The Most Successful Way of JEE Preparation

DPP

Daily Practice Problems

JEE Main & Advanced

Coordination Chemistry and *p*, *d* & *f* Blocks

with Metallurgy & Qualitative Analysis

CHEMISTRY Vol-7

GS Reddy

arihant
ARIHANT PRAKASHAN (SERIES), MEERUT

DPP
Daily Practice Problems

arihant
ARIHANT PRAKASHAN (SERIES), MEERUT

Administrative & Production Offices

Corporate Office 'Ramchhaya' 4577/15, Agarwal Road, Darya Ganj, New Delhi -110002
Tele: 011- 47630600, 43518550; Fax: 011- 23280316

Head Office Kalindi, TP Nagar, Meerut (UP) - 250002
Tel: 0121-2401479, 2512970, 4004199; Fax: 0121-2401648

Sales & Support Offices

Agra, Ahmedabad, Bengaluru, Bareilly, Chennai, Delhi, Guwahati, Hyderabad, Jaipur, Jhansi, Kolkata, Lucknow, Meerut, Nagpur & Pune

ISBN : 978-93-13193-46-3

Published by Arihant Publications (I) Ltd.

Production Team

Publishing Managers	Mahendra Singh Rawat & Keshav Mohan	*Inner Designer*	Mazher Chaudhary
Project Head	Sorabh Chaudhary	*Page Layouting*	Sunil Sharma, Hemant
Cover Designer	Shanu Mahad	*Proof Reading*	Salman

For further information about the products from Arihant, log on to www.arihantbooks.com or email to info@arihantbooks.com

/arihantpub /@arihantpub Arihant Publications /arihantpub

ABOUT DPP

Daily Practice Problems

Indian Institutes of Technology better known as IITs have always been the dream destination of each science group students who want to pursue their career as Engineers. These are the institutes which have got very high value not even in India but in the whole world. To get the admission in such institutes one need to face entrance Examinations; JEE Main & JEE Advanced, out of which JEE Main is considered as the preliminary exam & JEE Advanced is the final exam to get into the IITs. Approximately per year 12 lakh students enroll themselves for JEE Main and only the top 2.25 lakh in JEE Main merit can attempt for JEE Advanced, the fact clearly express the degree of toughness of JEE Main & JEE Advanced.

To have success in such an exam one needs to be very planned & systematic in its preparation. A very popular wise saying 'Rome was not built in a day' suggests that to achieve a big target like IITs we need to work continuously. For continuous work, one need to have proper planning that how much work he/she has to do on regular basis, i.e., on daily basis. Each student preparing for JEE has many books but unfortunately none of them gives the students a plan for daily studies, which is the most effective success key for JEE.

Keeping the daily practice needs of the students in mind we have come up with DPP Series (Daily Practice Problems) having approximately 26 books. The primary focus of this series is on achieving success through practice & proper planning. For that we have divided the whole syllabus of JEE Main & JEE Advanced; Physics, Chemistry & Mathematics, into different volumes (around 8 volumes of Mathematics & 9 volumes each of Physics & Chemistry), each volume has Daily Practice Problems Sheets having specific questions on various Topics of the individual chapter of the volume ensuring the complete Practice & Assessment of the topic.

DPP

Daily Practice Problems

In this way we have tried to give you an effective daily planning for JEE keeping in mind the daily hours that a student can devote to his practice, you are just required to practice one sheet of each subject on daily basis. This habit of daily practice will keep you away from exam stress and thoughts like; how much I have prepared/am I preparing properly or not/ do I need to study more on daily basis or what I am doing is ok, etc., which regularly tease your mind and can give you panic at any stage of your preparation.

Some special features that we have tried to incorporate in our Daily Practice Problems (DPP) series are discussed below

1. Micro Level coverage of all the topics of a Chapter/ Unit given in JEE Main & Advanced Syllabus.
2. Each DPP has various types of questions; Subjective Questions, Single/ Multiple Correct Options, Statements type, Comprehension based, Integer Value Answer type, Matching, etc., making you practice & ready for all the question formats of JEE.
3. New & innovative questions in each DPP challenging & improving your problem solving skills.
4. Along with Topical Coverage, Revisal DPPs for JEE Main & JEE Advanced with each Chapter, covering multidisciplinary questions proving comprehensive practice.
5. JEE Main & JEE Advanced Archive (collection of last 13 years' exams questions) with each chapter.
6. Complete Solutions for each DPP.

We hope that DPP series will prove to be very beneficial for the students preparing for JEE, by giving them daily practice in a systematic & planned manner. The suggestions/criticism from the students, teachers & parents about DPP series will be welcomed whole heartedly and we promise to update this series accordingly from time to time.

DPP
Daily Practice Problems

DAILY TASKS...

1 p-BLOCK ELEMENTS

DPP - 1 General Properties of 15th Group (Nitrogen and Phosphorus) 3-6
DPP - 2 Nitrogen and Phosphorus Compounds 7-10
DPP - 3 General Properties of 16th Group (Oxygen, Ozone and Sulphur) 11-15
DPP - 4 Oxygen and Sulphur Compounds 16-20
DPP - 5 General Properties of 17th Group (Halogens) 21-24
DPP - 6 Compounds of Halogens (F, Cl, Br and I) 25-29
DPP - 7 Interhalogens and Pseudo Halogens 30-33
DPP - 8 General Properties and Compounds of Xenon 34-37

- Revisal Problems for JEE Main 38-40
- Revisal Problems for JEE Advanced 41-43
- JEE Main & AIEEE Archive 44
- JEE Advanced & IIT JEE Archive 45-47

2 d & f-BLOCK ELEMENTS

DPP - 1 General Properties of *d*-Block Elements 51-54
DPP - 2 General Chemical and Magnetic Properties of *d*-Block Elements 55-58
DPP - 3 Oxides and Oxo Anions of Cr and Mn 59-63
DPP - 4 *f*-Block Elements-I (Lanthanides) 64-67
DPP - 5 *f*-Block Elements-II (Actinides) 68-71

- Revisal Problems for JEE Main 72-74
- Revisal Problems for JEE Advanced 75-77
- JEE Main & AIEEE Archive 78-79
- JEE Advanced & IIT JEE Archive 80

3 COORDINATION COMPOUNDS

DPP - 1 Introduction of Coordination Compounds and Werner's Theory 83-86
DPP - 2 IUPAC Nomenclature of Coordination Compounds 87-91
DPP - 3 Isomerism in Coordination Compounds-I 92-95
DPP - 4 Isomerism in Coordination Compounds-II 96-99
DPP - 5 Valence Bond Theory, Effective Atomic Number and Magnetic Properties 100-103

DPP
Daily Practice Problems

DPP - 6	Crystal Field Theory (CFT) and Colour of Complexes	104-107
DPP - 7	Bonding in Organometallic Compounds, Stability and Applications of Coordination Compounds	108-111

- Revisal Problems for JEE Main 112-114
- Revisal Problems for JEE Advanced 115-118
- JEE Main & AIEEE Archive 119-120
- JEE Advanced & IIT JEE Archive 121-122

4 METALLURGY

DPP - 1	Introduction of Metallurgy and Ores of Some Metals	125-128
DPP - 2	Metallurgy Process of Concentration of Ores	129-132
DPP - 3	Extraction of Crude Metals from Concentrated Ores	133-136
DPP - 4	Thermodynamic Principles of Metallurgy ,Ellingham's Diagram and Its Applications	137-141
DPP - 5	Refining of Crude Metals	142-145
DPP - 6	Metallurgy of Some Important Metals-I (Cu, Fe, Ag)	146-148
DPP - 7	Metallurgy of Some Important Metals-II (Al, Zn, Au)	149-152

- Revisal Problems for JEE Main 153-155
- Revisal Problems for JEE Advanced 156-158
- JEE Main & AIEEE Archive 159
- JEE Advanced & IIT JEE Archive 160-161

5 QUALITATIVE ANALYSIS

DPP - 1	Principles Involved in Qualitative Analysis of Inorganic Salts	165-167
DPP - 2	Qualitative Analysis of Acidic Radicals (Anions)	168-172
DPP - 3	Qualitative Analysis of Basic Radicals (Cations)	173-176
DPP - 4	Qualitative Analysis of Inorganic Salts (Acidic and Basic Radicals)	177-180
DPP - 5	Organic Qualitative Analysis	181-184
DPP - 6	Quantitative Analysis	185-188
DPP - 7	Principles Involved in Synthesis, Neutralisation and Kinetic Studies	189-192

- Revisal Problems for JEE Main 193-195
- Revisal Problems for JEE Advanced 196-201
- JEE Advanced & IIT JEE Archive 202-203

Answers with Explanations **205-280**

Last 3 Years' Questions JEE Main & Advanced **1-10**

1. p-Block Elements

DPP-1	General Properties of 15th Group (Nitrogen and Phosphorus)
DPP-2	Nitrogen and Phosphorus Compounds
DPP-3	General Properties of 16th Group (Oxygen, Ozone and Sulphur)
DPP-4	Oxygen and Sulphur Compounds
DPP-5	General Properties of 17th Group (Halogens)
DPP-6	Compounds of Halogens (F, Cl, Br and I)
DPP-7	Interhalogens and Pseudo Halogens
DPP-8	General Properties and Compounds of Xenon

- **Revisal Problems for JEE Main**
- **Revisal Problems for JEE Advanced**
- **JEE Main & AIEEE Archive**
- **JEE Advanced & IIT JEE Archive**

DPP

DAILY PRACTICE PROBLEMS

DPP-1 General Properties of 15th Group (Nitrogen and Phosphorus)

Subjective Questions

Working Space

Direction (Q. Nos. 1-4) *These questions are subjective in nature, need to be solved completely on notebook.*

1 Explain the following.

i. Catenation property of phosphorus is higher than nitrogen.

ii. 15th group elements have higher first ionisation enthalpies than 14th or 16th group elements.

iii. 15th group elements have low electron gain enthalpies than the surrounding groups.

2 i. What is the minimum and maximum oxidation states shown by nitrogen?

ii. Find the oxidation states of nitrogen in

a. NH_4NO_3 b. NH_4NO_2

c. NH_4CN d. HN_3

3 White phosphorus is highly reactive, red is inert and black is a conductor. Explain.

4 Explain the following.

i. Why nitrogen shows anomalous properties?

ii. Nitrogen is chemically inert.

iii. Nitrogen maximum valency is 4 while for other elements it is 6.

iv. For bismuth, + 3 is more stable than + 5.

Only One Option Correct Type

Direction (Q. Nos. 5-12) *This section contains 8 multiple choice questions. Each question has four choices (a), (b), (c) and (d), out of which ONLY ONE is correct.*

5 The formula of realgar and orpiment are

a. As_2S_4 and As_2S_3 b. $NaNO_3$ and KNO_3

c. Sb_2S_3 and As_2S_3 d. As_2S_3 and As_4S_4

6 Correct order of 2nd ionisation energy of C, N, O and F is

a. N > F > O > C b. O > F > N > C

c. F > O > N > C d. O > F > C > N

7 Which trend enthalpy is correct?

a. N—O > P—O b. N—F > P—F

c. N≡N > P≡P d. N—N > P—P

Working Space

8 Boiling point of liquid nitrogen is

a. 4 K b. 20 K c. 77 K d. 87 K

9 Oxidation of ammonia with CuO produce a gaseous chemical which can also be obtained by

a. heating ammonium nitrate
b. reacting excess ammonia with chlorine
c. heating potassium dichromate
d. catalytic oxidation of ammonia

10 Which of the molecular species has unpaired electrons?

a. N_2 b. N_2^+ c. NO^+ d. CN^-

11 Which is the correct order w.r.t the given property?

a. N > P > As > Sb > Bi (Atomic mass)
b. N > P > As > Sb = Bi (Electronegativity)
c. N > P > As > Sb = Bi (Covalent radii)
d. N = P > As > Sb > Bi (Density)

12 The ratio of bond pairs and lone pairs in a P_4 molecule is

a. 1 : 1 b. 2 : 3 c. 3 : 2 d. 4 : 1

One or More than One Options Correct Type

Direction (Q. Nos. 13-17) *This section contains 5 multiple choice questions. Each question has four choices (a), (b), (c) and (d), out of which ONE or MORE THAN ONE are correct.*

13 Regarding black phosphorus correct statements are

a. Most stable allotrope of phosphorus
b. Used in making safety matches
c. Good conductor of heat and electricity
d. It exists as P_4 units

14 Which are correct statements?

a. The melting point of antimony is higher than bismuth
b. Ionisation energy of C < O < N
c. In 15th group, all show allotropy except bismuth
d. Maximum covalency of nitrogen and phosphorus are 4 and 5 respectively

15 The correct statements regarding N and P are

a. Phosphorus is more abundant in earth's crust than nitrogen
b. Nitrogen has unique ability to form *pπ-pπ* bonding while phosphorus form *pπ-dπ* bonds
c. White P is soluble in water and CS_2 while red P is insoluble in both the solvents
d. Red and black phosphorus do not catch fire in air while white P readily catch fire in air

16 The correct statements among the given are

a. Antimony belongs to 15th group and 5th period
b. Electron gain enthalpy of P > N > S > O
c. Minimum and maximum oxidation number of phosphorus is − 3 and + 6
d. Fluoroapatite, formula is $Ca_9(PO_4)_6CaF_2$

17 Which of the following statements is/are correct?

a. Abundance in the earth's crust P > N > As > Sb
b. N_2 molecule has *pπ-pπ* bonding
c. Among the nitrogen halides, NF_3 is least stable
d. Both (a) and (b)

Working Space

Comprehension Type

Direction (Q. Nos. 18 and 19) *This section contains a paragraph, describing theory, experiments, data, etc. Two questions related to the paragraph have been given. Each question has only one correct answer among the four given options (a), (b), (c) and (d).*

Passage

Phosphorus was discovered by Brand (1669), Scheele isolated from bone ash and Lavoisier proved its elemental nature (1777). The principal minerals are phosphate rock, fluoroapatite and chloroapatite. Phosphorus is prepared by the direct reduction of phosphorite by carbon in the presence of silica. It exists in different allotropic forms such as yellow or white, red, α-black, β-black, etc. White P is most reactive, poisonous, glows in dark and readily catches fire due to unstable discrete P_4 molecules. Red P is inert, non-poisonous, does not glow etc., due to its polymeric structure. α-black, β-black allotropes are also chemically inert, do not ignite at normal temperature. It has layer structure like graphite and acts as conductor.

18 The allotrope of phosphorus with low ignition temperature is

a. white P b. red P c. α-black d. β-black

19 The allotrope of phosphorus that has layer lattice like graphite is

a. white P b. yellow P c. red P d. black P

Matching List Type

Direction (Q. Nos. 20 and 21) *Choices for the correct combination of elements from Column I and Column II are given as options (a), (b), (c) and (d), out of which one is correct.*

20 Match the Column I with Column II and mark the correct option from the codes given below.

	Column I		Column II
i.	Nitrogen	p.	Non-metal
ii.	Phosphorus	q.	Metal
iii.	Arsenic	r.	Metalloid
iv.	Bismuth	s.	Sublimation
		t.	Inert pair effect

Codes

	i	ii	iii	iv		i	ii	iii	iv
a.	p	p	r,s	q,t	b.	q	r	p	s,t
c.	q	p	s,t	r,p	d.	p	q	r	s

21 Match the Column I with Column II and mark the correct option from the codes given below.

	Column I		Column II
i.	$R_3P=O$	p.	−3 oxidation state
ii.	N_2	q.	$d\pi$-$p\pi$ bond
iii.	N_2O_3	r.	$p\pi$-$p\pi$ bond
iv.	Ca_3N_2	s.	Acidic
		t.	Basic

Codes

	i	ii	iii	iv
a.	p,s	p,q	s	r
b.	q	r	s	p
c.	p	q	r	s,t
d.	p,t	q,s	q	r

One Integer Value Correct Type

Working Space

Direction *(Q. Nos. 22-24) This section contains 3 questions. When worked out will result in an integer from 0 to 9 (both inclusive).*

22 How many of the given chemicals liberate dinitrogen on heating?

NH_4NO_2, NH_4NO_3, NaN_3, Li_3N, $(NH_4)_2Cr_2O_7$, $Pb(NO_3)_2$, Ba_3N_2

23 Number of chemical species having negative oxidation state for nitrogen among

NF_3, NCl_3, NH_2OH, NH_3, CH_3NH_2, NH_2^-, Li_3N, N_2O, HCN, HNC, NO_2^-

24 According to molecular orbital theory, number of electrons present in the antibonding orbitals of dinitrogen.

Statement Type

Direction *(Q. No. 25) This section is based on Statement I and Statement II. Select the correct answer from the codes given below.*

a. Both Statement I and Statement II are correct and Statement II is the correct explanation of Statement I
b. Both Statement I and Statement II are correct but Statement II is not the correct explanation of Statement I
c. Statement I is correct but Statement II is incorrect
d. Statement II is correct but Statement I is incorrect

25 **Statement I** N_2 is less reactive than P_4.

Statement II Nitrogen has more electron gain enthalpy than phosphorus.

DPP-2 Nitrogen and Phosphorus Compounds

Subjective Questions

Working Space

Direction *(Q. Nos. 1-4) These questions are subjective in nature, need to be solved completely on notebook.*

1 Answer the following.

i. Name the neutral oxides of nitrogen.

ii. Why NO and NO_2 are paramagnetic in the gaseous state while in the liquid or solid state, they are diamagnetic?

iii. Name the oxide of nitrogen used as anaesthetic.

iv. What is the nature of trioxides of 15th group elements?

2 Classify the following into normal and acidic salts.

NaH_2PO_3, NaH_2PO_4, NaH_2PO_2, Na_2HPO_3, Na_2HPO_4, $Na_2H_2P_2O_7$

3 Among the hydrides of 15th group elements

i. Why the enthalpy of vaporisation of NH_3 is higher than PH_3?

ii. Name the most stable and strong reducing agent.

iii. NH_3 has bond angle 107° while PH_3 has 94°.

4 i. What is aqua-regia and nitration mixture?

ii. When Au and Pt dissolves in aqua-regia, write the complex formed.

Only One Option Correct Type

Direction *(Q. Nos. 5-12) This section contains 8 multiple choice questions. Each question has four choices (a), (b), (c) and (d), out of which ONLY ONE is correct.*

5 The hybridisation of N in solid state for N_2O_5 is

a. sp and sp^2
b. Only sp^2
c. sp^2 and sp^2
d. Only sp^3

6 PH_3 becomes spontaneously inflammable due to the presence of

a. NH_3
b. N_2H_4
c. P_2H_4
d. All of the above

Working Space

7 Which is incorrect among the following options?

a. Reducing property of $H_3PO_4 > H_3PO_3 > H_3PO_2$
b. Oxidation state of nitrogen $N_2O < NO < N_2O_3 < N_2O_5$
c. Basicity $NH_3 > PH_3 > AsH_3 > SbH_3$
d. Boiling point of $SbH_3 > AsH_3 > NH_3 > PH_3$

8 Oxyacid of phosphorus having P—P bond is

a. $H_4P_2O_7$
b. $(HPO_3)_3$
c. $H_4P_2O_6$
d. None of these

9 In P_4O_6, the number of P—O bonds and lone pairs are

a. 12 and 4
b. 8 and 8
c. 12 and 16
d. 12 and 12

10 Which is the correct order of basic strength?

a. $NH_2OH < N_2H_4 < NH_3$
b. $N_2H_4 < NH_2OH < NH_3$
c. $NH_3 < N_2H_4 < NH_2OH$
d. $NH_2OH < NH_3 < N_2H_4$

11 Solid PCl_5 and solid PBr_5 exist respectively as

$[PCl_4]^+ [PCl_6]^-$ (I) $[PCl_4]^+ Cl^-$ (II) $[PBr_4]^+ [PBr_6]^-$ (III) $[PBr_4]^+ Br^-$ (IV)

a. I and III
b. I and IV
c. II and III
d. II and IV

12 Number of moles of NaOH needed to neutralise one mole each of H_3PO_2, H_3PO_3 and H_3PO_4 respectively are

a. 3, 3 and 3
b. 1, 2 and 3
c. 2, 2 and 3
d. 2, 1 and 3

One or More than One Options Correct Type

Direction *(Q. Nos. 13-17) This section contains 5 multiple choice questions. Each question has four choices (a), (b), (c) and (d), out of which ONE or MORE THAN ONE are correct.*

13 Which of the following can act as dehydrating agent?

a. P_4O_{10}
b. $POCl_3$
c. Conc. H_2SO_4
d. P_4O_6

14 NO_2 is obtained by heating

a. lead nitrate
b. potassium nitrate
c. lithium nitrate
d. silver nitrate

15 Phosphine is obtained when

a. red P is heated with NaOH
b. white P is heated with NaOH
c. Ca_3P_2 is reacted with water
d. by heating phosphorus acid

16 Regarding ammonia correct statements are

a. A nitride on hydrolysis produce NH_3
b. Haber's process of ammonia formation is exothermic
c. NH_3 dissolves AgCl forming a soluble complex
d. Ammonia is a Lewis base as well as Bronsted-Lowry base

17 Which is correctly matched?

a. Phosphinic acid (H_3PO_2) is a dibasic acid
b. Metaphosphoric acid is called glacial phosphoric acid
c. $H_4P_2O_5$ (diphosphorus acid) has P—O—P bond
d. H_3PO_4 has 7σ and 1π-bond

Working Space

Comprehension Type

Direction (Q. Nos. 18 and 19) This section contains a paragraph, describing theory, experiments, data, etc. Two questions related to the paragraph have been given. Each question has only one correct answer among the four given options (a), (b), (c) and (d).

Passage

In the all oxyacids of phosphorus, each phosphorus atom is in sp^3-hybridised state. All these acids contain P—OH bonds, the hydrogen atom of which are ionisable imparting acidic nature to the compound. The 'ous' acids (oxidation state of P is + 1 or + 3) also have P—H bonds in which hydrogens are not ionisable. The presence of P—H bonds in these acids imparts reducing properties. The structure of some oxyacids are drawn below:

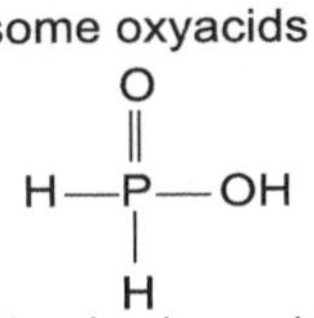

Hypophosphorus acid (*A*)

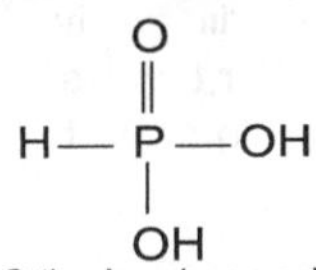

Orthophosphorus acid (*B*)

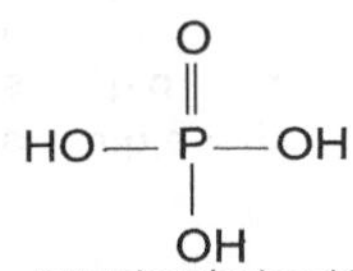

Orthophosphoric acid (*C*)

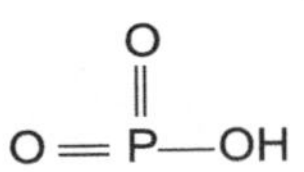

Metaphosphoric acid (*D*)

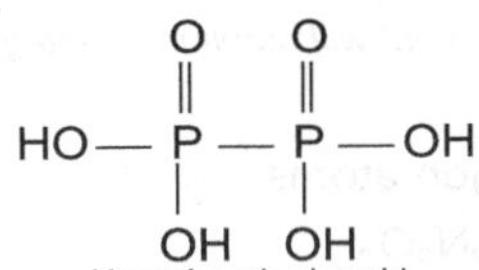

Hypophosphoric acid (*E*)

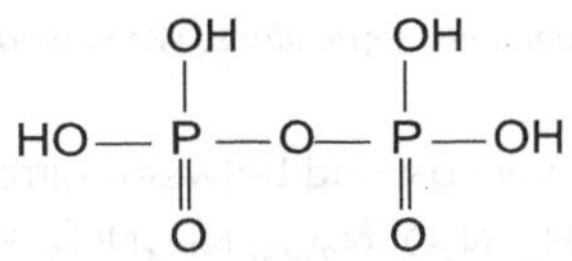

Pyrophosphoric acid (*F*)

18 Although metaphosphoric acid is written as a monomer, it exists as a polymer $(HPO_3)_n$. The number of P—O—P bonds in cyclic trimetaphosphoric acid is

a. zero
b. two
c. three
d. four

19 Which of the acids show reducing properties?

a. *A* and *C*
b. *A* and *B*
c. *A*, *B* and *D*
d. *C*, *D*, *E* and *F*

Matching List Type

Direction (Q. Nos. 20 and 21) Choices for the correct combination of elements from Column I and Column II are given as options (a), (b), (c) and (d), out of which one is correct.

20 Match the Column I with Column II and mark the correct option from the codes given below.

	Column I		Column II
i.	Haber's process	p.	Manufacture of ammonia
ii.	Le-blanc process	q.	Manufacture of sulphuric acid
iii.	Birkeland-Eyde process	r.	Manufacture of nitric acid
iv.	Solvay process	s.	Manufacture of sodium carbonate

Codes

	i	ii	iii	iv
a.	p	s	q	r
b.	s	r	q	p
c.	q	p	s	r
d.	s	r	p	q

Working Space

21 Match the Column I with Column II and mark the correct option from the codes given below.

	Column I		Column II
i.	HNO_3	p.	Oxidising agent
ii.	HNO_2	q.	Reducing agent
iii.	NH_3	r.	Acid
iv.	HN_3	s.	Ligand
		t.	Fractional oxidation state

Codes

	i	ii	iii	iv		i	ii	iii	iv
a.	p,r	p,q,r	q,s	r,t	b.	p,q	s,t	r,t	s
c.	q,r	p,t	s	r	d.	p,q,r	s,t	q,r	t

One Integer Value Correct Type

Direction *(Q. Nos. 22-24) This section contains 3 questions. When worked out will result in an integer from 0 to 9 (both inclusive).*

22 Number of chemical species having bond between nitrogen atoms

$$N_2O_4, N_2O_5, N_2H_4, N_2O, N_2O_3, NH_4NO_3, H_2N_2O_2$$

23 Number of metals that show passivity with conc. HNO_3 among Cr, Fe, Ni, Cu, Zn Al, Ag, Sn.

24 $2H_3PO_2 \longrightarrow H_3PO_4 + PH_3$ in this equivalent weight of the acid in the reactant side is obtained by dividing molecular weight with

Statement Type

Direction *(Q. No. 25) This section is based on Statement I and Statement II. Select the correct answer from the codes given below.*

a. Both Statement I and Statement II are correct and Statement II is the correct explanation of Statement I
b. Both Statement I and Statement II are correct but Statement II is not the correct explanation of Statement I
c. Statement I is correct but Statement II is incorrect
d. Statement II is correct but Statement I is incorrect

25 **Statement I** NF_3 is a weaker ligand than $N(CH_3)_3$.

Statement II NF_3 ionises to give F^- ions in aqueous solution.

DPP-3 General Properties of 16th Group
(Oxygen, Ozone and Sulphur)

Subjective Questions

Working Space

Direction (Q. Nos. 1-5) *These questions are subjective in nature, need to be solved completely on notebook.*

1 Explain the following.

i. Among the chalcogens, oxygen has lowest electron affinity.

ii. Oxygen is gas while sulphur is solid at room temperature.

iii. First ionisation energy of oxygen is less than nitrogen but second ionisation energy of O > N.

2 In *p*-block elements, variable oxidation state is of normal occurrence. Find the oxidation number of sulphur in the compounds given below.

i. $Na_2S_4O_6$

ii. SO_2Cl_2

iii. H_2S_2

iv. $S_2O_3^{2-}$

3 Give the reason of the following.

i. O_3 is diamagnetic while oxygen is paramagnetic.

ii. O—O bond lengths in ozone are equal.

iii. Bond length in ozone is higher than oxygen.

iv. Ozone estimation by iodometric method.

4 Explain the following.

i. Anomalous behaviour of oxygen.

ii. Oxygen common valency is 2 while for others show 2,4,6 covalency.

iii. Name two compounds in which oxygen has positive oxidation state.

iv. Give one example in which oxygen has fractional oxidation state.

5 Give reasons to the following.

i. Ozone destroys mercury meniscus.

ii. Oxygen and sulphur act as Lewis acids.

iii. Ozone acts as bleaching agent.

Only One Option Correct Type

Working Space

Direction *(Q. Nos. 6-15) This section contains 10 multiple choice questions. Each question has four choices (a), (b), (c) and (d), out of which ONLY ONE is correct.*

6 Which is incorrect statement?

a. Element with atomic number 52 belong to 5th period and 16th group
b. Number of electrons in the n and $(n-1)$ shells of selenium are 6 and 18
c. Among chalcogens oxygen and sulphur exist in free as well as in combined state
d. The differentiating electron in tellurium enters into $6p$ subshell

7 In earth's crust, sulphur exists as

a. elemental sulphur
b. sulphide ore
c. sulphate ores
d. All of these

8 The element used in the production of photocells and solar cells and also used in zerox machines is

a. sulphur
b. selenium
c. tellurium
d. polonium

9 The number of bond pairs and lone pairs in rhombic sulphur molecule are

a. 6 and 6
b. 8 and 8
c. 8 and 16
d. 16 and 8

10 Which has lowest bond energy (single bond)?

a. O—H
b. O—O
c. S—H
d. S—S

11 Harmful UV radiations emitted from the sun are prevented from reaching the earth by the presence of ozone in the

a. mesosphere
b. thermosphere
c. stratosphere
d. troposphere

12 Point out the incorrect statement among the following.

a. In OF_2, oxidation state of oxygen is + 2
b. Sulphur shows strong tendency to catenation while oxygen shows this tendency to a limited extent
c. The transition temperature of rhombic and monoclinic sulphur is 369° C
d. Oxygen is most abundant element in the earth's crust followed by sulphur in 16 th group

13 The correct order of thermal stability of the hydrides of group 16 elements is

a. $H_2Po > H_2Fe > H_2Se > H_2S > H_2O$
b. $H_2O < H_2S > H_2Se > H_2Te > H_2Po$
c. $H_2O = H_2S = H_2Se = H_2Te = H_2Po$
d. $H_2O > H_2S > H_2Se > H_2Te > H_2Po$

14 When O_2 changes to O_2^+

a. paramagnetic become diamagnetic
b. bond length increases
c. diamagnetic become paramagnetic
d. bond order increases

15 Increasing order of oxidation number of oxygen is found in

a. $KO_2 < BaO_2 < O_3 < O_2F_2$
b. $O_3 < KO_2 < BaO_2 < O_2F_2$
c. $BaO_2 < KO_2 < O_3 < O_2F_2$
d. $O_2F_2 < BaO_2 < KO_2 < O_3$

One or More than One Options Correct Type

Direction *(Q. Nos. 16-20) This section contains 5 multiple choice questions. Each question has four choices (a), (b), (c) and (d), out of which ONE or MORE THAN ONE are correct.*

16 Which of the following property do not increase down the group from oxygen to polonium?

a. Atomic size
b. Density
c. Electron affinity
d. Ionisation energy

Working Space

17 The correct statements among the following are

a. Rhombic sulphur is stable at room temperature
b. S_8 molecule is a planar ring
c. In S_8 molecule, sulphur has valence electrons 2 and oxidation state is 0
d. ΔH_f^o of rhombic sulphur is taken as 0

18 Which of the following statements regarding ozone are correct?

a. Conversion of ozone to oxygen is endothermic
b. O—O bonds in ozone have considerable double bond character
c. Pure ozone is pale blue gas and dark blue liquid
d. Nitric oxide and freons are responsible for the depletion of ozone layer

19 The correct statements among the following are

a. Bond lengths in O_2, O_2^-, O_2^+ are 121, 134,149 pm
b. Ozone is stronger oxidising agent than dioxygen
c. O_2 acts as reducing agent when it reacts with powerful oxidising agents like PtF_6
d. Ozone is much more reactive than oxygen

20 Which of the following do not stick to the poles of a magnet?

a. Liquid oxygen
b. Liquid nitrogen
c. Liquid SO_2
d. Liquid ozone

Comprehension Type

Direction (Q. Nos. 21 and 22) This section contains a paragraph, each describing theory, experiments, data, etc. Two questions related to the paragraphs have been given. Each question has only one correct answer among the four given options (a), (b), (c) and (d).

Passage

Sulphur and the rest of the elements of 16th group are less electronegative than oxygen. They can acquire ns^2np^6 by sharing two electrons with the atoms of other elements and thus, exhibit +2 oxidation state in their compounds. In addition to this, atoms have vacant *d*-orbitals in their valence shell due to which electrons can be promoted from the *s*- and *p*-orbitals of the same shell. As a result, they can show +4 and +6 oxidation states. The oxidation states of elements never exceed +6.

21 Oxidation states of sulphur in $H_2S_2O_7$ and $H_2S_2O_8$ respectively are

a. +6 and +7
b. +6 and +6
c. +5 and +6
d. +5 and +7

22 Like sulphur, oxygen does not show +4 and +6 oxidation states .The reason is

a. that oxygen is a gas while sulphur is solid
b. that oxygen has high ionisation energy in comparison to sulphur
c. that sulphur has high electron affinity in comparison to oxygen
d. that oxygen has no *d*-orbitals in its valence shell like sulphur

Matching List Type

Working Space

Direction *(Q. Nos. 23-25) Choices for the correct combination of elements from Column I and Column II are given as options (a), (b), (c) and (d), out of which one is correct.*

23 Match the Column I with Column II and mark the correct option from the codes given below.

	Column I (Allotropic form)		Column II (Structure)
i.	Engle's sulphur	p.	Rings, chair conformation, unstable
ii.	Sulphur	q.	Fibrous or rubber like
iii.	Rhombic sulphur	r.	Crystalline form-yellow crystals
iv.	Monoclinic sulphur	s.	Puckered S_8 rings crown conformation

Codes

	i	ii	iii	iv		i	ii	iii	iv
a.	s	q	r	p	b.	p	q	r	s
c.	q	p	s	r	d.	p	q	s	r

24 Match the Column I with Column II and mark the correct option from the codes given below.

	Column I (Molecular properties of dioxygen)		Column II (Values)
i.	Melting point (K)	p.	90.2
ii.	Boiling point (K)	q.	1.429
iii.	Density at STP (g/cm^3)	r.	120.7
iv.	Bond length (pm)	s.	493.4
v.	Bond energy (kJ mol^{-1})	t.	54.4

Codes

	i	ii	iii	iv	v
a.	p	q	t	r	s
b.	s	t	p	q	t
c.	t	p	q	r	s
d.	q	p	s	r	t

25 Match the Column I with Column II and mark the correct option from the codes given below.

	Column I		Column II
i.	Dioxygen	p.	Paramagnetic
ii.	Ozone	q.	Vulcanising agent
iii.	Rhombic sulphur	r.	Diamagnetic
iv.	Selenium	s.	Puckered ring
		t.	Photoconductor

Codes

	i	ii	iii	iv		i	ii	iii	iv
a.	p	r	q,r,s	t	b.	q	p,q	s	r
c.	s	p	r	q	d.	s	q	r	p

One Integer Value Correct Type

Direction (Q. Nos. 26-29) *This section contains 4 questions. When worked out will result in an integer from 0 to 9 (both inclusive).*

26 Number of chemicals that produce dioxygen on heating
$KClO_3$, $(NH_4)_2Cr_2O_7$, NH_4NO_3, $NaNO_3$, $KMnO_4$, $K_2Cr_2O_7$, H_2O_2, Pb_3O_4, NH_4NO_2

27 In the fully excited state, number of unpaired electrons in tellurium are

28 Number of chemical species having bent shape in
O_3, NO_2, NO_2^+, N_2O, SO_2, N_3^-, CS_2, C_3O_2, H_2S

29 Among the following compounds, the number of compounds which have oxidation states of S is +4 ?
PbS, SO_2, SF_6, $Na_2S_2O_3$, H_2SO_3

Statement Type

Direction (Q. No. 30) *This section is based on Statement I and Statement II. Select the correct answer from the codes given below.*

a. Both Statement I and Statement II are correct and Statement II is the correct explanation of Statement I
b. Both Statement I and Statement II are correct but Statement II is not the correct explanation of Statement I
c. Statement I is correct but Statement II is incorrect
d. Statement II is correct but Statement I is incorrect

30 **Statement I** Oxygen is more electronegative than sulphur, yet H_2S is acidic, while H_2O is neutral.
Statement II H—S bond is weaker than O—H bond.

DPP-4 Oxygen and Sulphur Compounds

Subjective Questions

Direction (Q. Nos. 1-5) These questions are subjective in nature, need to be solved completely on notebook.

Working Space

1 Give reason for the following statements.

i. H_2S is stronger acid than H_2O.

ii. SO_2 acts as Lewis acid as well as Lewis base.

iii. SO_3 acts as Lewis acid.

2 Explain the following.

i. Conc. H_2SO_4 reacts with CaF_2 liberating colourless vapours while with $CaBr_2$ reddish brown vapours are liberated.

ii. H_2S cannot be dried by passing through conc. H_2SO_4.

iii. For dilution of H_2SO_4, water should not be added to conc. H_2SO_4.

iv. When SO_2 reacts with NaOH.

3 Write molecular formula for the following compounds.

i. Oleum

ii. Vulcanising agent

iii. Marshall acid

iv. Sulphuric anhydride

v. Oil of vitriol

4 Explain the following.

i. An acidified $K_2Cr_2O_7$ paper turns green when exposed to SO_2.

ii. H_2S acts only as a reducing agent while SO_2 can act both as a reducing and oxidising agent.

iii. Sugar turns black on addition of conc. H_2SO_4.

iv. Copper reacts with hot conc. H_2SO_4 and dil. H_2SO_4.

5 Among the oxyacids of sulphur, name the acid(s)

i. having sulphur-sulphur double bond.

ii. having sulphur atoms in different oxidation states.

iii. having S—O—S bond.

iv. whose sodium salt is known as hypo.

Only One Option Correct Type

Direction (Q. Nos. 6-15) This section contains 10 multiple choice questions. Each question has four choices (a), (b), (c) and (d), out of which ONLY ONE is correct.

6 Which of the following statement is correct about the hydride of group 16 elements?

a. Order of acidic strength is $H_2O < H_2S < H_2Se < H_2Te$

b. Order of thermal stability is $H_2O > H_2S > H_2Se > H_2Te$

c. Order of melting and boiling points is $H_2O > H_2Te > H_2Se > H_2S$

d. All of the above

Working Space

7 The pair that contains mixed oxides is

a. Cl_2O_7 and Fe_3O_4
b. Pb_3O_4 and Fe_3O_4
c. C_3O_2 and Fe_3O_4
d. N_2O_5 and Cl_2O_7

8 The acid having a sulphur atom with a lone pair of electrons on it, is

a. H_2SO_4
b. H_2SO_5
c. $H_2S_2O_7$
d. $H_2S_2O_5$

9 Charring of cane sugar by sulphuric acid shows ...... property of the acid.

a. oxidation
b. complex formation
c. dehydrating
d. acidic

10 Which of the following statements is incorrect?

a. Oxidation state of sulphur in its compounds vary from –2 to +8
b. Tetrahalides of 16th group have *see-saw* shape
c. S_2Cl_2 has linear geometry
d. SCl_4 on hydrolysis yields sulphurous acid

11 $xSO_2 + yMnO_4^- + zH_2O \longrightarrow Mn^{2+} + SO_4^{2-} + H^+$. In this, three values of *x, y, z* are

a. 5, 2, 2
b. 4, 3, 2
c. 5, 3, 2
d. 6, 3, 2

12 Which of the following is wrong statement?

a. Dissolution of H_2SO_4, in H_2O is exothermic
b. H_2SO_4 is volatile liquid
c. In SO_2 and SO_3, hybridisation of S is same agent
d. H_2SO_4 is drying and dehydrating agent

13 Among the following oxyacids of sulphur, which one has more than one oxidation state for sulphur?

a. H_2SO_4
b. $H_2S_2O_3$
c. $H_2S_2O_7$
d. $H_2S_2O_6$

14 In SO_3 molecule, 3 oxygen atoms are attached to sulphur atom through double bonds. Hence, the molecules consist 3π-bonds. The types of pi-bonds present in SO_3 molecule are

a. two $p\pi$-$d\pi$ and one $p\pi$-$p\pi$
b. one $p\pi$-$d\pi$ and two $p\pi$-$p\pi$
c. one $p\pi$-$p\pi$ and one $p\pi$-$d\pi$
d. All $p\pi$-$p\pi$ bonds

15 Which of the following is known as Caro's acid?

a. $H_2S_2O_7$
b. $H_2S_2O_8$
c. $H_2S_2O_6$
d. H_2SO_5

Statement Type

Working Space

Direction *(Q. No. 16) This section is based on Statements I and Statements II. Select the correct answer from the codes given below.*

a. Both Statement I and Statement II are correct and Statement II is the correct explanation of Statement I
b. Both Statement I and Statement II are correct but Statement II is not the correct explanation of Statement I
c. Statement I is correct but Statement II is incorrect
d. Statement II is correct but Statement I is incorrect

16 **Statement I** In Caro's acid (H_2SO_5), the oxidation state of sulphur is found to be +8.

Statement II In Caro's acid, sulphur atom is attached to four oxygen atoms directly.

One or More than One Options Correct Type

Direction *(Q. Nos. 17-20) This section contains 4 multiple choice questions. Each question has four choices (a), (b), (c) and (d), out of which ONE or MORE THAN ONE are correct.*

17 The incorrect statements among the following are
a. Contact process is an example of heterogeneous catalysis
b. SO_2 decolourises the purple colour of acidified $KMnO_4$ solution
c. The two pi bonds in SO_2 are $d\pi$-$p\pi$
d. SO_2 is a reducing agent but not oxidising agent

18 The chemical species having sp^2-hybridisation for the central atom is

a. $SOCl_2$ b. SO_3^{2-}
c. SCl_2 d. SO_3

19 Which of the following is/are the correct statement(s) about sodium thiosulphate?
a. It is used as an antichlor
b. It is used in photography to remove AgBr
c. It is used in iodometry to estimate iodine
d. It has sulphur in +2 and +6 oxidation state

20 Which pair is isostructural?

a. SO_2 and SO_3^{2-} b. CO_2 and OCS
c. SO_2 and CS_2 d. SO_4^{2-} and PO_4^{3-}

Comprehension Type

Direction *(Q. Nos. 21 and 22) This section contains a paragraph, describing theory, experiments, data, etc. Two questions related to the paragraph have been given. Each question has only one correct answer among the four given options (a), (b), (c) and (d).*

Passage

Hydrated sodium thiosulphate, $Na_2S_2O_3 \cdot 5H_2O$ is called hypo. It can be prepared by Spring's reaction by adding iodine to a mixture of sodium sulphide and sodium sulphite.

$Na_2S + Na_2SO_3 + I_2 \longrightarrow 2NaI + Na_2S_2O_3$ (Spring's reaction)

It is an efflorescent substance and at 215°C, it is completely dehydrated to $Na_2S_2O_3$. Its reaction with iodine is used for the iodometric estimation of oxidising agents like $CuSO_4$, $CaOCl_2$, $KMnO_4$, MnO_2, etc., which on reaction with KI liberate I_2. It is used in the bleaching industry to destroy excess chlorine on fabrics. It is used in photography as fixer since, it dissolves unexposed AgBr on the film.

Working Space

21 Select the incorrect statement regarding hypo.

a. It is known as an antichlor
b. It is an efflorescent substance
c. AgBr with hypo forms a soluble complex, $Na_3[Ag(S_2O_3)_2]$
d. Thiosulphate ion $S_2O_3^{2-}$ has pyramidal shape

22 In $Na_2S_4O_6$, sulphur has different oxidation states. They are

a. +5 and +3
b. +5 and 0
c. 2.5 and 2.5
d. +4 and +6

Matching List Type

Direction *(Q. Nos. 23-25) Choices for the correct combination of elements from Column I and Column II are given as options (a), (b), (c) and (d), out of which one is correct.*

23 Match the Column I with Column II and mark the correct option from the codes given below.

	Column I (Oxyacids of sulphur)		Column II (Oxidation state of S-atom)
i.	H_2SO_3	p.	+4
ii.	H_2SO_4	q.	+5
iii.	H_2SO_5	r.	+6
		s.	+8

Codes

	i	ii	iii		i	ii	iii
a.	q	p	s	b.	p	r	r
c.	p	r	s	d.	r	r	s

24 Match the Column I with Column II and mark the correct option from the codes given below.

	Column I		Column II
i.	Ozone	p.	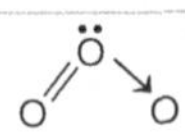
ii.	Sulphur dioxide	q.	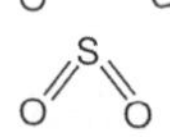
iii.	Oleum	r.	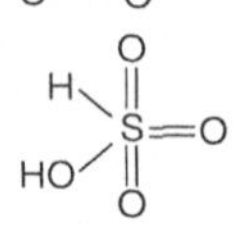
		s.	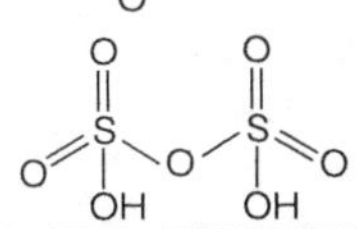

Codes

	i	ii	iii
a.	p	r	q
b.	q	p	s
c.	p	q	s
d.	r	p	q

Working Space

25 Match the Column I with Column II and mark the correct option from the codes given below.

	Column I		Column II
i.	SO_2	p.	Colourless gas
ii.	H_2S	q.	Oxidising agent
iii.	H_2SO_4	r.	Reducing agent
iv.	O_3	s.	Bleaching agent
		t.	Strong dibasic acid

Codes

	i	ii	iii	iv
a.	p,q,r,s	p,r	q,t	q,r,s
b.	p	q	r	s
c.	p,q	r,s	t	r,s
d.	p,q	r,s	q,r	s,t

One Integer Value Correct Type

Direction *(Q. Nos. 26-30) This section contains 5 questions. When worked out will result in an integer from 0 to 9 (both inclusive).*

26 The number of S—S bonds in sulphur trioxide trimer (S_3O_9) are

27 The number of metal sulphides precipitated by passing H_2S in acidic medium among ZnS, PbS, CuS, NiS, MnS, CdS, CoS, HgS, SnS.

28 The number of chemical species having sp^3-hybridisation.

SO_3, SO_2, $SOCl_2$, SO_2Cl_2, O_3, SF_2, SF_4, SO_3^{2-}, H_2SO_4

29 The number of covalent bonds present in H_2SO_5.

30 Number of molecules having non-zero dipole moment in O_3, SO_3, SF_4, SF_6, H_2S, CS_2, SO_2, H_2O and H_2O_2.

DPP-5 General Properties of 17th Group (Halogens)

Subjective Questions

Working Space

Direction (Q. Nos. 1-5) *These questions are subjective in nature, need to be solved completely on notebook.*

1 Which of the halogens (F,Cl, Br or I) provide an example of

i. liquid under ordinary conditions,
ii. undergoes sublimation,
iii. strongest oxidising agent,
iv. known as superhalogen,
v. commonly used as bleaching agent,
vi. undergoes disproportionation?

2 Explain the following.

i. F—F bond in fluorine is weaker than Cl— Cl bond in chlorine.
ii. Electron gain enthalpy of fluorine is less than chlorine.
iii. Fluorine forms only one oxyacid but other halogens form several acids.
iv. F_2 does not undergo disproportionation but other halogens do.
v. Why Cl_2 bleaches a substance permanently but SO_2 does it temporarily?

3 Give reasons for the following.

i. I_2 dissolves more in KI than in water.
ii. Among the halogens iodine is more electropositive element.
iii. All the halogens are coloured.
iv. F_2 is stronger oxidising agent than Cl_2.

4 Complete the following reactions.

i. $KMnO_4 + HCl \longrightarrow$
ii. $Cl_2 + \underset{\text{(cold and dil.)}}{NaOH} \longrightarrow$
iii. $F_2 + H_2O \longrightarrow$
iv. $I_2 + \underset{\text{(hot and conc.)}}{NaOH}$

5 Answer the following.

i. Why fluorine shows anomalous properties?
ii. Write the anomalous properties of fluorine.
iii. Why fluorine always shows –1 oxidation state and show monovalency while others show – 1 to + 7 and variable valency?

Only One Option Correct Type

Working Space

Direction (Q. Nos. 6-15) *This section contains 10 multiple choice questions. Each question has four choices (a), (b), (c) and (d), out of which ONLY ONE is correct.*

6 Which of the following property decreases down the group in the halogens?

a. Electropositive nature b. Density
c. Boiling point d. Ionisation enthalpy

7 Which of the following represents the decreasing order of van der Waals' forces in halogens?

a. $F_2 > Cl_2 > Br_2 > I_2$ b. $I_2 > Br_2 > Cl_2 > F_2$
c. $Br_2 > Cl_2 > F_2 > I_2$ d. $Cl_2 > F_2 > I_2 > Br_2$

8 The correct order of increasing order of radii ions Br^-, F^-, O^{2-} and S^{2-} is as follows

a. $Br^- < F^- < O^{2-} < S^{2-}$ b. $S^{2-} < O^{2-} < F^- < Br^-$
c. $F^- < O^{2-} < S^{2-} < Br^-$ d. $F^- < Br^- < O^{2-} < S^{2-}$

9 The correct order for the percentage of ionic character is

a. $HI > HBr > HCl > HF$ b. $HI < HBr < HCl < HF$
c. $HI = HBr < HCl < HF$ d. $HI < HBr < HCl = HF$

10 Which of the following statement is incorrect?

a. SRP values of halogens $X_2(g)/X^-(aq)$ $F_2 > Cl_2 > Br_2 > I_2$
b. Bond dissociation enthalpy of $Cl_2 > F_2 > Br_2 > I_2$
c. Boiling points of $I_2 > Br_2 > Cl_2 > F_2$
d. Reducing power of $I^- > Br^- > Cl^- > F^-$

11 Which option is incorrect among the following for the given property?

a. Dipole moment $(HF > HCl > HBr > HI)$
b. Bond length $(HI > HBr > HCl > HF)$
c. Acidic strength $(HI > HBr > HCl > HF)$
d. Thermal stability $(HF < HCl < HBr < HI)$

12 If chlorine gas is passed through hot NaOH solution. Choose the correct option for the change in oxidation number of chlorine.

a. 0 to −1or −5 b. 0 to −2 or −6
c. 0 to +1or +5 d. −1 to 0 or −5

13 Which of the following does not exist?

a. I_3^- b. Br_3^- c. ICl_2^- d. F_3^-

14 One gas bleaches the colour of flowers by reduction and another gas by oxidation. The gases respectively are

a. SO_2 and Cl_2 b. CO_2 and Cl_2 c. NO and Cl_2 d. H_2S and Br_2

15 Solid iodine is an example of

a. covalent solid b. ionic solid c. molecular solid d. metallic solid

One or More than One Options Correct Type

Direction (Q. Nos. 16-20) *This section contains 5 multiple choice questions. Each question has four choices (a), (b), (c) and (d), out of which ONE or MORE THAN ONE are correct.*

16 The fluoride ores are

a. carnalite b. cryolite
c. fluorspar d. chile saltpetre

Working Space

17 Which one of the following arrangements do not give the correct picture of the trends indicated against it?

a. $F_2 > Cl_2 > Br_2 > I_2$: Oxidising power
b. $F_2 > Cl_2 > Br_2 > I_2$: Electron gain enthalpy
c. $F_2 > Cl_2 > Br_2 > I_2$: Bond dissociation energy
d. $F > Cl > Br > I$: Electronegativity

18 Which of the following are insoluble in water?

a. AgCl b. AgBr c. AgF d. AgI

19 The source of iodine is

a. in free state b. in seaweeds c. as caliche d. as AgI

20 When chlorine reacts with hot, conc. NaOH, the products formed are

a. NaCl b. NaOCl c. $NaClO_3$ d. HCl

Comprehension Type

Direction (Q. Nos. 21 and 22) This section contains a paragraph, describing theory, experiments, data, etc. Two questions related to the paragraph have been given. Each question has only one correct answer among the four given options (a), (b), (c) and (d).

Passage

Group 17 elements are known as halogens (sea salt forming). Their general configuration is ns^2np^5 where, $n = 2$ to 7. Fluorine is the first member of the group differs in several ways from the rest of the group. Halogens are highly reactive elements having strong affinity for hydrogen. All the halogens form both ionic and covalent compounds .Fluorine is always monovalent and shows –1 oxidation state in its compounds .The other halogens exhibit 1, 3, 5, 7 valencies in their compounds and show – 1 to + 7 oxidation states.

21 Valence shell configuration that belongs to most reactive non-metal is

a. $(n-1)s^2p^6ns^1$ b. ns^2np^5 c. ns^2np^6 d. $(n-1)s^2p^6ns^2$

22 In which of the following chlorine has different oxidation states?

a. CrO_2Cl_2 b. $CaOCl_2$ c. Cl_2O_7 d. $Ca(OCl)_2$

Matching List Type

Direction (Q. No. 23 and 24) Choices for the correct combination of elements from Column I and Column II are given as options (a), (b), (c) and (d), out of which one is correct.

23 Match the Column I with Column II and mark the correct option from the codes given below.

	Column I		Column II
i.	Fluorine	p.	Decomposition of water
ii.	Chlorine	q.	Formation of polyhalides
iii.	Bromine	r.	Extraction of gold and platinum
iv.	Iodine	s.	Germicide
		t.	Steriliser of water

Codes

	i	ii	iii	iv		i	ii	iii	iv
a.	p,q	q,r	s	t	b.	p	q	r	s
c.	p	q,r,t	q	q,s	d.	q	s	r	t

Working Space

24 Match the Column I with Column II and mark the correct option from the codes given below.

	Column I		Column II
i.	Fluorine	p.	Bleaching
ii.	Chlorine	q.	Strongest oxidising agent
iii.	Bromine	r.	Show −1 oxidation state
iv.	Iodine	s.	Most negative electron gain enthalpy
		t.	Caliche
		u.	Oxidise only I^-

Codes

	i	ii	iii	iv		i	ii	iii	iv
a.	q,r	s,p	u	t	b.	t	u	s,r	q,p
c.	q,s	p,r	t	u	d.	u	s,r	q,p	t

One Integer Value Correct Type

Direction (Q. Nos. 25-29) This section contains 5 questions. When worked out will result in an integer from 0 to 9 (both inclusive).

25 The number of lone pair of electrons present on the central iodine in I_3^- ion is

26 In the MOT of F_2 molecule, number of electrons occupying antibonding orbitals are

27 The total number of positive oxidation states shown by fluorine is

28 When carnallite is dissolved in water the number of ions formed are

29 The number of water insoluble salts among

$AgF, AgCl, AgBr, AgI, BeF_2, CaF_2, KF, PbCl_2, HgCl_2$

Statement Type

Direction (Q. No. 30) This section is based on Statement I and Statement II. Select the correct answer from the codes given below.

a. Both Statement I and Statement II are correct and Statement II is the correct explanation of Statement I
b. Both Statement I and Statement II are correct but Statement II is not the correct explanation of Statement I
c. Statement I is correct but Statement II is incorrect
d. Statement II is correct but Statement I is incorrect

30 **Statement I** All halogens are coloured.

Statement II Halogens absorbs part of the light in the visible region.

DPP-6 Compounds of Halogens (F, Cl, Br and I)

Subjective Questions

Working Space

Direction *(Q. Nos. 1-5) These questions are subjective in nature, need to be solved completely on notebook.*

1 Explain the following.

i. Anhydrous HCl is a bad conductor of electricity while aqueous HCl is a good conductor.

ii. HF is least volatile while HCl is most volatile amongst hydrogen halides.

iii. HF is not stored in glass bottles but kept in wax bottles.

2 Arrange the following.

i. $HClO, HClO_2, HClO_3, HClO_4$ in the order of acidic strength.

ii. $HClO, HClO_2, HClO_3, HClO_4$ in the order of increasing oxidising power.

iii. $HClO, HClO_2, HClO_3, HClO_4$ in the increasing order of thermal stability.

iv. F^-, Cl^-, Br^-, I^- in the order of decreasing reducing nature.

v. HOF, HOCl, HOBr, HOI in the order of acidic strength.

vi. HF, HCl, HBr, HI in the order of acidic strength.

3 Explain the following.

i. What is euchlorine?

ii. In the preparation of HI from KI, phosphoric acid is preferred than sulphuric acid.

iii. HF has more percentage of ionic nature among the hydrogen halides.

4 Give reasons to the following.

i. Cl_2O is diamagnetic while ClO_2 is paramagnetic.

ii. Bond angle in Cl_2O is 111° while in OF_2 is 103°.

iii. HClO is a powerful oxidising, bleaching and disinfecting agent.

5 Explain the following.

i. Oxidation state of chlorine in bleaching powder.

ii. Available chlorine in bleaching powder.

iii. Bleaching powder loses its bleaching property when kept in an open bottle for a long time.

iv. How is available chlorine estimated in a bleaching powder sample?

Only One Option Correct Type

Working Space

Direction *(Q. Nos. 6-15) This section contains 10 multiple choice questions. Each question has four choices (a), (b), (c) and (d), out of which ONLY ONE is correct.*

6 Regarding the properties of hydrogen halides, which is wrong order?

a. Melting point order HI > HF > HBr > HCl
b. Boiling point order HF > HI > HBr > HCl
c. Acidic nature HI > HBr > HCl > HF
d. Percentage of ionic character HF > HI > HBr > HCl

7 The reaction $3ClO^-(aq) \longrightarrow 2Cl^-(aq) + ClO_3^-(aq)$ is an example of

a. oxidation reaction
b. reduction reaction
c. disproportionation reaction
d. decomposition reaction

8 Which is incorrect statement?

a. The stability of oxides formed by I > Br > Cl
b. Fluorine forms only one oxoacid while other halogens form different oxoacids
c. I_2O_5 is very good oxidising agent and used in the estimation of CO
d. ClO_2 is used as a bleaching agent for paper pulp and textiles

9 When gold and platinum dissolves in aqua-regia, the complexes formed are

a. $[AuCl_4]^-$ and $[PtCl_6]^{2-}$
b. $[AuCl_2]^-$ and $[PtCl_4]^{2-}$
c. $[AuCl_4]^{2-}$ and $[PtCl_6]^{2-}$
d. $[AuCl_3]^-$ and $[PtCl_6]^{2-}$

10 The type of hybrid orbitals used by the chlorine atom in ClO_2^- and ClO_3^- ions respectively are

a. sp^2, sp^3
b. sp^2, sp^2
c. sp^3, sp^3
d. sp^3, sp^2

11 In perchloric acid ($HClO_4$), the number of σ and π-bonds

a. 5σ and 4π-bonds
b. 6σ and 3π-bonds
c. 5σ and 3π-bonds
d. 5σ and 2π-bonds

12 The correct order of acidic strength is

a. $Cl_2O_7 > SO_2 > P_4O_{10}$
b. $SO_2 > Cl_2O_7 > P_4O_{10}$
c. $P_4O_{10} > SO_2 > Cl_2O_7$
d. $N_2O_5 > P_4O_{10} > SO_2$

13 Which of the following arrangements gives the correct order of increasing oxidation number of iodine?

a. $I_2 < HI < ICl < HIO_4$
b. $HI < ICl < I_2 < HIO_4$
c. $ICl < HI < HIO_4 < I_2$
d. $HI < I_2 < ICl < HIO_4$

14 Which of the following is incorrect?

a. ClO_4^- has $3d\pi$-$p\pi$ bonds
b. ClO_3^- and NO_3^- are isostructural
c. Cl_2O_7 is most acidic oxide
d. ClO^- is strong conjugate base

15 Correct order of bond angles are in

a. $H_2O > OF_2 > Cl_2O > ClO_2$
b. $ClO_2 > Cl_2O > H_2O > OF_2$
c. $OF_2 > H_2O > Cl_2O > ClO_2$
d. $OF_2 > OCl_2 > H_2O > ClO_2$

One or More than One Options Correct Type

Working Space

Direction (Q. Nos. 16-20) *This section contains 5 multiple choice questions. Each question has four choices (a), (b), (c) and (d), out of which ONE or MORE THAN ONE are correct.*

16 Paramagnetic oxide among the following is/are

a. ClO_2
b. Cl_2O
c. NO_2
d. NO

17 Which of the following does not show disproportionation reaction?

a. $CaF_2 + H_2SO_4 \longrightarrow CaSO_4 + 2HF$
b. $Cl_2 + 2OH^- \longrightarrow Cl + OCl^- + H_2O$
c. $Br^- + 3OCl^- \longrightarrow BrO_3^- + 3Cl^-$
d. $HOI + 2HCl \longrightarrow H^+ + ICl_2^- + H_2O$

18 NaCl, NaBr and NaI mixture on heating with conc. H_2SO_4 gives gases, respectively

a. HCl, Br_2, I_2
b. Cl_2, Br_2, I_2
c. Br_2, Cl_2, I_2
d. I_2, Br_2, Cl_2

19 Which of the following reactions is/are not feasible?

a. $I_2 + 2NaBr \longrightarrow Br_2 + 2NaI$
b. $I_2 + 2NaCl \longrightarrow Cl_2 + 2NaI$
c. $Br_2 + 2NaCl \longrightarrow Cl_2 + 2NaBr$
d. $Cl_2 + 2NaBr \longrightarrow Br_2 + 2NaCl$

20 Bleaching powder liberate chlorine by

a. reacting with CO_2
b. heating in the presence of $CoCl_2$
c. reacting with excess diluted acid
d. reacting with H_2S

Comprehension Type

Direction (Q. Nos. 21 and 22) *This section contains a paragraph, describing theory, experiments, data, etc. Two questions related to the paragraph have been given. Each question has only one correct answer among the four given options (a), (b), (c) and (d).*

Passage

A salt *A* when heated with $K_2Cr_2O_7$ and conc. H_2SO_4 liberates a gas which is absorbed in NaOH solution. The NaOH solution turns yellow. When this solution is acidified with acetic acid and lead acetate solution is added, a yellow precipitate *B* is formed. When *A* is mixed with MnO_2 and heated with conc. H_2SO_4, a gas *C* is evolved which turns a starch-iodide paper blue.

21 What is the colour of the gas which is evolved when salt *A* is heated with MnO_2 and H_2SO_4?

a. Violet
b. Brown
c. Greenish yellow
d. Colourless

22 When the salt *A* is heated with conc. H_2SO_4, colourless gas is evolved which gives dense fumes when exposed with ammonia. The gas is

a. chlorine
b. HCl
c. Both (a) and (b)
d. SO_2 and HCl

Matching List Type

Working Space

***Direction** (Q. Nos. 23-25) Choices for the correct combination of elements from Column I and Column II are given as options (a), (b), (c) and (d) out of which one is correct.*

23 Match the Column I with Column II and mark the correct option from the codes given below.

	Column I (Halogen acids)		Column II (Property)
i.	HF	p.	Monobasic
ii.	HCl	q.	Strong reducing agent
iii.	HBr	r.	Strong acid
iv.	HI	s.	React with glass

Codes

	i	ii	iii	iv
a.	q,r	r,s	p,s	q,s
b.	s	r	p	q
c.	q	r	s	p
d.	s	p	p	p,q,r

24 Match the Column I with Column II and mark the correct option from the codes given below.

	Column I (Reaction)		Column II (Nature of reaction) with respect to reactant which reacts with water
i.	$F_2 + H_2O \longrightarrow$	p.	Oxidised
ii.	$Cl_2 + H_2O \longrightarrow$	q.	Reduced
iii.	$Br_2 + H_2O \longrightarrow$	r.	Disproportionated
		s.	Neutralised

Codes

	i	ii	iii
a.	q	r	r
b.	p	q	r
c.	s	q	p
d.	s	r	p

25 Match the Column I with Column II and mark the correct option from the codes given below.

	Column I		Column II
i.	Bleaching powder	p.	Reducing agent
ii.	HCl	q.	Bleaching agent
iii.	$HClO_4$	r.	Oxidising agent
iv.	HI	s.	Strong acid
		t.	Colourless gas

Codes

	i	ii	iii	iv
a.	p	p	r,s	q,t
b.	q	r	p	s,t
c.	q,r	p,s,t	r,s	p,s,t
d.	p	q	r	s

Working Space

One Integer Value Correct Type

Direction (*Q. Nos.* 26-29) *This section contains* 4 *questions. When worked out will result in an integer from* 0 *to* 9 (both inclusive).

26 The total number of lone pairs in Cl_2O molecule are ....

27 Number of sigma bonds present in Cl_2O_7 are ....

28 Number of chemical species having sp^3-hybridisation among these
H_3O^+, Cl_2O, ClO_2^-, ClO_3^-, ClO_4^-, OF_2, Br_2O, NO_3^-

29 Aqua-regia is a mixture of one mole of conc. HNO_3 and x moles of conc. HCl. Here, x value is ....

Statement Type

Direction (*Q. No.* 30) *This section is based on Statement I and Statement II. Select the correct answer from the codes given below.*

a. Both Statement I and Statement II are correct and Statement II is the correct explanation of Statement I
b. Both Statement I and Statement II are correct but Statement II is not the correct explanation of Statement I
c. Statement I is correct but Statement II is incorrect
d. Statement II is correct but Statement I is incorrect

30 **Statement I** On dehydration with P_2O_5 at 263 K, $HClO_4$ gives Cl_2O_7.

Statement II This method is used in the preparation of Cl_2O_7.

DPP-7 Interhalogens and Pseudo Halogens

Subjective Questions

Working Space

Direction (Q. Nos. 1-5) *These questions are subjective in nature, need to be solved completely on notebook.*

1 Answer the following.

i. Which element forms highest number of interhalogen compounds?

ii. In the interhalogen compounds, what are the oxidation states shown by central halogen?

2 Interhalogens (diatomic $X—X'$ type) are more reactive than halogens (X_2). Why?

3 What are polyhalide ions and polyhalides? Give example.

4 Write two similarities between Cl^- and CN^- ion.

5 Identify the products in the following.

i. $KF + IF_5 \longrightarrow$

ii. $K[BrICl] \xrightarrow{\Delta}$

iii. $Rb[ICl_2] \xrightarrow{\Delta}$

iv. $I_2 + ICl + AlCl_3 \xrightarrow{\Delta}$

Only One Option Correct Type

Direction (Q. Nos. 6-15) *This section contains 10 multiple choice questions. Each question has four choices (a), (b), (c) and (d), out of which ONLY ONE is correct.*

6 Which of the following is not a pseudo halide ion?

a. CNO^-

b. CN^-

c. SCN^-

d. S^{2-}

7 The stability of interhalogen compounds follows the order

a. $IF_3 > BrF_3 > ClF_3$

b. $BrF_3 > IF_3 > ClF_3$

c. $ClF_3 > BrF_3 > IF_3$

d. $ClF_3 > IF_3 > BrF_3$

8 Match the following molecules or ions to their shapes. Which pairing is correct?

a. $[I_3]^-$: bent

b. BrF_5 : trigonal bipyramidal

c. ClF_3 : trigonal planar

d. $[BrF_4]^-$: square planar

9 In XA_5, the central atom has (both X and A are halogens)

a. 5 bond pairs and no lone pairs

b. 5 bond pairs and one lone pair

c. 6 bond pairs and no lone pairs

d. 4 bond pairs and one lone pair

Working Space

10 Which of the following pairs is isostructural?

a. ClF_3, NF_3 b. IF_5, PF_5
c. BrF_5, IF_5 d. ClO_3^-, NO_3^-

11 Which of the following pairs shows correct match between formula and structure of a species?

a. $[BrF_2]^+$: non-linear
b. $[ICl_4]^-$: tetrahedral
c. IF_5 : trigonal bipyramidal
d. BrF_3 : trigonal pyramidal

12 ClF on hydrolysis form

a. $HCl + HOF$ b. $HOCl + HF$
c. $Cl_2O + HF$ d. None of these

13 Which of the following is correct for $CsBr_3$?

a. It is a covalent compound
b. It contains Cs^{3+} and Br^- ions
c. It contains Cs^+ and Br_3^- ions
d. It contains Cs^+, Br^- and lattice Br_2 molecules

14 In the known interhalogen compounds, the maximum number of atoms are

a. 4 b. 5
c. 8 d. 7

15 Identify the interhalogen compound with zero dipole moment.

a. ICl_3 b. BrF_5
c. IF_7 d. IF_5

One or More than One Options Correct Type

Direction (Q. Nos. 16-20) *This section contains 5 multiple choice questions. Each question has four choices (a), (b), (c) and (d), out of which ONE or MORE THAN ONE are correct.*

16 Identify the products formed at the electrodes upon electrolysis of molten ICl.

a. I_2 at cathode b. Cl_2 at anode
c. I_2 and Cl_2 at anode d. Cl_2 at cathode

17 When $Cs[ICl_2]$ undergo thermal dissociation, the products formed are

a. CsCl b. CsI
c. ICl d. Cl_2

18 Pseudo halogens among the following are

a. $(CN)_2$ b. $(OCN)_2$
c. BrF_5 d. $(SeCN)_2$

19 Which has both ionic and covalent bonding?

a. KCN b. KI_3
c. $CsBr_3$ d. K_2O

20 Same hybridisation is present for the central atom in

a. I_3 b. ICl_2^-
c. ICl_4^- d. BrF_4^-

Comprehension Type

Working Space

Direction *(Q. Nos. 21 and 22) This section contains a paragraph, each describing theory, experiments, data, etc. Two questions related to the paragraph have been given. Each question has only one correct answer among the four given options (a), (b), (c) and (d).*

Passage

When two different halogens react with each other, the compounds formed are called interhalogen compounds.

The general formula of these compounds is given by XX', XX'_3, XX'_5 and XX'_7.

Here, X = halogen of large size and more electropositive.

X' = halogen of smaller size and more electronegative.

e.g. ClF, ClF_3, IF_5, IF_7, etc.

All are covalent and diamagnetic.

They are either volatile liquids or solids except ClF, BrF, ClF_3, IF_5 and IF_7.

Interhalogen compounds are more reactive than halogens (except fluorine). The reason is X—X' bond in interhalogens is weaker than X—X bond in halogens (except F—F bond).

Hydrolysis of interhalogens Interhalogen compounds undergo hydrolysis giving halide ion derived from the smaller halogen and a hypohalite (when XX'), halite (when XX'_3), halate (when XX'_5) and perhalate (when XX'_7) n ion derived from the larger halogen.

$$XX' + H_2O \longrightarrow HX + HOX$$

21 Which is more stable?

a. ClF
b. BrF
c. IF
d. BrCl

22 BrCl on hydrolysis produce

a. HOBr + HCl
b. HOCl + Br
c. HOBr + HOCl
d. HCl + HBr

Statement Type

Direction (Q. Nos. 23 and 24) *This section is based on Statement I and Statement II. Select the correct answer from the codes given below.*

a. Both Statement I and Statement II are correct and Statement II is the correct explanation of Statement I
b. Both Statement I and Statement II are correct but Statement II is not the correct explanation of Statement I
c. Statement I is correct but Statement II is incorrect
d. Statement II is correct but Statement I is incorrect

23 **Statement I** Interhalogens are ionic compounds.

Statement II Greater the difference in electronegativity of two different halogen atoms, more is the stability of their interhalogen compound.

24 **Statement I** Cyanide ions (CN^-) are also called pseudo halogens.

Statement II Cyanide ions make electrovalent bonds with electropositive elements like Na, K, etc.

Working Space

Matching List Type

Direction (Q. No. 25) *Choices for the correct combination of elements from Column I and Column II are given as options (a), (b), (c) and (d), out of which one is correct.*

25 Match the Column I with Column II and mark the correct option from the codes given below.

	Column I		Column II
i.	ClF_3	p.	Colourless liquid
ii.	BrF_3	q.	Yellow liquid
iii.	IF_3	r.	Yellow solid
iv.	IF_5	s.	Colourless gas
		t.	sp^3d

Codes

	i	ii	iii	iv		i	ii	iii	iv
a.	q	s	r	p	b.	s,t	q,t	r,t	s
c.	q,r	p,s	r,t	s,q	d.	p	s	q	r

One Integer Value Correct Type

Direction (Q. Nos. 26-30) *This section contains 5 questions. When worked out will result in an integer from 0 to 9 (both inclusive).*

26 How many pseudo halide ions are there, among the given ions?
ICl_2^-, Cl^-, CN^-, SCN^-, N_3^-, $SeCN^-$, F^-, OH^-

27 In ClF_3, if both lone pairs occupy axial positions then find the number of lone pair-bond pair repulsions.

28 How many of the given structures are isostructural with XeF_4
($XeOF_2$, BrF_4^-, ICl_4^-, IF_4^+, SF_4, PCl_4^+, IBr_2^-)?

29 Find the sum of bond pairs and non-bonding electron pairs in ICl molecule.

30 Find the total number of covalent bonds present in cyanogens.

DPP-8 General Properties and Compounds of Xenon

Subjective Questions

Direction (*Q. Nos. 1-5*) *These questions are subjective in nature, need to be solved completely on notebook.*

Working Space

1 With regard to noble gases name an element which is

i. most abundant in the universe
ii. most abundant in the atmosphere
iii. most reactive
iv. more adsorption on charcoal
v. easily liquefiable
vi. known as hidden and stranger
vii. radioactive element

2 Answer the following questions.

i. How is radon formed from radium?
ii. The oxidation state of xenon in perxenate ion $[XeO_6]^{4-}$.
iii. Name the principle used in the separation of noble gases from liquid air.
iv. Name the adsorbent used in Dewar's method of separation of individual noble gases from the mixture.
v. Name the chemicals that absorb CO_2, traces of oxygen and moisture associated with noble gas mixture.

3 Explain the following.

i. Why He and Ne do not form compounds?
ii. Why Xe, Kr form more number of compounds?
iii. What inspired Bartlett for carrying out reaction between Xe and PtF_6?
iv. Why is helium used in diving apparatus?

4 Give reasons for the following.

i. In XeF_2, the three lone pairs are positioned at equatorial position.
ii. In XeF_4, the two lone pairs are placed at axial position.
iii. $XeF_4 + SbF_5 \longrightarrow$
In this reaction, XeF_4 acts as Lewis base.
iv. All xenon compounds deviate from octet rule.

5 i. What are clathrates?
ii. Why He and Ne do not form clathrates with quinol?
iii. What are the uses of clathrates?

Working Space

Only One Option Correct Type

Direction (Q. Nos. 6-15) *This section contains 10 multiple choice questions. Each question has four choices (a), (b), (c) and (d), out of which ONLY ONE option is correct.*

6 Which of the following is/are correct statement(s) regarding noble gases?

i. He_2 molecule does not exist.

ii. $\frac{C_p}{C_V}$ value for noble gases is 1.4.

iii. Heaviest noble gas present in air is xenon.

iv. 6th period starts with caesium and ends with xenon.

a. i, iii and iv b. i, ii and iv c. ii, iii and iv d. All are correct

7 Which statement is false?

a. 118th element is a noble gas called ununoctium

b. Helium is lightest noble gas

c. Xenon is the most reactive among the rare gases

d. Noble gases Ne, Kr and Xe were discovered by Ramsay and Rayleigh

8 The ionisation potential of hydrogen atom is 13.6 eV. Then ionisation potential of He^+ ion is

a. 27.2 eV b. 6.8 eV c. 13.6 eV d. 54.4 eV

9 Which of the following is/are correct among the following?

i. $XeF_2 = sp^3d$, linear

ii. $XeF_4 = sp^3$, square planar

iii. $XeO_4 = sp^3$, tetrahedral

iv. $XeF_6 = sp^3d^2$, octahedral

a. ii and iii b. i and iii c. i, ii and iii d. Only iii

10 The oxidation number of oxygen in O_2PtF_6 is

a. zero b. +1/2 c. +1 d. −1/2

11 The structure of XeF_5^+ ion is

a. trigonal bipyramidal b. square pyramidal

c. octahedral d. pentagonal

12 The number of $(p\pi\text{-}d\pi)\,\pi$-bonds present in XeO_3 and XeO_4 respectively are

a. 3 and 4 b. 2 and 3 c. 4 and 2 d. 3 and 2

13 XeF_6 on reaction with CsF produce

a. $[XeF_5][CsF_2]$ b. $Cs[XeF_7]$ c. $[XeF_4][CsF_3]$ d. No reaction

14 Which of the following reaction occurs?

a. $XeF_2 + PF_5 \longrightarrow XeF^+[PF_6]^-$

b. $XeF_4 + SbF_5 \longrightarrow [XeF_3^+][SbF_6]^-$

c. $XeF_6 + H_2O \longrightarrow XeOF_4 + 2HF$

d. All of these

15 The wrong statement among the following is

a. Helium is used to fill observatory balloons because it is lighter than air and non-combustible

b. A mixture of 80% helium and 20% oxygen is used for respiration by deep sea divers

c. The noble gas used in atomic reactors is argon

d. Neon is used in discharge tubes and fluorescent bulbs for advertisement display purposes

One or More than One Options Correct Type

Working Space

Direction *(Q. Nos. 16-20) This section contains 5 multiple choice questions. Each question has four choices (a), (b), (c) and (d), out of which ONE or MORE THAN ONE are correct.*

16 Which of the following is/are correct regarding noble gases?

a. The commercial source of helium is natural gas

b. In 4th period, the elements with lowest and highest ionisation energies are K and Kr

c. Xenon is isoelectronic with La^{3+}

d. Radium when emits beta-particle, radon and helium are formed

17 The correct statements are

a. The correct order of occurrence (% by volume) in the air is Ar > Ne > Kr > He > Xe

b. In the separation of noble gases by Dewar's method, the principle used is fractional distillation

c. The boiling point order of noble gases is He < Ar < Kr < Xe

d. Xenon forms fluorides, oxides and oxyfluorides in excited states

18 XeO_3 can be prepared by

a. $XeF_2 \xrightarrow{\text{Hydrolysis}}$ b. $XeF_6 + SiO_2 \longrightarrow$

c. $XeF_4 \xrightarrow{\text{Hydrolysis}}$ d. $XeF_6 \xrightarrow{\text{Hydrolysis}}$

19 Pick up the wrong statement(s).

a. The ionisation energies of Xe and dioxygen are nearly same

b. In XeF_2, lone pairs are occupied in equatorial and axial sp^3d hybrid orbitals

c. $XeOF_4$, XeF_6 and XeO_3 have same number of lone pairs on xenon

d. XeF_4 and SF_4 are isostructural

20 The correct statements are

a. Solid argon is an example of molecular solid

b. The noble gas which do not form clathrates are He and Ne

c. Liquid He is used as cryogenic liquid

d. Polarisability of noble gases increases from He to Xe

Comprehension Type

Direction *(Q. Nos. 21 and 22) This section contains a paragraph, each describing theory, experiments, data, etc. Two questions related to the paragraph have been given. Each question has only one correct answer among the four given options (a), (b), (c) and (d).*

Passage

The noble gases have closed-shell electronic configuration and are monoatomic gases under normal conditions. The lower boiling points of the lighter noble gases are due to weak dispersion forces between the atoms and the absence of other interatomic interactions.

Direct reaction of xenon with fluorine leads to the formation of a series of compounds with oxidation numbers +2, + 4, + 6 and + 8. XeF_4 reacts violently with water to give XeO_3. The compounds of xenon exhibit rich stereochemistry and their geometries can be deduced considering the total number of electron pairs in the valence shell.

21 The noble gas compound without lone pair on xenon atom is

a. XeF_6 b. $XeOF_4$ c. XeO_2F_4 d. XeO_2F_2

22 Which of the following property decreases from helium to xenon?

a. Boiling point b. Density

c. Critical temperature d. Ionisation enthalpy

Working Space

Statement Type

Direction (Q. Nos. 23 and 24) *This section is based on Statement I and Statement II. Select the correct answer from the codes given below.*

a. Both Statement I and Statement II are correct and Statement II is the correct explanation of Statement I
b. Both Statement I and Statement II are correct but Statement II is not the correct explanation of Statement I
c. Statement I is correct but Statement II is incorrect
d. Statement II is correct but Statement I is incorrect

23 **Statement I** Radon is produced upon α-decay of radium atom.

Statement II α-particles which are equivalent to the atom, reduces the atomic number by 2 and atomic mass by 4 upon decay of mother nucleus.

24 **Statement I** The structure of XeF_4 is square pyramidal.

Statement II The hybridisation of XeF_4 is sp^3d^2.

Matching List Type

Direction (Q. No. 25) *Choices for the correct combination of elements from Column I and Column II are given as options (a), (b), (c) and (d) out of which one is correct.*

25 Match the Column I with Column II and mark the correct option from the codes given below.

	Column I		Column II
i.	He	p.	To detect defects in steel casting
ii.	Ar	q.	Used in airship balloons
iii.	Xe	r.	To provide inert atmosphere
iv.	Rn	s.	Used in flash bulbs
		t.	Thickness of metal and plastic sheets

Codes

	i	ii	iii	iv		i	ii	iii	iv
a.	q	r	s	p	b.	p	q	s	r
c.	q	p	r	s	d.	p	s	q	r

One Integer Value Correct Type

Direction (Q. Nos. 26-30) *This section contains 5 questions. When worked out will result in an integer from 0 to 9 (both inclusive).*

26 Number of lone pairs on the central atom in $XeOF_4$ are

27 Among XeF_2, XeF_4, XeF_6, XeO_3, XeO_4, XeO_2, $XeOF_4$ the number of compounds that have non-zero dipole moment.

28 In XeF_2, state the total number of electron pairs in its Lewis dot structure.

29 In argon, find the number of electrons with spin value $+\frac{1}{2}$.

30 Find the number of unpaired electrons in the fully excited xenon atom.

SET - 1

Revisal Problems for JEE Main

Only One Option Correct Type

***Direction** (Q. Nos. 1-30) This section contains 30 multiple choice questions. Each question has four choices (a), (b), (c) and (d), out of which ONLY ONE is correct.*

Working Space

1 The true statement for N_3^- is

a. It has non-linear structure
b. It is called pseudo halogen
c. It is isoelectronic with N_2O
d. It is known as nitride ion

2 One mole of each H_3PO_2, H_3PO_3 and H_3PO_4 will neutralise *x* moles of NaOH. *y* moles of $Ca(OH)_2$ and *z* moles of $Al(OH)_3$. Here *x*, *y*, *z* ratio is

a. 3 : 1.5 : 1
b. 1 : 2 : 3
c. 3 : 2 : 1
d. 1 : 1 : 1

3 When heated which metal nitrate would give the respective metal, oxide NO_2 and O_2?

a. $AgNO_3$
b. $Cu(NO_3)_2$
c. $Pb(NO_3)_2$
d. KNO_3

4 Which shape correctly describes the stated molecule?

a. NF_3, trigonal pyramidal
b. P_2H_4, planar
c. N_2O, bent
d. NO_2, linear

5 P_4O_{10} in cold and hot water produce respectively

a. HPO_3 and H_3PO_4
b. H_3PO_3 and H_3PO_4
c. H_3PO_2 and H_3PO_4
d. $H_4P_2O_7$ and H_3PO_4

6 $NH_4NO_3 \xrightarrow{\Delta} A + H_2O$, $NH_3 + O_2 \xrightarrow{\text{Catalyst}} B + H_2O$, $Pb(NO_3)_2 \xrightarrow{\Delta} C + PbO + O_2$

Among *A*, *B*, *C*, paramagnetic substance(s) is/are

a. Only *A*
b. Only *B*
c. *A* and *C*
d. *B* and *C*

7 Ammonium dichromate on heating '*X*' gas is liberated. The same gas is liberated by heating

a. NH_4NO_2
b. NH_4NO_3
c. $Mg_3N_2 + H_2O$
d. $Na_2O_2 + H_2O$

8 Which has more covalent bonds?

a. P_4O_{10}
b. Cl_2O_7
c. $H_4P_2O_7$
d. $H_2S_2O_7$

9 When sulphuric acid reacts with PCl_5, sulphuryl chloride is formed as the final product. This shows that sulphuric acid is

a. a dibasic acid
b. has great affinity for water
c. has two hydroxyl groups in its structure
d. is a derivative of sulphur dioxide

10 S—S bond is present in

I. $H_2S_2O_6$ II. $H_2S_2O_8$ III. $H_2S_2O_7$ IV. $H_2S_2O_5$

a. I and IV
b. I, II and IV
c. I, III and IV
d. All of these

Working Space

11 Which one of the following cannot act as an oxidising agent?

a. S^{2-}
b. SO_3^{2-}
c. SO_4^{2-}
d. $S_2O_8^{2-}$

12 The V-shape of SO_2 is due to the presence of

a. two σ and π-bonds
b. two σ and two π-bonds
c. two σ bonds and one lone pair of electrons
d. two σ and two π-bonds, and one lone pair of electrons

13 Among SF_4, XeF_4, BrF_4^- and ICl_4^- the number of species having two lone pairs of electrons on the central atom according to VSEPR theory is

a. 2
b. 3
c. 4
d. 0

14 Liquid oxygen and liquid nitrogen are allowed to flow between the poles of an electromagnet. Choose the correct observation.

a. Both will be attracted but to oppositely pole pieces
b. Both will be attracted but to same poles
c. Liquid O_2 will be attracted but liquid N_2 will be repelled to the same degree
d. Liquid O_2 will be attracted but liquid N_2 unaffected

15 Among the species, O_2^+, O_2 and O_2^- the order of first ionisation energy is

a. $O_2^+ < O_2 < O_2^-$
b. $O_2^- < O_2^+ < O_2$
c. $O_2^- < O_2 < O_2^+$
d. $O_2^+ < O_2^- < O_2$

16 The amount of $Na_2S_2O_3 \cdot 5H_2O$ required to completely reduce 100 mL of 0.25 N iodine solution is

a. 6.20 g
b. 9.30 g
c. 3.10 g
d. 7.75 g

17 The anion forms an oxyacid of chlorine has the same number of $p\pi$- $d\pi$ bonds as that of SO_3 (gas), oxidation state of chlorine and the structure of anion is

a. + 5, angular
b. +3, angular
c. +5, pyramidal
d. +7, tetrahedral

18 The products formed when slaked lime reacts with chlorine on heating

a. $CaCl_2$ and $Ca(ClO_3)_2$
b. $CaCl_2$ and $CaOCl_2$
c. $CaCl_2$ and $Ca(OCl)_2$
d. $CaCl_2$ and O_2

19 Which one among CO_3^{2-}, SO_3, XeO_3 and NO_3^- have planar shape?

a. CO_3^{2-}, SO_3 and XeO_3
b. SO_3, XeO_3 and NO_3^-
c. CO_3^{2-}, XeO_3 and NO_3^-
d. CO_3^{2-}, SO_3 and NO_3^-

20 Which of the following molecules has a measurable dipole moment?

a. BrF_3
b. XeF_4
c. CF_4
d. PCl_5

21 When Cl_2 gas reacts with hot and concentrated sodium hydroxide solution, the oxidation number of chlorine changes from

a. 0 to −1 and 0 to +3
b. 0 to +1 and 0 to −3
c. 0 to +1 and 0 to −5
d. 0 to −1 and 0 to +5

22 The geometry of the central atom in ClF_4^+ is

a. square planar
b. square pyramidal
c. octahedral
d. *see-saw*

23 If *A* represents the central atom, in which molecule F-*A*-F angle is the smallest?

a. BF_3
b. CF_4
c. NF_3
d. XeF_2

Working Space

24 In the preparation of fluorine, electrolysis of molten KHF_2 is preferred than molten fluorspar. The reason is

a. KHF_2 is more abundant
b. KHF_2 is more ionic
c. KHF_2 has less melting point than CaF_2
d. KHF_2 has more melting point than CaF_2

25 XeF_6 on complete hydrolysis yields '*X*'. The molecular formula of *X* and its geometry respectively are

a. XeO_2 and linear
b. XeO_3 and trigonal planar
c. XeO_3 and pyramidal
d. XeO_4 and tetrahedral

26 The ratio of the heat capacities $\frac{C_p}{C_V}$, for one mole of a gas is 1.6. The gas is

a. He
b. CO_2
c. H_2
d. CH_4

27 Which pair have identical shapes?

a. CF_4 and SF_4
b. PF_5 and IF_5
c. XeF_2 and CS_2
d. PCl_3 and ClF_3

28 Which classification is correct?

a. XeF_2 is a fluoride donor
b. XeF_6 is a fluoride acceptor but not a fluoride donor
c. AsF_5 donates fluoride ions to XeF_6
d. SbF_5 is always a fluoride acceptor and always forms SbF_6^-

29 Which of the following is the life saving mixture of asthma patient?

a. Mixture of helium and oxygen
b. Mixture of neon and oxygen
c. Mixture of xenon and nitrogen
d. Mixture of argon and oxygen

30 Match the shape to the formula. Which pairing is incorrect?

a. XeO_2F_2 = *see-saw*
b. XeO_3 = trigonal planar
c. $[XeF_3]^+$ = T shape
d. $[XeF_5]^-$ = pentagonal planar

SET - 2

Revisal Problems
for JEE Advanced

Only One Option Correct Type

Working Space

Direction *(Q. Nos. 1-10) This section contains 10 multiple choice questions. Each question has four choices (a), (b), (c) and (d), out of which ONLY ONE is correct.*

1 The number of unpaired electrons present in the species $[Fe(H_2O)_5NO]^{2+}$ which is formed during 'brown ring' test is
a. 2 b. 3 c. 4 d. 5

2 In which one of the following pairs the species have similar geometry?
a. CO_2 and SO_2 b. BH_3 and NH_3 c. CO_3^{2-} and SO_3^{2-} d. SO_4^{2-} and ClO_4^-

3 O_2 is converted into O_2^+ using
a. PtF_6 b. KF c. Br_2 d. $Na_2S_2O_3$

4 Which of the following pairs has the highest difference in their first ionisation energy?
a. Xe and Cs b. Kr and Rb c. Ar and K d. Ne and Na

5 Among the compounds I to IV, those which hydrolyse easily

I. NCl_3 II. NF_3 III. $BiCl_3$ IV. PCl_3
a. I and II b. I, III and IV c. II, III and IV d. I, II and III

6 Among the following pairs, those in which both species have similar structure are
a. N_3^- and XeF_2 b. $[ClF_2]^+$ and $[ICl_2]^-$ c. $[ICl_4]^-$ and $[SF_4]$ d. XeO_3 and SO_3

7 The oxoacid of sulphur that does not contain S—S bond is
a. pyrosulphurous acid ($H_2S_2O_5$) b. dithionous acid ($H_2S_2O_4$)
c. dithionic acid ($H_2S_2O_6$) d. pyrosulphuric acid ($H_2S_2O_7$)

8 Among the following NO, SO_2, SO_3, CaO, Cl_2O, Cl_2O_7, CO, N_2O_3, P_4O_{10} the number of anhydrides of acids are
a. 3 b. 4 c. 5 d. 6

9 How many H^+ ions are required when the equation below is balanced with the smallest whole number coefficients?

$$Cu(s) + NO_3^-(aq) + H^+(aq) \longrightarrow Cu^{2+}(aq) + NO(g) + H_2O(l)$$

a. 2 b. 4 c. 6 d. 8

10 In the reaction, $SO_2 + 2H_2S \longrightarrow 3S + 2H_2O$
a. sulphur is reduced and there is no oxidation
b. sulphur is reduced and hydrogen is oxidised
c. sulphur is oxidised and hydrogen is reduced
d. sulphur is both oxidised and reduced

One or More than One Options Correct Type

Direction *(Q. Nos. 11-20) This section contains 10 multiple choice questions. Each question has four choices (a), (b), (c) and (d), out of which ONE or MORE THAN ONE are correct.*

11 Among the following, the species having same bond order are
a. NO b. O_2^+ c. O_2^- d. F_2^-

12 The oxidation number of sulphur atoms in $Na_2S_4O_6$ are
a. +6 b. +5 c. 0 d. +3

Working Space

13 Which of the following statement is correct?

a. ONCl and ONO^- are not isoelectronic
b. O_3 molecule is bent
c. Ozone is violet-black solid
d. Ozone is diamagnetic gas

14 Among PbS, CuS, MnS, NiS, CoS, HgS, Ag_2S, Bi_2S_3 and SnS_2 the black coloured sulphides are

a. CuS
b. CdS
c. MnS
d. Ag_2S

15 Choose the correct statement from the following.

a. Fluoride ion has higher hydration energy than other halide ions
b. More volatile hydride is HF
c. Fluorine has a lesser tendency to form cation than anion
d. Except fluorine other halogens form polyhalide ions

16 The number of lone pairs are identical in the pairs

a. XeF_4, ClF_3
b. XeO_2F_2, ICl_4^-
c. XeO_4, ICl_4^-
d. ICl_4^-, XeF_4

17 Which of the following is oxidised by ozone?

a. $KMnO_4$
b. K_2MnO_4
c. KI
d. $FeSO_4$

18 Identify the correct statements.

a. N—O bond in NO_2^- is shorter than the N—O bond in NO_3^-
b. NH_3 has higher proton affinity than PH_3
c. Both SF_4 and SF_6 undergo hydrolysis easily
d. Sulphur in vapour state exhibit paramagnetism

19 Which of the following statements is/are correct?

a. HI cannot be prepared by heating KI with conc. H_2SO_4
b. BrO_3^- and XeO_3 are isostructural
c. Bi^{5+} is stronger oxidising agent among Bi^{5+} and Sb^{5+}
d. Both NH_3 and N_3^- ion are isostructural

20 Regarding S_8 molecule correct statements are

a. This gas have 8 bond pairs and 16 lone pairs
b. Sulphur is sp^3-hybridised
c. It has planar cyclic structure
d. Sulphur is divalent and it has '0' oxidation state

Comprehension Type

***Direction** (Q. Nos. 21 and 22) This section contains a paragraph, each describing theory, experiments, data, etc. Two questions related to the paragraph have been given. Each question has only one correct answer among the four given options (a), (b), (c) and (d).*

Passage

The reactions of Cl_2 gas with cold dilute and hot conc. NaOH in water give sodium salts of two (different) oxoacids of chlorine, *P* and *Q* respectively. The Cl_2 gas reacts with SO_2 gas in the presence of charcoal, to give product *R*. *R* reacts with white phosphorus to give a compound *S*. On hydrolysis, *S* gives an oxoacid of phosphorus, *T*.

21 *P* and *Q* respectively are the sodium salts of

a. hypochlorous and chloric acids
b. hypochlorous and chlorous acids
c. chloric and perchloric acids
d. chloric and hypochlorous acids

22 *R*, *S* and *T* respectively are

a. $SOCl_2$, PCl_5 and H_3PO_4
b. SO_2Cl_2, PCl_3 and H_3PO_4
c. SO_2Cl_2, PCl_5 and H_3PO_2
d. SO_2Cl_2, PCl_5 and H_3PO_4

Working Space

Statement Type

Direction (Q. No. 23) This section is based on Statement I and Statement II. Select the correct answer from the codes given below.

a. Both Statement I and Statement II are correct and Statement II is the correct explanation of Statement I
b. Both Statement I and Statement II are correct but Statement II is not the correct explanation of Statement I
c. Statement I is correct but Statement II is incorrect
d. Statement II is correct but Statement I is incorrect

23 **Statement I** The most stable octaatomic interhalogen compound is IF_7.

Statement II The difference between electronegativities of iodine and fluorine is greatest among the halogens.

Matching List Type

Direction (Q. No. 24) Choices for the correct combination of elements from Column I and Column II are given as options (a), (b), (c) and (d), out of which one is correct.

24 Match the following reactions given in Column I with the corresponding products given in Column II.

	Column I		Column II
i.	Cu+ dil. HNO_3	p.	NO
ii.	Cu+ conc. HNO_3	q.	NO_2
iii.	Zn+ dil. HNO_3	r.	N_2O
iv.	Zn+ conc. HNO_3	s.	$Cu(NO_3)_2$
		t.	$Zn(NO_3)_2$

Codes

	i	ii	iii	iv		i	ii	iii	iv
a.	p,s	q,s	r,t	q,t	b.	p	q	r	s
c.	p,s	q	s	t	d.	p	r	q	s

One Integer Value Correct Type

Direction (Q. Nos. 25-30) This section contains 6 questions. When worked out will result in an integer from 0 to 9 (both inclusive).

25 How many moles of water reacts with one mole of PCl_5 for complete hydrolysis?

26 The number of P— O— P bonds in the cyclic trimer of metaphosphoric acid is

27 Find the number of chemical species having pyramidal shape among these species H_3O^+, SO_2, H_2S, OSF_2, SO_3, SO_3^{2-}, $S_2O_3^{2-}$, NF_3.

28 Among the following the number of metals which are rendered passive in the presence of conc. HNO_3,

Al, Sn, Pb, Fe, Cr, Zn, Ni, Hg, Cu.

29 Find the number of chemical species among following that have $d\pi$-$p\pi$ bonds in their structure NO_3^-, SO_3, XeO_3, H_3PO_4, ClO_4^-, CO_3^{2-}, SO_4^{2-}, $(CN)_2$, $XeOF_2$

30 How many orbitals are involved in the hybridisation of iodine in IF_5?

JEE Main & AIEEE Archive

(Compilation of Last 13 Years' Questions)

1 Among the following oxoacids, the correct decreasing order of acidic strength is **(JEE Main 2014)**

a. $HOCl > HClO_2 > HClO_3 > HClO_4$
b. $HClO_4 > HOCl > HClO_2 > HClO_3$
c. $HClO_4 > HClO_3 > HClO_2 > HOCl$
d. $HClO_2 > HClO_4 > HClO_3 > HOCl$

2 Which one of the following properties is not shown by NO? **(JEE Main 2014)**

a. It is diamagnetic in gaseous state
b. It is a neutral oxide
c. It combines with oxygen to form nitrogen dioxide
d. Its bond order is 2.5

3 Which of the following is correct statement? **(JEE Main 2013)**

a. ONCl and ONO^- are not isoelectronic
b. O_3 molecule is bent
c. Ozone is violet-black in solid state
d. Ozone is diamagnetic gas

4 Which of the following on thermal decomposition yields a basic as well as acidic oxide? **(AIEEE 2012)**

a. $NaNO_3$ b. $KClO_3$ c. $CaCO_3$ d. NH_4NO_3

5 Which of the following statements is wrong? **(AIEEE 2011)**

a. The stability of hydrides increases from NH_3 to BiH_3 in group 15 of the periodic table
b. Nitrogen cannot form $d\pi$-$p\pi$ bond
c. Single N—N bond is weaker than the single P—P bond
d. N_2O_4 has two resonance structures

6 Which of the following statements regarding sulphur is incorrect? **(AIEEE 2011)**

a. S_2 molecule is paramagnetic
b. The vapour at 200°C consists mostly of S_8 rings
c. At 600°C, the gas mainly consists of S_2 molecules
d. The oxidation state of sulphur is never less than +4 in its compounds

7 In which of the following arrangements, the sequence is not strictly according to the property written against it? **(AIEEE 2009)**

a. $CO_2 < SiO_2 < SnO_2 < PbO_2$
increasing oxidising power
b. $HF < HCl < HBr < HI$
increasing acidic strength
c. $NH_3 > PH_3 < AsH_3 < SbH_3$
increasing basic strength
d. $B < C < O < N$
increasing first ionisation enthalpy

8 Which one of the following reaction of xenon compounds is not feasible? **(AIEEE 2009)**

a. $XeO_3 + 6HF \longrightarrow XeF_6 + 3H_2O$
b. $3XeF_4 + 6H_2O \longrightarrow 2Xe + XeO_3 + 12HF + 1.5O_2$
c. $2XeF_2 + 2H_2O \longrightarrow 2Xe + 4HF + O_2$
d. $XeF_6 + RbF \longrightarrow Rb[XeF_7]$

9 Regular use of which of the following fertilisers increases the acidity of soil? **(AIEEE 2007)**

a. Potassium nitrate b. Urea
c. Superphosphate of lime d. Ammonium sulphate

10 The number of hydrogen atom(s) attached to phosphorus atom in hypophosphorous acid is **(AIEEE 2005)**

a. three b. one
c. two d. zero

11 The correct order of the thermal stability of hydrogen halides (H — X) is **(AIEEE 2005)**

a. $HI > HCl > HF > HBr$ b. $HCl < HF > HBr < HI$
c. $HF > HCl > HBr > HI$ d. $HI > HBr > HCl > HF$

12 Which one of the following statements regarding helium is incorrect? **(AIEEE 2004)**

a. It is used to fill gas balloons instead of hydrogen because it is lighter and non-inflammable
b. It is used as a cryogenic agent for carrying out experiments at low temperatures
c. It is used to produce and sustain powerful superconducting magnets
d. It is used in gas-cooled nuclear reactors

13 The substance used in Holmes signals of the ship is a mixture of **(AIEEE 2003)**

a. $CaC_2 + Ca_3P_2$ b. $Ca_3(PO_4)_2 + Pb_3O_4$
c. $H_3PO_4 + CaCl_2$ d. $NH_3 + HOCl$

14 Concentrated hydrochloric acid when kept in open air, sometimes produces a cloud of white fumes. The explanation for it is that **(AIEEE 2003)**

a. concentrated hydrochloric acid emits strongly smelling HCl gas all the time
b. oxygen in air reacts with the emitted HCl gas to form a cloud of chlorine gas
c. strong affinity of HCl gas for moisture in air results in forming of droplets of liquid solution which appears like a cloudy smoke
d. due to strong affinity for water, concentrated hydrochloric acid pulls moisture of air towards itself. This moisture forms droplets of water and hence, the cloud

15 PCl_3 and PCl_5 both exist; NCl_3 exists but NCl_5 does not exist. It is due to **(AIEEE 2002)**

a. lower electronegativity of P than N
b. lower tendency of N to form covalent bond
c. availability of vacant *d*-orbital in P but not in N
d. statement is itself incorrect

JEE Advanced & IIT JEE Archive

(Compilation of Last 10 Years' Questions)

1 Under ambient conditions, the total number of gases released as products in the final step of the reaction scheme, shown below, is

(2014 Adv., Only One Option Correct Type)

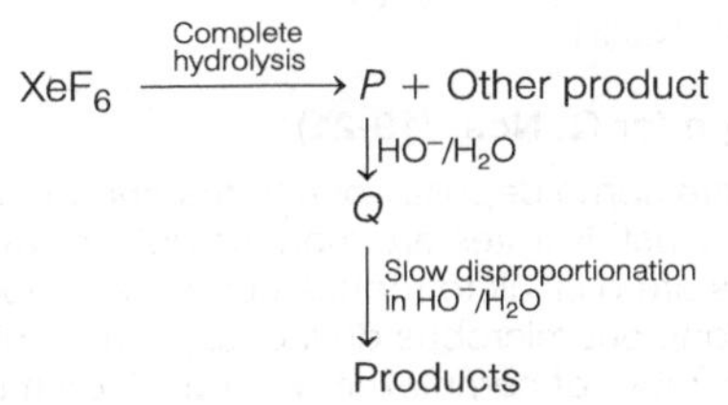

a. 0 b. 1
c. 2 d. 3

2 The product formed in the reaction of $SOCl_2$ with white phosphorus is

(2014 Adv., Only One Option Correct Type)

a. PCl_3 b. SO_2Cl_2
c. SCl_2 d. $POCl_3$

3 The correct statement(s) about O_3 is(are)

(2013 Adv., One or More than One Options Correct Type)

a. O—O bond lengths are equal
b. Thermal decomposition of O_3 is endothermic
c. O_3 is diamagnetic in nature
d. O_3 has a bent structure

Passage for Q. Nos. (4-5)

The reactions of Cl_2 gas with cold-dilute and hot-concentrated NaOH in water give sodium salts of two (different) oxoacids of chlorine, *P* and *Q*, respectively. The Cl_2 gas reacts with SO_2 gas in the presence of charcoal, to give a product *R*. *R* reacts with white phosphorus to give a compound *S*. On hydrolysis, *S* gives an oxoacid of phosphorus *T*.

(2013 Adv., Comprehension Type)

4 *P* and *Q* respectively, are the sodium salts of
a. hypochlorous and chloric acids
b. hypochlorous and chlorous acids
c. chloric and perchloric acids
d. chloric and hypochlorous acids

5 *R*, *S* and *T* respectively, are
a. SO_2Cl_2, PCl_5 and H_3PO_4
b. SO_2Cl_2, PCl_3 and H_3PO_3
c. $SOCl_2$, PCl_3 and H_3PO_2
d. $SOCl_2$, PCl_5 and H_3PO_4

6 The unbalanced chemical reactions given in Column I show missing reagent or condition (?) which are provided in Column II. Match Column I with Column II and select the correct answer using the codes given below.

(2013 Adv., Matching Type)

	Column I		Column II
i.	$PbO_2 + H_2SO_4 \xrightarrow{?} PbSO_4 + O_2 +$ other product	p.	NO
ii.	$Na_2S_2O_3 + H_2O \xrightarrow{?} NaHSO_4 +$ other product	q.	I_2
iii.	$N_2H_4 \xrightarrow{?} N_2 +$ other product	r.	Warm
iv.	$XeF_2 \xrightarrow{?} Xe +$ other product	s.	Cl_2

Codes

	i	ii	iii	iv		i	ii	iii	iv
a.	s	q	r	p	b.	r	q	p	s
c.	p	s	q	r	d.	r	s	q	p

Passage for Q. Nos. (7-8)

Bleaching powder and bleach solution are produced on a large scale and used in several household products. The effectiveness of bleach solution is often measured by iodometry. **(2012, Comprehension Type)**

7 25 mL of household bleach solution was mixed with 30 mL of 0.50 M KI and 10 mL of 4 N acetic acid. In the titration of the liberated iodine, 48 mL of 0.25 N $Na_2S_2O_3$ was used to reach the end point. The molarity of the household bleach solution is

a. 0.48 M b. 0.96 M c. 0.24 M d. 0.024 M

8 Bleaching powder contains a salt of an oxoacid as one of its components. The anhydride of that oxoacid is

a. Cl_2O b. Cl_2O_7 c. ClO_2 d. Cl_2O_6

9 Which ordering of compounds is according to the decreasing order of the oxidation state of nitrogen? **(2012, Only One Option Correct Type)**

a. HNO_3, NO, NH_4Cl, N_2
b. HNO_3, NO, N_2, NH_4Cl
c. HNO_3, NH_4Cl, NO, N_2
d. NO, HNO_3, NH_4Cl, N_2

10 The reaction of white phosphorus with aqueous NaOH gives phosphine along with another phosphorus containing compound. The reaction type; the oxidation states of phosphorus in phosphine and the other product respectively are **(2012, Only One Option Correct Type)**

a. redox reaction; − 3 and − 5
b. redox reaction; 3 and + 5
c. disproportionation reaction; − 3 and + 5
d. disproportionation reaction; − 3 and + 3

11 The shape of XeO_2F_2 molecule is **(2012, Only One Option Correct Type)**

a. trigonal bipyramidal b. square planar
c. tetrahedral d. *see-saw*

12 Extra pure N_2 can be obtained by heating **(2011, Only One Option Correct Type)**

a. NH_3 with CuO b. NH_4NO_3
c. $(NH_4)_2Cr_2O_7$ d. $Ba(N_3)_2$

13 Reaction of Br_2 with Na_2CO_3 in aqueous solution gives sodium bromide and sodium bromate with evolution of CO_2 gas. The number of sodium bromide molecules involved in the balanced chemical equation is **(2011, Interger Type)**

14 Among the following, the number of compounds that can react with PCl_5 to give $POCl_3$ is O_2, CO_2, SO_2, H_2O, H_2SO_4, P_4O_{10}. **(2011, Interger Type)**

15 All the compounds listed in Column I react with water. Match the result of the respective reactions with the appropriate options listed in Column II. **(2010, Matching Type)**

	Column I		Column II
i.	$(CH_3)_2SiCl_2$	p.	Hydrogen halide formation
ii.	XeF_4	q.	Redox reaction
iii.	Cl_2	r.	Reacts with glass
iv.	VCl_5	s.	Polymerisation
		t.	O_2 formation

Codes

	i	ii	iii	iv
a.	p,s	p,q,r,t	p,q,t	p
b.	q,r,t	p,r,s	q,t	r,s
c.	p,q,t	q,s	q	r,s,t
d.	p	q,r	r,t	r,t

16 The reaction of P_4 with *X* leads selectively to P_4O_6. The *X* is **(2009, Only One Option Correct Type)**

a. dry O_2
b. a mixture of O_2 and N_2
c. moist O_2
d. O_2 in the presence of aqueous NaOH

17 The nitrogen oxide(s) that contain(s) N—N bond(s) is(are) **(2009, One or More than One Options Correct Type)**

a. N_2O b. N_2O_3
c. N_2O_4 d. N_2O_5

18 Aqueous solution of $Na_2S_2O_3$ on reaction with Cl_2 gives **(2008, Only One Option Correct Type)**

a. $Na_2S_4O_6$
b. $NaHSO_4$
c. NaCl
d. NaOH

Passage for Q. Nos. (19-21)

There are some deposits of nitrates and phosphates in earth's crust. Nitrates are more soluble in water. Nitrates are difficult to reduce under the laboratory conditions but microbes do it easily. Ammonia forms large number of complexes with transition metal ions. Hybridisation easily explains the ease of sigma donation capability of NH_3 and PH_3. Phosphine is a flammable gas and is prepared from white phosphorus. **(2008, Comprehension Type)**

19 Among the following, the correct statement is

a. Phosphates have no biological significance in humans
b. Between nitrates and phosphates, phosphates are less abundant in earth's crust
c. Between nitrates and phosphates, nitrates are less abundant in earth's crust
d. Oxidation of nitrates is possible in soil

20 Among the following, the correct statement is

a. Between NH_3 and PH_3, NH_3 is a better electron donor because the lone pair of electrons occupies spherical 's' orbital and is less directional
b. Between NH_3 and PH_3, PH_3 is a better electron donor because the lone pair of electrons occupies sp^3- orbital and is more directional
c. Between NH_3 and PH_3, NH_3 is a better electron donor because the lone pair of electrons occupies sp^3- orbital and is more directional
d. Between NH_3 and PH_3, PH_3 is a better electron donor because the lone pair of electrons occupies spherical 's' orbital and is less directional

21 White phosphorus on reaction with NaOH gives PH_3 as one of the products. This is a

a. dimerisation reaction
b. disproportionation reaction
c. condensation reaction
d. precipitation reaction

22 The percentage of *p*-character in the orbitals forming P—P bonds in P_4 is **(2007, Only One Option Correct Type)**

a. 25 b. 33
c. 50 d. 75

Passage for Q. Nos. (23-25)

The noble gases have closed-shell electronic configuration and are monoatomic gases under normal conditions. The low boiling points of the lighter noble gases are due to weak dispersion forces between the atoms and the absence of other interatomic interactions.

The direct reaction of xenon with fluorine leads to a series of compounds with oxidation numbers $+2, +4$ and $+6$. XeF_4 reacts violently with water to give XeO_3. The compounds of xenon exhibit rich stereochemistry and their geometries can be deduced considering the total number of electron pairs in the valence shell.

(2007, Comprehension Type)

23 Argon is used in arc welding because of its

a. low reactivity with metal
b. ability to lower the melting point of metal
c. flammability
d. high calorific value

24 The structure of XeO_3 is

a. linear
b. planar
c. pyramidal
d. *T*-shaped

25 XeF_4 and XeF_6 are expected to be

a. oxidising
b. reducing
c. unreactive
d. strongly basic

26 Which of the following is not oxidised by O_3? **(2005, Only One Option Correct Type)**

a. KI
b. $FeSO_4$
c. $KMnO_4$
d. K_2MnO_4

27 Which gas is evolved when PbO_2 is treated with conc. HNO_3? **(2005, Only One Option Correct Type)**

a. NO_2
b. O_2
c. N_2
d. N_2O

28 A pale blue liquid which obtained by equimolar mixture of two gases at $-30°C$ is **(2005, Only One Option Correct Type)**

a. N_2O
b. N_2O_3
c. N_2O_4
d. N_2O_5

29 Which of the following isomers of phosphorus is thermodynamically most stable? **(2005, Only One Option Correct Type)**

a. Red
b. White
c. Black
d. Yellow

2. *d* & *f*-Block Elements

DPP-1	General Properties of *d*-Block Elements
DPP-2	General Chemical and Magnetic Properties of *d*-Block Elements
DPP-3	Oxides and Oxo Anions of Cr and Mn
DPP-4	*f*-Block Elements-I (Lanthanides)
DPP-5	*f*-Block Elements-II (Actinides)

- **Revisal Problems for JEE Main**
- **Revisal Problems for JEE Advanced**
- **JEE Main & AIEEE Archive**
- **JEE Advanced & IIT JEE Archive**

DPP
DAILY PRACTICE PROBLEMS

DPP-1 General Properties of *d*-Block Elements

Subjective Questions

Working Space

Direction (Q. Nos. 1-5) *These questions are subjective in nature, need to be solved completely on notebook.*

1 Answer the following regarding the transition metals.
 i. Metal, more abundant in earth's crust.
 ii. Metal having lowest melting point.
 iii. Metal having highest melting point.
 iv. First man-made element.
 v. Best conductor of electricity.
 vi. Lightest (low density) and heaviest element (high density).

2 Why Zn, Cd and Hg have low melting points unlike other *d*-block elements?

3 Why do the 4*d* and 5*d* series of elements resemble each other more closely than the 3*d* series of elements?

4 Give reasons for the following.
 i. Fe^{3+} is more stable than Fe^{2+}.
 ii. Silver is a transition metal while zinc is not, why?

5 Give answers of the following.
 i. Name the mineral that has both Fe and Cu.
 ii. Name the metals present in insulin, haemoglobin and vitamin B_{12}.
 iii. % of gold in 21 carats gold.

Only One Option Correct Type

Direction (Q. Nos. 6-15) *This section contains 10 multiple choice questions. Each question has four choices (a), (b), (c) and (d), out of which ONLY ONE is correct.*

6 Atomic number of second transition series lies from
 a. 38 to 47 b. 39 to 48 c. 40 to 49 d. 41 to 50

7 Which has lower standard reduction potential (SRP) value?
 a. Mn^{2+}/Mn b. Zn^{2+}/Zn c. Fe^{2+}/Fe d. Ni^{2+}/Ni

8 An element has configuration $4d^5 5s^2$. The element belongs to
 a. 3rd group and 4th period b. 5th group and 5th period
 c. 7th group and 5th period d. 7th group and 7th period

9 The electronic configuration of palladium is
 a. [Ar] $4f^{14} 5d^9 6s^1$ b. [Kr] $4d^8 5s^2$
 c. [Kr] $5d^8 6s^2$ d. [Kr] $4d^{10} 5s^0$

Working Space

10 Which has lowest and highest first ionisation enthalpy in 3*d* series?

a. Sc and Zn
b. Zn and Sc
c. Cu and Zn
d. Cr and Zn

11 Electronic configuration of a transition element *X* in +3 oxidation state is [Ar] $3d^5$ and *Y* in +2 state is [Ar] $3d^8$. What are the atomic numbers of the elements?

a. 23 and 26
b. 26 and 28
c. 21 and 25
d. 24 and 26

12 The electronic configuration of the element which is just above the element with atomic number 42 in the same group

a. [Ar] $3d^4 4s^2$
b. [Ar] $3d^3 4s^1$
c. [Ar] $3d^5 4s^1$
d. [Ar] $3d^{10} 4s^1$

13 All the following statements about the 3*d*-series elements are true except that

a. scandium has lowest density
b. enthalpy of atomisation of zinc is lowest
c. first ionisation energy of Sc is lowest and zinc is highest
d. scandium has lowest melting point

14 The atomic radii of Cu and Ag are 1.17 Å and 1.34 Å. The atomic radius of gold is expected to be

a. 1.41 Å
b. 1.51 Å
c. 1.34 Å
d. 1.12 Å

15 The reduction potentials of M^{2+}/M follow the trend

a. V < Fe < Ni < Cu
b. Fe < Cu < V < Ni
c. Cu < V < Ni < Fe
d. Ni < Fe < Cu < V

One or More than One Options Correct Type

***Direction** (Q. Nos. 16-20) This section contains 5 multiple choice questions. Each question has four choices (a), (b), (c) and (d), out of which ONE or MORE THAN ONE are correct.*

16 Identify the pairs in which one covalent radii of elements are almost similar.

a. Na and Ta
b. Mo and W
c. La and Lu
d. Sc and Y

17 The metals that do not liberate H_2 gas with dilute acids.

a. Zn
b. Cu
c. Ag
d. Au

18 Which of the following have pseudo noble gas configuration?

a. Cu^{2+}
b. Zn^{2+}
c. Ag^{+}
d. Au^{3+}

19 Set of metals having ccp crystal structure

a. V, Nb and Ta
b. Cu, Ag and Au
c. Ni, Pd and Pt
d. Fe, Ru and Os

20 The pair of ions that have same electronic configuration

a. Fe^{2+} and Mn^{2+}
b. Fe^{2+} and Co^{3+}
c. Fe^{3+} and Co^{3+}
d. V^{2+} and Cr^{3+}

Comprehension Type

Working Space

***Direction** (Q. Nos. 21 and 22) This section contains a paragraph, describing theory, experiments, data, etc. Two questions related to the paragraph have been given. Each question has only one correct answer among the four given options (a), (b), (c) and (d).*

Passage

The atomic and ionic radii for transition elements are smaller than their corresponding *s*-block elements and are greater than their corresponding *p*-block elements. The atomic and ionic radii transition elements for a given series show decreasing trend for first elements constant in the middle and slight increase towards the end. This is due to the combined effect of increasing effective nuclear charge and increasing screening effect along the period.

The atomic and ionic radii of 4*d* series are greater than 3*d* series. However, the elements of 4*d* series and 5*d* series have almost similar sizes due to lanthanide contraction. Transition elements have low atomic volumes, high densities, high melting points.

21 Which pair of elements has nearly same atomic size?

a. Sc and Y
b. V and Zr
c. Zr and Hf
d. La and Y

22 The metal with high melting point in 3*d* series is

a. Cu
b. Zn
c. Cr
d. Fe

Matching List Type

***Direction** (Q. Nos. 23 and 24) Choices for the correct combination of elements from Column I and Column II are given as options (a), (b), (c) and (d), out of which one is correct.*

23 Match the Column I with Column II and mark the correct option from codes given below.

	Column I		Column II
i.	Nb	p.	$4d^7 5s^1$
ii.	Mo	q.	$4d^6 5s^1$
iii.	Tc	r.	$4d^5 5s^1$
iv.	Ru	s.	$4d^4 5s^1$
		t.	$4d^3 5s^2$

Codes

	i	ii	iii	iv		i	ii	iii	iv
a.	p	r	s	q	b.	s	r	q	p
c.	r	s	p	q	d.	p	s	r	q

24 Match the Column I with Column II and mark the correct option from codes given below.

	Column I (Ion)		Column II (Magnetic moment)
i.	Cu^{2+}	p.	2.9 BM
ii.	Fe^{2+}	q.	4.0 BM
iii.	Ni^{2+}	r.	2.8 BM
iv.	Co^{2+}	s.	5.0 BM

Codes

	i	ii	iii	iv		i	ii	iii	iv
a.	r	s	p	q	b.	p	q	r	s
c.	s	q	p	r	d.	q	r	s	p

Working Space

One Integer Value Correct Type

***Direction** (Q. Nos. 25-29) This section contains 5 questions. When worked out will result in an integer from 0 to 9 (both inclusive).*

25 The number of elements that do not have completely filled *d* subshell among the elements Ni, Pd, Pt, Cu, Ag, Au, Zn, Cd and Hg are

26 Total spin of the electrons in a chromium atom is

27 Number of electrons with $l = 2$ and $s = \frac{1}{2}$ in zinc atom are

28 Number of elements that generally do not show variable valencies Cu, Zn Au, Cd, Sc, La, Pt, Co.

29 In 4*d* series, number of elements that have ns^2 configuration in the valence shell are

Statement Type

***Direction** (Q. No. 30) This section is based on Statement I and Statement II. Select the correct answer from the codes given below.*

a. Both Statement I and Statement II are correct and Statement II is the correct explanation of Statement I
b. Both Statement I and Statement II are correct but Statement II is not the correct explanation of Statement I
c. Statement I is correct but Statement II is incorrect
d. Statement II is correct but Statement I is incorrect

30 **Statement I** The highest oxidation state of osmium is +8.

Statement II Osmium is a 5-*d* block element.

DPP-2 General Chemical and Magnetic Properties of *d*-Block Elements

Subjective Questions

Working Space

Direction *(Q. Nos. 1-5) These questions are subjective in nature, need to be solved completely on notebook.*

1 Why Cr^{2+} is reducing and Mn^{3+} is oxidising when both have d^4 configuration?

2 $E^{\circ}_{(M^{2+}/M)}$ value for copper is positive (+ 0.34 V). What is the possible reason for this?

3 Which is a stronger reducing agent Cr^{2+} or Fe^{2+} and why?

4 Name the elements with highest oxidation number in 3*d*, 4*d* and 5*d* series.

5 E° for Mn^{3+}/Mn^{2+} couple is more positive than for Fe^{3+}/Fe^{2+} couple, why?

Only One Option Correct Type

Direction *(Q. Nos. 6-15) This section contains 10 multiple choice questions. Each question has four choices (a), (b), (c) and (d), out of which ONLY ONE is correct.*

6 Of the following outer electronic configurations of atoms, the highest oxidation state is achieved by which one of them?

a. $(n-1)d^3ns^2$
b. $(n-1)d^5ns^1$
c. $(n-1)d^8ns^2$
d. $(n-1)d^5ns^2$

7 Among the following series of transition metal ions, the one where all metal ions have $3d^2$ electronic configuration is (Atomic number, Ti = 22, V = 23, Cr = 24, Mn = 25) **[AIEEE 2004]**

a. $Ti^{3+}, V^{2+}, Cr^{3+}, Mn^{4+}$
b. $Ti^{+}, V^{4+}, Cr^{6+}, Mn^{7+}$
c. $Ti^{4+}, V^{3+}, Cr^{2+}, Mn^{3+}$
d. $Ti^{2+}, V^{3+}, Cr^{4+}, Mn^{5+}$

8 For the four successive transition elements (Cr, Mn, Fe and Co), the stability of + 2 oxidation state will be there in which of the following order?

a. Mn > Cr > Fe > Co
b. Fe > Mn > Co > Cr
c. Co > Mn > Fe > Cr
d. Cr > Mn > Co > Fe

9 The catalytic activity of transition metals and their compounds is mainly due to

a. their magnetic behaviour
b. their ful-filled *d*-orbitals
c. their ability to adopt variable oxidation states
d. their chemical reactivity

Working Space

10 Many copper (I) compounds are unstable in aqueous solution and undergo disproportionation as $2Cu^+ \longrightarrow Cu + Cu^{2+}$. This is due to

a. free energy change of reaction is positive
b. free energy of reaction is negative
c. second ionisation energy of Cu is higher than first ionisation energy
d. hydration energy of Cu^+ is less than Cu^{2+}

11 Gun metal is an alloy of

a. Cu and Al
b. Cu and Sn
c. Cu, Zn and Sn
d. Cu, Zn and Ni

12 In which of the following pairs both the ions are coloured in aqueous solution? (Atomic number, Sc = 21, Ti = 22, Ni = 28, Cu = 29, Co = 27)

a. Sc^{3+} and Ti^{3+}
b. Sc^{3+} and Co^{2+}
c. Ni^{2+} and Cu^+
d. Ni^{2+} and Ti^{3+}

13 $[Ti(H_2O)_6]^{3+}$ absorbs green and yellow region part of visible light. Then the transmitted colour of the compound is

a. purple
b. green
c. blue
d. red

14 Compound of metal ion M^{x+} ($Z = 26$) has a spin only magnetic moment of $\sqrt{35}$ BM The number of unpaired electrons and x value in the compound are

a. 2 and 4
b. 3 and 4
c. 2 and 5
d. 5 and 3

15 In which of the following the stability of two oxidation states is correctly represented?

a. $Ti^{3+} > Ti^{4+}$
b. $Mn^{2+} > Mn^{3+}$
c. $Fe^{2+} > Fe^{3+}$
d. $Cu^+ > Cu^{2+}$

One or More than One Options Correct Type

***Direction** (Q. Nos. 16-20) This section contains 5 multiple choice questions. Each question has four choices (a), (b), (c) and (d), out of which ONE or MORE THAN ONE are correct.*

16 Which one of the following arrangements represent the correct order of the property stated against it?

a. $V^{2+} < Cr^{2+} < Mn^{2+} < Fe^{2+}$: paramagnetic behaviour
b. $Ni^{2+} < Co^{2+} < Fe^{2+} < Mn^{2+}$: ionic size
c. $Co^{3+} < Fe^{3+} < Cr^{3+} < Sc^{3+}$: stability in aqueous solution
d. Sc < Ti < Cr < Mn : number of oxidation states

17 The metals present in German silver alloy are

a. Cu
b. Cr
c. Zn
d. Ni

18 Magnetic moment is 3.87 BM for the ion(s)

a. Co^{3+}
b. V^{2+}
c. Cr^{3+}
d. Mn^{4+}

19 +8 oxidation state is/are shown by

a. Fe
b. Ru
c. Os
d. W

20 Ions having same colour in aqueous solution are

a. Cu^{2+} and Fe^{2+}
b. Ni^{2+} and Fe^{2+}
c. Fe^{3+} and Cr^{3+}
d. V^{4+} and Cr^{2+}

Working Space

Comprehension Type

Direction (Q. Nos. 21 and 22) *This section contains a paragraph, describing theory, experiments, data, etc. Two questions related to the paragraph have been given. Each question has only one correct answer among the four given options (a), (b), (c) and (d).*

Passage

Transition metals and their compounds have paramagnetic properties due to the presence of unpaired electrons in $(n-1)d$-orbitals. The paramagnetic behaviour is expressed in terms of magnetic moment which is because of the spin of the unpaired electrons (n). It is given as

Magnetic moment = $\sqrt{n(n+2)}$ BM

Majority of transition metal compounds are coloured both in solid state as well as in aqueous solution. This is also due to the presence of unpaired electrons in $(n-1)$ *d*-orbitals, *d*-orbitals splitting and *d-d* transition of electrons absorbing suitable visible light.

21 Which of the following exhibit colour due to charge transfer phenomenon, but not due to *d-d* transition?

a. $Cr_2O_7^{2-}$ b. MnO_4^- c. CrO_4^{2-} d. All of these

22 Which pair of ions are expected to be diamagnetic?

a. Cu^+ and Cu^{2+} b. Ag^+ and Au^+ c. Ag^+ and Au^{3+} d. Zn^{2+} and Cu^{2+}

Matching List Type

Direction (Q. No. 23-25) *Choices for the correct combination of elements from Column I and Column II are given as options (a), (b), (c) and (d), out of which one is correct.*

23 Match the Column I with Column II and mark the correct option from the given codes.

	Column I (Alloys)		Column II (Composition)
i.	Type metal	p.	Sn (20%), Cu(80%)
ii.	Bell metal	q.	Sn (82%), Sb(12%), Cu(6%)
iii.	White metal	r.	Sn (28%), Pb(22%), Bi(50%)
iv.	Rose metal	s.	Sn (5%), Pb (80%), Sb(15%)

Codes

	i	ii	iii	iv		i	ii	iii	iv
a	s	p	q	r	b	s	p	s	q
c	p	q	s	r	d	q	p	r	s

24 Match the Column I with Column II and mark the correct option from the given codes.

	Column I (Ions)		Column II (Magnetic moment in BM)
i.	Sc^{3+}	p.	3.87
ii.	Ti^{4+}	q.	4.90
iii.	Co^{2+}	r.	2.83
iv.	Cr^{2+}	s.	0

Codes

	i	ii	iii	iv		i	ii	iii	iv
a	r	s	p	q	b	r	q	p	s
c	p	q	r	s	d	s	s	p	q

Working Space

25 Match the Column I with Column II and mark the correct option from the given codes.

	Column I		Column II
i.	*Zeigler-Natta* catalyst	p.	Pt/PtO used in reduction
ii.	Contact process	q.	Polymerisation of ethylene
iii.	Fenton's reagent	r.	Vanadium pentoxide
iv.	Adams catalyst	s.	Oxidation of alcohol to aldehyde
		t.	Nickel

Codes

	i	ii	iii	iv		i	ii	iii	iv
a.	q	r	s	p	b.	p	q	r	s
c.	q	p	s	r	d.	s	p	q	r

One Integer Value Correct Type

Direction *(Q. Nos. 26-29) This section contains 4 questions. When worked out will result in an integer from 0 to 9 (both inclusive).*

26 Number of alloys that contain nickel among the following solder, gun metal, German silver, nichrome, monel metal, constanton, bell metal, duralumin, type metal, invar, alnico.

27 Green vitriol is $FeSO_4 \cdot xH_2O$ and white vitriol is $ZnSO_4 \cdot yH_2O$, Then, the values of x and y are

28 Oxidation state of chromium in CrO_5 is $+x$. Here, value of x is

29 Number of compounds in which metal has zero oxidation state

WO_3, $Ni(CO)_4$, MoO_3, $Fe(CO)_5$, $Cr(CO)_6$, $[Pt(NH_3)_2Cl_2]$, $Co_2(CO)_8$ and $Mn_2(CO)_{10}$

Statement Type

Direction *(Q. No. 30) This section is based on Statement I and Statement II. Select the correct answer from the codes given below.*

a. Both Statement I and Statement II are correct and Statement II is the correct explanation of Statement I
b. Both Statement I and Statement II are correct but Statement II is not the correct explanation of Statement I
c. Statement I is correct but Statement II is incorrect
d. Statement II is correct but Statement I is incorrect

30 **Statement I** $Cr_2O_7^{2-}$, CrO_4^{2-}, MnO_4^-, MnO_4^{2-}, Cu_2O and Ni-DMG complex are coloured.

Statement II It is due to charge transfer transition.

DPP-3 Oxides and Oxo Anions of Cr and Mn

Subjective Questions

Working Space

Direction (Q. Nos. 1-5) *These questions are subjective in nature, need to be solved completely on notebook.*

1 Answer the following.

i. Mn^{3+} disproportionate to Mn^{2+} and MnO_2 in acidic medium.

ii. $KMnO_4$ is heated with KOH.

iii. MnO_4^{2-} disproportionate into MnO_2 and MnO_4^- in acidic or alkaline medium.

2 Pyrolusite on heating with KOH in the presence of air gives dark green compound *A*. The solution of *A* on treatment with H_2SO_4 gives a purple coloured compound *B*, which gives the following:

i. KI on reaction with alkaline solution of (*B*) changes into a compound (*C*).

ii. With conc. H_2SO_4 compounds (*B*) gives (*D*) which can decompose to yield (*E*) and oxygen.

Identify the compounds *A*, *B*, *C*, *D* and *E* and write the equations involved.

3 Explain the following.

i. How the colour of $K_2Cr_2O_7$ solution depend on pH of the solution?

ii. Why $K_2Cr_2O_7$ is preferred to $Na_2Cr_2O_7$ in volumetric analysis?

iii. Acidified solution of dichromate ions forms a deep blue colour with H_2O_2. Why?

iv. Acidified $K_2Cr_2O_7$ solution turns green when sodium sulphite is added to it.

4 Answer the following.

i. Philosopher's wool

ii. Rinmann's green

iii. Rutile

iv. Which is more basic TiO_2 or ZrO_2?

v. Which is more stable RuO_4 or OsO_4?

vi. Calomel

vii. Corrosive sublimate

5 If 40 mL $KMnO_4$ oxidises 100 mL 0.05 M oxalic acid, find the molarity of $KMnO_4$.

Only One Option Correct Type

Direction (Q. Nos. 6-15) *This section contains 10 multiple choice questions. Each question has four choices (a), (b), (c) and (d), out of which ONLY ONE is correct.*

6 Which is most basic oxide of chromium?

a. CrO b. CrO_2

c. CrO_3 d. Cr_2O_3

Working Space

7 The basic character of the transition metal monoxides follows the order (Atomic Number, Ti = 22, V = 23, Cr = 24, Fe = 26)

a. TiO > VO > CrO > FeO
b. VO > CrO > TiO > FeO
c. CrO > VO > FeO > TiO
d. TiO > FeO > VO > CrO

8 $MnO_4^- + S^{2-} + H^+ \longrightarrow Mn^{2+} + S + H_2O$. When this equation is balanced, the total number of coefficients on the left hand side are

a. 15
b. 23
c. 17
d. 14

9 $MnO_4^- + I^- \longrightarrow MnO_2 + IO_3^-$ when this equation is balanced, the number of OH^- ions added

a. $4OH^-$ ions left side
b. $2OH^-$ ions left side
c. $4OH^-$ ions right side
d. $2OH^-$ ions right side

10 One mole of each $KMnO_4$ and $K_2Cr_2O_7$ can oxidise .... moles of ferrous ion.

a. 5 and 5
b. 5 and 6
c. 6 and 5
d. 6 and 6

11 MnO_4^- is of intense pink colour, though Mn is in (+7) oxidation state. It is due to

a. oxygen gives colour to it
b. charge transfer when Mn gives its electron to oxygen
c. character transfer when oxygen gives its electron to Mn making it Mn (+ VI) hence, coloured
d. None of the above is correct

12 Which of the following is incorrect statement?

a. Both MnO_4^-, MnO_4^{2-} have tetrahedral shape
b. MnO_4^- is purple while MnO_4^{2-} ion is green
c. Both MnO_4^- and MnO_4^{2-} are diamagnetic
d. MnO_4^- is diamagnetic while MnO_4^{2-} ion is paramagnetic

13 Which is the correct statement about $Cr_2O_7^{2-}$ structure?

a. It has neither Cr—Cr bonds nor O—O bonds
b. It has one Cr—Cr bond and six O—O bonds
c. It has no Cr—Cr bonds and has six O—O bonds
d. It has one Cr—Cr bond and seven Cr—O bonds

14 Pyrolusite in MnO_2 is used to prepare $KMnO_4$. Steps are

$$MnO_2 \xrightarrow{\text{I}} MnO_4^{2-} \xrightarrow{\text{II}} MnO_4^-$$

Here, I and II are

a. fused with KOH/air, electrolytic oxidation
b. fused with KOH/air, electrolytic reduction
c. fuse with conc. HNO_3/air, electrolytic oxidation
d. All are correct

Working Space

15 Out of Cr (VI) as $Cr_2O_7^{2-}$ and CrO_4^{2-}, which is better oxidising agent?

a. CrO_4^{2-}, basic medium
b. $Cr_2O_7^{2-}$, basic medium
c. $Cr_2O_7^{2-}$, acidic medium
d. CrO_3, basic medium

One or More than One Options Correct Type

Direction (Q. Nos. 16-20) This section contains 5 multiple choice questions. Each question has four choices (a), (b), (c) and (d), out of which ONE or MORE THAN ONE are correct.

16 Which of the following statements is correct when a mixture of NaCl and $K_2Cr_2O_7$ is gently warmed with conc. H_2SO_4?

a. A deep red vapours are evolved
b. Chlorine gas is evolved
c. Chromyl chloride vapours are evolved
d. The vapours when passed into NaOH solution gives a yellow solution of Na_2CrO_4

17 Amphoteric oxides in the following is/are

a. ZnO
b. V_2O_5
c. MnO_2
d. Mn_2O_7

18 Which of the following on heating produce oxygen gas?

a. $KMnO_4$
b. KNO_3
c. $(NH_4)_2Cr_2O_7$
d. $K_2Cr_2O_7$

19 Which of the following statements is not true?

a. On passing H_2S through acidified $K_2Cr_2O_7$ solution, a milky colour is observed
b. $Na_2Cr_2O_7$ is preferred over $K_2Cr_2O_7$ in volumetric analysis
c. $K_2Cr_2O_7$ solution in acidic medium is orange
d. $K_2Cr_2O_7$ solution becomes yellow on increasing the pH beyond 7

20 $MnO_4^- + NO_2^- \longrightarrow Mn^{2+} + NO_3^-$ when this redox reaction is balanced in acidic medium, we get

a. $6H^+$ on the left hand side
b. three water molecules on right hand side
c. total coefficients of all reactants is 12
d. total coefficients of all products is 10

Comprehension Type

Direction (Q. Nos. 21 and 22) This section contains a paragraph, describing theory, experiments, data, etc. Two questions related to the paragraph have been given. Each question has only one correct answer among the four given options (a), (b), (c) and (d).

Passage

Oxides are generally formed by the reaction of metals with oxygen at high temperatures. All the metals except scandium form *MO* oxides which are ionic. The highest oxidation number in the oxides, coincides with the group number and is attained in Sc_2O_3 to Mn_2O_7. Beyond 7th group, higher oxides coinciding with the group number do not form. As the oxidation number of metal increases covalent character and acidity increases. Mn_2O_7, CrO_3 are acidic and their lower oxides are basic and amphoteric. Stability of higher oxides increases down the group.

Working Space

21 Which of the following reacts with both acids and bases?

a. ZnO
b. V_2O_5
c. CrO_3
d. Both (a) and (b)

22 Which of the following undergo disproportionation?

a. MnO_4^-
b. MnO_4^{2-}
c. V_2O_5
d. CuO

Matching List Type

***Direction** (Q. Nos. 23 and 24) Choices for the correct combination of elements from Column I and Column II are given as options (a), (b), (c) and (d), out of which one is correct.*

23 Match the Column I with Column II and mark the correct option from the codes given below.

	Column I		Column II
i.	K_2CrO_4	p.	Baeyer's reagent
ii.	$KMnO_4$	q.	Transition element in + 6 oxidation state
iii.	$K_2Cr_2O_7$	r.	Oxidising agent in acidic medium
iv.	K_2MnO_4	s.	Manufactured from pyrolusite
		t.	Manufactured from chromite ore

Codes

	i	ii	iii	iv
a.	q,t	p,r	q,r,t	q,s
b.	q,p	p,s	q,p,s	q,t
c.	p	q	r	s
d.	q	s	p	t

24 Match the Column I with Column II and mark the correct option from the codes given below.

	Column I		Column II
i.	$KMnO_4$ in acidic medium	p.	Mol. wt/6
ii.	$KMnO_4$ in alkaline medium	q.	Mol. wt/5
iii.	$KMnO_4$ in neutral solution	r.	Mol.wt/3
iv.	$K_2Cr_2O_7$ in acidic medium	s.	Mol. wt/1

Codes

	i	ii	iii	iv
a.	s	r	q	p
b.	q	s	r	p
c.	r	s	p	q
d.	p	s	q	r

One Integer Value Correct Type

Working Space

Direction (Q. Nos. 25-28) This section contains 4 questions. When worked out will result in an integer from 0 to 9 (both inclusive).

25 The number of moles of FeC_2O_4 that reduces two moles of $KMnO_4$ in acidic medium are

26 50 cc of 0.04 M $K_2Cr_2O_7$ in acidic medium oxidises a sample of H_2S gas to sulphur. Volume of 0.03 M $KMnO_4$ required to oxidise the same amount of H_2S gas to sulphur in acidic medium is $10 \times x$. Here, the value of x is

27 The number of sigma bonds in $Cr_2O_7^{2-}$ ion are

28 If molarity of $K_2Cr_2O_7$ solution is 0.5 M, its normality value is

Statement Type

Direction (Q. Nos. 29 and 30) This section is based on Statement I and Statement II. Select the correct answer from the codes given below.

a. Both Statement I and Statement II are correct and Statement II is the correct explanation of Statement I
b. Both Statement I and Statement II are correct but Statement II is not the correct explanation of Statement I
c. Statement I is correct but Statement II is incorrect
d. Statement II is correct but Statement I is incorrect

29 **Statement I** When a solution of potassium chromate is treated with an excess of dilute nitric acid chromate undergoes oxidation.

Statement II Dichromate ions are produced in above process.

30 **Statement I** The oxidation state of chromium in the final product formed by the reaction between KI and acidified potassium dichromate solution has green colour.

Statement II The product has Cr^{3+} ions.

DPP-4 f-Block Elements-I (Lanthanides)

Subjective Questions

Working Space

Direction *(Q. Nos. 1-5) These questions are subjective in nature, need to be solved completely on notebook.*

1 Explain why Ce^{4+} is good oxidising agent whereas Sm^{2+} is good reducing agent?

2 Name the members of lanthanide series that show + 4 and + 2 oxidation states.

3 Even though Ce^{3+} $(4f^1)$ and Yb^{3+} $(4f^{13})$ have unpaired electron, they are colourless in solution. Why?

4 Why lanthanides are called rare-earths?

5 i. Name the main mineral from which lanthanides are extracted.
ii. What is lanthanoid contraction? Why this phenomenon is considered as extraordinary in nature?

Only One Option Correct Type

Direction *(Q. Nos. 6-15) This section contains 10 multiple choice questions. Each question has four choices (a), (b), (c) and (d), out of which ONLY ONE is correct.*

6 Set of continuous atomic numbers of elements are present in the same group as well as same period.

a. 56, 57, 58, 59
b. 88, 89, 90, 91
c. 101, 102, 103, 104
d. 68, 69, 70, 71

7 The element with atomic number 62 belongs to

a. III A group and 6th period
b. III B group and 7th period
c. IV B group and 7th period
d. III B group and 6th period

8 The radius of lanthanum is 1.06 Å. Then which of the following is closest to the radius of Lu^{3+} ion?

a. 1.06 Å
b. 1.4 Å
c. 1.6 Å
d. 0.85 Å

9 + 4 ion of which has half-filled 4f subshell?

a. Tb
b. Lu
c. Gd
d. Eu

10 Lanthanide for which + II and + III oxidation states are common is

a. La
b. Nd
c. Ce
d. Eu

11 An alloy of lanthanides

a. misch metal
b. nichrome
c. wood metal
d. type metal

Working Space

12 Across the lanthanide series, the basicity of the lanthanide hydroxides

a. increases
b. decreases
c. first increases and then decreases
d. first decreases and then increases

13 Gadolinium belongs to 4*f* series. Its atomic number is 64. Which of the following is the correct electronic configuration of gadolinium?

a. [Xe] $4f^7 5d^1 6s^2$
b. [Xe] $4f^6 5d^2 6s^2$
c. [Xe] $4f^8 6d^2$
d. [Xe] $4f^9 5s^1$

14 Lanthanides contraction causes

a. small increase in basic strength of oxides of lanthanides
b. great variation in chemical properties of lanthanides
c. small increase in electronegativities of lanthanides
d. small decrease in standard electrode potential value of lanthanides

15 Which one of the following statement(s) is/are incorrect for lanthanides?

a. Lanthanide metal ions are mostly coloured
b. Lanthanides have high melting and boiling points
c. Lanthanides have low densities
d. Lanthanides have high ionisation energy

One or More than One Options Correct Type

***Direction** (Q. Nos. 16-20) This section contains 5 multiple choice questions. Each question has four choices (a), (b), (c) and (d), out of which ONE or MORE THAN ONE are correct.*

16 Which of the following statements are incorrect regarding lanthanides?

a. All ions having unpaired electrons are coloured
b. The first and second ionisation energies are comparable with calcium
c. Melting point range is 1000 to 1200 K but samarium has 1623 K
d. Ce^{4+} is a strong reducing agent

17 Observe the following statements. Which is correct statement regarding lanthanides?

a. Cerium shows + 4 and + 3 oxidation states
b. Element with atomic numbers 66 and 99 belong to *f*-block
c. Lanthanides belong to 3rd group and 7th period
d. Monazite is the mineral that constitutes the major source of the lanthanides

18 The colourless ions among the lanthanides are

a. La^{3+}
b. Gd^{3+}
c. Lu^{3+}
d. Ce^{3+}

19 The correct statements among the following are

a. Reducing power decreases from La to Lu
b. La to Lu $E°$ values become less negative
c. Basic strength increases from $La(OH)_3$ to $Lu(OH)_3$
d. Promethium is artificial radioactive element

20 Which of the following pairs have same atomic size?

a. Zr and Hf
b. Nb and Ta
c. Mo and W
d. Y and La

Working Space

Comprehension Type

Direction *(Q. Nos. 21 and 22) This section contains a paragraph, describing theory, experiments, data, etc. Two questions related to the paragraph have been given. Each question has only one correct answer among the four given options (a), (b), (c) and (d).*

Passage

The observed oxidation state of lanthanides either in solution or in insoluble compounds form tripositive cations. Based on the ionisation enthalpies and hydration energy factors, it is concluded that tripositive species is more stable than di or tetrapositive species in aqueous solution. In general, ions having half-filled (Tb^{4+}, Eu^{2+}, Gd^{3+}) or completely filled 4*f* orbitals (Yb^{2+}, Lu^{3+}) or noble gas configuration (La^{3+}, Ce^{4+}) are stable. This is supported by their SRP values. Ions with + 4 oxidation state act as oxidising and ions having + 2 oxidation state act as reducing agents. Ions having same number of unpaired electrons in *f*-orbitals possess same colour.

21 Observe the following equations.

$Sm^{3+}(aq) + e^- \longrightarrow Sm^{2+}(aq), E° = -1.55 \text{ V}$

$Eu^{3+}(aq) + e^- \longrightarrow Eu^{2+}(aq), E° = -0.43 \text{ V}$

$Yb^{3+}(aq) + e^- \longrightarrow Yb^{2+}(aq), E° = -1.55 \text{ V}$

Based on the above equations, the correct reducing strength is in the order of

a. $Sm^{2+} > Yb^{2+} > Eu^{2+}$

b. $Eu^{2+} > Yb^{2+} > Sm^{2+}$

c. $Eu^{2+} > Sm^{2+} > Yb^{2+}$

d. $Sm^{2+} > Eu^{2+} > Yb^{2+}$

22 Which is correct statement?

a. Ce^{4+} is more stable than Ce^{3+} in solution

b. Ce^{3+} is more sable than Ce^{4+} in solution

c. Ce^{4+} and Ce^{3+} are equally stable

d. Ce^{4+} has xenon configuration

Matching List Type

Direction *(Q. No. 23) Choices for the correct combination of elements from Column I and Column II are given as options (a), (b), (c) and (d), out of which one is correct.*

23 Match the Column I with Column II and mark the correct option from the codes given below.

	Column I		Column II
i.	Promethium	p.	Lanthanide
ii.	Misch metal	q.	Actinide
iii.	Uranium	r.	An alloy
iv.	Plutonium	s.	Nuclear fuel
		t.	Artificial element

Codes

	i	ii	iii	iv
a.	p,t	r	q,s	q,s
b.	r,s	q	s	q,s
c.	p,s	r	q	t,p
d.	p,r	q	r	q,s

Working Space

One Integer Value Correct Type

Direction (Q. Nos. 24-28) *This section contains 5 questions. When worked out will result in an integer from 0 to 9 (both inclusive).*

24 Number of species that have half-filled 4*f* subshell among Gd, Lu, Eu, Yb, Eu^{2+}, Gd^{3+}, Eu^{3+}, Ce^{4+}, Tb^{4+}.

25 Number of unpaired electrons in ytterbium is/are

26 Number of diamagnetic lanthanide ions among La^{3+}, Ce^{3+} Ce^{4+}, Gd^{3+}, Yb^{2+}, Yb^{3+}, Lu^{2+}, Lu^{3+} are

27 Number of good reducing agents among Ce^{4+}, Eu^{2+}, Sm^{2+}, Yb^{2+}, Tb^{4+}, La^{3+} Ce^{3+} are

28 The highest oxidation state shown by any transition element is

Statement Type

Direction (Q. Nos. 29 and 30) *This section is based on Statement I and Statement II. Select the correct answer from the codes given below.*

a. Both Statement I and Statement II are correct and Statement II is the correct explanation of Statement I
b. Both Statement I and Statement II are correct and Statement II is not the correct explanation of Statement I
c. Statement I is correct but Statement II is incorrect
d. Statement II is correct but Statement I is incorrect

29 **Statement I** The second and third rows of transition elements resemble each other much more than they resemble the first row.

Statement II Due to lanthanide contraction, the atomic radii of the second and the third row transition elements.

30 **Statement I** Many trivalent lanthanide ions are coloured both in solid state and in aqueous solution.

Statement II Colour of these ions is due to the presence of *f*-electrons.

DPP-5 f-Block Elements-II (Actinides)

Subjective Questions

Working Space

Direction (Q. Nos. 1-5) These questions are subjective in nature, need to be solved completely on notebook.

1 i. What are inner-transition elements? Give their general outer electronic configuration.

ii. Is actinium d-block or f-block element?

2 Actinide contraction is greater from element to element than lanthanide contraction. Why?

3 Explain the oxidation states shown by actinide series of elements.

4 i. What are transuranium elements?

ii. Name the first transuranium element.

iii. What are its oxidations states?

5 i. Which actinides were the first to be discovered? Which other actinides were discovered in nature?

ii. What evidence shows that actinides are very reactive, for the most part?

Only One Option Correct Type

Direction (Q. Nos. 6-15) This section contains 10 multiple choice questions. Each question has four choices (a), (b), (c) and (d), out of which ONLY ONE is correct.

6 Which one of the following is an electronic configuration of thorium?

a. [Rn] $5f^1 6d^1 7s^2$
b. [Rn] $5f^2 6d^2 7s^0$
c. [Rn] $5f^2 6d^0 7s^2$
d. [Rn] $5f^0 6d^2 7s^2$

7 Which of the following is general electronic configuration of actinides?

a. [Rn] $5f^{0\text{-}14} 6d^{0\text{-}1} 7s^2$
b. [Rn] $5f^{1\text{-}14} 6d^{0\text{-}1} 7s^2$
c. [Rn] $5f^{0\text{-}14} 6d^2 7s^2$
d. [Rn] $5f^1 6d^2 7s^0$

8 The actinoids exhibit more member of oxidation states in general than the lanthanoids. This is because

a. the $5f$-orbitals are more buried than the $4f$-orbitals
b. there is a similarity between $4f$ and $5f$-orbitals in their angular part of the wave function
c. the actinides are more reactive than the lanthanides
d. the $5f$-orbitals extend farther from the nucleus than the $4f$-orbitals

Working Space

9 Identify the incorrect statement among the following.
a. All f-block elements belong to 3rd group
b. Element curium ($Z = 96$) has $5f^7\,6d^1\,7s^2$
c. The actinides tarnish readily in air
d. $4f$ and $5f$-orbitals are equally buried

10 In which $5f$ subshell is half-filled?
a. Am and Cm b. Cm and Bk c. Cm and Cf d. No and Lr

11 Actinides
a. are all synthetic elements
b. include element 104
c. are highly electropositive and reactive
d. do not show variable oxidation states

12 Which of the following is not an actinide?
a. Curium b. Californium c. Uranium d. Rutherfordium

13 The compound used in enrichment of uranium for nuclear power plant is
a. U_3O_8 b. UF_6 c. UCl_4 d. $UO_2(NO_2)_2$

14 Which one of the following shows oxidation state upto + 7?
a. Am b. Pu c. U d. Cm

15 The general electronic configuration of actinides is $(n-2)f^{1\text{-}14}\ (n-1)d^{0\text{-}1}\ ns^2$. Which one of the following actinides has one electron in $6d$-orbital?
a. Cm ($Z = 96$) b. Lr ($Z = 103$)
c. Np ($Z = 93$) d. All of these

One or More than One Options Correct Type

Direction (Q. Nos. 16-20) *This section contains 5 multiple choice questions. Each question has four choices (a), (b), (c) and (d), out of which ONE or MORE THAN ONE are correct.*

16 Commonly used nuclear fuels are
a. U^{235} b. Pu^{239}
c. U^{238} d. Pu^{243}

17 Which are the correct statements?
a. All actinides are radioactive
b. All actinides are synthetic in nature
c. Uranium is the heaviest natural element
d. 100th element is curium

18 Pick up the incorrect statements.
a. Oxides and hydroxides of actinides are less basic than lanthanides
b. Actinides show +2, + 4, + 5, + 6, + 7 oxidation states besides +3
c. Actinides form oxo anions like UO_2^{2+}, UO_2^{+}, etc
d. Actinides have less tendency to form complexes

19 Which species have half-filled $5f$ subshell?
a. Am b. Am^{2+}
c. Cm d. Np

20 Which are correct regarding actinides?
a. Highly reactive metals b. These react easily with HCl
c. These react with alkalis d. Act as strong reducing agents

Working Space

Comprehension Type

Direction *(Q. Nos. 21 and 22) This section contains a paragraph, describing theory, experiments, data, etc. Two questions related to the paragraph have been given. Each question has only one correct answer among the four given options (a), (b), (c) and (d).*

Passage

Rutherford and Soddy found that every uranium ore that they examined contained minute quantities of radium and an appreciable amount of non-radioactive lead. From this, they concluded that uranium is the starting substance in a series of radioactive transformations which pass through radium and ultimately ends with stable lead. These transformations are accompanied by spontaneous emissions of alpha and beta particles resulting in the formation of new atoms. All the nuclei from the initial element to the final stable element constitute a series called disintegration series. Mass number changes by 4 units per each alpha particle emission.

There are four disintegration series. They are

Series	Starting	End product
$4n$ (Thorium series)	Thorium-232	Lead-208
$(4n+1)$ Neptunium series	Neptunium-237	Bismuth-209
$(4n+2)$ Uranium series	Uranium-238	Lead-206
$(4n+3)$ Actinium series	Uranium-235	Lead-207

21 The number of alpha and beta particles emitted in the nuclear reaction $^{228}_{90}Th \longrightarrow \, ^{212}_{83}Bi$ are

a. 4α and 1β b. 3α and 7β
c. 8α and 1β d. 4α and 7β

22 Lead isotope is the end product in all disintegration series except

a. uranium series b. thorium series
c. actinium series d. neptunium series

Matching List Type

Direction *(Q. No. 23) Choices for the correct combination of elements from Column I and Column II are given as options (a), (b), (c) and (d), out of which one is correct.*

23 Match the Column I with Column II and mark the correct option from the codes given below.

	Column I		Column II
i.	Actinide that has f^7 in $+3$ state	p.	Neptunium
ii.	Element that has f^{14} in $+3$ state	q.	Curium
iii.	Ion that is colourless in +3 state	r.	Actinium
iv.	Element that shows $+7$ oxidation state	s.	Lawrencium

Codes

	i	ii	iii	iv
a.	q	s	r	p
b.	p	q	s	r
c.	q	p	r	s
d.	p	s	q	r

One Integer Value Correct Type

Direction (Q. Nos. 24-28) *This section contains 5 questions. When worked out will result in an integer from 0 to 9 (both inclusive).*

24 The number of electrons present in 5*f* subshell of berkelium are

25 The number of electrons present in 5*f* subshell of Th^{4+} ion are

26 Number of elements that show + 6 oxidation state among Th, U, Np, Pu, Am, Cm, Bk, Cf, Es, Fm are

27 The number of unpaired electrons in the element with atomic number 100 are

28 ${}^{238}_{92}U \longrightarrow {}^{206}_{82}Pb$, in this transformation, number of alpha particles (α) liberated are

Statement Type

Direction (Q. Nos. 29 and 30) *This section is based on Statement I and Statement II. Select the correct answer from the codes given below.*

a. Both Statement I and Statement II are correct and Statement II is the correct explanation of Statement I
b. Both Statement I and Statement II are correct and Statement II is not the correct explanation of Statement I
c. Statement I is correct but Statement II is incorrect
d. Statement II is correct but Statement I is incorrect

29 **Statement I** Actinides show a large number of oxidation state whereas lanthanides show a limited number of oxidation state.

Statement II Energy gap between 4*f*, 5*d* and 6*s* subshells is small whereas that between 5*f*, 6*d* and 7*s* subshell is large.

30 **Statement I** Actinides form relatively less stable complexes as compared to lanthanides.

Statement II Actinides can utilise their 5*f*-orbitals along with 6*d*-orbitals in bonding but lanthanides do not use their 4*f*-orbitals for bonding.

SET - 1

Revisal Problems for JEE Main

Only One Option Correct Type

Direction (Q. Nos. 1-30) *This section contains 30 multiple choice questions. Each question has four choices (a), (b), (c) and (d), out of which ONLY ONE is correct.*

Working Space

1 Which of the following does not relevant with transition elements?

a. Melting point of transition elements are high
b. Some ions of transition elements possess paramagnetic properties
c. All transition elements dissolve in acids
d. Transition elements possess various oxidation states

2 In the following statements about the transition elements are true except that

a. all of the transition elements are predominantly metallic
b. in aqueous solution, many of their simple ions are coloured
c. most of the transition elements show pronounced catalytic activity
d. most of the transition elements show + 1 and + 2 oxidation states

3 The ground state electronic configuration of the Mn ion in MnO_2 is

a. [Ar] $3d^3$ b. $3d^1 4s^2$ c. $3d^9 4s^2$ d. $3d^5$

4 Rutile is an ore of

a. manganese b. chromium c. titanium d. chromium

5 Arrange the transition metals Ti, Ni, Mn and La in order of increasing atomic radii.

a. La < Ti < Mn < Ni b. La < Ni < Ti < Mn
c. La < Mn < Ni < Ti d. Ni < Mn < Ti < La

6 For the same transition metal ion, the colour of its compounds will depend upon

a. temperature of the reaction
b. pressure of the reaction
c. nature of ligands or Lewis base attached to the metal ion
d. concentration of the ligands

7 Isoelectronic pair of ions is

a. Sc^{2+} and V^{3+} b. Mn^{3+} and Fe^{2+}
c. Mn^{2+} and Fe^{3+} d. Ni^{3+} and Fe^{2+}

8 Which oxide is expected to be most acidic?

a. TiO_2 b. CrO_3 c. CoO d. Fe_2O_3

9 The enthalpies of hydration of Ca^{2+}, Mn^{2+} and Zn^{2+} follow the order

a. $Mn^{2+} > Ca^{2+} > Zn^{2+}$ b. $Zn^{2+} > Ca^{2+} > Mn^{2+}$
c. $Mn^{2+} > Zn^{2+} > Ca^{2+}$ d. $Zn^{2+} > Mn^{2+} > Ca^{2+}$

10 Which aqueous solution is coloured?

a. $Zn(NO_3)_2$ b. $LiNO_3$
c. $Co(NO_3)_2$ d. Potash alum

11 Addition of high proportions of manganese makes steel useful in making rails of rail road because manganese

a. can show highest oxidation state of +7 b. can remove oxygen and sulphur
c. helps the formation of oxides of iron d. give hardness to steel

12 Which of the following set is diamagnetic?

a. Hg_2Cl_2 and $HgCl_2$ b. $CuCl_2$ and CuCl
c. $ZnCl_2$ and $CuCl_2$ d. $TiCl_3$ and $TiCl_4$

Working Space

13 Which of the following reactions are disproportionation reactions?

i. $Cu^+ \longrightarrow Cu^{2+} + Cu$

ii. $3MnO_4^{2-} + 4H^+ \longrightarrow 2MnO_4^- + MnO_2 + 2H_2O$

iii. $2KMnO_4 \longrightarrow K_2MnO_4 + MnO_2 + O_2$

iv. $2MnO_4^- + 3Mn^{2+} + 2H_2O \longrightarrow 5MnO_2 + 4H^+$

a. i, ii
b. i, ii, iii
c. ii, iii, iv
d. i, iv

14 Which of the following is amphoteric oxide among Mn_2O_7, CrO_3, Cr_2O_3, CrO, V_2O_5, V_2O_4?

a. Cr_2O_3 and V_2O_4
b. Mn_2O_7 and CrO_3
c. CrO and V_2O_5
d. V_2O_5 and CrO_3

15 Interstitial compounds are formed when small atoms are trapped inside the crystal lattice of metals. Which of the following is not the characteristic property of interstitial compounds?

a. They have high melting points in comparison to pure metals
b. They are very hard
c. They retain metallic conductivity
d. They are chemically very reactive

16 Experimentally, it was found that a metal oxide has formula $M_{0.98}O$. Metal M present as M^{2+} and M^{3+} in its oxide. Fraction of the metal which exists as M^{3+} would be

a. 7.0%
b. 4.08%
c. 6.05%
d. 5.08%

17 In which of the following the hybridisation is different from others?

a. MnO_4^-
b. $Ni(CO)_4$
c. $Ni(CN)_4^{2-}$
d. $MnBr_4^{2-}$

18 Which of the following pairs of transition metal ions are the stronger oxidising agents in aqueous solution?

a. V^{2+} and Cr^{2+}
b. Ti^{2+} and Cr^{2+}
c. Mn^{3+} and Co^{3+}
d. V^{2+} and Fe^{2+}

19 It is possible to change the $Cr_2O_7^{2-}$ ion into the CrO_4^{2-} ion by the addition of

a. H_2SO_4
b. H_2O_2
c. NaOH
d. H_3PO_4 and H_2O_2

20 The number of moles of dichromate required to oxidise one mole of ferrous and one mole of ferric oxalate respectively are

a. 0.5 and 1.0
b. 1.0 and 0.5
c. 0.6 and 1.2
d. 0.5 and 0.33

21 MnO_4^- oxidises

i. oxalate ion in acidic medium at 333 K and

ii. HCl

For balanced equations, the ratios (MnO_4^- : $C_2O_4^{2-}$ in (i) and MnO_4^- : HCl in (ii)) respectively are

a. 1 : 5 and 2 : 5
b. 2 : 5 and 1 : 5
c. 2 : 5 and 1 : 8
d. 5 : 2 and 1 : 8

22 Which statement given below is correct regarding dichromate ion?

a. In the reaction between $Cr_2O_7^{2-}$ and H_2O_2,oxidising agent is H_2O_2
b. A strong acid changes the CrO_4^{2-} ion to the $Cr_2O_7^{2-}$ ion
c. In an acidic medium, the $Cr_2O_7^{2-}$ ion behaves as reducing agent
d. In the reaction between $Cr_2O_7^{2-}$ and H_2O_2, the solution colour changes from orange of dichromate ion through green to the blue of the Cr^{3+} ion

Working Space

23 Sodium hydrogen sulphite ($NaHSO_3$) is added to a neutral permanganate ion solution. In the reaction that follows, the permanganate ion will be reduced to

a. MnO_2
b. MnO_4^-
c. MnO_4^{2-}
d. Mn^{2+}

24 $KMnO_4$ acts as an oxidising agent in acidic medium. The number of moles of $KMnO_4$ needed to react with one mole of sulphide ions in acidic solution is

a. 2/5
b. 3/5
c. 4/5
d. 1/5

25 $X + KOH + O_2 \longrightarrow Y + H_2O$

$Y + H_2SO_4 \longrightarrow Z + K_2SO_4 + MnO_2 + H_2O$. Here X, Y and Z are

a. $X = K_2MnO_4$, $Y = MnO_2$, $Z = KMnO_4$
b. $X = K_2MnO_4$, $Y = MnO_4$, $Z = KMnO_4$
c. $X = MnO_2$, $Y = K_2MnO_4$, $Z = KMnO_4$
d. $X = MnO_2$, $Y = KMnO_4$, $Z = K_2MnO_4$

26 Lanthanide for which + 2 and + 3 oxidation states are common is

a. La
b. Nd
c. Ce
d. Eu

27 The radioactive lanthanide is

a. promethium (Pm)
b. lutetium (Lu)
c. ytterbium (Yb)
d. samarium (Sm)

28 Which of the following ions will exhibit colour in aqueous solutions?

a. $La^{3+}(Z = 57)$
b. $Ti^{3+}(Z = 22)$
c. $Lu^{3+}(Z = 71)$
d. $Sc^{3+}(Z = 21)$

29 Which one of the following pairs consists of a good oxidising and good reducing agent, respectively?

a. Ce (IV) and La (III)
b. La (III) and Eu (II)
c. Ce (IV) and Eu (II)
d. Ln (III) and Ce (III)

30 The pair of lanthanides with the highest third ionisation energy is

a. Eu and Gd
b. Eu and Yb
c. Dy and Yb
d. Lu and Yb

Revisal Problems for JEE Advanced

SET - 2

Only One Option Correct Type

Working Space

Direction (Q. Nos. 1-10) *This section contains 10 multiple choice questions. Each question has four choices (a), (b), (c) and (d), out of which ONLY ONE is correct.*

1 Consider the four sulphates $FeSO_4$, $MnSO_4$, $CoSO_4$ and $ZnSO_4$. Four separate capsules of these materials have been made containing 0.1 mole of the substance and are suspended by a thread. Then a strong magnet is brought close to each capsule. Order of sulphates according to the strength of the response that the magnet will produce on these capsules is

a. $MnSO_4 > FeSO_4 > CoSO_4 > ZnSO_4$
b. $FeSO_4 > MnSO_4 > CoSO_4 > ZnSO_4$
c. $ZnSO_4 > CoSO_4 > FeSO_4 > MnSO_4$
d. $CoSO_4 > FeSO_4 > MnSO_4 > ZnSO_4$

2 A non-stoichiometric oxide of silver has composition $Ag_{1.8}O$. What percentage of Ag is present in the atom of Ag^{2+}?

a. 11 b. 14
c. 20 d. 25

3 Which of the following will not act as an oxidising agent?

a. CrO_3 b. MoO_3
c. WO_3 d. CrO_4^{2-}

4 Which of the following is the correct order of magnetic moment (in BM) for Mn^{2+}, Cr^{2+} and V^{2+}?

a. $Mn^{2+} > Cr^{2+} > V^{2+}$ b. $V^{2+} > Cr^{2+} > Mn^{2+}$
c. $Mn^{2+} > V^{2+} > Cr^{2+}$ d. $Cr^{2+} > V^{2+} > Mn^{2+}$

5 For a given sample of Fe^{2+} solution, if 30 mL of 0.1 M dichromate solution is required, the volume of 0.1 M permanganate required is

a. 30 mL b. 36 mL
c. 40 mL d. 45 mL

6 Which one of the following elements shows maximum number of different oxidation numbers in its compounds?

a. Eu b. La
c. Am d. Gd

7 Oxidation state of mercury in amalgam, calomel and corrosive sublimate is

a. 0, +1 and +2 b. +1, +1 and +2
c. +1, +2 and +1 d. 0, +1 and +1

8 Which of the following electrode reactions has more negative SRP value?

a. Mn^{3+}/Mn^{2+} b. Fe^{3+}/Fe^{2+}
c. Cr^{3+}/Cr^{2+} d. V^{3+}/V^{2+}

9 Among the following the strongest oxidising agent is

a. $[WO_4]^{2-}$ b. $[CrO_4]^{2-}$
c. $[MoO_4]^{2-}$ d. $[ReO_4]^{2-}$

10 Acidified $K_2Cr_2O_7$ is treated with hydrogen sulphide. In this reaction, the oxidation state of chromium

a. increases from +3 to +6 b. remains unchanged
c. decreases from +6 to +3 d. decreases from +6 to +2

One or More than One Options Correct Type

Working Space

Direction (Q. Nos. 11-20) *This section contains 10 multiple choice questions. Each question has four choices (a), (b), (c) and (d), out of which ONE or MORE THAN ONE are correct.*

11 Which of the following are soluble in water?

a. AgF
b. $AgNO_3$
c. $AgClO_4$
d. AgBr

12 Which are correct statements?

a. In any transition series, magnetic moment of M^{2+} ions first increases and then decreases
b. Metals of 4*d* and 5*d* series have greater enthalpies of atomisation than 3*d* series
c. In any series, metals with highest oxidation numbers lie towards the end of the series
d. In 3*d* series $E°$ value is positive for Cu^{2+}/Cu while for others $E°$ value is negative

13 Which is correct regarding structure of dichromate ion?

a. It has 8 sigma bonds
b. All Cr—O bond lengths are equal
c. It has 7 sigma bonds
d. All Cr—O bond lengths are not equal

14 Which are correctly matched?

a. Baeyer's reagent = cold alkaline $KMnO_4$
b. Tollen's reagent = ammoniacal $AgNO_3$
c. Nessler's reagent = rhodium complex
d. Bordeaux mixture = $CuSO_4 + Cu(OH)_2$

15 Among the following metals, which have relatively higher second ionisation potential?

a. V ($Z = 23$)
b. Cr ($Z = 24$)
c. Mn ($Z = 25$)
d. Cu ($Z = 29$)

16 Find the correct statements.

a. In 3*d* series, enthalpy of atomisation is lowest for zinc and highest for vanadium
b. Cr^{2+} is stronger reducing agent than Fe^{2+}
c. Metallic radius of zinc is less than copper
d. Density of copper is greater than zinc

17 Which are correctly matched?

a. MnO_2 = dry battery cells
b. TiO_2 = pigment industry
c. $TiCl_4$ = in *Zeigler-Natta* catalyst
d. AgBr = photographic plates

18 Find incorrect statements.

a. The maximun oxidation state of Cr is +6
b. All Cr—O bond lengths in CrO_4^{2-} are equal but in $Cr_2O_7^{2-}$ they are not equal
c. Mercurous ion exists as Hg_2^{2+} but not as Hg^+
d. German silver contains silver metal

19 $KMnO_4$ in an acidic medium oxidises

a. H_2O_2
b. NO_2^-
c. SO_4^{2-}
d. $S_2O_3^{2-}$

20 The species having same number of unpaired electrons

a. Eu^{2+}
b. Gd^{3+}
c. Tb^{4+}
d. Lu^{3+}

Comprehension Type

Direction (Q. Nos. 21 and 22) *This section contains a paragraph, each describing theory, experiments, data, etc. Two questions related to the paragraph have been given. Each question has only one correct answer among the four given options (a), (b), (c) and (d).*

Passage

A violet compound of manganese *A* decomposes on heating to give compounds *B* and *C* along with the liberation of a gaseous substance *D*. Compound *C* reacts with KOH in the presence of KNO_3 to give compound *B*. On heating *C* with conc. H_2SO_4 and NaCl, greenish yellow gas *E* is liberated and a compound of manganese *F* is formed along with other products.

Working Space

21 The greenish yellow gas liberated in the above reactions is

a. O_2 b. Cl_2 c. F_2 d. SO_2

22 The compounds *A* and *C* in the above reactions are

a. $KMnO_4$ and MnO_2 b. K_2MnO_4 and MnO_2
c. $KMnO_4$ and K_2MnO_4 d. $KMnO_4$ and Mn_2O_3

Matching List Type

Direction (Q. Nos.23 and 24) *Choices for the correct combination of elements from Column I and Column II are given as options (a), (b), (c) and (d), out of which one is correct.*

23. Match the following reagents given in Column I with the Column II. Mark the correct option from the codes given below.

	Column I		Column II
i.	Lucas reagent	p.	$CuSO_4$ solution + lime
ii.	Bordeaux mixture	q.	CrO_2Cl_2
iii.	Etard's reagent	r.	$[Cu(NH_3)_4]SO_4$
iv.	Schweitzer reagent	s.	Conc. HCl + anhydrous $ZnCl_2$

Codes

	i	ii	iii	iv		i	ii	iii	iv
a.	s	p	q	r	b.	p	q	s	r
c.	q	p	r	s	d.	r	s	q	p

24. Match the following characteristics given in Column I with the corresponding elements given in Column II. Mark the correct option from the codes given below.

	Column I		Column II
i.	Element with low enthalpy of atomisation	p.	Copper
ii.	Element with highest second ionisation energy	q.	Zinc
iii.	Element with high melting point	r.	Fe
iv.	Most abundant element	s.	Cr

Codes

	i	ii	iii	iv		i	ii	iii	iv
a.	s	p	r	q	b.	q	p	r	s
c.	p	r	q	s	d.	q	p	s	r

One Integer Value Correct Type

Direction (Q. Nos. 25-30) *This section contains 6 questions. When worked out will result in an integer from 0 to 9 (both inclusive).*

25 What is van't Hoff factor if $Hg_2(NO_3)_2$ is 100% ionised in aqueous solution?

26 The number of electrons in 4*f* subshell of +4 oxidation state of terbium are

27 The calculated magnetic moment of Fe^{n+} ion is 4.90 BM. The sum of the number of electrons present in 3*d* subshell of Fe^{n+} ion and the value of *n* is

28 In solid $CuSO_4 \cdot 5H_2O$ copper is coordinated to .... water molecules.

29 $MnO_4^- + I^- \longrightarrow MnO_2 + IO_3^-$ when this equation is balanced in basic medium, the sum of the coefficients in the balanced equation is

30 Among the following ion, the number of ions which are colourless

$La^{3+}, Pr^{3+}, Lu^{3+}, Ce^{3+}, Tm^{3+}, Yb^{3+}, Er^{3+}$

JEE Main & AIEEE Archive

(Compilation of Last 13 Years' Questions)

1 Which of the following arrangements does not represent the correct order of the property stated against it? **(JEE Main 2013)**

a. $V^{2+} < Cr^{2+} < Mn^{2+} < Fe^{2+}$: paramagnetic behaviour
b. $Ni^{2+} < Co^{2+} < Fe^{2+} < Mn^{2+}$: ionic size
c. $Co^{3+} < Fe^{3+} < Cr^{3+} < Sc^{3+}$: stability in aqueous solution
d. Sc < Ti < Cr < Mn : number of oxidation states

2 Four successive members of the first row transition elements listed below with atomic numbers. Which one of them is expected to have the highest $E^\circ_{M^{3+}/M^{2+}}$ value? **(JEE Main 2013)**

a. Cr ($Z = 24$) b. Mn ($Z = 25$)
c. Fe ($Z = 26$) d. Co ($Z = 27$)

3 Iron exhibits +2 and +3 oxidation states. Which of the following statements about iron is incorrect? **(AIEEE 2012)**

a. Ferrous oxide is more basic in nature than the ferric oxide
b. Ferrous compounds are relatively more ionic than the corresponding ferric compounds
c. Ferrous compounds are less volatile than the corresponding ferric compounds
d. Ferrous compounds are more easily hydrolysed than the corresponding ferric compounds

4 In context of the lanthanides, which of the following statements is not correct? **(AIEEE 2011)**

a. There is a gradual decrease in the radii of the members with increasing atomic number in the series
b. All the members exhibit +3 oxidation state
c. Because of similar properties, the separation of lanthanoids is not easy
d. Availability of 4*f* electrons results in the formation of compounds in +4 state for all the members of the series

5 The outer electronic configuration of Gd (Atomic number 64) is **(AIEEE 2011)**

a. $4f^3, 5d^5, 6s^2$ b. $4f^8, d^0, 6s^2$
c. $4f^4, 5d^4, 6s^2$ d. $4f^7, 5d^1, 6s^2$

6 The correct order of $E^\circ_{M^{2+}/M}$ values with negative sign for the four successive elements Cr, Mn, Fe and Co is **(AIEEE 2010)**

a. Mn > Cr > Fe > Co
b. Cr > Fe > Mn > Co
c. Fe > Mn > Cr > Co
d. Cr > Mn > Fe > Co

7 Knowing that the chemistry of lanthanides (Ln) is dominated by its +3 oxidation state, which of the following statements is incorrect? **(AIEEE 2009)**

a. Because of the large size of the Ln (III) ions, the bonding in its compounds is predominantly ionic in character
b. The ionic sizes of Ln (III) decrease in general with increasing atomic number
c. Ln (III) compounds are generally colourless
d. Ln (III) hydroxide are mainly basic in character

8 In context with the transition elements, which of the following statements is incorrect? **(AIEEE 2009)**

a. In addition to the normal oxidation state, the zero oxidation state is also shown by these elements in complexes
b. In the highest oxidation states, the transition metal shows basic character and form cationic complexes
c. In the highest oxidation states of the first five transition elements (Sc to Mn), all the 4*s* and 3*d* electrons are used for bonding
d. Once the d^5 configuration is exceeded, the tendency to involve all the 3*d* electrons in bonding decreases

9 Larger number of oxidation states are exhibited by the actinides than those by the lanthanides, the main reason being **(AIEEE 2008)**

a. 4*f*-orbitals more diffused than the 5*f*-orbitals
b. lesser energy difference between 5*f* and 6*d* than between 4*f* and 5*d*-orbitals
c. more energy difference between 5*f* and 6*d* than between 4*f* and 5*d*-orbitals
d. more reactive nature of the actinides than the lanthanides

10 Identify the incorrect statement among the following. **(AIEEE 2007)**

a. *d*-block elements show irregular and erratic chemical properties among themselves
b. La and Lu have partially filled *d*-orbitals and no other partially filled orbitals
c. The chemistry of various lanthanoids is very similar
d. 4*f* and 5*f*-orbitals are equally shielded

11 The actinides exhibit more number of oxidation states in general than the lanthanides. This is because **(AIEEE 2007)**

a. the 5*f* orbitals are more buried than the 4*f* orbitals
b. there is a similarity between 4*f* and 5*f*-orbitals in their angular part of the wave function
c. the actinides are more reactive than the lanthanides
d. the 5*f* orbitals extend farther from the nucleus than the 4*f* orbitals

12 Lanthanides contraction is caused due to (AIEEE 2006)

a. the appreciable shielding on outer electrons by 4*f* electrons from the nuclear charge
b. the appreciable shielding on outer electrons by 5*d* electrons from the nuclear charge
c. the same effective nuclear charge from Ce to Lu
d. the imperfect shielding on outer electrons by 4*f* electrons from the nuclear charge

13 On heating, mixture of Cu_2O and Cu_2S will give (AIEEE 2005)

a. Cu_2SO_3 b. $CuO + CuS$
c. $Cu + SO_3$ d. $Cu + SO_2$

14 Calomel (Hg_2Cl_2) on reaction with ammonium hydroxide gives (AIEEE 2005)

a. HgO b. Hg_2O
c. $NH_2—Hg—Hg—Cl$ d. $HgNH_2Cl$

15 Cerium (Z = 58) is an important member of the lanthanides. Which of the following statements about cerium is incorrect?

a. The common oxidation states of cerium are +3 and +4 (AIEEE 2004)
b. The +3 oxidation state of cerium is more stable than the +4 oxidation state
c. The +4 oxidation state of cerium is not known in solutions
d. Cerium (IV) acts as an oxidising agent

16 Of the following outer electronic configurations of atoms, the highest oxidation state is achieved by which one of them? (AIEEE 2004)

a. $(n-1)d^8ns^2$ b. $(n-1)d^5ns^1$
c. $(n-1)d^3ns^2$ d. $(n-1)d^5ns^2$

17 Which of the following groups of transition metals is called coinage metals? (AIEEE 2003)

a. Cu, Ag, Au b. Ru, Rh, Pd
c. Fe, Co, Ni d. Os, Ir, Pt

18 Which one of the following nitrates will leave behind a metal on strong heating? (AIEEE 2003)

a. Ferric nitrate
b. Copper nitrate
c. Manganese nitrate
d. Silver nitrate

19 For making good quality mirrors, plates of float glass are used. These are obtained by floating molten glass over a liquid metal which does not solidify before glass. The metal used can be (AIEEE 2003)

a. mercury b. tin
c. sodium d. magnesium

20 Which one of the following statements is correct? (AIEEE 2003)

a. Manganese salts give a violet borax bead test in the reducing flame
b. From a mixed precipitate of AgCl and AgI, ammonia solution dissolves only AgCl
c. Ferric ions give a deep green precipitate on adding potassium ferrocyanide solution
d. On boiling a solution having K^+, Ca^{2+} and HCO_3^- ions, we get a precipitate of $K_2Ca(CO_3)_2$

21 A red solid is insoluble in water. However, it becomes soluble if some KI is added to water. Heating the red solid in a test tube results in liberation of some violet coloured fumes and droplets of a metal appear on the cooler parts of the test tube. The red solid is (AIEEE 2003)

a. $(NH_4)_2Cr_2O_7$ b. HgI_2
c. HgO d. Pb_3O_4

22 What would happen when a solution of potassium chromate is treated with an excess of dilute nitric acid? (AIEEE 2003)

a. Cr^{3+} and $Cr_2O_7^{2-}$ are formed
b. $Cr_2O_7^{2-}$ and H_2O are formed
c. CrO_4^{2-} is reduced to +3 state of Cr
d. None of the above

23 Most common oxidation states of Ce (cerium) are (AIEEE 2002)

a. + 3, + 4
b. + 2, + 3
c. + 2, + 1
d. + 3, + 5

24 Which of the following ions has the maximum magnetic moment? (AIEEE 2002)

a. Mn^{2+} b. Fe^{2+}
c. Ti^{2+} d. Cr^{2+}

JEE Advanced & IIT JEE Archive

(Compilation of Last 10 Years' Questions)

1 Consider the following list of reagents, acidified $K_2Cr_2O_7$, alkaline $KMnO_4$, $CuSO_4$, H_2O_2, Cl_2, O_3, $FeCl_3$, HNO_3 and $Na_2S_2O_3$. The total number of reagents that can oxidise aqueous iodide to iodine is

(2014 Adv., Integer Type)

2 The colour of light absorbed by an aqueous solution of $CuSO_4$ is

(2012, Only One Option Correct Type)

a. orange-red b. blue-green
c. yellow d. violet

3 Which of the following halides react(s) with $AgNO_3(aq)$ to give a precipitate that dissolves in $Na_2S_2O_3(aq)$?

(2012, One or More than One Options Correct Type)

a. HCl b. HF
c. HBr d. HI

4 Reduction of the metal centre in aqueous permanganate ion involves

(2011, One or More than One Options Correct Type)

a. three electrons in neutral medium
b. five electrons in neutral medium
c. three electrons in alkaline medium
d. five electrons in acidic medium

5 Which of the following pairs is expected to exhibit same colour in solution?

(2005, Only One Option Correct Type)

a. $VOCl_2$; $FeCl_2$ b. $CuCl_2$; $VOCl_2$
c. $MnCl_2$; $FeCl_2$ d. $FeCl_2$; $CuCl_2$

6 Which of the following will not be oxidised by O_3?

(2005, Only One Option Correct Type)

a. KI b. $FeSO_4$
c. $KMnO_4$ d. K_2MnO_4

7 $\underset{\text{(White fumes with pungent smell)}}{(B)} \xleftarrow{\text{Moist air}} \underset{\text{(}M\text{ = Transition element, colourless)}}{MCl_4} \xrightarrow{\text{Zn}} \underset{\text{(Purple colour)}}{(A)}$ Identify the metal M and hence MCl_4. Explain the difference in colours of MCl_4 and A.

(2005, Subjective Type)

JEE Archive

3. Coordination Compounds

DPP-1	Introduction of Coordination Compounds and Werner's Theory
DPP-2	IUPAC Nomenclature of Coordination Compounds
DPP-3	Isomerism in Coordination Compounds-I
DPP-4	Isomerism in Coordination Compounds-II
DPP-5	Valence Bond Theory, Effective Atomic Number and Magnetic Properties
DPP-6	Crystal Field Theory (CFT) and Colour of Complexes
DPP-7	Bonding in Organometallic Compounds, Stability and Applications of Coordination Compounds

- **Revisal Problems for JEE Main**
- **Revisal Problems for JEE Advanced**
- **JEE Main & AIEEE Archive**
- **JEE Advanced & IIT JEE Archive**

DPP
DAILY PRACTICE PROBLEMS

DPP-1 Introduction of Coordination Compounds and Werner's Theory

Subjective Questions

Working Space

Direction *(Q. Nos. 1-5) These questions are subjective in nature, need to be solved completely on notebook.*

1 Alums are double salts. How many ions are produced by one mole of an alum molecule in solution?

2 In $CoCl_3 \cdot 5NH_3$ and $K_4[Fe(CN)_6]$, which satisfy primary and secondary valency?

3 Write two defects in Werner's theory.

4 i. $IrCl_3 \cdot 3NH_3$ complex is a non-conductor in aqueous solution. Why?
ii. What are homo and heteroleptic complexes? Give example.

5 $CoCl_2$ solution can be used as invisible ink. On heating, it gives very legible blue colour which then slowly disappears. Why?

Only One Option Correct Type

Direction *(Q. Nos. 6-15) This section contains 10 multiple choice questions. Each question has four choices (a), (b), (c) and (d), out of which ONLY ONE is correct.*

6 Primary and secondary valency of Pt in $[Pt(en)_2Cl_2]$ are

a. +4 and 4
b. +4 and 6
c. +6 and 4
d. +2 and 6

7 If zeise's salt has the formula $[Pt(C_2H_4)Cl_3]^-$. In this, platinum primary and secondary valency are

a. +1and 3
b. +1and 4
c. +3 and 4
d. +4 and 6

8 The two complexes $PtCl_4 \cdot 2NH_3$ and $PtCl_4 \cdot 2KCl$ do not give precipitate of AgCl when treated with $AgNO_3$. The structures of these complexes are

a. $[Pt(NH_3)_2Cl_4]$ and $K_2[PtCl_6]$
b. $[Pt(NH_3)_2]Cl_4$ and $K_2[PtCl_6]$
c. $[Pt(NH_3)_2Cl_2]Cl_2$ and $K_2[PtCl_4]Cl_2$
d. $[Pt(NH_3)_2Cl_4]$ and $K_2[PtCl_5]Cl$

Working Space

9 A cobaltamine has the formula $CoCl_3 \cdot xNH_3$. This when reacted with $AgNO_3$ solution, one third of the chloride is precipitated. It can have the structure

a. $[Co(NH_3)_6]Cl_3$
b. $[Co(NH_3)_5]Cl_2$
c. $[Co(NH_3)_4Cl_2]Cl$
d. $[Co(NH_3)_5H_2O]Cl_3$

10 Which is the incorrect?

a. $[Co(NH_3)_4Cl_2]^+$ is heteroleptic complex
b. In $[Fe(C_2O_4)_3]^{3+}$ and $[Co(en)_3]^{3+}$, the coordination number of both Fe and Co is 6
c. Ferric alum is a double salt
d. Mohr's salt is a double salt and do not give test for Fe^{2+} ions

11 Ligands which can get attached to the central metal ion through more than one atom are called

a. ambident ligands
b. polydentate ligands
c. chelate ligands
d. neutral ligands

12 The anion acetylacetonate (acac) forms chelate with Co^{3+}. The ring of the chelate is

a. five membered
b. four membered
c. six membered
d. three membered

13 Oxidation states of iron in the complexes $[Fe(H_2O)_5NO]^{2+}$ and $[Fe(CN)_5NO]^{2-}$ are

a. +1and +2
b. +1and +1
c. +2 and +2
d. +2 and +1

14 Complexes in which the oxidation number of a metal is zero is/are

a. $Cr(C_6H_6)_2$
b. $Mn_2(CO)_{10}$
c. $[Ni(CO)_4]$
d. All of these

15 An example for bidentate and negatively charged ligand is

a. acetylacetonate (acac)
b. propylene diamine (pn)
c. ethylene diamine (en)
d. bipyridyl (bipy)

One or More than One Options Correct Type

Direction *(Q. Nos. 16-20) This section contains 5 multiple choice questions. Each question has four choices (a), (b), (c) and (d), out of which ONE or MORE THAN ONE are correct.*

16 An example of double salt is/are

a. hypo
b. bleaching powder
c. common alum
d. carnallite

17 Type of bonding in $K_4[Fe(CN)_6]$ is/are

a. ionic
b. covalent
c. metallic
d. coordinate covalent

18 Regarding the complex $K_3[Fe(CN)_6]$, incorrect statement(s) is/are

a. CN^- ion is Lewis base
b. Fe^{3+} ion is Lewis acid
c. It is a cationic complex
d. It produces 10 ions in solution

Working Space

19 Which of the following statements is/are correct?

a. $[Co(en)_2NO_2Cl]Br$ is cationic complex
b. $[Co(en)_3]Cl_3$ produces 3 ions in solution
c. $[Fe(CO)_5]$ is neutral complex
d. $[Cu(NH_3)_4]SO_4$ is deep blue colour

20 A freshly prepared aqueous solution of $Pd(NH_3)_2Cl_2$ does not conduct electricity. It suggests that

a. the structure of the compound involves covalent bonding only
b. the chlorine atoms must be in coordination sphere
c. the van't Hoff factor of the compound would be unity
d. on adding excess aqueous $AgNO_3$ to 0.1 L of 0.1 M solution of the compound, 0.02 mole of AgCl would be obtained

Comprehension Type

Direction (Q. Nos. 21 and 22) This section contains a paragraph, describing theory, experiments, data, etc. Two questions related to the paragraph have been given. Each question has only one correct answer among the four given options (a), (b), (c) and (d).

Passage

Complex compounds are addition compounds formed by the stoichiometric combination of two or more simple salts but do not decompose into constituent ions completely. The first such complex prepared by Tassaert is hexamine cobalt (III) chloride. Later many such compounds were prepared and their properties were studied. The chloramines complexes of cobalt (III) chromium (III) not only exhibit a spectrum of colours but also differ in the reactivity of their chlorides. Moreover, greater the number of ions produced by a complex in solution, greater is the electrical conductivity. This type of information was obtained for several series of complexes.

21 Coordination number of Co in $CoCl_3 \cdot 5H_2O$ is six. The volume of 0.1 N $AgNO_3$ needed to precipitate the chlorine in 200 mL of 0.01 M solution of complex is

a. 140 mL b. 40 mL c. 80 mL d. 20 mL

22 The number of ions per mole of the complex $CoCl_3 \cdot 5NH_3$ in aqueous solution will be

a. 3 b. 9 c. 2 d. 4

Matching List Type

Direction (Q. No. 23) Choices for the correct combination of elements from Column I and Column II are given as options (a), (b), (c) and (d), out of which one is correct.

23 Match the Column I with Column II and mark the correct option from the codes given below.

	Column I (Molar conductance)		Column II (Complex)
i.	229	p.	$[Pt(NH_3)_5Cl]Cl_3$
ii.	0	q.	$[Pt(NH_3)_2Cl_4]$
iii.	404	r.	$[Pt(NH_3)_4Cl_2]Cl_2$
iv.	523	s.	$[Pt(NH_3)_6]Cl_4$

Codes

	i	ii	iii	iv		i	ii	iii	iv
a.	r	q	p	s	b.	p,q	p	q	r,s
c.	p	q,r	p	s	d.	p	s	r	p,s

One Integer Value Correct Type

Working Space

Direction *(Q. Nos. 24-28) This section contains 5 questions. When worked out will result in an integer from 0 to 9 (both inclusive).*

24 Formula of Mohr's salt is $FeSO_4 \cdot (NH_4)_2SO_4 \cdot xH_2O$. Here x value is

25 When $AgNO_3$ solution is added in excess to 1 M solution of $CoCl_3 \cdot xNH_3$, one mole of AgCl is formed. The value of x is

26 How many moles of AgCl would be obtained when 100 mL of 0.1 M $CoCl_3 \cdot 3NH_3$ is treated with excess $AgNO_3$?

27 Oxidation state of cobalt in $Co(CO)_6$ is

28 In the complex $PtCl_4 \cdot 5NH_3$ if coordination number of platinum is 6, number of chloride ions precipitated by adding AgCl are

Statement Type

Direction *(Q. Nos. 29 and 30) This section is based on Statement I and Statement II. Select the correct answer from the codes given below.*

a. Both Statement I and Statement II are correct and Statement II is the correct explanation of Statement I
b. Both Statement I and Statement II are correct but Statement II is not the correct explanation of Statement I
c. Statement I is correct but Statement II is incorrect
d. Statement II is correct but Statement I is incorrect

29 **Statement I** Oxidation state of Fe in $Fe(CO)_5$ is zero.

Statement II EAN of Fe in this complex is 36.

30 **Statement I** The oxidation number of platinum in Zeise's salt is +4.

Statement II Zeise's salt is ionic complex.

DPP-2 IUPAC Nomenclature of Coordination Compounds

Subjective Questions

Working Space

Direction (Q. Nos. 1-5) *These questions are subjective in nature, need to be solved completely on notebook.*

1 Write the structure to the following.

i. Potassium tetrafluoroargentate(III)

ii. Calcium dioxalatodiamminecobaltate(II)

iii. *Bis* benzene chromium(0)

iv. Sodium *bis* (thiosulphato)argentite (I)

2 Write the IUPAC names to the following.

i. $[Pt(NH_3)_4]\,[PtCl_4]$ (Magnus's green)

ii. $NH_4[Cr(NH_3)_2(NCS)_4]$ (Reinecke's salt)

iii. $K\,[Pt(\eta^2 - C_2H_4)Cl_3]$ (Zeise's salt)

iv. $[Fe(bipy)_3]^{2+}$

3 Write the IUPAC name of the following.

i. $PtCl_4 \cdot 5NH_3$ (gives 4 ions in solution)

ii. $PtCl_4 \cdot 2KCl$ (gives 3 ions in solution)

iii. $IrCl_3 \cdot 3NH_3$ (no Cl^- ion is precipitated)

iv. $CoCl_3 \cdot 4NH_3$ (one Cl is precipitated by $AgNO_3$)

4 In the IUPAC system, the name given to the ligand

i. $CH_3COCH_2COCH_3 \rightleftharpoons CH_3-\overset{O^-}{\overset{|}{C}}=CH-\overset{O}{\overset{\|}{C}}-CH_3$

ii. $H_2N-CH_2-COO^-$

iii. 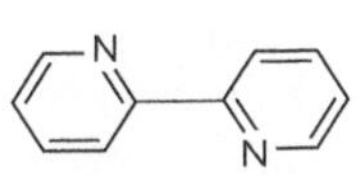

iv. $H_2NCH_2CH_2NHCH_2CH_2NH_2$

5 Explain the following.

i. Chelate complex

ii. Dimethylglyoxime(dmg)

iii. Tetradentate

Only One Option Correct Type

Working Space

Direction *(Q. Nos. 6-15) This section contains 10 multiple choice questions. Each question has four choices (a), (b), (c) and (d), out of which ONLY ONE is correct.*

6 According to IUPAC nomenclature, sodium nitroprusside is named as

a. sodium nitroferricyanide
b. sodium nitroferrocyanide
c. sodium pentacyanonitrosylferrate (III)
d. sodium pentacyanonitrosyliumferrate (II)

7 The formula of the complex hexamminecobalt (III) chloride sulphate is

a. $[Co(NH_3)_6]ClSO_4$
b. $[Co(NH_3)_6Cl]SO_4$
c. $[Co(NH_3)_6ClSO_4]$
d. None of these

8 The IUPAC name of the compound $[Cr(NH_3)_5(NCS)][ZnCl_4]$ is

a. pentammine isothiocyanatochromium (III) tetrachlorozincate (II)
b. pentaammine thiocyanatochromium (III) tetrachlorozincate (II)
c. pentammine isothiocyanatochromium (0) tetrachlorozincate (IV)
d. pentammine thiocyanatochromium (0) tetrachlorozincate (IV)

9 The IUPAC name of the complex $[(CO)_5Mn-Mn(CO)_5]$ is

a. *bis* (pentacarbonyl dimanganese)
b. *bis* (pentacarbonyldimanganate(II)
c. decacarbonyldimanganate(0)
d. *bis* (pentacarbonyldimanganese(0)

10 The donor sites of EDTA ligand are

a. O-atoms only
b. N-atoms only
c. two N-atoms and four O-atoms
d. three N-atoms of EDTA ion

11 The IUPAC name of $[Ni(PPh_3)_2Cl_2]^{2+}$ is

a. *bis* dichloro(triphenylphosphine) nickel (II) ion
b. dichloro *bis* (triphenylphosphine) nickel (II)ion
c. dichloro triphenylphosphine nickel (II) ion
d. triphenylphosphine nickel (II)dichloride

12 Formula of brown ring complex is

a. $[Fe(H_2O)(NO_2)_5]SO_4$
b. $[Fe(H_2O)_3(NO_2)_3]SO_4$
c. $[Fe(H_2O)_5(NO)]SO_4$
d. $[Fe(H_2O)_4(NO)_2]SO_4$

13 The IUPAC name of the compound $K[SbCl_5Ph]$ is

a. Potassium chlorophenylantimonate (V)
b. Potassium pentachloro(phenyl) antimonate (V)
c. Potassium pentachlorobenzylantimonate (V)
d. None of the above

14 $K_2[OsCl_5N]$ is named as

a. potassium pentachloroazoosmate(VI)
b. potassium pentachloronitridoosmate(VI)
c. potassium pentachloroazidoosmate(VI)
d. potassium pentachloronitroniumosmate(VI)

Working Space

15 What is the IUPAC name of compound $NaBH_4$?

a. Sodium boronhydride
b. Sodium tetrahydridoboron (III)
c. Sodium tetrahydridoborate (III)
d. Sodium tetrahydridoborate (I)

One or More than One Options Correct Type

Direction (Q. Nos. 16-20) This section contains 5 multiple choice questions. Each question has four choices (a), (b), (c) and (d), out of which ONE or MORE THAN ONE are correct.

16 Which of the following complexes have correct name?

a. $K[Pt(NH_3)Cl_5]$ = potassiumamminepentachloridoplatinate (IV)
b. $[Ag(CN)_2]^-$ = dicyanidoargentate(I) ion
c. $K_3[Cr(C_2O_4)_3]$ = tripotassium trioxalatochromate(III)
d. $Na_2[Ni(EDTA)]$ = sodium ethylenediaminetetraacetatonickel(II)

17 Which is/are the correctly named?

a. acac = Acetylacetonato, chelating ,uninegative
b. bipy = 2, 2′ bipyridine , bidentate, neutral
c. DMG = bidentate and neutral
d. gly = glycinato, bidentate, uninegative

18 The correct name of ligand(s) in IUPAC name is/are

a. $S_2O_3^{2-}$ = thiosulphato
b. C_5H_5N = pyridyl
c. $(C_5H_5)^-$ = cyclopentadienyl
d. $C_2O_4^{2-}$ = acetato

19 Which is/are the incorrect IUPAC name?

a. Formula of potassium tetrafluoroargentite is $K_2[AgF_4]$
b. IUPAC name for $[AuCl_4(H_2O)_2]^-$ diaquatetrachloroaurate (I)
c. $[Cr(C_6H_6)_2]$ is named as *bis* (benzene) chromium (0)
d. $[AlH_4]^-$ is named as tetrahydrido aluminium (III) ion

20 The ligand(s) which gives chelate complexes is/are

a. SCN^-
b. $NH_2—\overset{+}{N}H_3$
c. EDTA
d. $C_2O_4^{2-}$

Comprehension Type

Direction (Q. Nos. 21 and 22) This section contains a paragraph, describing theory, experiments, data, etc. Two questions have been related to the paragraph have been given. Each question has only one correct answer among the four given options (a), (b), (c) and (d).

Passage

In naming complexes, cation is named first followed by anion. Non-ionic or neutral complexes are given a one word name.

The cationic and neutral complex ions are named by writing 'the number and name of the ligands followed by the name of the central metal ion with its oxidation number in roman numerals within the parenthesis. If the complex is anionic, name of the metals ends with 'ate'. When the complex contain two or more metal atoms linked with the help of bridging ligand then each bridging ligand is started with μ

Working Space

21 Name of $[CoBr(NH_3)_4NO_2]$ is

a. tetramminebromonitrocobalt (II)
b. bromotetramminenitritocobalt(III)
c. tetramminebromonitrocobalt (III)
d. tetramminebromine nitritocobalt (II)

22 The IUPAC name of $[Ni(NH_3)_4]\,[NiCl_4]$ is

a. tetrachloronickel (II) tetramminenickel (II)
b. tetramminenickel (II) tetrachloronickel (II)
c. tetramminenickel (II) tetrachloronickelate (II)
d. tetrachloronickel (II) tetramminenickelate (0)

Matching List Type

Direction *(Q. Nos. 23 and 24) Choices for the correct combination of elements from Column I and Column II are given as options (a), (b), (c) and (d), out of which one is correct.*

23 Match the Column I with Column II and mark the correct option from the codes given below.

	Column I		Column II
i.	Sodium argentothiosulphate	p.	Neutral complex
ii.	$PtCl_4 \cdot 5NH_3$	q.	Cationic complex
iii.	Na_2ZnCl_4	r.	Anionic complex
iv.	$[Pt(NH_3)_2Br_2Cl_2]$	s.	Heteroleptic complex
		t.	Homoleptic complex

Codes

	i	ii	iii	iv
a.	p	p,s	q	r,s
b.	q	r,s	p,r,s	q
c.	q,t	q,s	r,t	p,s
d.	p	q,s	r,t	q

24 Match the Column I with Column II and mark the correct option from the codes given below.

	Column I		Column II
i.	$[CoCl(ONO)(en)_2]^+$	p.	Cobalt (III)
ii.	$[CoBr(H_2O)(NH_3)_4](NO_3)_2$	q.	Cobalt (III) ion
iii.	$[Ni(CO)_4]$	r.	Zero charge
iv.	$(C_6H_5)_3P$	s.	Homoleptic complex
		t.	Neutral complex

Codes

	i	ii	iii	iv
a.	p	p,s	q	r,s
b.	q	r,s	p,r,s	q
c.	q,t	q,s	r,t	p,s
d.	q	p	s,t	r

Working Space

One Integer Value Correct Type

Direction (*Q. Nos. 25-29*) *This section contains 5 questions. When worked out will result in an integer from 0 to 9 (both inclusive).*

25 Ethylenediamminetetraacetate ion is a polydentate ligand and negatively charged. The magnitude of negative charge is

26 The oxidation state of the central metal in the complex ion $[Fe(EDTA)]^-$.

27 The complex potassium dicyanodioxalatonickelate (II) in solution produce .... ions.

28 Number of EDTA molecules required to form an octahedral complex.

29 The number of complexes in which IUPAC names contain metal name ending with 'ate'
$K[Co(EDTA)]$, $[Pt(NH_3)_4]Br_4$, $[Co(NO_2)_6]^{3-}$, $[FeF_6]^{3-}$, $[Pt(en)Cl_2Br_2]$ $[Pt(en_2)]^{2+}$,
$K[Co(EDTA)$ $[Pt(py)_4]$ $[PtCl_4]$

Statement Type

Direction (*Q. No. 30*) *This section is based on Statement I and Statement II. Select the correct answer from the codes given below.*

a. Both Statement I and Statement II are correct and Statement II is the correct explanation of Statement I
b. Both Statement I and Statement II are correct but Statement II is not the correct explanation of Statement I
c. Statement I is correct but Statement II is incorrect
d. Statement II is correct but Statement I is incorrect

30 **Statement I** IUPAC name of the complex $[(NH_3)_5—Cr—OH— Cr(NH_3)_5]Cl_5$ is pentammine chromium-μ-hydroxopentammine chromium (III) chloride.

Statement II IUPAC name of the complex $[Ni(dmg)_2]$ is *bis* (dimethylglyoximato) nickel (III).

DPP-3 Isomerism in Coordination Compounds-I

Subjective Questions

Working Space

Direction *(Q. Nos. 1-5) These questions are subjective in nature, need to be solved completely on notebook.*

1 $[Pt(gly)_2]$ show geometrical but $[Pt(en)_2]^{2+}$ do not show. Why?

2 *Fac* and *mer* isomers of an $[Ma_3b_3]$ octahedral complex are optically inactive. Why?

3 How many isomers are possible for a complex having the formula $[Mabcdef]$ having octahedral geometry?

4 How many geometrical isomers are possible for the following?

i. $[Pt(NH_3)_3Cl]^+$

ii. $[Pt(NH_3)_3Cl_5]^-$

iii. $[Pt(NH_3)_2NO_2Cl]$

iv. $[Pt(NH_3)_4ClBr]^{2+}$

5 Draw the structure of EDTA as ligand.

Only One Option Correct Type

Direction *(Q. Nos. 6-15) This section contains 10 multiple choice questions. Each question has four choices (a), (b), (c) and (d), out of which ONLY ONE is correct.*

6 *Tris* (ethylenediamine) cobalt (III) cation, $[Co(en)_3]^{3+}$ can have

a. three stereoisomers, all chiral and optically active
b. two chiral stereoisomers (enantiomers) and one achiral
c. three stereoisomers, all achiral
e. two stereoisomers, both achiral

7 Which of the following types of octahedral complexes will exhibit geometrical isomerism (where, M = metal, a, b = achiral ligands)?

a. $[Ma_6]$
b. $[Ma_5b]$
c. $[Ma_4b_2]$
d. $[M(aa)_3]$

8 Which of the following can exhibit linkage isomerism?

a. NO_2^-
b. H_2O
c. NH_3
d. $H_2NCH_2CH_2NH_2$

Working Space

9 What type of isomerism is shown by $[Co(NH_3)_4Br_2]Cl$?

a. Geometrical only b. Ionisation only
c. Geometrical and ionisation d. Ionisation and optical

10 Which of the following does not show optical isomerism?

a. $[Co(en)_3]^{3+}$ b. $[Co(en)_2Cl_2]^+$
c. $[Co(NH_3)_3Cl_3]$ d. $[Co(en)(NH_3)_2Cl_2]^+$

11 The ionisation isomer of $[Cr(H_2O)_4Cl(NO_2)]Cl$ is

a. $[Cr(H_2O)_4(O_2N)]Cl_2$ b. $[Cr(H_2O)_4Cl_2](NO_2)$
c. $[Cr(H_2O)_4Cl(ONO)]Cl$ d. $[Cr(H_2O)_4Cl_2(NO_2)]\cdot H_2O$

12 Which of the following complex species is not expected to exhibit optical isomerism?

a. $[Co(en)_3]^{3+}$ b. $[Co(en)_2Cl_2]^+$
c. $[CoNH_3Cl_3]$ d. $[Co(en)(NH_3)_2Cl_2]^+$

13 The complex $[Co(NH_3)_6][Cr(C_2O_4)_3]$ and $[Cr(NH_3)_6][Co(C_2O_4)_3]$ exhibit

a. coordination isomerism b. geometrical isomerism
c. ionisation isomerism d. linkage isomerism

14 Which of the following exhibit *cis-trans* isomerism?

a. $[PtCl_2(NH_3)_2]$ b. $[PdCl_2BrI]$
c. $[Pt(NH_3)(py)(Cl)(Br)]$ d. All of these

15 Which of the following is considered to be an anticancer species?

a. 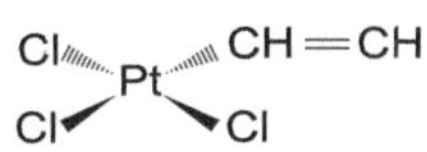

b. 

c. 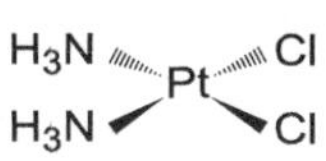

d. 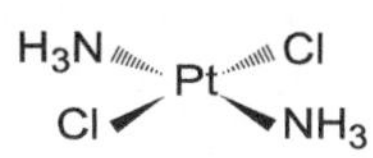

One or More than One Option Correct Type

Direction (Q. Nos. 16-20) *This section contains 5 multiple choice questions. Each question has four choices (a), (b), (c) and (d), out of which ONE or MORE THAN ONE are correct.*

16 Linkage isomerism is/are shown by

a. NO_2^- b. CN^- c. SCN^- d. H_2O

17 Which is/are correct pair?

a. $K[Ag(CN)_2]$ = linkage isomerism
b. $[Ag(NH_3)_2][CuCl_2]$ = ionisation isomerism
c. $[CrCl(NH_3)_5]I_2$ = ionisation
d. $[PtCl_4I_2]$ = geometrical

18 Coordination isomerism is/are shown by

a. $[Ag(NH_3)_2][CuCl_2]$ b. $[Al(H_2O)_6][Co(CN)_6]$
c. $[Pt(NH_3)_4(OH)_2]SO_4$ d. $[Co(NH_3)_5Cl]SO_4$

19 Which of the following show geometrical isomerism?

a. $[Pt(NH_2CH_2COO)_2]$ b. $[Pt(NH_3)(NH_2OH)(NO_2)(C_6H_6N)]$
c. $[Cu(NH_3)_4]Cl_2$ d. $[Pt(NH_3)Cl_2]$

20 Which of the following statements is/are correct?

a. Geometrical isomerism is not observed in complexes of CN 4 having tetrahedral geometry

b. Square planar complexes with CN 4 and 6 generally show geometrical isomerism

c. Square planar complexes with Ma_3b or Mab_3 exhibit *cis-trans* isomerism

d. The platinumglycinato complex $[Pt(gly)_2]$ show geometrical isomerism

Comprehension Type

***Direction** (Q. Nos. 21 and 22) This section contains a paragraph, describing theory, experiments, data etc. Two questions related to the paragraph have been given. Each has only one correct answer among the four given options (a) ,(b), (c) and (d).*

Passage

Consider the following isomers of $[Co(NH_3)_4Br_2]^+$. The black sphere represents Co, grey sphere represents NH_3 and unshaded sphere represents Br.

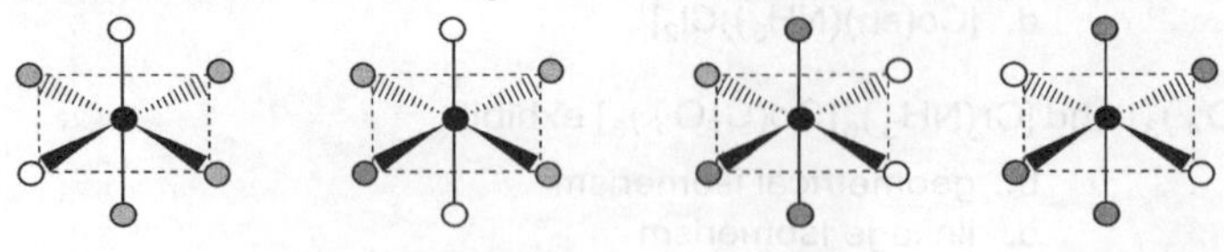

(a) (b) (c) (d)

21 The oxidation state and coordination number of cobalt in the complex $[Co(NH_3)_4Br_2]^+$ are

a. +2 and 6
b. +3 and 6
c. +3 and +6
d. +1and 6

22 Which of the structures is identical?

a. Structure (a) = structure (b) and structure (c) = structure (d)

b. Structure (a) = structure (c) and structure (b) = structure (d)

c. Structure (a) = structure (d) and structure (b) = structure (c)

d. None of the above

Matching List Type

***Direction** (Q. Nos. 23 and 24) Choices for the correct combination of elements from Column I and Column II are given as options (a), (b), (c) and (d) out of which one is correct.*

23 Match the Column I with Column II and mark the correct option from the codes given below.

	Column I (Complex species)		Column II (Isomerism)
i.	$[Co(NH_3)_4Cl_2]^+$	p.	Optical
ii.	*cis*- $[Co(en)_2Cl_2]^+$	q.	Ionisation
iii.	$[Co(NH_3)_5(NO_2)]Cl_2$	r.	Coordination
iv.	$[Co(NH_3)_6][Cr(CN)_6]$	s.	Geometrical
		t.	Linkage

Codes

	i	ii	iii	iv
a.	p	q	r	s
b.	t	r	q	p
c.	s	p	t	r
d.	q	p	t	s

Working Space

24 Match the Column I with Column II and mark the correct option from the codes given below.

	Column I		Column II
i.	$[CoCl_3(NH_3)_3]$	p.	*cis* form is optically acitve
ii.	$[Co(ox)_3]^{3-}$	q.	*trans* form is optically inactive
iii.	$[CrCl_2(ox)_2]^{3-}$	r.	Show meridian form
iv.	$[RhCl_3(py)_3]$	s.	Show facial form
			Show both geometrical and optical isomerism

Codes

	i	ii	iii	iv		i	ii	iii	iv
a.	p,q	q	p,t	r,s	b.	r	r,s	q,p	p,q
b.	p,t	s	p,t	p,r,t	d.	r,s	p,q,t	p,q,t	r,s

One Integer Value Correct Type

Direction (Q. Nos. 25-29) *This section contains 5 questions. When worked out will result in an integer from 0 to 9 (both inclusive).*

25 Total number of geometrical isomers for the complex $[RhCl(CO)(PPh_3)(NH_3)]$ is

26 The total number of isomers of $[Co(en)_2Cl_2]^+$ is

27 The total number of possible isomers for the complex compound $[Cu^{II}(NH_3)_4][Pt^{II}Cl_4]$

28 The number of geometrical isomers possible for $Cr(NH_3)_3Cl_3$ are

29 For square planar complex of platinum (II), $[Pt(NH_3)(Br)(Cl)py]^{2+}$, how many isomeric forms are possible?

Statement Type

Direction (Q. No. 30) *This section is based on Statement I and Statement II. Select the correct answer from the codes given below.*

a. Both Statement I and Statement II are correct and Statement II is the correct explanation of Statement I
b. Both Statement I and Statement II are correct but Statement II is not the correct explanation of Statement I
c. Statement I is correct but Statement II is incorrect
d. Statement II is correct but Statement I is incorrect

30 **Statement I** Complexes of MX_6 and MX_5L type do not show geometrical isomerism. Assume that X and L are unidentate.

Statement II Geometrical isomerism is not shown by complexes of coordination number 6.

DPP-4 Isomerism in Coordination Compounds-II

Subjective Questions

Working Space

Direction *(Q. Nos. 1-5) These questions are subjective in nature, need to be solved completely on notebook.*

1 How many geometrical isomers are possible for
i. $[Rh(en)_3]^{3+}$ and
ii. $[CoCl_2Br_2]^{2-}$?

2 $[CoCl_2Br_2]^{2-}$ do not show geometrical but $[PdBr_2Cl_2]^{2-}$ show geometrical isomerism. Why?

3 What type of isomerism is shown by the complex $K[Cr(H_2O)_2(C_2O_4)_2]$?

4 Which platinum complex is known as anticancer drug?

5 Give the size of ring structure formed by the following chelating ligands.
i. Propylenediamine (pn) ii. Trimethylenediamine (tn)
iii. Ethylenediamine (en) iv. Oxalate ion

Only One Option Correct Type

Direction *(Q. Nos. 6-15) This section contains 10 multiple choice questions. Each question has four choices (a), (b), (c) and (d), out of which ONLY ONE is correct.*

6 Which type of isomerism is shown by $[Co(NH_3)_4Br_2Cl]$?
a. Geometrical and ionisation
b. Optical and ionisation
c. Geometrical and optical
d. Geometrical only

7 The complex that exists as a pair of enantiomers is
a. *trans*-$[Co(H_2NCH_2CH_2NH_2)_2Cl_2]^+$
b. $[Co(H_2NCH_2CH_2NH_2)_3]^{3+}$
c. $Pt(PPh_3)(Cl)(Br)(CH_3)]^-$
d. *cis*-$[Co(NH_3)_4Cl_2]^+$

8 Which of the following will have three stereoisomeric forms?
i. $[Cr(NO_3)_3(NH_3)_3]$ ii. $K_3[Co(C_2O_4)_3]$
iii. $K_3[Co(C_2O_4)_2Cl_2]$ iv. $[Co(en)_2ClBr]$
a. iii and iv b. i and iv
c. ii and iii d. i and ii

Working Space

9 A coordination compound of cobalt has the molecular formula containing five ammonia molecules, one nitro group and two chlorine atoms for one cobalt atom. One mole of this compound produces three ions in an aqueous solution. The aqueous solution on treatment with an excess of $AgNO_3$ gives two moles of AgCl as a precipitate. The formula of the complex and the isomerism shown by this is

a. $[Co(NH_3)_5NO_2]Cl_2$, ionisation isomerism
b. $[Co(NH_3)_5NO_2]Cl_2$, linkage isomerism
c. $[Co(NH_3)_5NO_2]Cl_2$, both linkage and ionisation isomerism
d. $[Co(NH_3)_5Cl_2]NO_2$, ionisation isomerism

10 Which of the following will give maximum number of isomers?

a. $[Ni(C_2O_4)(en)_2]^{2-}$
b. $[Ni(en)(NH_3)_4]^{2+}$
c. $[Cr(SCN)_2(NH_3)_4]^{+}$
d. $[Co(NH_3)_4Cl_2]$

11 Which of the following coordination compounds would exhibit optical isomerism?

a. Pentamminenitrocobalt (III) iodide
b. Diamminedichloroplatinum (II)
c. *trans*-dicyano*bis*(ethylenediamine) chromium (III) chloride
d. *trans*-(ethylenediamine) cobalt (III) bromide

12 Which one of the following is expected to exhibit optical isomerism? (en = ethylenediamine)

a. *cis*-$[Pt(NH_3)_2Cl_2$
b. *trans*-$[Pt(NH_3)_2Cl_2]$
c. *cis*-$[Co(en)_2Cl_2]$
d. *trans*-$[Co(en)_2Cl_2]$

13 How the isomeric complexes $[Co(NH_3)_6][Cr(NO_2)_6]$ and $[Cr(NH_3)_6][Co(NO_2)_6]$ can be distinguished from one another by

a. conductivity measurement
b. measuring magnetic moment
c. electrolysis of their aqueous solution
d. optical measurement

14 The complexes $[Co(NH_3)_6][Cr(CN)_6]$ and $[Cr(NH_3)_6][Co(CN)_6]$ are the examples of which type of isomerism? **[AIEEE 2011]**

a. Linkage isomerism
b. Ionisation isomerism
c. Coordination isomerism
d. Geometrical isomerism

15 The isomerism shown by $[Co(EDTA)]^-$ ion is

a. geometrical
b. optical
c. Both (a) and (b)
d. None of these

One or More than One Options Correct Type

Direction *(Q. Nos. 16-20) This section contains 5 multiple choice questions. Each question has four choices (a), (b), (c) and (d), out of which ONE or MORE THAN ONE are correct.*

16 $[Co(NH_3)_4(NO_2)_2]Cl$ exhibits

a. linkage isomerism
b. ionisation isomerism
c. geometrical isomerism
d. optical isomerism

17 The complex $[Co(NH_3)_5Br]SO_4$ and $[Co(NH_3)_5SO_4]Br$ can be identified by

a. $BaCl_2$
b. $AgNO_3$
c. $PbCl_2$
d. H_2SO_4

Working Space

18 The pair(s) of coordination complexes/ions exhibiting the same kind of isomerism is/are

a. $[Cr(NH_3)_5Cl]^{2+}$ and $[Cr(NH_3)_4Cl_2]^+$
b. $[Cr(NH_3)_4Cl_2]^+$ and $[Pt(NH_3)_2(H_2O)Cl]^+$
c. $[CoBr_2Cl_2]^{2-}$ and $[PtBr_2Cl_2]^{2-}$
d. $[Pt(NH_3)_3NO_3]Cl$ and $[Pt(NH_3)_3Cl]Br$

19 Which of the following complexes show geometrical isomerism?

a. Tetrahedral
b. Octahedral
c. Square planar
d. Linear

20 The compound(s) that exhibit(s) geometrical isomerism is/are

a. $[Pt(en)Cl_2]$
b. $[Pt(en)_2Cl_2]Cl_2$
c. $[Pt(NH_3)_2Cl_2]$
d. $[Pt(en)_2]Cl_2$

Comprehension Type

***Direction** (Q. Nos. 21 and 22) This section contains a paragraph, each describing theory, experiments, data, etc. Two questions related to the paragraph have been given. Each question has only one correct answer among the four given options (a), (b), (c) and (d).*

Passage

A metal complex having the composition $Cr(NH_3)_4Cl_2Br$ has been isolated in two forms *A* and *B*. The form *A* reacts with $AgNO_3$ to give a white precipitate readily soluble in dilute aqueous ammonia whereas *B* gives a pale yellow precipitate soluble in conc. ammonia.

21 The formula of complex *A*

a. $[Cr(NH_3)_4Br]Cl_2$
b. $[Cr(NH_3)_4ClBr]Cl$
c. $[Cr(NH_3)_4Cl_2]Br$
d. Both (a) and (b)

22 The isomerism shown by complex *A* is

a. ionisation
b. geometrical
c. linkage
d. Both (a) and (b)

Matching List Type

***Direction** (Q. Nos. 23 and 24) Choices for the correct combination of elements from Column I and Column II are given as options (a), (b), (c) and (d) out of which one is correct.*

23 Match the Column I with Column II and mark the correct option from the codes given below.

	Column I		Column II
i.	$[Pd(PPh_3)_2(SCN)_2]$	p.	Ionisation isomerism
ii.	$[Co(NH_3)_5(NO_3)]SO_4$	q.	Linkage isomerism
iii.	$[Cr(H_2O)_5Cl]Cl_2 \cdot H_2O$	r.	Coordination isomerism
iv.	$[Cu(NH_3)_4][Pt(CN)_4]$	s.	Hydrate isomerism

Codes

	i	ii	iii	iv		i	ii	iii	iv
a.	p	s	q	r	b.	q	p	s	r
c.	s	q	p	r	d.	q	p	r	s

Working Space

24 Match the Column I with Column II and mark the correct option from the codes given below.

	Column I		Column II
i.	$[CoCl_2(en)_2]^+$	p.	Shows geometrical isomerism only
ii.	$[Cr(ox)_3]^{3-}$	q.	Shows optical isomerism only
iii.	$[RhCl_3(py)_3]$	r.	Shows optical as well as geometrical
iv.	$[ZnCl_2(NH_3)_2]$	s.	Shows no isomerism

Codes

	i	ii	iii	iv		i	ii	iii	iv
a.	r	q,s	p	s	b.	p	s	r	q
c.	s	r	p	q,s	d.	r	p	q	s

One Integer Value Correct Type

Direction (*Q. Nos.* 25 -29) *This section contains* 5 *questions. When worked out will result in an integer from* 0 *to* 9 (*both inclusive*).

25 The total number of isomers expected for $[Pt(NCS)(en)_2]^{2+}$ are

26 Number of isomers possible for $[Co(NH_3)_5SCN]Cl$ are

27 $[Fe(EDTA)]^-$ exhibits optical isomerism. In this complex, the number of rings formed are

28 The number of geometrical isomers possible for the complex $[CoCl_2Br_2]^-$ are

29 The total number of possible isomers for the complex compound $[Cu^{II}(NH_3)_4][Pt^{II}Cl_4]$ are

Statement Type

Direction (*Q. No.* 30) *This section is based on Statement I and Statement II. Select the correct answer from the codes given below.*

a. Both Statement I and Statement II are correct and Statement II is the correct explanation of Statement I
b. Both Statement I and Statement II are correct but Statement II is not the correct explanation of Statement I
c. Statement I is correct but Statement II is incorrect
d. Statement II is correct but Statement I is incorrect

30 **Statement I** Ambidentate ligands lead to linkage isomerism.

Statement II The ionisation sphere is different in different linkage isomers.

DPP-5 Valence Bond Theory, Effective Atomic Number and Magnetic Properties

Subjective Questions

Working Space

Direction *(Q. Nos. 1-5) These questions are subjective in nature, need to be solved completely on notebook.*

1 i. What is effective atomic number (EAN)?
ii. Calculate the EAN value of Mn in $Mn_2(CO)_{10}$.

2 Using VBT explain the geometry and magnetic behaviour of $[Fe(CO)_5]$.

3 The EAN of a metal carbonyl $[M(CO)_x] = 36$. Atomic number of metal is 24. Find the value of x.

4 The species $[CuCl_4]^{2-}$ exists but $[CuI_4]^{2-}$ does not exist. Why?

5 $Fe^{2+} \xrightarrow[\text{Excess}]{\text{SCN}} A \xrightarrow[\text{Excess}]{F^-} B$. In this what are A and B? Give the IUPAC name of A. Find the spin only magnetic moment of B.

Only One Option Correct Type

Direction *(Q. Nos. 6-15) This section contains 10 multiple choice questions. Each question has four choices (a), (b), (c) and (d), out of which ONLY ONE is correct.*

6 Which one of the following is wrongly matched?
a. $[Ni(CO)_4]$ = neutral complex
b. $[Cu(NH_3)_4]^{2+}$ = square planar
c. $[Co(en)_3]^{3+}$ = follow EAN rule
d. $[Fe(CN)_6]^{3+} = sp^3d^2$

7 In which of the following the central atom has sp^3d^2-hybridisation?
a. $[CoF_6]^{3-}$
b. $[Co(NH_3)_6]^{3+}$
c. $[Fe(CN)_6]^{3-}$
d. $[Cr(NH_3)_6]^{3+}$

8 Which is the diamagnetic?
a. $[CoF_6]^{3-}$
b. $[Ni(CN)_4]^{2-}$
c. $[NiCl_4]^{2-}$
d. $[Fe(CN)_6]^{3-}$

9 A magnetic moment of 1.73 BM will be shown by one among following.
a. $[Ni(CN)_4]^{2-}$
b. $TiCl_4$
c. $[CoCl_6]^{4-}$
d. $[Cu(NH_3)_4]^{2+}$

10 Which one of the following complex species does not involve inner orbital hybridisation?
a. $[CoF_6]^{3-}$
b. $[Co(NH_3)_6]^{3+}$
c. $[Fe(CN)_6]^{3-}$
d. $[Cr(NH_3)_6]^{3+}$

Working Space

11 Which of the following complex has zero magnetic moment (spin only)?

a. $[Ni(NH_3)_6]Cl_2$
b. $Na_3[FeF_6]$
c. $[Cr(H_2O)_6]SO_4$
d. $K_4[Fe(CN)_6]$

12 The number of unpaired electrons calculated in $[Co(NH_3)_6]^{3+}$ and $[CoF_6]^{3-}$ are

a. 4 and 4
b. 0 and 2
c. 2 and 4
d. 0 and 4

13 The tetrachloro complexes of Ni(II) and Pd(II) respectively are (atomic number of Ni and Pd are 28 and 46 respectively).

a. diamagnetic and diamagnetic
b. diamagnetic and paramagnetic
c. paramagnetic and diamagnetic
d. paramagnetic and paramagnetic

14 Which one of the following is an inner orbital complex as well as diamagnetic in behaviour? (Atomic number, Zn = 30, Cr = 24, Co = 27, Ni = 28)

a. $[Zn(NH_3)_6]^{2+}$
b. $[Cr(NH_3)_6]^{3+}$
c. $[Co(NH_3)_6]^{3+}$
d. $[Ni(NH_3)_6]^{2+}$

15 The pair of compounds having the same hybridisation for the central atom is

a. $[Cu(NH_3)_4]^{2+}$ and $[Ni(NH_3)_4]^{2+}$
b. $[NiCl_4]^{2-}$ and $[PtCl_4]^{2-}$
c. $[Cu(NH_3)_4]^{2+}$ and $[Zn(NH_3)_4]^{2+}$
d. $[Co(NH_3)_6]^{3+}$ and $[Co(H_2O)_6]^{3+}$

One or More than One Options Correct Type

Direction (Q. Nos. 16-20) *This section contains 5 multiple choice questions. Each question has four choices (a), (b), (c) and (d), out of which ONE or MORE THAN ONE are correct.*

16 The formation of complex which involves *d*-orbitals of outer shell are called

a. low spin complex
b. high spin complex
c. inner orbital complex
d. outer orbital complex

17 sp^3-hybridisation is found in

a. $[ZnCl_4]^{2-}$
b. $[Cu(NH_3)_4]^{2+}$
c. $[CuCl_4]^{2-}$
d. $[Ni(CO)_4]$

18 Regarding the complex $[Co(H_2O)_6]^{2-}$, correct statement(s) is/are

a. Paramagnetic complex
b. Inner orbital complex
c. Outer orbital complex
d. Magnetic moment is 3.873 BM

19 For which the EAN value is equal to the atomic number of a noble gas?

a. $K_2[HgI_4]$
b. $[Pd(NH_3)Cl_2]$
c. $[CdI_4]^{2-}$
d. $Co_2(CO)_8$

20 $[Fe(H_2O)_6]^{2+}$ and $[Fe(CN)_6]^{4-}$ differ in

a. magnetic moment
b. colour
c. hybridisation
d. geometry

Comprehension Type

Working Space

Direction (Q. Nos. 21 and 22) This section contains a paragraph, describing theory, experiments, data, etc. Two questions related to the paragraph have been given. Each question has only one correct answer among the four given options (a), (b), (c) and (d).

Passage

When crystals of $CuSO_4 \cdot 4NH_3$ are dissolved in water, there is hardly any evidence for the presence of Cu^{2+} ions or ammonia molecules. A new ion $[Cu(NH_3)_4]^{2+}$ is furnished in which ammonia molecules are directly linked with the metal ion. Similarly, aqueous solution of $Fe(CN)_2 \cdot 4KCN$ does not give tests of Fe^{2+} and CN^- ions but give test for new ion $[Fe(CN)_6]^{4-}$ called ferrocyanide ion.

21 The hybridisation and geometry and magnetic property of $[Cu(NH_3)_4]^{2+}$ ion are

a. sp^3, tetrahedral, diamagnetic   b. sp^3, tetrahedral, paramagnetic
c. dsp^2, square planar, diamagnetic   d. dsp^2, square planar, paramagnetic

22 The hybridisation, geometry and magnetic property of $[Fe(CN)_6]^{4-}$ are

a. d^2sp^3, octahedral, diamagnetic   b. d^2sp^3, octahedral, paramagnetic
c. sp^3d^2, octahedral, diamagnetic   d. sp^3d^2, octahedral, paramagnetic

Matching List Type

Direction (Q. Nos. 23 and 24) Choices for the correct combination of elements from Column I and Column II are given as options (a), (b), (c) and (d) out of which one is correct.

23 Match the Column I with Column II and mark the correct option from the codes given bleow.

	Column I		Column II
i.	$[Zn(NH_3)_4]^{2+}$	p.	sp^3
ii.	$[MnBr_4]^{2-}$	q.	Octahedral
iii.	$K_4[Fe(CN)_6]$	r.	sp^3d^2
iv.	$[Fe(H_2O)_6]Cl_3$	s.	Tetrahedral
		t.	d^2sp^3

Codes

	i	ii	iii	iv		i	ii	iii	iv
a.	p,s	p,s	q,t	q,r	b.	r,s	q	r	p
c.	q	p,t	p,q	q,r	d.	s	s,t	p,r,s	r,s,t

24 Match the Column I with Column II and mark the correct option from the codes given bleow.

	Column I (Complex ion)		Column II (Hybridisation and number of unpaired electrons)
i.	$[Cr(H_2O)_6]^{3+}$	p.	dsp^2,1
ii.	$[Co(CN)_4]^{2-}$	q.	sp^3d^2, 5
iii.	$[Ni(NH_3)_6]^{2+}$	r.	d^2sp^3, 3
iv.	$[MnFe]^{4-}$	s.	sp^3d^2, 2

Codes

	i	ii	iii	iv		i	ii	iii	iv
a.	r	p	q	s	b.	r	q	s	p
c.	s	r	q	p	d.	s	p	q	r

Working Space

One Integer Value Correct Type

Direction (Q. Nos. 25-29) *This section contains 5 questions. When worked out will result in an integer from 0 to 9 (both inclusive).*

25 How many of the following species obey Sidgwick EAN rule?

$K_3[Fe(CN)_6]$, $[Ru(CO)_5]$, $[Cr(NH_3)_6]^{3+}$, $[Co(NH_3)_6]^{3+}$, $[Ni(NH_3)_6]^{2+}$, $[Fe(CO)_5]$, $[W(CO)_6]$

26 How many of the following are diamagnetic?

$[Zn(OH)_4]^{2-}$, $[Ni(NH_3)_6]^{2+}$, $K_4[Fe(CN)_6]$, $K_3[Fe(CN)_6]$, $[Cu(NH_3)_4]^{2+}$, $[PdBr_4]^{2-}$,

$[Ni(CO)_4]$, $[CoF_6]^{3-}$

27 The number of unpaired electrons in the complex $[Mn(acac)_3]$ is (Atomic number of Mn = 25)

28 The magnetic moment of $[Ru(H_2O)_6]^{2+}$ corresponds to the presence of ...... unpaired electrons.

29 The number of unpaired electrons in $[Fe(dipy)_3]^{2+}$ are

Statement Type

Direction (Q. No. 30) *This section is based on Statement I and Statement II. Select the correct answer from the codes given below.*

a. Both Statement I and Statement II are correct and Statement II is the correct explanation of Statement I
b. Both Statement I and Statement II are correct but Statement II is not the correct explanation of Statement I
c. Statement I is correct but Statement II is incorrect
d. Statement II is correct but Statement I is incorrect

30 **Statement I** Both $[Ni(CN)_4]^{2-}$ and $[NiCl_4]^{2-}$ have same shape and same magnetic behaviour.

Statement II Both are square planar and diamagnetic.

DPP-6 Crystal Field Theory (CFT) and Colour of Complexes

Subjective Questions

Direction (Q. Nos. 1-5) *These questions are subjective in nature, need to be solved completely on notebook.*

Working Space

1 $[Ni(H_2O)_6]^{2+}$ is green while $[Ni(NH_3)_6]^{2+}$ complex is violet. Explain.

2 i. The value of Δ for $[RhCl_6]^{3-}$ is 243 kJ/mol. What wavelength of light will promote an electron from the t_{2g} to the e_g set?

ii. The crystal field stabilisation energy (CFSE) value in terms of wave number for $[Ti(H_2O)_6]^{3+}$ that has absorption maximum at 492 nm.

3 What will be the correct order for the wavelengths of absorption in the visible region of the following $[Ni(H_2O)_6]^{2+}$, $[Ni(NH_3)_6]^{2+}$, $[Ni(NO_2)_6]^{4-}$?

4 Explain, why nearly all tetrahedral complexes are high spin?

5 What are weak field ligands and strong field ligands? Give examples.

Only One Option Correct Type

Direction (Q. Nos. 6-15) *This section contains 10 multiple choice questions. Each question has four choices (a), (b), (c) and (d), out of which ONLY ONE is correct.*

6 Among the following complexes the one which shows zero crystal field stabilisation energy is

a. $[Co(H_2O)_6]^{2+}$ b. $[Co(H_2O)_6]^{3+}$ c. $[Mn(H_2O)_6]^{3+}$ d. $[Fe(H_2O)_6]^{3+}$

7 The wavelength of light absorbed is highest in

a. $[Co(NH_3)_5Cl]^{2+}$ b. $[Co(NH_3)_5H_2O]^{3+}$

c. $[Co(NH_3)_6]^{3+}$ d. $[Co(en)_3]^{3+}$

8 The CFSE for octahedral $[CoCl_6]^{4-}$ is 18000 cm^{-1}. Then the CFSE for tetrahedral $[CoCl_4]^{2-}$ will be

a. 18000 cm^{-1} b. 16000 cm^{-1}

c. 8000 cm^{-1} d. 20000 cm^{-1}

9 The crystal field splitting energy (Δ_o) of

I. $[CoBr_6]^{3-}$ II. $[CoF_6]^{3-}$ III. $[Co(NCS)_6]^{3-}$ IV. $[Co(CN)_6]^{3-}$ is in the order of

a. I < II < III < IV b. II < IV < I < III

c. III < I < IV < II d. IV < III < II < I

Working Space

10 Crystal field splitting energy for high spin d^4 octahedral complex is

a. $-1.2\Delta_o$ b. $-0.6\Delta_o$ c. $-0.8\Delta_o$ d. $-1.6\Delta_o$

11 The correct order of ligand field strength is **[2010 ISAT]**

a. $H_2O < Cl^- < CO < NH_3$ b. $CO < NH_3 < Cl^- < H_2O$

c. $H_2O < CO < NH_3 < Cl^-$ d. $Cl^- < H_2O < NH_3 < CO$

12 The crystal field splitting energies (CFSE) of high spin and low spin d^6 metal complexes in octahedral complex in terms of Δ_o respectively are

a. -0.4 and -2.4 b. -2.4 and -0.4 c. -0.4 and 0.0 d. -2.4 and 0.0

13 The Δ_t of the following complexes follows the order

I. $[CoCl_4]^{2-}$ II. $[CoBr_4]^{2-}$ III. $[Co(NCS)_4]^{2-}$

a. III > I > II b. I > II > III c. II > I > III d. III > II > I

14 In which of the following coordination entities, the magnitude Δ_o (CFSE in octahedral field) will be maximum?

a. $[Co(H_2O)_6]^{3+}$ b. $[Co(NH_3)_6]^{3+}$ c. $[Co(CN)_6]^{3-}$ d. $[Co(C_2O_4)_3]^{3-}$

15 The increasing order of wavelength of absorption for the complex ions

I. $[Cr(NH_3)_6]^{3+}$ II. $[CrCl_6]^{3-}$ III. $[Cr(H_2O)_6]^{3+}$ IV. $[Cr(CN)_6]^{3-}$ is

a. IV < II < III < I b. IV < III < II < I

c. IV < I < III < II d. II < III < I < IV

One or More than One Options Correct Type

Direction (Q. Nos. 16-20) *This section contains 5 multiple choice questions. Each question has four choices (a), (b), (c) and (d), out of which ONE or MORE THAN ONE are correct.*

16 Which of the following metal ions cannot form both high spin and low spin octahedral complexes?

a. Ti^{3+} b. Co^{2+} c. Fe^{2+} d. Cu^{2+}

17 Which is/are the correct statements?

a. CFSE of $[Co(NH_3)_6]^{3+} < [Rh(NH_3)_6]^{3+} < [Ir(NH_3)_6]^{3+}$

b. CFSE of $[Co(NH_3)_6]^{3+} < [Co(en)_3]^{3+}$

c. $\Delta_o < P$ low spin state is more stable

d. $[Ni(H_2O)_6]^{2+}$ complex is yellow while $[Ni(en)_3]^{2+}$ is violet

18 CFSE is zero for the complexes with

a. d^5 weak field b. d^5 strong field

c. d^{10} weak field d. d^{10} strong field

19 CFSE value depends on

a. nature of ligands

b. charge on the metal ion

c. whether the metal is $3d$, $4d$ or $5d$ series

d. geometry of complex

20 Octahedral complex of Cr (III) will be

a. sp^3d^2 in case of weak field ligand b. d^2sp^3 in case of strong ligand

c. d^2sp^3 always d. always coloured complexes

Working Space

Comprehension Type

Direction *(Q. Nos. 21 and 22) This section contains a paragraph, describing theory, experiments, data, etc. Two questions related to the paragraph have been given. Each question has only one correct answer among the four given options (a), (b), (c) and (d).*

Passage

In octahedral complexes due to repulsion between the ligands and *d*-orbitals , there is splitting of *d*-orbitals into two sets, i.e. two orbitals of higher energy called e_g and three orbitals of lower energy called t_{2g}. The difference of energy between the two sets of *d*-orbitals is called crystal field stabilisation energy denoted by Δ_o. For any given metal cation, the magnitude of Δ_o depends on the nature of ligands.

21 The CFSE for d^7 configuration for strong ligand field is

a. $-1.6\,\Delta_o$ b. $-2.0\,\Delta_o$ c. $-2.4\,\Delta_o$ d. $-1.8\,\Delta_o$

22 In the following complexes of manganese, the distribution of electrons in *d*-orbitals of manganese

i. $[Mn(H_2O)_6]^{2+}$ ii. $[Mn(CN)_6]^{4-}$

a. $t_{2g}^3 e_g^2$ in both b. $t_{2g}^5 e_g^0$ in both

c. $t_{2g}^3 e_g^2$ in (i) and $t_{2g}^5 e_g^0$ in (ii) d. $t_{2g}^5 e_g^0$ in (i) and $t_{2g}^3 e_g^2$ in (ii)

Matching List Type

Direction *(Q. No. 25) Choices for the correct combination of elements from Column I and Column II are given as options (a), (b), (c) and (d) out of which one is correct.*

23 Match the Column I with Column II and mark the correct option from the codes given below.

	Column I (Complex)		Column II (Δ / cm^{-1})
i.	$[Co(NH_3)_6]^{3+}$	p.	$-0.8\Delta_o$
ii.	$[Co(NH_3)_6]^{2+}$	q.	$-0.4\Delta_o$
iii.	$[Co(H_2O)_6]^{3+}$	r.	$-2.4\Delta_o$
iv.	$[Co(H_2O)_6]^{2+}$	s.	$-1.8\Delta_o$

Codes

	i	ii	iii	iv		i	ii	iii	iv
a.	r,s	p	q	s	b.	p	q	p	s
c.	r	s	q	p	d.	q	p	q	q

24 Match the Column I with Column II and mark the correct option from the codes given below.

	Column I		Column II
i.	$[Ti(H_2O)_6]^{3+}$	p.	$t_{2g}^3 e_g^2$
ii.	$[Fe(F)_6]^{3-}$	q.	$-0.4\Delta_o$
iii.	$[Fe(CN)_6]^{4-}$	r.	$t_{2g}^6 e_g^0$
iv.	$[Zn(H_2O)_6]^{2+}$	s.	$-0.0\Delta_o$
		t.	$\Delta_o < P$

Codes

	i	ii	iii	iv		i	ii	iii	iv
a.	q	p,s	r,t	s	b.	p	q	p	s
c.	r	s	q	p	d.	q	p	q	q

One Integer Value Correct Type

Direction (Q. Nos. 25-29) *This section contains 5 questions. When worked out will result in an integer from 0 to 9 (both inclusive).*

25 Number of unpaired electrons in d^6, low spin octahedral complex.

26 The number of ligands which have strong crystal field splitting than H_2O among SCN^-, NCS^-, $EDTA^{4-}$, $C_2O_4^{2-}$, NO_2^-, Br^-, PPh_3, F^-

27 The total number of unpaired electrons in the two complexes $[Cr(H_2O)_6]^{2+}$ and $[Cr(CN)_6]^{4-}$ having octahedral geometry are

28 Number of unpaired electrons in t_{2g} and e_g orbitals in weak octahedral ligand fields with d^7 configuration.

29 In tetrahedral complexes having weak field ligand, pairing starts with ...... electron.

Statement Type

Direction (Q. No. 30) *This section is based on Statement I and Statement II. Select the correct answer from the codes given below.*

a. Both Statement I and Statement II are correct and Statement II is the correct explanation of Statement I
b. Both Statement I and Statement II are correct but Statement II is not the correct explanation of Statement I
c. Statement I is correct but Statement II is incorrect
d. Statement II is correct but Statement I is incorrect

30 **Statement I** The value of CFSE for the octahedral complex $[Co(NH_3)_6]^{3+}$ is $-2.4\Delta_0$.

Statement II The ammonia molecule is strong field ligand.

Working Space

DPP-7 Bonding in Organometallic Compounds, Stability and Applications of Coordination Compounds

Subjective Questions

Direction (*Q. Nos.* 1-5) *These questions are subjective in nature, need to be solved completely on notebook.*

Working Space

1 Give one example of organometallic compounds having

i. σ-bonded complex

ii. π-bonded complex

iii. σ and π-bonded complex.

2 Gold and platinum dissolve in aqua-regia forming soluble complexes. Give the formula and write the IUPAC names.

3 What is synergic effect?

4 What is meant by chelate effect? Give an example.

5 i. Which is more stable $K_4[Fe(CN)_6]$ or $K_3[Fe(CN)_6]$ and why?

ii. Explain why $[Cu(en)_2]^{2+}$ is less table than $[Fe(EDTA)]^-$?

Only One Option Correct Type

Direction (*Q. Nos.* 6-15) *This section contains* 10 *multiple choice questions. Each question has four choices (a), (b), (c) and (d), out of which ONLY ONE is correct.*

6 A filter paper moistened with ammoniacal sodium nitroprusside solution turns violet on contact with a drop of alkaline Na_2S solution .The violet colour is due to the formation of

a. $[Fe(SCN)_5NO]^-$

b. $[Fe(SCN)_5NO]^{2-}$

c. $[Fe(CN)_5NOS]^{4-}$

d. $[Fe(CN)_5NOS]^{3-}$

7 Wilkinson's catalyst is

a. $[Rh(CO)_2 I_2]^-$

b. $[(Ph_3P)_3RhCl]$

c. $[Co_2(CO)_8]$

d. $[(Ph_3P)_2Rh(CO)Cl]$

Working Space

8 Tollen's reagent contains

a. AgOH
b. $AgNO_3$
c. $[Ag(NO_3)_2]^+$
d. $[Ag(NH_3)_2]^+$

9 The compound formed by dissolving gold and platinum in aqua-regia is

a. $[AuCl_4]^{3-}$ and $[PtCl_6]^{2-}$
b. $[AuCl_4]^-$ and $[PtCl_6]^{4-}$
c. $[AuCl_4]^-$ and $[PtCl_6]^{2-}$
d. $[AuCl_4]^-$ and $[PtCl_4]^{2-}$

10 Among the following, which is not the π-bonded organometallic compound?

a. $(CH_3)_4Sn$
b. $K[PtCl_3(\eta^2 - C_2H_4)]$
c. $Fe(\eta^5 - C_5H_5)_2$
d. $Cr(\eta^6 - C_6H_6)_2$

11 Which of the following biomolecules contain a non-transition metal ion?

a. Vitamin B-12
b. Chlorophyll
c. Haemoglobin
d. Insulin

12 In black and white photography, the developed film is fixed by washing with hypo solution which dissolves the undecomposed AgBr to form a complex ion formed is

a. $[Ag(S_2O_3)_2]^{3-}$
b. $[Ag(S_2O_3)_2]^{2-}$
c. $[Ag(S_2O_3)_3]^{2-}$
d. $[Ag(S_2O_3)_2]^-$

13 In the estimation of hardness of water, the reagent used is

a. hypo solution
b. $KMnO_4$
c. EDTA solution
d. $K_2Cr_2O_7$

14 Which of the following is powerful σ-donor and π-acceptor ligand?

a. Cl^-
b. CO
c. NH_3
d. OH^-

15 Fe^{2+} and Fe^{3+} can be distinguished by

a. $K_3[Fe(CN)_6]$
b. $K_4[Fe(CN)_6]$
c. [(KSCN)]
d. All of these

One or More than One Options Correct Type

Direction (Q. Nos. 16-20) *This section contains 5 multiple choice questions. Each question has four choices (a), (b), (c) and (d), out of which ONE or MORE THAN ONE are correct.*

16 Which of the following forms a metal carbonyl corresponding $[M(CO)_6]$?

a. Cr
b. W
c. Fe
d. Ni

17 In which of the following porphyrin acts as ligand?

a. Haemoglobin
b. Chlorophyll
c. Vitamin B-12
d. Insulin

18 Which are π-bonded organometallic compounds?

a. $[Cr(\eta^6 - (C_6H_6)]$
b. $[Fe(CO)_5]$
c. $K[PtCl_3\eta^2 - C_2H_4]$
d. $[Fe(\eta^5 - C_5H_5)_2]$

Working Space

19 Which is the correct?

a. $[Fe(NCS)_6]^{3-}$ = blood red

b. $[Fe(H_2O)_6]^{2+}$ = pale green

c. $[Cu(H_2O)_4]^{2+}$ = blue

d. $K_2[HgI_4]$ = scarlet red

20 How many of the following combinations in an aqueous medium will give a blue colour or precipitate?

a. $Fe^{2+} + [Fe(CN)_6]^{3-}$

b. $Fe^{3+} + [Fe(CN)_6]^{4-}$

c. $Co^{2+} + SCN^-$

d. $Fe^{3+} + SCN^-$

Comprehension Type

***Direction** (Q. Nos. 21 and 22) This section contains a paragraph, describing theory, experiments, data, etc. Two questions related to the paragraph have been given. Each question has only one correct answer among the four given options (a), (b), (c) and (d).*

Passage

Coordination compounds play a vital role in our lives. The importance of these compounds can be realised from the fact that life would not have been possible without the existence of chlorophyll in plants and haemoglobin in the blood of the animals. The field of such compounds has expanded very fast in recent years and coordination compounds are playing important roles in analytical chemistry, polymerisation reactions, metallurgy and refining of metals, etc.

21 The reagent used for identifying nickel ion is

a. EDTA

b. dimethylglyoxime

c. potassium ferrocyanide

d. sodium nitroprusside

22 Which is used in cancer chemotherapy?

a. *cis*-platin

b. Zeise's salt

c. Ferrocene

d. Fischer salt

Matching List Type

***Direction** (Q. Nos. 23 and 24) Choices for the correct combination of elements from Column I and Column II are given as options (a), (b), (c) and (d) out of which one is correct.*

23 Match the Column I with Column II and mark the correct option from the codes given below.

	Column I		Column II
i.	$[Fe(\eta^5 - C_5H_5)_2]$	p.	Nessler's reagent
ii.	$K_2[Co(NO_2)_6]$	q.	Fischer salt
iii.	$[Pt(NH_3)_4]\ [PtCl_4]$	r.	Ferrocene
iv.	$K_2[HgI_4]$	s.	Magnus's green

Codes

	i	ii	iii	iv
a.	p,q	p	q,r	s
b.	r	q	s	p
c.	p	q	r	r,s
d.	s	p	q	r

Working Space

24 Match the Column I with Column II and mark the correct option from the codes given below.

	Column I		Column II
i.	$[Cr(CO)_6]$	p.	EAN = 36
ii.	$Fe_2(CO)_9$	q.	Three CO—CO bond
iii.	$Co_4(CO)_{12}$	r.	Zero bridged CO group
iv.	$Fe_3(CO)_{12}$	s.	One Fe—Fe bond
		t.	Two Fe—Fe bond

Codes

	i	ii	iii	iv
a.	p,r	s	q	t
b.	r	q	s	p
c.	p	q	r	r,s
d.	s	p	q	r

One Integer Value Correct Type

Direction (*Q. Nos.* 25-29) *This section contains* 5 *questions. When worked out will result in an integer from* 0 *to* 9 *(both inclusive).*

25 The number of terminal carbonyl groups present in $Fe_2(CO)_9$ are

26 In $Fe_2(CO)_9$, the number of CO present between iron atoms ......

27 In $[Co_2(CO)_8]$, the number of CO molecules lying between the metal atoms are

28 The number of $M—M$ bonds in $[Ir_4(CO)_{12}]$ are

29 In $Mn_2(CO)_{10}$, the number of CO molecules in between the metal atoms are

Statement Type

Direction (*Q. No.* 30) *This section is based on Statement I and Statement II. Select the correct answer from the codes given below.*

a. Both Statement I and Statement II are correct and Statement II is the correct explanation of Statement I
b. Both Statement I and Statement II are correct but Statement II is not the correct explanation of Statement I
c. Statement I is correct but Statement II is incorrect
d. Statement II is correct but Statement I is incorrect

30 **Statement I** $Mn(\pi— C_3H_5)(CO)_4$, obey effective atomic number rule.

Statement II π-allyl ligand act as a $3e^-$ donor.

SET - 1

Revisal Problems for JEE Main

Only One Option Correct Type

Working Space

***Direction** (Q. Nos. 1-28) This section contains 28 multiple choice questions. Each question has four choices (a), (b), (c) and (d), out of which ONLY ONE option is correct.*

1 Which has higher van't Hoff factor value in aqueous solution?

a. Potash alum b. Mohr's salt
c. Potassium ferrocyanide d. Carnalite

2 A chelating ligand has two or more than two donor atoms to bind a single metal ion. Which of the following is not a chelating agent?

a. Thiosulphato b. Glycinato
c. Oxalate d. Ethane-1, 2-diamine

3 Pick up a poor electrolytic conductor complex in solution.

a. $K_2[PtCl_6]$ b. $K_4[Fe(CN)_6]$
c. $[Co(NH_3)_4]SO_4$ d. $[Co(NH_3)_3(NO_2)_3]$

4 How many EDTA molecules are required to make an octahedral complex with a Ca^{2+} ion?

a. Six b. Three c. One d. Two

5 Which of the following is an outer orbital complex?

a. $[Co(NH_3)_6]^{3+}$ b. $[Mn(CN)_6]^{4-}$
c. $[Fe(CN)_6]^{4-}$ d. $[Ni(NH_3)_6]^{2+}$

6 The structure of the complexes $[Cu(NH_3)_4]$ (ClO_4) and $[Cu(NH_3)_4]ClO_4$ in solution respectively are

a. square planar and tetrahedral b. octahedral and square pyramidal
c. octahedral and trigonal bipyramidal d. tetrahedral and square pyramidal

7 EAN value of the complex is not equal to 36 in

a. $[Ni(H_2O)_6]^{2+}$ b. $[Zn(OH)_4]^{2-}$
c. $[Fe(CO)_5]$ d. $[Cr(CO)_6]$

8 The correct statement is

a. $[Ni(NH_3)_6]^{2+}$ is an inner orbital complex
b. $[CrBr_2(en)_2]^-$ exhibits linkage isomerism
c. $[MnBr_4]^{2-}$ is tetrahedral
d. $[Co(NH_3)_6]^{3+}$ is paramagnetic

9 How many moles of AgCl would be obtained when 100 moles of 0.1 M $CoCl_3 \cdot 5NH_3$ is treated with excess $AgNO_3$?

a. 0.01 b. 0.02 c. 0.03 d. 0.04

10 The IUPAC name of $[Ni(NH_3)_4][NiCl_4]$ is

a. tetrachloronickel(II) tetraamminenickel (II)
b. tetraamminenickel(II) tetrachloronickel (II)
c. tetraamminenickel (II) tetrachloronickelate (II)
d. tetrachloronickelate (II) tetramminenickelate(II)

11 Which is the incorrect?

a. $[Co(en)_2NO_2Cl]Br$ is cationic complex
b. $[Co(en)_3]Cl_3$ shows geometrical isomerism
c. $[Fe(CO)_5]$ is neutral complex
d. $[Cu(NH_3)_4]SO_4$ is deep blue colour

Working Space

12 Extent of crystal field splitting in octahedral complexes of the given metal with particular weak field ligand are such that

a. Fe(III) < Cr (III) < Rh (III) < Ir (III)
b. Cr (III) < Fe (III) < Rh (III) < Ir (III)
c. Ir (III) < Rh (III) < Fe (III) < Cr (III)
d. Fe (III) = Cr (III) < Rh (III) < Ir (III)

13 Which of the following is correct arrangement of ligands in terms of the Dq values of their complexes with any particular 'hard' metal ion?

a. $Cl^- < F^- < NCS^- < NH_3 < CN^-$
b. $NH_3 < F^- < Cl^- < NCS^- < CN^-$
c. $Cl^- < F^- < NCS^- < CN^- < NH_3$
d. $NH_3 < CN^- < NCS^- < Cl^- < F^-$

14 Cr^{3+} form four complexes with four different ligands which are $[Cr(Cl)_6]^{3-}$, $[Cr(H_2O)_6]^{3+}$, $[Cr(NH_3)_6]^{3+}$ and $[Cr(CN)_6]^{3-}$. The order of CFSE (D_o) in these complexes is in the order

a. $[CrCl_6]^{3-} = [Cr(H_2O)_6]^{3+} = [Cr(NH_3)_6]^{3+} = [Cr(CN)_6]^{3-}$
b. $[CrCl_6]^{3-} < [Cr(H_2O)_6]^{3+} < [Cr(NH_3)_6]^{3+} < [Cr(CN)_6]^{3-}$
c. $[CrCl_6]^{3-} > [Cr(H_2O)_6]^{3+} > [Cr(NH_3)_6]^{3+} > [Cr(CN)_6]^{3-}$
d. $[CrCl_6]^{3-} < [Cr(H_2O)_6]^{3+} = [Cr(NH_3)_6]^{3+} < [Cr(CN)_6]^{3-}$

15 The IUPAC name of $[Fe(O_2)(CN)_4Cl]^{4-}$ is

a. chlorotetracyanodioxoferrate (II) ion
b. chlorotetracyanoperoxoferrate (II) ion
c. chlorotetracyano superoxoferrate (II) ion
d. None of the above

16 The value of 'spin only' magnetic moment for one of the following configuration in octahedral field is 3.87 BM. The correct one is

a. d^5 (in strong field ligand)
b. d^6 (in weak field ligand)
c. d^3 (in weak as well as in strong field ligand)
d. d^9 (in strong field ligand)

17 There are four complexes of Ni. Select the complex(es) which will be attracted by magnetic field.

I. $[Ni(CN)_4]^{2-}$ II. $[NiCl_4]^{2-}$ III. $Ni(CO)_4$ IV. $[Ni(H_2O)_6]^{2+}$

a. Only I
b. Only II
c. II, III and IV
d. II and IV

18 An octahedral complex with molecular formula $5NH_3 \cdot Cl \cdot SO_4$ has two isomers *A* and *B*. The solution of *A* gives a white precipitate with $AgNO_3$ solution and the solution of *B* gives white precipitate with $BaCl_2$ solution .The type of isomerism exhibited by the complex is

a. linkage isomerism
b. ionisation isomerism
c. coordination isomerism
d. geometrical isomerism

19 Both geometrical and optical isomerism are shown by

a. $[Co(en)_2Cl_2]^+$
b. $[Co(NH_3)_5Cl]^{2+}$
c. $[Co(NH_3)_4Cl_2]^+$
d. $[Co(O)_3]^{3-}$

20 Low spin complex of d^6 cation in an octahedral field will have which of the following energy?

a. $\frac{-12}{5}\Delta_o + P$
b. $\frac{-12.5}{5}\Delta_o + 3P$
c. $\frac{-2}{5}\Delta_o + 2P$
d. $\frac{-2}{5}\Delta_o + P$

21 The crystal field stabilisation energy (CFSE) and the spin only magnetic moment in Bohr magneton (BM) for the complex $K_3[Fe(CN)_6]$ respectively, are

a. $0.0\Delta_o$ and $\sqrt{35}$ BM
b. $-2.0\Delta_o$ and $\sqrt{3}$ BM
c. $-0.4\Delta_o$ and $\sqrt{24}$ BM
d. $-2.4\Delta_o$ and 0.0 BM

22 If the bond length of CO bond in carbon monoxide is 1.128 Å, then what is the value of CO bond length in $Fe(CO)_5$?

Working Space

a. 1.15
b. 1.128
c. 1.72
d. 1.11

23 Crystal field theory does not explain which of the following property of coordination compounds?

a. Magnetic property
b. Colour
c. Structure of coordination compounds
d. The covalent character of the bond between metal and ligand

24 The correct order of ligand field strength is

a. $H_2O < Cl < CO < NH_3$
b. $CO < NH_3 < Cl < H_2O$
c. $H_2O < CO < NH_3 < Cl$
d. $Cl < H_2O < NH_3 < CO$

25 What is(are) number of unpaired electron(s) in the square planar $[Pt(CN)_4]^{2-}$ ions?

a. zero
b. 1
c. 4
d. 6

26 In decacarbonyl dimanganese (0), the number of CO groups bonded in between the Mn atoms are

a. 2
b. 1
c. 0
d. 4

27 The ligand that shows synergic effect is

a. H_2O
b. NH_3
c. CO
d. C_5H_5N

28 For the reactions and their equilibrium constants given below:

$$CuCl_4^{2-} + Br^- \rightleftharpoons CuCl_3Br^{2-} + Cl^-, K_1$$

$$CuCl_3Br^{2-} + Br^- \rightleftharpoons CuCl_2Br_2^{2-} + Cl^-, K_2$$

$$CuCl_2Br_2^{2-} + Br^- \rightleftharpoons CuClBr_3^{2-} + Cl^-, K_3$$

$$CuClBr_3^{2-} + Br^- \rightleftharpoons CuBr_4^{2-} + Cl^-, K_4$$

The equilibrium constant, K for the reaction $CuCl_4^{2-} + 3Br^- \rightleftharpoons CuClBr_3^{2-} + 3Cl^-$ is

a. $K_1 \times K_2 \times K_3$
b. $K_1 \times K_2 \times K_3 \times K_4$
c. $K_1 + K_2 + K_3$
d. $1/K_1 \times K_2 \times K_3$

Statement Type

***Direction** (Q. No. 29) This section is based on Statement I and Statement II. Select the correct answer from the codes given below.*

a. Both Statement I and Statement II are correct and Statement II is the correct explanation of Statement I
b. Both Statement I and Statement II are correct but Statement II is not the correct explanation of Statement I
c. Statement I is correct but Statement II is incorrect
d. Statement II is correct but Statement I is incorrect

29 **Statement I** $[Fe(CN)_6]^{3-}$ ion shows magnetic moment corresponding to two unpaired electrons.

Statement II Because it has d^2sp^3 type hybridisation.

Revisal Problems for JEE Advanced

SET - 2

Only One Option Correct Type

Working Space

Direction *(Q. Nos. 1-10) This section contains 10 multiple choice questions. Each question has four choices (a), (b), (c) and (d), out of which ONLY ONE option is correct.*

1 Consider the following complex ions P, Q and R. $P = [FeF_6]^{3-}$, $Q = [V(H_2O)_6]^{2+}$ and $R = [Fe(H_2O)_6]^{2+}$. The correct order of the complex ions according to their spin only magnetic moment value (in BM) is

a. $R < Q < P$
b. $Q < R < P$
c. $R < P < Q$
d. $Q < P < R$

2 Which of the following complexes has a shape different from others?

a. $[Zn(NH_3)_4]^{2+}$
b. $[NiCl_2(PPh_3)_2]$
c. $[Cu(NH_3)_4]^{2+}$
d. $[Cu(CN)_4]^{3-}$

3 The species with metal ion having d^5 configuration is

a. $K_4[Fe(CN)_6]$
b. $[Co(NH_3)_6]PO_4$
c. $K_4[Mn(CN)_6]$
d. $[Co(NH_3)_5SO_4]NO_3$

4 Consider the two complexes (A) $[Ni(H_2O)_6]^{2+}$ and (B) $[Ni(NH_3)_6]^{2+}$, the correct statement is

a. Complex A is diamagnetic and complex B is paramagnetic
b. Complex A is paramagnetic and complex B is diamagnetic
c. Both are paramagnetic
d. Both are diamagnetic

5 The spin only magnetic moment of $[Mn(CN)_6]^{4-}$ and $[MnBr_4]^{2-}$ in Bohr magneton

a. 5.92 and 5.92 BM
b. 4.89 and 1.73 BM
c. 1.73 and 5.92 BM
d. 1.73 and 1.73 BM

6 Chocolate brown precipitate is formed with

a. Cu with $[Fe(CN)_6]^{3-}$
b. Cu^{2+} with $[Fe(CN)_6]^{4-}$
c. Fe^{2+} with $[Fe(CN)_6]^{4-}$
d. Ni^{2+} with DMG

7 Which of the following complex formation has more –ve free energy change $(-\Delta G)$?

a. $[Ni(NH_3)_2(H_2O)_4]^{2+}$
b. $[Ni(NH_3)_4(H_2O)_2]^{2+}$
c. $[Ni(en)(H_2O)_4]^{2+}$
d. $[Ni(en)_2(H_2O)_2]^{2+}$

8 What is the magnetic moment and hybridisation of the brown ring complex formed in the test of nitrate ion (NO_3^-)?

a. $\sqrt{3}$ BM, sp^3d^2
b. $\sqrt{3}$ BM, d^2sp^3
c. $\sqrt{15}$ BM, sp^3d^2
d. $\sqrt{15}$ BM, d^2sp^3

9 For the coordination compound, $Na_2[Cd(en)(CN)_4]$, what are the coordination number (CN) and oxidation number (ON) of the central metal ion? (Note : The letters (en) represent the ligand ethylenediamine)

a. CN = 6, ON = +2
b. CN = 5, ON = +2
c. CN = 4, ON = +3
d. CN = 4, ON = +2

10 The crystal field splitting energy of octahedral (Δ_o) and tetrahedral (Δ_t) complexes is related as

a. $\Delta_t = \frac{1}{2}\Delta_o$
b. $\Delta_t = \frac{4}{9}\Delta_o$
c. $\Delta_t = \frac{3}{5}\Delta_o$
d. $\Delta_t = \frac{2}{5}\Delta_o$

Working Space

One or More than One Options Correct Type

Direction (Q. Nos. 11-20) *This section contains 10 multiple choice questions. Each question has four choices (a), (b), (c) and (d), out of which ONE or MORE THAN ONE are correct.*

11 Which of the following statement is correct ?

a. $[AgF_4]^-$, square planar and diamagnetic
b. $[Pd(CN)_4]^{2-}$, square planar and diamagnetic
c. $[AuCl_4]^-$, tetrahedral and paramagnetic
d. $[PdCl_2(PPh_3)_2]^-$, tetrahedral and paramagnetic

12 Which of the following complexes have metal-metal bond?

a. $Co_2(CO)_8$
b. $Mn_2(CO)_{10}$
c. $Fe_2(CO)_9$
d. $Ir_4(CO)_{12}$

13 Which of the following statements is correct?

a. $[Ag(NH_3)_2]^+$ is linear
b. Oxidation state of sulphur in sodium nitroprusside is +2
c. $[Fe(CO)_3$ is triangle pyramidal
d. In $[Fe(H_2O)_6]^{3+}$, complex Fe is d^2sp^3-hybridised

14 An aqueous pink solution of cobalt (II) chloride changes to deep blue colour on addition of excess of HCl. This is because

a. $[Co(H_2O)_6]^{2+}$ is transformed into $[CoCl_6]^{4-}$
b. $[Co(H_2O)_6]^{2+}$ is transformed into $[CoCl_4]^{2-}$
c. tetrahedral complexes have smaller crystal field splitting than octahedral complexes
d. tetrahedral complexes have large crystal field splitting than octahedral complexes

15 One of the most general reactions exhibited by coordination compounds is that of substitution, or replacement of one ligand by another.
When tetrachloroplatinate (II) ion reacted with ammonia yield *X* and tetrammine platinum (II) ion reacted with chloride ion yield *Y*, here *X* and *Y*

a. *X* is *cis*-diamminedichloroplatinum (II)
b. *X* is *trans*-diamminedichloroplatinum (II)
c. *Y* is *trans*-diamminedichloroplatinum (II)
d. *Y* is *cis*-diamminedichloroplatinum

16 For which of the following complex, EAN value is not equal to the atomic number of a rare gas?

a. $[IrCl_6]^{3-}$
b. $[Ag(NH_3)_2]^+$
c. $[Mn_2(CO)_{10}]$
d. $[Ni(en)_3]^{2+}$

17 In which of the following pairs both the complexes show optical isomerism?

a. *cis* $[Cr(C_2O_4)_2Cl_2]^{3-}$ and $[Co(NH_3)_4Cl_2]$
b. $[Co(en)_3]Cl_3$ and *cis* $[Co(en)_2Cl_2]^+$
c. $[Co(NO_2)_3(NH_3)_3]$ and *cis* $[Pt(en)_2Cl_2]^+$
d. $[Co(en)_2(C_2O_4)]^+$ and $[Co(EDTA)]^-$

18 Which is the symmetrical bidentate ligand?

a. Glycinate ion
b. Oxalate ion
c. Ethane-1, 2-diamine
d. 2, 2′-dipyridine

19 For which of the following CFSE is zero in an octahedral field?

a. d^5 configuration in strong ligand field
b. d^5 configuration in weak ligand field
c. d^{10} configuration in strong ligand field
d. d^{10} configuration in weak ligand field

20 Which is an example of bridging ligand?

a. OH^-
b. NH_2^-
c. CO
d. NH_3

Working Space

Comprehension Type

Direction (Q. Nos. 21 and 22) *This section contains a paragraph, describing theory, experiments, data, etc. Two questions related to the paragraphs have been given. Each question has only one correct answer among the four given options (a), (b), (c) and (d).*

Passage

In coordination chemistry, there are variety of methods applied to find out the structure of complexes. An isomer of the complex $Co(en)_2(H_2O)Cl_2Br$ on reaction with conc. H_2SO_4 (dehydrating agent), it suffers loss in weight and on reaction with $AgNO_3$ solution, it gives a white precipitate which is soluble in NH_3.

21 The correct formula of the complex is

a. $[CoClBr(en)_2]\ H_2O$
b. $[CoCl(en)_2H_2O]BrCl$
c. $[CoBr(en)_2H_2O]Cl_2$
d. $[CoBrCl(en)_2]ClH_2O$

22 If one mole of original complex is reacted with excess lead nitrate solution, then number of moles of $PbCl_2$ formed will be

a. 0.5 b. 1 c. 0 d. 3

Matching List Type

Direction (Q. Nos. 23-25) *Choices for the correct combination of elements from Column I and Column II are given as options (a), (b), (c) and (d), out of which one is correct.*

23 Match the following reactions given Column I with the corresponding products given in Column II.

	Column I (Compound)		Column II (Oxidation state of Co)
i.	$[Co(NCS)(NH_3)_5]SO_3$	p.	+4
ii.	$[Co(NH_3)_4Cl_2]SO_4$	q.	0
iii.	$Na_4[Co(S_2O_3)_3]$	r.	+1
iv.	$Co(CO)_8]$	s.	+2
		t.	+3

Codes

	i	ii	iii	iv
a.	p	q	s	t
b.	s	r	q	p
c.	t	p	s	q
d.	s	p	q	r

24 Match the following reactions given Column I with the corresponding products given in Column II.

	Column I		Column II
i.	$[CoF_3(H_2O)_3]$	p.	Diamagnetic, low spin complex
ii.	$[Cr(C_2O_4)_3]^{3-}$	q.	Paramagnetic, outer orbital complex
iii.	$[AuCl_4]^-$	r.	Paramagnetic, inner orbital complex
iv.	$[Fe(CN)_6]^{3-}$	s.	Diamagnetic, high spin complex
		t.	Paramagnetic and low spin complex

Codes

	i	ii	iii	iv
a.	r	q	p	s
b.	p,q	p	q	r,s
c.	p	q,r	p	s
d.	q	r	p	t

25 Match the Column I with Column II and mark the correct option from the codes given below.

Working Space

	Column I (Coordination compounds)		Column II (Central metal atoms)
i.	Chlorophyll	p.	Rhodium
ii.	*Ziegler-Natta* catalyst	q.	Cobalt
iii.	Wilkinson's catalyst	r.	Lead
iv.	Vitamin-B_{12}	s.	Titanium
		t.	Magnesium

Codes

	i	ii	iii	iv		i	ii	iii	iv
a.	t	s	p	q	b.	r	s	t	p
c.	s	r	q	p	d.	r	s	p	q

One Integer Value Correct Type

***Direction** (Q. Nos. 26-30) This section contains 5 questions. When worked out will result in an integer from 0 to 9 (both inclusive).*

26 Among the following complexes (*K* to *P*)

$K_3[Fe(CN)_6]$ (*K*), $[Co(NH_3)_6]Cl_3$(*L*), $Na_3[Co(oxalate)_3]$ (*M*),

$[Ni(H_2O)_6]Cl_2$ (*N*), $K_2[Pt(CN)_4]$ (*O*) and $[Zn(H_2O)_6](NO_3)_2$ (*P*)

The number of diamagnetic complexes are

27 For a d^4 ion with high spin complex, CFSE is $-x$ Dq. Here, x is

28 Number of unpaired electrons in t_{2g} set of *d*-orbitals in the complex $[Co(H_2O)_3F_3]$ are

29 The number of optically active isomers for the complex of formula $[Ma_2b_2cd]$.

30 The number of ions formed, when *bis* (ethane-1, 2-diamine) copper (II) sulphate is dissolve in water will be

Statement Type

***Direction** (Q. No. 31) This section is based on Statement I and Statement II. Select the correct answer from the codes given below.*

a. Both Statement I and Statement II are correct and Statement II is the correct explanation of Statement I
b. Both Statement I and Statement II are correct and Statement II is not the correct explanation of Statement I
c. Statement I is correct but Statement II is incorrect
d. Statement II is correct but Statement I is incorrect

31 **Statement I** Toxic metals are removed by the chelating ligands.

Statement II Chelate complexes tend to be more stable.

JEE Main & AIEEE Archive

(Compilation of Last 13 Years' Questions)

1 The octahedral complex of a metal ion M^{3+} with four monodentate ligands L_1, L_2, L_3 and L_4 absorb wavelengths in the region of red, green, yellow and blue, respectively. The increasing order of ligand strength of the four ligands is **(JEE Main 2014)**

a. $L_4 < L_3 < L_2 < L_1$ b. $L_1 < L_3 < L_2 < L_4$
c. $L_3 < L_2 < L_4 < L_1$ d. $L_1 < L_2 < L_4 < L_3$

2 Which of the following complex species is not expected to exhibit optical isomerism? **(JEE Main 2013)**

a. $[Co(en)_3]^{3+}$
b. $[Co(en)_2Cl_2]^+$
c. $[Co(NH_3)_3Cl_3]$
d. $[Co(en)(NH_3)Cl_2]^+$

3 Which among the following will be named as dibromidobis- (ethylenediamine) chromium (III) bromide? **(AIEEE 2012)**

a. $[Cr(en)_3]Br_3$ b. $[Cr(en)_2Br_2]Br$
c. $[Cr(en)Br_4]^-$ d. $[Cr(en)Br_2]Br$

4 Which of the following facts about the complex $[Cr(NH_3)_6]Cl_3$ is wrong? **(AIEEE 2011)**

a. The complex involves d^2sp^3 hybridisation and is octahedral in shape
b. The complex is paramagnetic
c. The complex is an outer orbital complex
d. The complex gives white precipitate with silver nitrate solution

5 The magnetic moment (spin only) of $[NiCl_4]^{2-}$ is **(AIEEE 2011)**

a. 1.82 BM b. 5.46 BM
c. 2.82 BM d. 1.41 BM

6 Among the ligands NH_3, en, CN^- and CO, the correct order of their increasing field strength is **(AIEEE 2011)**

a. $CO < NH_3 < en < CN^-$
b. $NH_3 < en < CN^- < CO$
c. $CN^- < NH_3 < CO < en$
d. $en < CN^- < NH_3 < CO$

7 Which one of the following complex ions has geometrical isomers? **(AIEEE 2011)**

a. $[Co(en)_3]^{3+}$
b. $[Ni(NH_3)_5Br]^+$
c. $[Co(NH_3)_2(en)_2]^{3+}$
d. $[Cr(NH_3)_4(en)]^{3+}$

8 Which one of the following has an optical isomer? (en = ethylenediamine) **(AIEEE 2010)**

a. $[Zn(en)(NH_3)_2]^{2+}$ b. $[Co(en)_3]^{3+}$
c. $[Co(H_2O)_4(en)]^{3+}$ d. $[Zn(en)_2]^{2+}$

9 Which of the following has an optical isomer? **(AIEEE 2009)**

a. $[Co(NH_3)_3Cl]^+$ b. $[Co(en)(NH_3)_2]^{2+}$
c. $[Co(H_2O)_4(en)]^{3+}$ d. $[Co(en)_2(NH_3)_2]^{3+}$

10 Which of the following pairs represents linkage isomers? **(AIEEE 2009)**

a. $[Cu(NH_3)_4][PtCl_4]$ and $[Pt(NH_3)_4][CuCl_4]$
b. $[Pd(PPh_3)_2(NCS)_2]$ and $[Pd(PPh_3)_2(SCN)_2]$
c. $[Co(NH_3)_5]NO_3SO_4$ and $[Co(NH_3)_5SO_4]NO_3$
d. $[PtCl_2(NH_3)_4]Br_2$ and $[PtBr_2(NH_3)_4]Cl_2$

11 The coordination number and the oxidation state of the element 'E' in the complex $[E(en)_2(C_2O_4)]NO_2$ (where, (en) is ethylene diamine) respectively are **(AIEEE 2008)**

a. 6 and 2 b. 4 and 2
c. 4 and 3 d. 6 and 3

12 In which of the following octahedral complexes of Co (atomic number 27), will the magnitude of Δ be the highest? **(AIEEE 2008)**

a. $[Co(CN)_6]^{3-}$ b. $[Co(C_2O_4)_3]^{3-}$
c. $[Co(H_2O)_6]^{3+}$ d. $[Co(NH_3)_6]^{3+}$

13 Which one of the following has a square planar geometry? **(AIEEE 2007)**

(Atomic number, Co = 27, Ni = 28, Fe = 26, Pt = 78)

a. $[CoCl_4]^{2-}$ b. $[FeCl_4]^{2-}$
c. $[NiCl_4]^{2-}$ d. $[PtCl_4]^{2-}$

14 In $Fe(CO)_5$, the Fe—CO bond possesses **(AIEEE 2006)**

a. π-character only
b. both σ and π-characters
c. ionic character
d. σ-character only

15 The IUPAC name for the complex $[Co(NO_2)(NH_3)_5]Cl_2$ is **(AIEEE 2006)**

a. nitrito-N-pentamminecobalt (III) chloride
b. nitrito-N-pentamminecobalt (II) chloride
c. pentammine nitrito-N-cobalt (II) chloride
d. pentammine nitrito-N-cobalt (III) chloride

16 Nickel (Z = 28) combines with a uninegative monodentate ligand X^- to form a paramagnetic complex $[NiX_4]^{2-}$. The number of unpaired electron(s) in the nickel and geometry of this complex ion, respectively are **(AIEEE 2006)**

a. one, tetrahedral
b. two, tetrahedral
c. one, square planar
d. two, square planar

17 The value of the spin only magnetic moment for one of the following configurations is 2.84 BM. The correct one is **(AIEEE 2005)**

a. d^5 (in strong ligand field)
b. d^3 (in weak as well as in strong fields)
c. d^4 (in weak ligand field)
d. d^4 (in strong ligand field)

18 Which of the following factors may be regarded as the main cause of lanthanide contraction? **(AIEEE 2005)**

a. Greater shielding of 5*d* electron by 4*f* electrons
b. Poorer shielding of 5*d* electron by 4*f* electrons
c. Effective shielding of one of 4*f* electron by another in the subshell
d. Poor shielding of one of 4*f* electron by another in the subshell

19 The lanthanide contraction is responsible for the fact that **(AIEEE 2005)**

a. Zr and Zn have the same oxidation state
b. Zr and Hf have about the same radius
c. Zr and Nb have similar oxidation state
d. Zr and Y have about the same radius

20 The IUPAC name of the coordination compound $K_3[Fe(CN)_6]$ is **(AIEEE 2005)**

a. tripotassium hexacyanoiron (II)
b. potassium hexacyanoiron (II)
c. potassium hexacyanoferrate (III)
d. potassium hexacyanoferrate (II)

21 Which one of the following cyano complexes would exhibit the lowest value of paramagnetic behaviour? **(AIEEE 2005)**
(Atomic number of Cr = 24, Mn = 25, Fe = 26, Co = 27)

a. $[Co(CN)_6]^{3-}$ b. $[Fe(CN)_6]^{3-}$
c. $[Mn(CN)_6]^{3-}$ d. $[Cr(CN)_6]^{3-}$

22 Which of the following compounds shows optical isomerism? **(AIEEE 2005)**

a. $[Co(CN)_6]^{3-}$ b. $[Cr(C_2O_4)_3]^{3-}$
c. $[ZnCl_4]^{2-}$ d. $[Cu(NH_3)_4]^{2+}$

23 The coordination number of a central metal atom in a complex is determined by **(AIEEE 2004)**

a. the number of ligands around a metal ion bonded by sigma bonds
b. the number of ligands around a metal ion bonded by pi bonds
c. the number of ligands around a metal ion bonded by sigma and pi bonds both
d. the number of only anionic ligands bonded to the metal ion

24 Which one of the following has largest number of isomers? **(AIEEE 2004)**
(*R* = alkyl group, en = ethylenediamine)

a. $[Ru(NH_3)_4Cl_2]^+$ b. $[Co(NH_3)_5Cl]^{2+}$
c. $[Ir(PR_3)_2H(CO)]^{2+}$ d. $[Co(en)_2Cl_2]^+$

25 The correct order of magnetic moments (spin only values in BM) among the following is (atomic number of Mn = 25, Fe = 26, Co = 27) **(AIEEE 2004)**

a. $[MnCl_4]^{2-} > [CoCl_4]^{2-} > [Fe(CN)_6]^{4-}$
b. $[MnCl_4]^{2-} > [Fe(CN)_6]^{4-} > [CoCl_4]^{2-}$
c. $[Fe(CN)_6]^{4-} > [MnCl_4]^{2-} > [CoCl_4]^{2-}$
d. $[Fe(CN)_6]^{4-} > [CoCl_4]^{2-} > [MnCl_4]^{2-}$

26 Which one of the following complexes is an outer orbital complex? **(AIEEE 2004)**
(Atomic number of Mn = 25, Fe = 26, Co = 27, Ni = 28)

a. $[Fe(CN)_6]^{4-}$ b. $[Mn(CN)_6]^{4-}$
c. $[Co(NH_3)_6]^{3+}$ d. $[Ni(NH_3)_6]^{2+}$

27 Ammonia forms the complex ion $[Cu(NH_3)_4]^{2+}$ with copper ions in the alkaline solutions but not in acidic solutions. What is the reason for it? **(AIEEE 2003)**

a. In acidic solutions, hydration protects copper ions
b. In acidic solutions, protons coordinate with ammonia molecules forming NH_4^+ ions and NH_3 molecules are not available
c. In alkaline solutions, insoluble $Cu(OH)_2$ is precipitated which is soluble in excess of any alkali
d. Copper hydroxide is an amphoteric substance

28 One mole of the complex compound $Co(NH_3)_5Cl_3$ gives 3 moles of ions on dissolution in water. One mole of the same complex reacts with two moles of $AgNO_3$ solution to yield two moles of AgCl(*s*). The structure of the complex is **(AIEEE 2003)**

a. $[Co(NH_3)_5Cl]\,Cl_2$
b. $[Co(NH_3)_3Cl_3]\cdot 2NH_3$
c. $[Co(NH_3)_4Cl_2]Cl\cdot NH_3$
d. $[Co(NH_3)_4Cl]\,Cl_2\cdot NH_3$

29 In the coordination compound, $K_4[Ni(CN)_4]$ the oxidation state of nickel is **(AIEEE 2003)**

a. − 1 b. 0
c. + 1 d. + 2

30 Type of isomerism shown by $[Cr(NH_3)_5NO_2]\,Cl_2$ is **(AIEEE 2002)**

a. optical b. ionisation
c. geometrical d. linkage

31 A square planar complex is formed by the hybridisation of which of the following atomic orbitals? **(AIEEE 2002)**

a. s, p_x, p_y, p_z b. s, p_x, p_y, p_z, d
c. d, s, p_x, p_y d. s, p_x, p_y, p_z, d, d

32 The most stable ion is **(AIEEE 2002)**

a. $[Fe(OH)_5]^{3-}$ b. $[FeCl_6]^{3-}$
c. $[Fe(CN)_6]^{3-}$ d. $[Fe(H_2O)_6]^{3+}$

JEE Advanced & IIT JEE Archive

(Compilation of Last 10 Years' Questions)

1 Match each coordination compound in Column I with an appropriate pair of characteristics from Column II and select the correct answer using the codes given below the columns (en = $H_2NCH_2CH_2NH_2$; atomic numbers : Ti = 22; Cr = 24; Co = 27; Pt = 78)
(2014 Adv. Matching Type)

	Column I		Column II
i.	$[Cr(NH_3)_4Cl_2]Cl$	p.	Paramagnetic and exhibits ionisation isomerism
ii.	$[Ti(H_2O)_5Cl](NO_3)_2$	q.	Diamagnetic and exhibits *cis-trans* isomerism
iii.	$[Pt(en)(NH_3)Cl]NO_3$	r.	Paramagnetic and exhibits *cis-trans* isomerism
iv.	$[Co(NH_3)_4(NO_3)_2]NO_3$	s.	Diamagnetic and exhibits ionisation isomerism

Codes

	i	ii	iii	iv		i	ii	iii	iv
a.	s	q	r	p	b.	r	p	s	q
c.	q	p	r	s	d.	p	r	s	q

2 $EDTA^{4-}$ is ethylenediaminetetraacetate ion. The total number of N— Co — O bond angles in $[Co(EDTA)]^-$ complex ion is **(2013 Adv. Integer Type)**

3 Consider the following complex ions, P, Q and R.
$P = [FeF_6]^{3-}$, $Q = [V(H_2O)_6]^{2+}$ and $R = [Fe(H_2O)_6]^{2+}$
The correct order of the complex ions, according to their spinonly magnetic moment values (in BM) is **(2013 Adv. Only One Option Correct Type)**

a. $R < Q < P$
b. $Q < R < P$
c. $R < P < Q$
d. $Q < P < R$

4 The pair(s) of coordination complexes/ions exhibiting the same kind of isomerism is(are)
(2013 Adv. One or More than One Options Correct Type)

a. $[Cr(NH_3)_5Cl]Cl_2$ and $[Cr(NH_3)_4Cl_2]Cl$
b. $[Co(NH_3)_4Cl_2]^+$ and $[Pt(NH_3)_2(H_2O)Cl]^+$
c. $[CoBr_2Cl_2]^{2-}$ and $[PtBr_2Cl_2]^{2-}$
d. $[Pt(NH_3)_3(NO_3)]Cl$ and $[Pt(NH_3)_3Cl]Br$

5 As per IUPAC nomenclature, the name of the complex $[Co(H_2O)_4(NH_3)_2]Cl_3$ is
(2012, Only One Option Correct Type)

a. tetraaquadiaminecobalt (III) chloride
b. tetraaquadiamminecobalt (III) chloride
c. diaminotetraaquacobalt (III) chloride
d. diamminetetraaquacobalt (III) chloride

6 $NiCl_2\{P(C_2H_5)_2(C_6H_5)\}_2$ exhibits temperature dependent magnetic behaviour (paramagnetic/diamagnetic) the coordination geometries of Ni^{2+} in the paramagnetic and diamagnetic states respectively are
(2012, Only One Option Correct Type)

a. tetrahedral and tetrahedral
b. square planar and square planar
c. tetrahedral and square planar
d. square planar and tetrahedral

7 Geometrical shapes of the complexes formed by the reaction of Ni^{2+} with Cl^-, CN^- and H_2O respectively, are **(2011, Only One Option Correct Type)**

a. octahedral, tetrahedral and square planar
b. tetrahedral, square planar and octahedral
c. square planar, tetrahedral and octahedral
d. octahedral, square planar and octahedral

8 Among the following complexes (*K-P*),
(2011, Only One Option Correct Type)

$K_3[Fe(CN)_6]$ (*K*), $[Co(NH_3)_6]Cl_3$ (*L*), $Na_3[Co(ox)_3]$ (*M*), $[Ni(H_2O)_6]Cl_2$ (*N*), $K_2[Pt(CN)_4]$ (*O*) and $[Zn(H_2O)_6](NO_3)_2$ (*P*)
the diamagnetic complexes are

a. *K, L, M, N* b. *K, M, O, P*
c. *L, M, O, P* d. *L, M, N, O*

9 The volume (in mL) of 0.1 M $AgNO_3$ required for complete precipitation of chloride ions present in 30 mL of 0.01 M solution of $[Cr(H_2O)_5Cl]Cl_2$, as silver chloride is close to **(2011, Integer Type)**

10 Total number of geometrical isomers for the complex $[RhCl(CO)(PPh_3)(NH_3)]$ is
(2010, Integer Type)

11 The ionisation isomer of $[Cr(H_2O)_4Cl(NO_2)]Cl$ is
(2010, Only One Option Correct Type)

a. $[Cr(H_2O)_4(O_2N)]Cl_2$
b. $[Cr(H_2O)_4Cl_2](NO_2)$
c. $[Cr(H_2O)_4Cl(ONO)]Cl$
d. $[Cr(H_2O)_4Cl_2(NO_2)]H_2O$

12 The correct structure of ethylenediaminetetraacetic acid (EDTA) is
(2010, Only One Option Correct Type)

a. 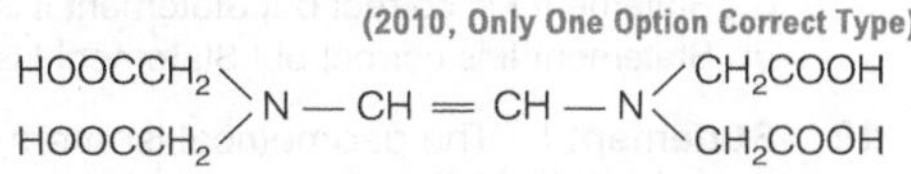

b. 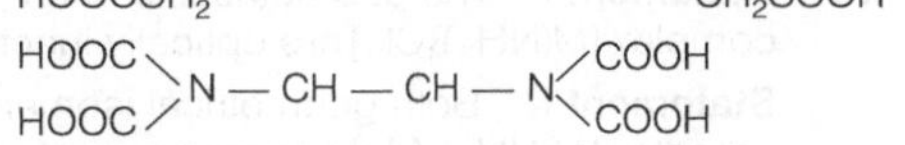

c. 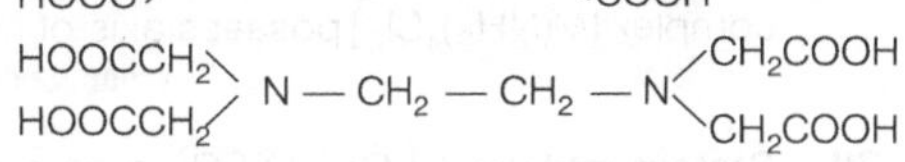

d. 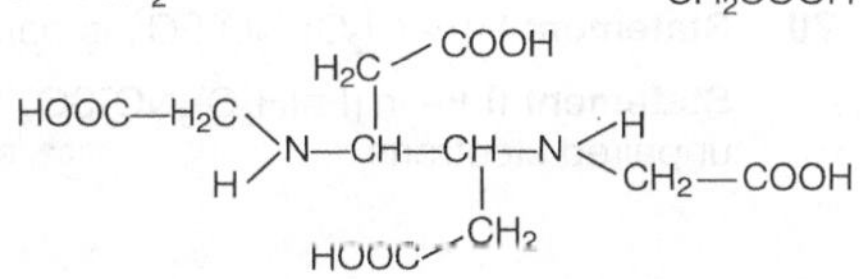

13 The complex showing a spin-only magnetic moment of 2.82 BM is (2010, Only One Option Correct Type)

a. $Ni(CO)_4$ b. $[NiCl_4]^{2-}$
c. $Ni(PPh_3)_4$ d. $[Ni(CN)_4]^{2-}$

14 The compound(s) that exhibit(s) geometrical isomerism is(are) (2009, One or More than One Options Correct Type)

a. $[Pt(en)Cl_2]$
b. $[Pt(en)_2]Cl_2$
c. $[Pt(en)_2Cl_2]Cl_2$
d. $[Pt(NH_3)_2]Cl_2$

15 The spin only magnetic moment value (in Bohr magneton units) of $Cr(CO)_6$ is (2009, Only One Option Correct Type)

a. 0 b. 2.84
c. 4.90 d. 5.92

16 Both $[Ni(CO)_4]$ and $[Ni(CN)_4]^{2-}$ are diamagnetic. The hybridisation of nickel in these complexes respectively, are (2008, Only One Option Correct Type)

a. sp^3, sp^3 b. sp^3, dsp^2
c. dsp^2, sp^3 d. dsp^2, dsp^2

17 The IUPAC name of $[Ni(NH_3)_4][NiCl_4]$ is (2008, Only One Option Correct Type)

a. tetrachloronickel (II)-tetraamminenickel (II)
b. tetraamminenickel (II)-tetrachloronickel (II)
c. tetraamminenickel (II)-tetrachloronickelate (II)
d. tetrachloronickel (II)-tetraamminenickelate (0)

18 Among the following, the coloured compound is (2008, Only One Option Correct Type)

a. CuCl
b. $K_3[Cu(CN)_4]$
c. CuF_2
d. $[Cu(CH_3CN)_4]BF_4$

Direction *(Q. Nos. 19 and 20) This section is based on Statement I and Statement II. Select the correct answer from the codes given below.*

a. Both Statement I and Statement II are correct and Statement II is the correct explanation of Statement I
b. Both Statement I and Statement II are correct and Statement II is not the correct explanation of Statement I
c. Statement I is correct but Statement II is incorrect
d. Statement II is correct but Statement I is incorrect

19 **Statement I** The geometrical isomers of the complex $[M(NH_3)_4Cl_2]$ are optically inactive.

Statement II Both geometrical isomers of the complex $[M(NH_3)_4Cl_2]$ possess axis of symmetry. (2008, Statement Type)

20 **Statement I** $[Fe(H_2O)_5NO]SO_4$ is paramagnetic.

Statement II Fe in $[Fe(H_2O)_5NO]SO_4$ has three unpaired electrons. (2008, Statement Type)

21 Match the complexes in Column I with their properties listed in Column II. (2007, Matching Type)

	Column I		Column II
i.	$[Co(NH_3)_4(H_2O)_2]Cl_2$	p.	Geometrical isomers
ii.	$[Pt(NH_3)_2Cl_2]$	q.	Paramagnetic
iii.	$[Co(H_2O)_5Cl]Cl$	r.	Diamagnetic
iv.	$[Ni(H_2O)_6]Cl_2$	s.	Metal ion with +2 oxidation state

Codes

	i	ii	iii	iv		i	ii	iii	iv
a.	r	s	p	q	b.	q	r	s	q
c.	p	s	q	r	d.	p	q	r	s

22 Among the following metal carbonyls, the C—O bond order is lowest in (2007, Only One Option Correct Type)

a. $[Mn(CO)_6]^+$ b. $[Fe(CO)_5]$
c. $[Cr(CO)_6]$ d. $[V(CO)_6]^-$

23 If the bond length of CO bond in carbon monoxide is 1.128 Å, then what is the value of CO bond length in $Fe(CO)_5$? (2006, Only One Option Correct Type)

a. 1.15 Å b. 1.128 Å c. 1.72 Å d. 1.118 Å

Passage for Q. Nos. (24-26)

The coordination number of Ni^{2+} is 4.

$NiCl_2$ + KCN (excess) $\longrightarrow$ *A* (cyano complex)
$NiCl_2$ + conc. HCl (excess) $\longrightarrow$ *B* (chloro complex)

24 The IUPAC name of *A* and *B* are (2006, Comprehension Type)

a. potassium tetracyanonickelate (II), potassium tetrachloronickelate (II)
b. tetracyanopotassiumnickelate (II), tetrachloropotassiumnickelate (II)
c. tetracyanonickel (II), tetrachloronickel (II)
d. potassium tetracyanonickel (II), potassium tetrachloronickel (II)

25 Predict the magnetic nature of *A* and *B*.

a. Both are diamagnetic
b. *A* is diamagnetic and *B* is paramagnetic with one unpaired electron
c. *A* is diamagnetic and *B* is paramagnetic with two unpaired electrons
d. Both are paramagnetic

26 The hybridisation of *A* and *B* are

a. dsp^2, sp^3 b. sp^3, sp^3
c. dsp^2, dsp^2 d. sp^3d^2, d^2sp^3

27 Which kind of isomerism is shown by $Co(NH_3)_4Br_2Cl$? (2005, Only One Option Correct Type)

a. Geometrical and ionisation
b. Optical and ionisation
c. Geometrical and optical
d. Geometrical only

4. Metallurgy

DPP-1	Introduction to Metallurgy and Ores of Some Metals
DPP-2	Metallurgy Process of Concentration of Ores
DPP-3	Extraction of Crude Metals from Concentrated Ores
DPP-4	Thermodynamic Principles of Metallurgy, Ellingham's Diagram and Its Applications
DPP-5	Refining of Crude Metals
DPP-6	Metallurgy of Some Important Metals I (Cu, Fe, Ag)
DPP-7	Metallurgy of Some Important Metals II (Al, Zn, Au)

- **Revisal Problems for JEE Main**
- **Revisal Problems for JEE Advanced**
- **JEE Main & AIEEE Archive**
- **JEE Advanced & IIT JEE Archive**

DPP
DAILY PRACTICE PROBLEMS

DPP-1 Introduction to Metallurgy and Ores of Some Metals

Subjective Questions

Working Space

Direction (Q. Nos. 1-5) *These questions are subjective in nature, need to be solved completely on notebook.*

1 How does the term "Ore" differ from "Mineral" ? Give an example.

2 Metals like Cu, Ag, Zn, Hg and Pb occur in nature as sulphides rather than oxides. Why ?

3 What are ruby, emerald and sapphire?

4 In general which metal do you expect to occur in the native state in nature? Give examples.

5 What is the relative abundance of elements in

i. earth crust? ii. human body? iii. sea water?

Only One Option Correct Type

Direction (Q. Nos. 6-15) *This section contains 10 multiple choice questions. Each question has four choices (a), (b), (c) and (d), out of which ONLY ONE is correct.*

6 Which represents the correct abundance order of elements in the earth crust?

a. O > Si > Fe > Al b. O > Si > Al > Fe c. O > C > Si > Al d. Al > Fe > O > Si

7 Which of the following contains both copper and iron?

a. Cuprite b. Chalcocite
c. Chalcopyrite d. Malachite

8 Regarding cryolite incorrect statement is

a. Its formula is $3NaF \cdot AlF_3$
b. The non-metal is fluorine
c. It is used in aluminium extraction from bauxite
d. It is an ore of Al

9 In chromite ore, the oxidation state of iron and chromium respectively are

a. +3, + 6 b. +3, + 2
c. +2, + 3 d. +2, + 6

10 Sulphide ores are common for the metals

a. Ag, Cu and Pb
b. Ag, Cu and Sn
c. Ag, Mg and Pb
d. Al, Cu and Pb

Working Space

11 Oxidation states of the metal in the minerals haematite and magnetite respectively are

a. II, III in haematite and III in magnetite
b. II, III in haematite and II in magnetite
c. II in haematite and II, III in magnetite
d. III in haematite and II, III in magnetite

12 Among the following statements the incorrect one is

a. Calamine and siderite are carbonate ores
b. Argentite and cuprite are oxides
c. Zinc blende and pyrites are sulphide ores
d. Malachite and azurite are ores of copper

13 A mineral is called an ore if

a. the percentage of metal content is more than 50 %
b. it is possible to isolate metal from it
c. a metal can be profitably extracted from it
d. if it is free from any earthly impurities

14 Which is the incorrect match?

a. Calamine = $ZnCO_3$
b. Siderite = Fe_3O_4
c. Sphalerite = ZnS
d. Malachite = $CuCO_3 \cdot Cu(OH)_2$

15 Molecular formula of Glauber salt is

a. $MgSO_4 \cdot 7H_2O$
b. $FeSO_4 \cdot 7H_2O$
c. $CuSO_4 \cdot 5H_2O$
d. $Na_2SO_4 \cdot 10H_2O$

One or More than One Options Correct Type

Direction (Q. Nos. 16-20) *This section contains 5 multiple choice questions. Each question has four choices (a), (b), (c) and (d), out of which ONE or MORE THAN ONE are correct.*

16 Which of the following is the correct combination?

a. Borax = $Na_2B_4O_7 \cdot 10H_2O$
b. Colemanite = $Ca_2B_6O_{11} \cdot 5H_2O$
c. Anglesite = $PbCO_3$
d. Chile salt petre = $NaNO_3$

17 Which of the following pairs is/are carbonate minerals?

a. Calamine, siderite
b. Magnesite, dolomite
c. Malachite, azurite
d. Calcite, witherite

18 Which is not the name of a mineral?

a. Zypsum
b. Plaster of Paris
c. Litharge
d. Galena

19 Which of the following is a halide ore?

a. Feldspar
b. Fluorspar
c. Sylvine
d. Carnallite

20 Which is not an example of ore?

a. Ruby silver
b. Horn silver
c. German silver
d. Quick silver

Working Space

Comprehension Type

Direction (Q. Nos. 21and 22) This section contains a paragraph, describing theory, experiments, data, etc. Two questions related to the paragraph have been given. Each question has only one correct answer among the four given options (a), (b), (c) and (d).

Passage

The natural substances in which metals occur in the earth are called minerals. A mineral has a definite composition. It may be a single compound or a complex mixture . It is usually associated with number of impurities. The minerals from which the metals can be conveniently extracted are called ores .Thus, all ores are minerals but all minerals are not ores. Depending on the composition ores are classified into different types such as oxides, sulphides, carbonates, halides, silicates etc.

21 The impurities associated with minerals are called

a. Gangue b. Flux c. Slag d. Depressant

22 Metals that exist in their native state are

a. Cu and Fe b. Al and Cu c. Au and Pt d. Pd and Pt

Matching List Type

Direction (Q. No. 23 and 24) Choices for the correct combination of elements from Column I and Column II are given as options (a), (b), (c) and (d), out of which one is correct.

23 Match the Column I with Column II and mark the correct option from the codes given below.

	Column I		Column II
i.	China clay	p.	$MgSO_4 \cdot 7H_2O$
ii.	Feldspar	q.	$PbCO_3$
iii.	Cerussite	r.	$PbSO_4$
iv.	Epsom salt	s.	$KAlSi_3O_8$
		t.	$[Al_2(OH)_4Si_2O_5]$

Codes

	i	ii	iii	iv		i	ii	iii	iv
a.	t	s	q	p	b.	s	t	p	q
c.	p	q	s	t	d.	s	p	q	t

24 Match the Column I with Column II and mark the correct option from the codes given below.

	Column I		Column II
i.	Gypsum	p.	Silver glance
ii.	Thortveitite	q.	$CaCO_3$
iii.	Halides	r.	Sulphates
iv.	Calcite	s.	Silicate
		t.	Carbonates

Codes

	i	ii	iii	iv		i	ii	iii	iv
a.	r	s	p	q,t	b.	s	t	p	q
c.	p	q	s	t	d.	s	p	q	t

One Integer Value Correct Type

Working Space

Direction *(Q. Nos. 25-29) This section contains 5 questions. When worked out will result in an integer from 0 to 9 (both inclusive).*

25 The number of carbonate minerals among the following

Cassiterite, Siderite, Anglesite, Azurite, Calamine, Cryolite, Calcite, Magnetite , Magnesite

26 The different number of non-metals present in fluorapatite are

27 Number of oxide ores among Galena, Pyrolusite, Rutile, Cinnabar, Cassiterite, Haematite, Sphalerite, Salt petre, Cuprite

28 How many of the given mineral contain silicon in them? Pyrargyrite (Ruby silver), Kaolinite (China clay) ,Feldspar, Willemite, Limonite, Mica, Beryl

29 Number of ores that have more than one metal or non-metal among: Horn silver, Quick silver, Chromite, Limonite, Cinnabar, Cryolite, Caliche, Diaspore, Fluorspar

Statement Type

Direction *(Q. No. 30) This section is based on Statements I and Statements II. Select the correct answer from the codes given below.*

a. Both Statement I and Statement II are correct and Statement II is the correct explanation of Statement I
b. Both Statement I and Statement II are correct and Statement II is not the correct explanation of Statement I
c. Statement I is correct but Statement II is incorrect
d. Statement II is correct but Statement I is incorrect

30 **Statement I** The copper pyrite ($CuFeS_2$) is the ore of copper.

Statement II The clay ($Al_2O_3 \cdot 2SiO_2 \cdot 2H_2O$) is the ore of aluminium.

DPP-2 Metallurgy Process of Concentration of Ores

Subjective Questions

Working Space

Direction (Q. Nos. 1-5) *These questions are subjective in nature, need to be solved completely on notebook.*

1 What are the three major steps involved in the extraction of a metal after its ore is mined?

2 An ore gave SO_2 on heating with oxygen. How will you concentrate this ore?

3 In metallurgy what do the terms gangue, flux, slag stand for ? Give examples.

4 How does NaCN act as a depressant in preventing ZnS from forming the froth?

5 Name the different type of fluxes used in the concentration of the ores.

Only One Option Correct Type

Direction (Q. Nos. 6-15) *This section contains 10 multiple choice questions. Each question has four choices (a), (b), (c) and (d), out of which ONLY ONE is correct.*

6 Flux is often added to remove impurities from an iron ore in a blast furnace. In the reaction slag and flux are

a. $CaSiO_3$ and SiO_2
b. $CaSiO_3$ and CaO
c. $CaO + SiO_3$
d. SiO_2 and $CaSiO_3$

7 Froth floatation process is based on

a. Specific gravity of the particles
b. Magnetic properties of the particles
c. Wetting properties of the particles
d. Electrical properties of the particles

8 Which one of the following ores is concentrated by chemical leaching method?

a. Galena
b. Copper pyrite
c. Cinnabar
d. Argentite

9 Which of the ore dressing process requires finest size of ore?

a. Froth floatation
b. Gravity separation
c. Magnetic separation
d. All of these

10 Wolframite ($FeWO_4$) impurities present in cassiterite (SnO_2) can be separated by the process of

a. Froth floatation
b. Electromagnetic
c. Zone refining
d. Gravity separation

Working Space

11 In the metallurgical process, the flux used for removing basic impurities is

a. Silica
b. Sodium chloride
c. Lime stone
d. Sodium carbonate

12 Gravity method of ore concentration depends on

a. Preferential wetting properties of particles
b. Difference in the specific gravity of ore and gangue particles
c. Fusion temperature of gangue and ore particles
d. Magnetic properties of ore and gangue particles

13 Sulphide ores of metals are usually concentrated by froth floatation process. Which one of the following sulphide ores offers an exception and is concentrated by leaching ?

a. Sphalerite
b. Argentite
c. Galena
d. Copper pyrites

14 The slag formed in the extraction of iron and copper respectively are

a. $CaSiO_3$ and $FeSiO_3$
b. $FeSiO_3$ and $CaSiO_3$
c. $MgSiO_3$ and $FeSiO_3$
d. $CaCO_3$ and SiO_2

15 In the froth floatation process for benefaction of the ores, the ore particles float because

a. they are light
b. their surface is not easily wetted by water
c. they are insoluble
d. they bear electric charge

One or More than One Options Correct Type

Direction (Q. Nos. 16-20) *This section contains 5 multiple choice questions. Each question has four choices (a), (b), (c) and (d), out of which ONE or MORE THAN ONE are correct.*

16 The ore that is concentrated by froth floatation process is

a. Cinnabar
b. Galena
c. Zinc blende
d. Calamine

17 In froth floatation process the particles of the ore come on the froth because

a. They become water repellent
b. They are light
c. They have lower molecular weight
d. Impurities are more polar nature

18 Purification of bauxite is called

a. Baeyer's process
b. Hall process
c. Serpeck process
d. Hoope's process

19 The common impurities found in bauxite are

a. Fe_2O_3
b. $CaCO_3$
c. SiO_2
d. CaO

20 Which of the following chemicals are involved in froth floatation process ?

a. Collectors
b. Depressant
c. Flux
d. Activators

Working Space

Comprehension Type

Direction (Q. Nos. 21 and 22) *This section contains a paragraph, describing theory, experiments, data, etc. Two questions related to the paragraph have been given. Each question has only one correct answer among the four given options (a), (b), (c) and (d).*

Passage

Most of the ores available in nature contain large amount of impurities i.e. gangue. The pretreatment of ores, based on physical properties and without bringing out any major chemical change in the ore is called ore dressing or ore concentration.

21 Which of the following is not related to ore concentration?

a. Levigation b. Smelting c. Zone refining d. Electrolysis

22 In the purification of bauxite by leaching, the chemicals employed in Baeyer's and Hall's process respectively are

a. Na_2CO_3 and NaOH b. NaOH and Na_2CO_3
c. NaOH and $C + N_2$ d. Na_2CO_3 and NaCN

Matching List Type

Direction (Q. No. 23 and 24) *Choices for the correct combination of elements from Column I and Column II are given as options (a), (b), (c) and (d), out of which one is correct.*

23 Match the Column I with Column II and mark the correct option from the codes given below.

	Column I		Column II
i.	Wilfley table method	p.	$FeCr_2O_4$
ii.	Magnetic separation	q.	$CuFeS_2$
iii.	Froth floatation	r.	SnO_2
iv.	Wet process	s.	$Al_2O_3 \cdot 2H_2O$

Codes

	i	ii	iii	iv		i	ii	iii	iv
a.	r	p	q	s	b.	p	q	r	r
c.	s	r	q	p	d.	p	s	q	r

24 Match the Column I with Column II and mark the correct option from the codes given below.

	Column I		Column II
i.	Electromagnetic separation	p.	Al_2O_3
ii.	Froth floatation	q.	Fe_3O_4
iii.	Baeyer's process	r.	Collectors
iv.	Leaching	s.	Argentite
		t.	$CuFeS_2$

Codes

	i	ii	iii	iv
a.	q	r,t	p,q	s
b.	p	q	r	s
c.	q	r,t	p	s
d.	p	p	s	q

Working Space

One Integer Value Correct Type

Direction (Q. Nos. 25-29) *This section contains 5 questions. When worked out will result is an integer from 0 to 9 (both inclusive).*

25 The number of ores concentrated by froth floatation process among cassiterite, galena, zinc blende, chromite, haematite, copper iron pyrites

26 The number of ores concentrated by leaching or wet process among calamine, tinstone, bauxite, argentite, horn silver, galena, cinnabar, native gold

27 In the leaching of Ag_2S with NaCN the equation is

$Ag_2S + 4NaCN \longrightarrow 2Na[Ag(CN)_2] + Na_2S$. In this change in oxidation state of silver.

28 Number of ores concentrated by Levigation (hydraulic washing)among pitche blende, tinstone, chromite, galena, copper pyrites, argentite.

29 The number of magnetic substances among : tinstone, wolframite, chromite, magnetite, pyrolusite, rutile, magnesite, sphalerite.

Statement Type

Direction (Q. No. 30) *This section is based on Statements I and Statements II. Select the correct answer from the codes given below.*

a. Both Statement I and Statement II are correct and Statement II is the correct explanation of Statement I
b. Both Statement I and Statement II are correct and Statement II is not the correct explanation of Statement I
c. Statement I is correct but Statement II is incorrect
d. Statement II is correct but Statement I is incorrect

30 **Statement I** Cresol and aniline are froth stabilisers.

Statement II Collectors enhance the non-wettability of the ore particles.

DPP-3 Extraction of Crude Metals from Concentrated Ores

Subjective Questions

Working Space

Direction (Q. Nos. 1-5) *These questions are subjective in nature, need to be solved completely on notebook.*

1 How does roasting differ from calcination? Give equations.

2 i. In metallurgy, what does the term 'Roasting' stand for ?
ii. Name the byproduct formed during roasting and give one of its uses.

3 Aluminium metal is frequently used as a reducing agent for the extraction of metals such as chromium, manganese etc. Why ?

4 Copper can be extracted by hydrometallurgy but not zinc. Why ?

5 Explain, Goldschmidt aluminothermic process.

Only One Option Correct Type

Direction (Q. Nos. 6-15) *This section contains 10 multiple choice questions. Each question has four choices (a), (b), (c) and (d), out of which ONLY ONE is correct.*

6 Which of the following is not correctly matched ?
a. Smelting process Cu and Fe
b. Leaching: Extraction of Au, Ag
c. Cyanide process : Extraction of Sn and Pb
d. Electrolyte reduction : Extraction of Al

7 The purpose of adding cryolite is
a. to increase the electrical conductivity of pure aluminium
b. to lower the melting point of Al_2O_3
c. to remove the impurities as slag
d. to increase the Al% in the yield

8 In Mac Arthur Forest method, silver is extracted from the solution of $Na[Ag(CN)_2]$ by the use of
a. Fe b. Mg c. Cu d. Zn

9 In the extraction of copper, the metal formed in the Bessemer's convertor is due to the reaction
a. $Cu_2S \longrightarrow 2Cu + S$
b. $2Cu_2O \longrightarrow 4Cu + O_2$
c. $2Cu_2S + 3O_2 \longrightarrow 2Cu_2O + 2SO_2$
d. $2Cu_2O + Cu_2S \longrightarrow 6Cu + SO_2$

Working Space

10 The nitrate of which metal leaves metallic globule on strongly heating ?

a. $Cu(NO_3)_3$ b. $AgNO_3$ c. $NaNO_3$ d. $Pb(NO_3)_2$

11 In the extraction of gold, the cyanide solution acts as a

a. leaching agent that dissolves the impurities
b. leaching agent that acts as a complex forming reactant
c. flux
d. reducing agent that converts the metal oxide or sulphide into the metal

12 The matte is impure substance obtained during the extraction of

a. iron b. copper c. lead d. aluminium

13 During the extraction of gold the following reactions takes place

$Au + CN^- + H_2O \xrightarrow{O_2} X, X + Zn \longrightarrow Y + Au$

Here, X and Y respectively are

a. $[Au(CN)_2]^-$ and $[Zn(CN)_4]^{2-}$ b. $[Au(CN)_4]^{3-}$ and $[Zn(CN)_4]^{2-}$
c. $[Au(CN)_4]^{2-}$ and $[Zn(CN)_4]^{2-}$ d. $[Au(CN)_2]^-$ and $[Zn(CN)_6]^{4-}$

14 Identify the reaction that does not take place in a blast furnace

a. $CaCO_3 \longrightarrow CaO + CO_2$ b. $CaO + SiO_2 \longrightarrow CaSiO_3$
c. $2Fe_2O_3 + 3C \longrightarrow 4Fe + 3CO$ d. $CO_2 + C \longrightarrow 2CO$

15 Roasting of sulphides gives the gas X as a byproduct. This is a colourless gas with choking smell of burnt sulphur and causes great damage to respiratory organs as a result of acid rain. Its aqueous solution is acidic, acts as reducing agent and its acid is never isolated. The gas X is

a. H_2S b. SO_2 c. CO_2 d. SO_3

One or More than One Options Correct Type

Direction *(Q. Nos. 16-20) This section contains 5 multiple choice questions. Each question has four choices (a), (b), (c) and (d), out of which ONE or MORE THAN ONE are correct.*

16 In the following processes, which are not sulphatising roasting ?

a. $2ZnS + 3O_2 \longrightarrow 2ZnO + 2SO_2$
b. $ZnS + 2O_2 \longrightarrow ZnSO_4$
c. $2Cu_2S + 3O_2 \longrightarrow 2Cu_2O + 2SO_2$
d. $Ag_2S + 2NaCl + 2O_2 \longrightarrow 2AgCl + Na_2SO_4$

17 The metallurgical process involved in the preparation of wrought from haematite are

a. Gravity separation b. Smelting
c. Puddling process d. Bessemerisation

18 Which is correctly matched?

a. Pyrometallurgy = Fe, Zn and Cu b. Electrometallurgy = Al, Mg and Na
c. Hydrometallurgy = Ag and Au d. Smelting = Fe, Cu, Ag

19 Which of the following metal sulphides are converted into metals by self reduction?

a. Cu_2S and PbS b. PbS and HgS c. PbS and ZnS d. FeS_2 and Cu_2S

20 Which of the following are correct process ?

a. $2Fe + Al_2O_3 \longrightarrow 2Al + Fe_2O_3$
b. $ZnO + C \longrightarrow Zn + CO$
c. $Cr_2O_3 + 2Al \longrightarrow 2Cr + Al_2O_3$
d. $2Na[Au(CN)_2] + Zn \longrightarrow Na_2[Zn(CN)_4] + 2Au$

Working Space

Comprehension Type

Direction (Q. Nos. 21 and 22) *This section contains a paragraph, describing theory, experiments, data, etc. Two questions related to the paragraph have been given. Each question has only one correct answer among the four given options (a), (b), (c) and (d).*

Passage

Copper is extracted from copper pyrites. After roasting, the ore is mixed with silica and coke and then smelted in a blast furnace. The matte obtained from the blast furnace is charged into silica lined convertor. Some silica is also added and a hot air blast is blown into the mixture to obtain blister copper which is purified by electro refining.

21 The chemical formula of copper pyrites is

a. $CuFeS_2$ b. Cu_2S c. Cu_2O d. $CuCO_3Cu(OH)_2$

22 Coke is added during smelting

a. to reduce FeO to Fe
b. to reduce Cu_2O to Cu
c. to check the oxidation of FeO to Fe_2O_3
d. None of these

Matching List Type

Direction (Q. Nos. 23 and 24) *Choices for the correct combination of elements from Column I and Column II are given as options (a), (b), (c) and (d), out of which one is correct.*

23 Match the Column I with Column II and mark the correct option from the codes given below.

	Column I		Column II
i.	Cr_2O_3	p.	Reduction with CO after roasting
ii.	Fe_2O_3	q.	Reduction with Al
iii.	Cu_2S	r.	Self reduction after roasting
iv.	ZnS	s.	Reduction with CO

Codes

	i	ii	iii	iv		i	ii	iii	iv
a.	r	s	q	p	b.	q	s	r	p
c.	p	q	s	r	d.	s	p	q	r

24 Match the Column I with Column II and mark the correct option from the codes given below.

	Column I		Column II
i.	Lithium	p.	Roasting of sulphide.
ii.	Copper	q.	Electrolysis of Al_2O_3 with Na_3AlF_6.
iii.	Aluminium	r.	Electrolysis of fused LiCl/KCl.
iv.	Magnesium	s.	Carnallite
		t.	Malachite

Codes

	i	ii	iii	iv		i	ii	iii	iv
a.	t	s	q	p	b.	s	t	p	q
c.	r	p,t	q	s	d.	s	p	q	t

Working Space

One Integer Value Correct Type

Direction (*Q. Nos.* 25-29) *This section contains 5 questions. When worked out will result is an integer from* 0 *to* 9 *(both inclusive).*

25 Among Ca, Pb, Sn, Ag, Mg, Au, Zn, Cu, Al, Na, Hg the number of metals extracted by pyrometallurgy are

26 Among Ca, Pb, Sn, Ag, Mg, Au, Zn, Cu, Al, Na, Hg, K, the number of metals extracted by electrometallurgy are

27 $Au + CN^- + H_2O + O_2 \xrightarrow[\text{Balanced equation}]{} [Au(CN)_2]^- + OH^-$ in this how many CN^- ions are involved in the balanced equation?

28 In the electrolytic extraction of Al at cathode Al metal and at anode F_2 gas is liberated. If the molar ratio of Al and F_2 is $2 : x$, here value of is x

29 In Goldschmidt aluminothermic process, thermit mixture contains ...... parts Fe_2O_3 and one part Al powder.

Statement Type

Direction (*Q. No.* 30) *This section is based on Statement I and Statement II. Select the correct answer from the codes given below.*

a. Both Statement I and Statement II are correct and Statement II is the correct explanation of Statement I
b. Both Statement I and Statement II are correct and Statement II is not the correct explanation of Statement I
c. Statement I is correct but Statement II is incorrect
d. Statement II is correct but Statement I is incorrect

30 **Statement I** Cast iron is obtained by remelting pig iron with scrap iron and coke wing hot air blast.

Statement II In the extraction of silver, silver is extracted as cationic complex.

DPP-4 Thermodynamic Principles of Metallurgy Ellingham's Diagram and Its Applications

Subjective Questions

Working Space

Direction (*Q. Nos.* 1-5) *These questions are subjective in nature, need to be solved completely on notebook.*

1 What are Ellingham diagrams ?

2 i. Out of C and CO which is a better reducing agent at 673 K?

ii. Out of C and CO which is a better reducing agent for ZnO?

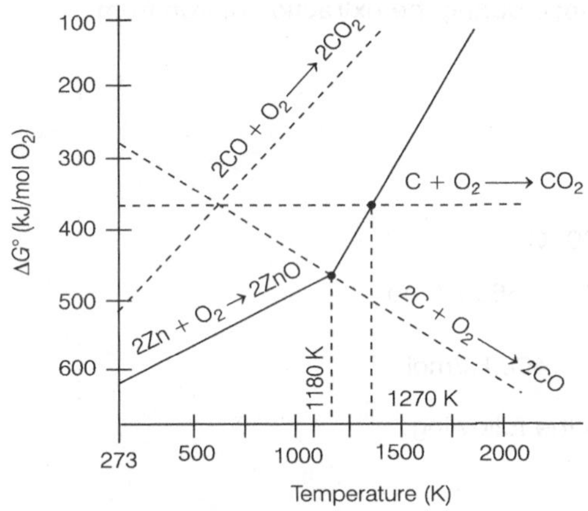

3 i. Suggest a condition under which Mg could reduce alumina.

ii. Predict condition under which Al might be expected to reduce MgO.

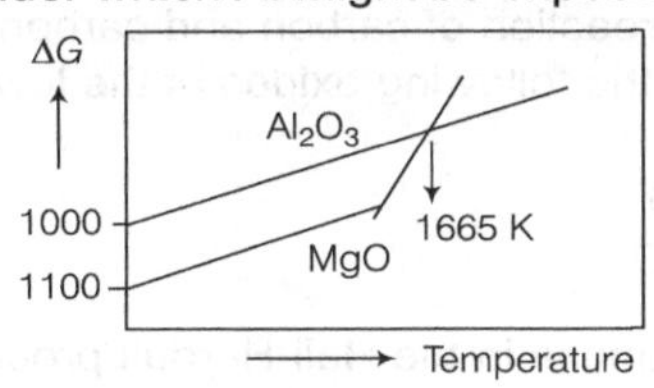

4 Why is the reduction of a metal oxide easier if the metal formed is in liquid state at the temperature of reduction ?

5 Why carbon cannot be used to reduce chromium oxide ?

Only One Option Correct Type

Working Space

Direction *(Q. Nos. 6-14) This section contains 9 multiple choice questions. Each question has four choices (a), (b), (c) and (d), out of which ONLY ONE is correct.*

6 Ellingham diagram represents a graph of

a. ΔG vs T
b. ΔG° vs T
c. ΔS° vs T
d. ΔG vs P

7 ΔG° vs T plot in the Ellingham diagram slopes down for the reaction.

a. $Mg + \frac{1}{2}O_2 \longrightarrow MgO$

b. $2Ag + \frac{1}{2}O_2 \longrightarrow Ag_2O$

c. $C + \frac{1}{2}O_2 \longrightarrow CO$

d. $CO + \frac{1}{2}O_2 \longrightarrow CO_2$

8 When compared to ΔG° for the formation of Al_2O_3 the ΔG° for the formation of Cr_2O_3 is

a. Higher
b. Lower
c. Same
d. Unpredicted

9 Which metal has a greater tendency to form metal oxide?

a. Cr
b. Fe
c. Al
d. Ca

10 The main reaction occurring in blast furnace during the extraction of iron from haematite is

a. $Fe_2O_3 + 3CO \longrightarrow 2Fe + 3CO_2$
b. $FeO + SiO_2 \longrightarrow FeSiO_3$
c. $Fe_2O_3 + 3C \longrightarrow 2Fe + 3CO$
d. $CaO + SiO_2 \longrightarrow CaSiO_3$

11 Consider the following reactions at 10000° C,

(A) $Zn(s) + \frac{1}{2}O_2(g) \longrightarrow ZnO(s); \quad \Delta G^\circ = -360$ kJ/mol

(B) $C(s) + \frac{1}{2}O_2(g) \longrightarrow CO(s); \quad \Delta G^\circ = -460$ kJ/mol

Then choose the correct statement from the following.

a. Zinc can be oxidised by CO
b. ZnO can be reduced by CO
c. ZnO can be reduced by C
d. None of the above

12 According to Ellingham diagrams the oxidation reaction of carbon and carbon monoxide may be used to reduce which one of the following oxides at the lowest temperature?

a. Al_2O_3
b. ZnO
c. MgO
d. Cu_2O

13 The minimum voltage required to electrolyse alumina in the Hall-Heroult process is [Given, $\Delta G_f^\circ(Al_2O_3) = -1520$ kJ/mol and $\Delta G_f^\circ(CO_2) = 394$ kJ/mol]

a. 1.575 V
b. 1.60 V
c. 1.312 V
d. -2.62 V

14 Based on the given information $\Delta G_f^\circ(CaO) = -604.2$ kJ/mol and $\Delta G_f^\circ(Al_2O_3) = -1582$ kJ/mol, which of the following is feasible?

a. $3CaO + 2Al \longrightarrow Al_2O_3 + 3Ca$
b. $3Ca + Al_2O_3 \longrightarrow 2Al + 3CaO$
c. Both (a) and (b)
d. None of the above

One or More than One Options Correct Type

Direction (Q. Nos. 15-19) *This section contains 5 multiple choice questions. Each question has four choices (a), (b), (c) and (d), out of which ONE or MORE THAN ONE are correct.*

15 For the metallurgical process of which of the calcined ore can be reduced by carbon?

a. Haematite
b. Calamine
c. Iron pyrites
d. Zincite

16 For which oxide formation ΔG° is more negative than Cr_2O_3?

a. FeO
b. ZnO
c. Al_2O_3
d. MgO

17 In the extraction of iron from haematite in blast furnace reducing agent

a. At low temperature CO is effective
b. At high temperature carbon is more effective
c. Both carbon and CO are effective at any temperature
d. Carbon is effective at low temperature and CO at high temperature

18 Based on the given information, $\Delta G_f^\circ(CaO) = -604.2$ kJ/mol and $\Delta G_f^\circ(Al_2O_3) = -1582$ kJ/mol, which of the following is correct?

a. $3CaO + 2Al \longrightarrow Al_2O_3 + 3Ca$, is spontaneous
b. $3CaO + 2Al \longrightarrow Al_2O_3 + 3Ca$, is non-spontaneous
c. ΔG° for the reaction $3CaO + 2Al \longrightarrow Al_2O_3 + 3Ca$, is $+230.6$
d. ΔG° for the reaction $3CaO + 2Al \longrightarrow Al_2O_3 + 3Ca$, is -977.8

19 Consider the Ellingham diagram given below. Which metal(s) in the diagram can be extracted at 1100 K using coke as a reducing agent?

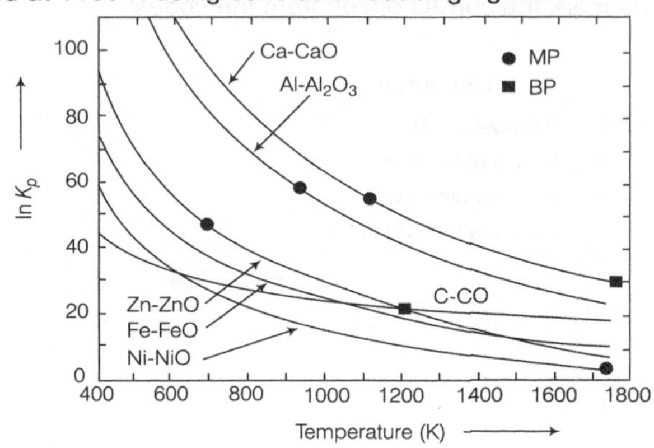

a. Ca, Al
b. Ca, Al, Zn, Fe
c. Zn, Fe
d. Ni

Comprehension Type

Working Space

Direction *(Q. Nos. 20 and 21) This section contains a paragraph, describing theory, experiments, data, etc. Two questions related to the paragraph have been given. Each question has only one correct answer among the four given options (a), (b), (c) and (d).*

Passage

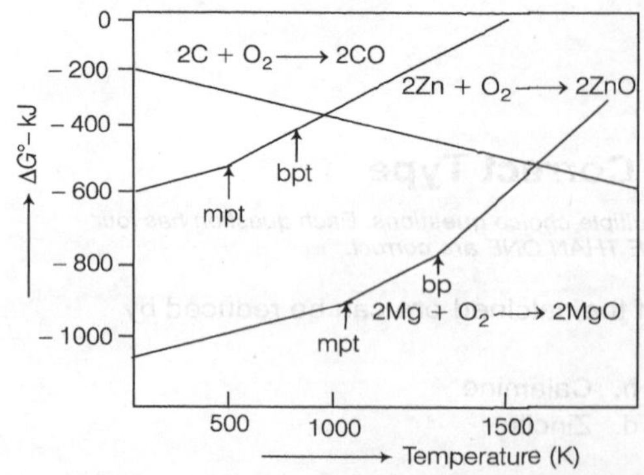

20 To make the reduction $ZnO + C \longrightarrow Zn + CO$ spontaneous, temperature should be

a. < 1000°C
b. > 1000°C
c. < 500°C
d. > 500°C but < 1000°C

21 At 1100°C, which reaction is spontaneous to a maximum extent ?

a. $MgO + C \longrightarrow Mg + CO$
b. $ZnO + C \longrightarrow Zn + CO$
c. $MgO + Zn \longrightarrow Mg + ZnO$
d. $ZnO + Mg \longrightarrow Zn + MgO$

Matching List Type

Direction *(Q. No. 22 and 23) Choices for the correct combination of elements from Column I and Column II are given as options (a), (b), (c) and (d), out of which one is correct.*

22 Match the Column I with Column II and mark the correct option from the codes given below.

	Column I		Column II
i.	CaO	p.	Reduced by Al
ii.	HgO	q.	Most stable oxide
iii.	ZnO	r.	Least stable oxide
iv.	Cr_2O_3	s.	Reduced by carbon

Codes

	i	ii	iii	iv
a.	q	r	s	p
b.	q	s	r	p
c.	p	q	r	s
d.	s	p	q	r

Working Space

23 Match the Column I with Column II and mark the correct option from the codes given below.

	Column I		Column II
i.	Reduced by Al	p.	At equilibrium
ii.	$\Delta G = 0$	q.	Spontaneous
iii.	$\Delta G = +$ ve	r.	FeO
iv.	$\Delta S = +$ ve	s.	non-spontaneous
		t.	CrO

Codes

	i	ii	iii	iv		i	ii	iii	iv
a.	q	p,t	r	p,s	b.	r,t	p	s	q
c.	p	p	q	r,s	d.	q	s	p	r

One Integer Value Correct Type

Direction (Q. Nos. 24-28) *This section contains 5 questions. When worked out will result in an integer from 0 to 9 (both inclusive).*

24 In the Gibbs equation at equilibrium the value of ΔG is

25 The reaction $3CaO + 2Al \longrightarrow Al_2O_3 + 3Ca$ is non-spontaneous. The number of electrons involved are

26 The given reaction $2Al + Cr_2O_3 \longrightarrow Al_2O_3 + 2Cr$ is spontaneous. The number of electrons involved are

27 Among PbO, ZnO, CaO, Al_2O_3, SnO_2, Cr_2O_3, WO_3, HgO the oxides reduced by carbon are

28 $Cu^{2+} + Fe \longrightarrow Cu + Fe^{2+}$ for this reaction to calculate ΔG° the formula $\Delta G^\circ = -nFE^\circ$ is used. Here, value of n is

Statement Type

Direction (Q. Nos. 29 and 30) *This section is based on Statement I and Statement II. Select the correct answer from the codes given below.*

a. Both Statement I and Statement II are correct and Statement II is the correct explanation of Statement I
b. Both Statement I and Statement II are correct and Statement II is not the correct explanation of Statement I
c. Statement I is correct but Statement II is incorrect
d. Statement II is correct but Statement I is incorrect

29 **Statement I** Carbon is good reducing agent for sulphides.

Statement II Carbon is poor reducing agent for iron oxide.

30 **Statement I** Carbon can reduce Fe_2O_3 to Fe at a temperature below 983 K.

Statement II Iron has higher affinity toward oxygen than carbon.

DPP-5 Refining of Crude Metals

Subjective Questions

Working Space

Direction (*Q. Nos. 1-5*) *These questions are subjective in nature, need to be solved completely on notebook.*

1 Sodium is prepared by the electrolysis of molten NaCl but not by the electrolysis of its aqueous solution. Why?

2 What is the principle used in zone refining?

3 Answer the following.

i. $Ni + 4CO \xrightarrow{xK} [Ni(CO)_4] \xrightarrow{yK} Ni + 4CO$ here values of x, y are

ii. $Ti(s) + 2I_2(g) \xrightarrow{aK} TiI_4(g) \xrightarrow{bK} Ti + 2I_2(g)$ values of a, b are

4 What is the difference between electrolysis of $AgNO_3$ using

i. Pt electrodes? ii. silver electrodes?

5 What is Bessemerisation?

Only One Option Correct Type

Direction (*Q. Nos. 6-15*) *This section contains 10 multiple choice questions. Each question has four choices (a), (b), (c) and (d), out of which ONLY ONE is correct.*

6 Poling process is used

a. for the removal of Cu_2O from Cu
b. for the removal of Al_2O_3 from Al
c. for the removal of Fe_2O_3 from Fe
d. In all of these

7 In the electrolytic refining of zinc

a. the impure metal is at the cathode
b. graphite is at the anode
c. acidified zinc sulphate is the electrolyte
d. the metal ion gets reduced at the anode

8 The super pure silicon can be obtained by

a. fractional distillation
b. fractional crystallisation
c. zone refining
d. solvent extraction

9 Purification of aluminium is called

a. Baeyer's process
b. Bosch process
c. Castner's process
d. Hoopes process

10 The processes used in the refining of aluminium and zinc metals respectively are

a. Hoopes' process and fractional distillation
b. Hoopes' process and cupellation
c. Poling and fractional distillation
d. cupellation and fractional distillation

Working Space

11 van Arkel method of purification of metals involves converting the metal to
a. volatile stable compound
b. volatile unstable compound
c. non-volatile stable compound
d. non-volatile unstable compound

12 The most electropositive metals are extracted from their ores by
a. high temperature reduction with carbon
b. self reduction
c. displacement method
d. electrolysis of fused ionic salts

13 Which method of purification is represented by the following equation?

$$\text{Ti}(s) + 2\text{I}_2(g) \xrightarrow{523\text{K}} \text{TiI}_4(g) \xrightarrow{1700\text{K}} \text{Ti}(s) + 2\text{I}_2(g)$$

a. Cupellation
b. Poling
c. van Arkel
d. Zone refining

14 Which of the following process is used in the extractive metallurgy of magnesium?
a. Fused salt electrolysis
b. Self reduction
c. Aqueous solution electrolysis
d. Thermite reduction

15 In the electrolytic refining of copper using $CuSO_4$ as electrolyte the anode reaction is
a. $4OH^- \rightarrow O_2 + 2H_2O + 4e^-$
b. $Cu^{2+} + 2e^- \rightarrow Cu$
c. $Cu \rightarrow Cu^{2+} + 2e^-$
d. $2SO_4^{2-} \rightarrow S_2O_8^{2-} + 2e^-$

One or More than One Options Correct Type

Direction (Q. Nos. 16-20) *This section contains 5 multiple choice questions. Each question has four choices (a), (b), (c) and (d), out of which ONE or MORE THAN ONE are correct.*

16 Which of the following is related to metal purification?
a. Mond's process
b. van Arkel process
c. Zone refining
d. Mac Arthur Forest process

17 Which of the following are correctly matched?
a. Zone refining : Ultra pure Ge
b. Mond's process : Pt
c. Cupellation : Ag
d. Poling : Cu

18 Regarding blister copper incorrect statements are
a. Pure copper
b. An ore of copper
c. An alloy of copper
d. Copper containing 2% impurities

19 Regarding electrolytic refining of silver which is correct?
a. Impure silver metal as anode
b. Pure silver as cathode
c. Impure metal as cathode
d. $AgNO_3$ solution as electrolyte

20 In the chromatography method of purification of metals the adsorbent used in the column
a. alumina
b. slaked lime
c. silica gel
d. alum

Comprehension Type

Working Space

Direction *(Q. Nos. 21and 22) This section contains a paragraph, describing theory, experiments, data, etc. Two questions related to the paragraph have been given. Each question has only one correct answer among the four given options (a), (b), (c) and (d).*

Passage

Impure metals are purified by different methods in order to remove the undesired impurities. These methods depend on the nature of metal and nature of impurities. Liquation is used for purifying metals with low melting points like Bi, Sn, Pb, Hg etc. Distillation is used for volatile metals like Zn, Cd, Hg. Cupellation is used to purify silver metal with lead impurities. Electrolytic refining (Cu, Ag, Mg, Al) Zone refining (Si, Ge, Ga) Mond's process (Ni) van Arkel process (Zr, Ti).

21 In the extraction of nickel by Mond's process, the metal is obtained by

a. electrochemical reduction
b. thermal decomposition
c. chemical reduction by aluminium
d. chemical reduction by carbon

22 Suggest a method of purification for obtaining highly pure solar grade silicon used as a semiconductor.

a. Electrochemical refining
b. Mond's process
c. Zone refining method
d. van Arkel method

Matching List Type

Direction *(Q. No. 23 and 24) Choices for the correct combination of elements from Column I and Column II are given as options (a), (b), (c) and (d), out of which one is correct.*

23 Match the Column I with Column II and mark the correct option from the codes given below.

	Column I		Column II
i.	Iron, copper	p.	Poling
ii.	Zr and Ti	q.	Bessemerisation
iii.	Pb and Sn	r.	van Arkel
iv.	Cu and Sn	s.	Liquation

Codes

	i	ii	iii	iv
a.	q	r	s	p
b.	p	s	q	r
c.	q	p	p	s
d.	p	q	r	r

24 Match the Column I with Column II and mark the correct option from the codes given below.

	Column I		Column II
i.	$Ni(CO)_4$	p.	Liquation
ii.	Less fusible impurities	q.	Mond's process
iii.	Blister copper	r.	Bessemerisation
iv.	Pig iron	s.	Poling

Codes

	i	ii	iii	iv		i	ii	iii	iv
a.	q	s	p	r	b.	q	p	s	r
c.	r	p	q	r	d.	p	q	r	s

Working Space

One Integer Value Correct Type

Direction (*Q. Nos.* 25-29) *This section contains* 5 *questions. When worked out will result in an integer from* 0 *to* 9 *(both inclusive).*

25 In auto reduction process of copper, the part of cuprous sulphide is oxidised with cuprous oxide to form a free copper metal. The balanced equation is as follows

$$Cu_2S + 2Cu_2O \longrightarrow xCu + SO_2$$

The value of x is

26 The central atom in the stable volatile complex formed in the Mond's method has the oxidation number

27 Number of ligands in the stable volatile complex formed during the extraction of Ni by Mond's process.

28 Sea water contains 1297.3g of Mg^{2+} per metric ton. How much of slaked lime in kg, must be added to 1.0 metric ton of sea water to precipitate all of the Mg^{2+} ion?

29 The percentage impurity in 23 carat gold is nearly

Statement Type

Direction (*Q. No.* 30) *This section is based on Statement I and Statement II. Select the correct answer from the codes given below.*

a. Both Statement I and Statement II are correct and Statement II is the correct explanation of Statement I
b. Both Statement I and Statement II are correct and Statement II is not the correct explanation of Statement I
c. Statement I is correct but Statement II is incorrect
d. Statement II is correct but Statement I is incorrect

30 **Statement I** In electrorefining process cathode is made up of a pure strip of the same metal.

Statement II Blister copper contains 50% impurities.

DPP-6 Metallurgy of Some Important Metals-I (Cu, Fe, Ag)

Subjective Questions

Direction (Q. Nos. 1-5) *These questions are subjective in nature, need to be solved completely on notebook.*

Working Space

1 Among Fe, Cu and Ag which one liberates hydrogen with dilute acid? Give reason.

2 Inorganic compound *A* gave the following reactions.
   i. The compound on heating gave a residue and a gaseous mixture of NO_2 and O_2.
   ii. Aqueous solution of *A* gave a white precipitate with sodium chloride solution. The precipitate dissolves in ammonia solution. Find *A* and give reactions.

3 How a silver halide forms a complex with hypo solution?

4 What is passivity? Does iron exhibit this property explain?

5 What happens when green vitriol is heated?

Only One Option Correct Type

Direction (Q. Nos. 6-15) *This section contains 10 multiple choice questions. Each question has four choices (a), (b), (c) and (d), out of which ONLY ONE is correct.*

6 Which one of the following is the major impurity in pig iron?
a. Graphite b. Sulphur c. Silicon d. Phosphorus

7 In smelting the furnace most suitable is
a. Blast furnace b. Reverberatory furnace
c. Open hearth furnace d. Muffle furnace

8 Silver from argentiferous lead is obtained by
a. Pattinson's process b. Parke's process
c. Cyanide process d. Both (a) and (b)

9 Which of the following has lowest carbon content?
a. Cast iron b. Wrought iron c. Stainless steel d. Chrome steel

10 During the process of electrolytic refining of copper, some metals present as impurity settle as anode mud. These are
a. Ag and Au b. Fe and Ni c. Sn and Ag d. Pb and Zn

11 $2CuCO_3 \cdot Cu(OH)_2$ is the formula of
a. Azurite b. Malachite c. Siderite d. Chalcopyrite

12 The temperature of the slag zone in the metallurgy of iron using blast furnace is
a. 400–700°C b. 1200–1500°C c. 800–1000°C d. 1500–1600°C

Working Space

13 The Puddler's candles are produced due to the burning of

a. carbon b. carbon monoxide
c. hydrogen d. None of these

14 In the extraction of copper, the metal formed in the Bessemers convertor is due to the reaction

a. $Cu_2S \longrightarrow 2Cu + S$ b. $2Cu_2O \longrightarrow 4Cu + O_2$
c. $2Cu_2S + 3O_2 \longrightarrow 2Cu_2O + 2SO_2$ d. $2Cu_2O + Cu_2S \longrightarrow 6Cu + SO_2$

15 Native silver metal forms a water soluble complex with a dilute aqueous solution of NaCN in the presence of

a. nitrogen b. oxygen c. carbon dioxide d. argon

One or More than One Options Correct Type

Direction *(Q. Nos. 16-20) This section contains 5 multiple choice questions. Each question has four choices (a), (b), (c) and (d), out of which ONE or MORE THAN ONE are correct.*

16 Which one of the following reactions, take place during the extraction of iron?

a. $FeO + SiO_2 \longrightarrow FeSiO_3$ b. $Fe_2O_3 + 3O_2 \longrightarrow 2Fe + 3CO$
c. $FeO + CO \longrightarrow Fe + CO_2$ d. $CO_2 + C \longrightarrow 2CO$

17 Which are the ores of iron?

a. Limonite b. Cassiterite c. Magnetite d. Green vitriol

18 The statements that are correct among the given are

a. a furnace lined with haematite is used to convert cast iron to wrought iron
b. collectors enhance the wettability of mineral particles during froth floatation
c. in vapour phase refining, metal should form a volatile compound
d. copper from its low grade ores is extracted by hydrometallurgy

19 Which of the following process is not based on the formation of complex compounds?

a. Belzean process b. Mac Arthur Forest process
c. Poling process d. Mond's process

20 Which of the following are white in colour?

a. $CuSO_4 \cdot 5H_2O$ b. Cu_2I_2 c. Cu_2O d. $CuSO_4$

Comprehension Type

Direction *(Q. Nos. 21-22) This section contains a paragraph, describing theory, experiments, data, etc. Two questions related to the paragraph have been given. Each question has only one correct answer among the four given options (a), (b), (c) and (d).*

Passage

Steel is the most important commercial variety of iron. The % of carbon in this form of iron is midway between that of cast iron and wrought iron i.e. 0.25 to 2 %. There are many varieties of steel depending on the amount of carbon present in it such as mild steels, hard steels and special steels. Steel mixed with small amount of Ni, Co, Cr, W, Mo, Mn etc., acquire special properties. Such products are called alloy steels.

21 Which of the following is used to make permanent magnets?

a. Invar b. Alnico c. Stainless steel d. Manganese steel

22 Steel used in making pendulums, measuring tapes

a. invar b. alnico c. stainless steel d. manganese steel

Working Space

Matching List Type

Direction *(Q. No. 23 and 24) Choices for the correct combination of elements from Column I and Column II are given as options (a), (b), (c) and (d), out of which one is correct.*

23 Match the Column I with Column II and mark the correct option code.

	Column I		Column II
i.	Fool's gold	p.	Fe + Ni
ii.	Stainless steel	q.	FeS_2
iii.	German silver	r.	Fe + Cr + Ni
iv.	Invar steel	s.	Cu + Zn + Ni

Codes

	i	ii	iii	iv		i	ii	iii	iv
a.	q	p	s	p	b.	p	q	s	r
c.	q	s	p	r	d.	r	p	q	s

24 Match the Column I with Column II and mark the correct option code.

	Column I		Column II
i.	Silver	p.	Cast iron
ii.	Blast furnace	q.	Carbon
iii.	Smelting	r.	Cyanide process
iv.	Chalcocite	s.	Native metal
		t.	Copper

Codes

	i	ii	iii	iv		i	ii	iii	iv
a.	r,s	r	q	p,s	b.	r,s	p	r	t
c.	p,q	s,t	q	p,s	d.	p	q,t	q,s	p

One Integer Value Correct Type

Direction *(Q. Nos. 25-29) This section contains 5 questions. When worked out will result is an integer from 0 to 9 (both inclusive).*

25 The percentage of copper in duralumin is

26 The percentage of silver in German silver is

27 Mohr's salt is a double salt with the formula $FeSO_4(NH_4)_2SO_4xH_2O$. Here, value of x is

28 Number of unpaired electrons in the complex $[Ag(CN)_2]^-$ are

29 The number of silver containing compounds among Nessler's reagent, Tollen's reagent, Baeyer's reagent, Corrosive sublimate, Horn silver, Ruby silver, Calomel.

Statement Type

Direction *(Q. No. 30) This section is based on Statement I and Statement II. Select the correct answer from the codes given below.*

a. Both Statement I and Statement II are correct and Statement II is the correct explanation of Statement I
b. Both Statement I and Statement II are correct and Statement II is not the correct explanation of Statement I
c. Statement I is correct but Statement II is incorrect
d. Statement II is correct but Statement I is incorrect

30 **Statement I** Silver is obtained from its cyanide complex during leaching.
Statement II Copper can be extracted by poling process when it contain Cu_2O as impurities.

DPP-7 Metallurgy of Some Important Metals-II (Al, Zn, Au)

Subjective Questions

Working Space

Direction *(Q. Nos. 1-5) These questions are subjective in nature, need to be solved completely on notebook.*

1 Why ZnO is called philosopher's wool ?

2 What is meant by galvanising ?

3 What is Rinmann's green ?

4 How is the purity of gold expressed ?

5 i. Why Al metal is corrosion resistance ?
ii. Why Al metal is the most commonly used material in large power transmission lines?

Only One Option Correct Type

Direction *(Q. Nos. 6-15) This section contains 10 multiple choice questions. Each question has four choices (a), (b), (c) and (d), out of which ONLY ONE is correct.*

6 In the electrolytic refining of zinc

a. the impure metal is at the cathode
b. graphite is at the anode
c. acidified zinc sulphate is the electrolyte
d. the metal ion gets reduced at the anode

7 When compared to $\Delta G°$ for the formation of Al_2O_3 the $\Delta G°$ for the formation of Cr_2O_3 is

a. higher b. lower c. same d. unpredicted

8 During the extraction of gold the following reactions take place

$Au + CN^- + H_2O \xrightarrow{O_2} X$, $X + Zn \longrightarrow Y + Au$. Here, X and Y respectively are

a. $[Au(CN)_2]^-$ and $[Zn(CN)_4]^{2-}$ b. $[Au(CN)_4]^{3-}$ and $[Zn(CN)_4]^{2-}$
c. $[Au(CN)_4]^{2-}$ and $[Zn(CN)_4]^{2-}$ d. $[Au(CN)_2]^-$ and $[Zn(CN)_6]^{4-}$

9 Gold is extracted by hydrometallurgical process based on its property

a. of being electropositive
b. of being less reactive
c. to form complexes which are water soluble
d. to form salts which are water soluble

Working Space

10 During Hoope's process for electrolytic refining of Al, the middle layer is of
a. pure Al
b. impure aluminium
c. cryolite and BaF_2
d. alloys of Al, Ca and Si

11 Thermite is a mixture of iron oxide and
a. zinc powder
b. potassium metal
c. sodium shavings
d. aluminium powder

12 Which is an alloy of aluminium?
a. Magnalium b. Duralumin c. Brass d. Both (a) and (b)

13 Which one of the following metals has the largest abundance in the earth's crust?
a. Aluminium b. Calcium c. Magnesium d. Sodium

14 The processes used in the refining of aluminium and zinc metals respectively are
a. Hoope's process and fractional distillation
b. Hoope's process and cupellation
c. Poling and fractional distillation
d. cupellation and fractional distillation

15 In sulphatizing roasting of ZnS products are
a. $ZnO + SO_2$
b. $ZnO + ZnSO_4 + SO_2$
c. $ZnCl_2$
d. $Zn + SO_2$

One or More than One Options Correct Type

Direction *(Q. Nos. 16-20) This section contains 5 multiple choice questions. Each question has four choices (a), (b), (c) and (d), out of which ONE or MORE THAN ONE are correct.*

16 Purification of alumina is called
a. Baeyer's process
b. Hall process
c. Mond's process
d. Hoope's process

17 Among the following statements the correct one is
a. calamine and siderite are carbonate ores
b. argentite and cuprite are oxides
c. zinc blende and pyrites are sulphide ores
d. malachite and azurite are ores of copper

18 The alloys having zinc metal are
a. German silver b. Bell metal c. Gun metal d. Brass

19 Which is a zinc compound?
a. White vitriol
b. Fool's gold
c. Philosopher's wool
d. Calomel

20 Which is not a mineral of Al?
a. Spinel b. Feldspar c. Corundum d. Magnalium

Comprehension Type

Direction *(Q. Nos. 21 and 22) This section contains a paragraph, describing theory, experiments, data, etc. Two questions related to the paragraph have been given. Each question has only one correct answer among the four given options (a), (b), (c) and (d).*

Passage

The chief ore of zinc is ZnS.The ore is purified by floatation process and then heated in air which converts ZnS into ZnO. ZnO can be reduced with coke in Belgian process to get zinc which is purified by electrolysis.

Alternately zinc can be extracted from calamine .This after purification by levigation, subjected to calcination to get oxide. This on reduction by Belgian method followed by electrolytic refining yields pure zinc.

Working Space

21 Zinc oxide is converted into zinc by

a. reduction with carbon
b. reduction with CO
c. heating to high temperature
d. Any one of them

22 Which of the following graph in the given figure is the Ellingham diagram of ZnO?

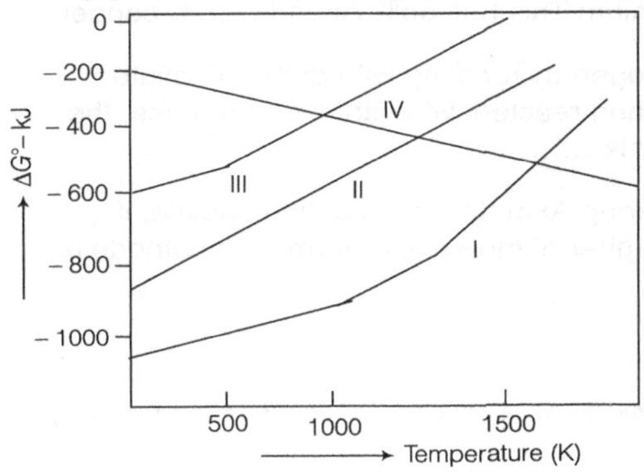

a. Only I
b. Only II
c. Only III
d. Only IV

Matching List Type

Direction *(Q. Nos. 23 and 24) Choices for the correct combination of elements from Column I and Column II are given as options (a), (b), (c) and (d), out of which one is correct.*

23 Match the Column I with Column II and mark the correct option from the codes given below.

	Column I		Column II
i.	Gold	p.	Pyrometallurgy
ii.	Al from red bauxite	q.	Hydrometallurgy
iii.	Zinc	r.	Baeyer and electrometallurgy
iv.	Al from white bauxite	s.	Serpeck and electrometallurgy

Codes

	i	ii	iii	iv		i	ii	iii	iv
a.	q	r	p	s	b.	p	q	r	s
c.	s	p	q	r	d.	q	s	p	r

24 Match the Column I with Column II and mark the correct option from the codes given below.

	Column I		Column II
i.	Cyanide process	p.	Zinc smelting
ii.	Belgian process	q.	$Na[Au(CN)_2]$
iii.	Parkes process	r.	Smelting aluminium
iv.	Hall process	s.	Liquid-liquid extraction

Codes

	i	ii	iii	iv		i	ii	iii	iv
a.	p	q	s	r	b.	s	q	p	r
c.	q	p	s	r	d.	r	s	q	p

Working Space

One Integer Value Correct Type

***Direction** (Q. Nos. 25-29) This section contains 5 questions. When worked out will result in an integer from 0 to 9 (both inclusive).*

25 Gibbsite formula is $Al_2O_3 \cdot xH_2O$. Here, value of x is

26 Gold is soluble in aqua-regia which is conc. HNO_3 and conc. HCl in 1: x ratio. Then the value of x is

27 Duralumin is an important alloy of aluminium.This has 95% Al and ......% copper.

28 Al is amphoteric metal and liberate hydrogen by reacting with both with dilute acids and bases . 2 moles of Al metal when reacted with either acid or base, the number of moles of dihydrogen liberated is ......

29 During the Hall-Heroult process of preparing Al metal from purified bauxite, if 3 moles of O_2 is formed at anode the number of moles of Al formed at cathode is

Statement Type

***Direction** (Q. No. 30) This section is based on Statement I and Statement II. Select the correct answer from the codes given below.*

a. Both Statement I and Statement II are correct and Statement II is the correct explanation of Statement I
b. Both Statement I and Statement II are correct but Statement II is not the correct explanation of Statement I
c. Statement I is correct but Statement II is incorrect
d. Statement II is correct but Statement I is incorrect

30 **Statement I** Many gemstones are impure forms of alumina (Al_2O_3).

Statement II In cyanide process gold (Au) is treated with a conc. solution of sodium or potassium cyanide.

SET - 1

Revisal Problems
for JEE Main

Only One Option Correct Type

Working Space

Direction (Q. Nos. 1-30) This section contains 30 multiple choice questions. Each question has four choices (a), (b), (c) and (d), out of which ONLY ONE is correct.

1 The most electropositive metals are isolated from their ores by

a. high temperature reduction with carbon
b. self-reduction
c. thermal decomposition
d. electrolysis of fused ionic salts

2 Which of the following does not involve heating?

a. Calcination
b. Roasting
c. Smelting
d. Leaching

3 Which is incorrectly matched?

a. Chalcocite = $CuFeS_2$
b. Horn silver = AgCl
c. Sylvine = KCl
d. Ruby silver = $3Ag_2S \cdot Sb_2S_3$

4 Magnesium is extracted from sea water by Dow process. In this process to separate Mg, the process carried is

a. Na_2CO_3 is added to sea water to get $MgCO_3$
b. $Ca(OH)_2$ is added to get $Mg(OH)_2$
c. Na_2SO_4 is added to get $MgSO_4$
d. NaCl is added to get $MgCl_2$

5 Roasted iron ore is mixed with coke and lime stone and smelted in blast furnace. Which among the following does not represent the correct answer about the reactions taking place in the blast furnace?

a. In the lower part, called zone of fusion temperature (1773 K) coke burns to CO_2 producing 97000 cal of heat
b. In the middle part, lime stone decomposes to give CO_2 and lime which acts as a reducing agent
c. In the middle part, of the furnace called zone of heat absorption (temperature1073–1273 K) CO_2 is reduced to CO
d. At the top is the zone of reduction (temperature 673–973 K). Here iron oxide is reduced to spongy iron by CO

6 Zone refining is based on the principle of

a. fractional distillation
b. partition coefficient
c. fractional crystallisation
d. chromatographic separation

7 Addition of high proportions of manganese makes steel useful in making rails of railroads, because manganese

(I) gives hardness to steel
(II) helps the formation of oxides of iron
(III) can remove oxygen and sulphur
(IV) can show highest oxidation state of +7

a. (I), (II) and (III)
b. (I) and (III)
c. (II) and (IV)
d. (I),(III) and (IV)

8 Among the following group of oxides, the group that cannot be reduced by carbon to give the respective metals

a. Cu_2O, SnO_2
b. Fe_2O_3, ZnO
c. CuO, K_2O
d. PbO, ZnO

Working Space

9 Selection of temperature to carry out a reduction process depends so as to make

a. ΔG negative
b. ΔG positive
c. ΔH negative
d. ΔH positive

10 The common impurities present in the bauxite ore are

a. Fe_2O_3 and CuO
b. Fe_2O_3 and PbO
c. Fe_2O_3 and SiO_2
d. SiO_2 and CuO

11 The major role of fluorspar (CaF_2) which is added in small quantity in the electrolytic reduction of alumina dissolved in fused cryolite (Na_3AlF_6) is/are

a. as a catalyst
b. to make the fused mixture for reduction
c. to lower the temperature of melting
d. to decrease the rate of oxidation of carbon at the anode

12 Which of the following is not the correct process ?

a. $Fe + Al_2O_3 \longrightarrow 2Al + Fe_2O_3$
b. $ZnO + C \longrightarrow Zn + CO$
c. $Cr_2O_3 + 2Al \longrightarrow 2Cr + Al_2O_3$
d. $2Ca_3(PO_4)_2 + 6SiO_2 + 10C \longrightarrow 6CaSiO_3 + P_4 + 10CO$

13 In the extraction of Ag and Au using a KCN solution, cyanide ions react with metal ions as

a. an oxidising agent
b. a reducing agent
c. a complexing agent
d. Both (b) and (c)

14 Which of the following are correctly matched ?

a. Zone refining : Ultra pure Ge
b. Leaching : Extraction of Au, Ag
c. Cyanide process : Extraction of Pb
d. Electrolyte reduction : Extraction of Al

15 Which of the following minerals do not contain lead ?

a. Anglesite
b. Cerussite
c. Heavy spar
d. Galena

16 Spiegel (or) spiegeleisen, used in the manufacture of steel by the Bessemer process, is an alloy of

a. Iron, chromium and carbon
b. Iron, nickel and carbon
c. Iron, tungsten and carbon
d. Iron, manganese and carbon

17 Among the following statements the incorrect one is

a. calamine and siderite are carbonates
b. argentite and cuprite are oxides
c. zinc blende and iron pyrites are sulphides
d. malachite and azurite are ores of copper

18 In the commercial electrochemical process for aluminium extraction the electrolyte used is

a. $Al(OH)_3$ in NaOH
b. an aqueous solution of $Al_2(SO_4)_3$
c. a molten mixture of Al_2O_3 and Na_3AlF_6
d. a molten mixture of Al_2O_3 and $Al(OH)_3$

19 The metal that cannot be obtained by electrolysis of an aqueous solution of its salt is

a. Ag
b. Ca
c. Au
d. Cu

20 When a metal is precipitated by adding a more electropositive metal, then the process is called

a. Hydrometallurgy
b. Aluminothermic process
c. Self-reduction
d. Poling

21 In the extraction of lead the Parke's process is used to remove the metal

a. Cu
b. Ag
c. Zn
d. Mn

22 In which of the following isolations no reducing agent is required?

a. Iron from haematite
b. Aluminium from bauxite
c. Mercury from cinnabar
d. Zinc from zinc blende

Working Space

23 Which of the following reaction(s) does not occur in Bessemers convertor ?

a. $2Cu_2S + 5O_2 \longrightarrow CuSO_4 + 2CuO$
b. $2Cu_2S + 3O_2 \longrightarrow 2Cu_2O + 2SO_2 \uparrow$
c. $Cu_2S + 2Cu_2O \longrightarrow 6Cu + SO_2$
d. $FeO + SiO_2 \longrightarrow FeSiO_3$

24 Refactory materials are generally used in furnaces because

a. they are chemically inert
d. they can withstand high temperature
c. they do not contain impurities
d. they decrease melting point of ore

25 In the leaching of Ag_2S with NaCN, a stream of air is also passed. It is because of

a. reversible nature of reaction between Ag_2S and NaCN
b. to oxidise Na_2S formed into Na_2SO_4 and S
c. Both (a) and (b)
d. Both are suitable

26 Which is used to extract zinc?

a. Bauxite or cassiterite
b. Magnesite or haematite
c. Malachite or calamine
d. Sphalerite or calamine

27 Identify *x*, *y*, *z* in the following process; Metal sulphide $\xrightarrow{x}$ metal oxide $\xrightarrow{y}$ impure metal $\xrightarrow{z}$ pure metal

a. Roasting, smelting, electrolysis
b. Roasting, calcination, smelting
c. Calcination, auto-reduction, bessemerisation
d. None of the above

28 Element carbon is most likely to reduce Fe_2O_3 in which of the following temperature ranges ?

a. 500–1000 K
b. 500–1500 K
c. Below 1000 K
d. Above 1000 K

29 The reverberatory furnace used for preparation of wrought iron, its hearth is lined with

a. silicon dioxide
b. iron oxide
c. magnesium oxide
d. aluminium oxide

30 Th electrolyte used in Castner's process of sodium extraction is

a. aqueous Na_2CO_3
b. aqueous NaOH
c. $NaCl + CaCl_2$
d. fused anhydrous NaOH

SET - 2

Revisal Problem
for JEE Advanced

Only One Option Correct Type

Direction (Q. Nos. 1-10) *This section contains 10 multiple choice questions. Each question has four choices (a), (b), (c) and (d), out of which ONLY ONE is correct.*

Working Space

1 Observe the following statements regarding purification of bauxite
i. During Hall's process, silica, is removed as Si (vapour).
ii. Bauxite ore contaminated with iron oxide is purified in Baeyer's process.
iii. During Serpeck's process silica, is removed as Si (vapour).
The correct answer is
a. I, II and III b. Both I and II c. Both I and III d. Both II and III

2 Puddling process is used to convert pig iron into
a. Cast iron b. Wrought iron c. Cast iron d. Steel

3 The impurity present in ruby and sapphire are
a. Fe and Cr b. Cr and Co c. Cr and Al d. Co and Fe

4 Which of the following is froth stabilizer in froth floatation process?
a. Pine oil and xanthate b. Aniline and xanthate
c. Xanthate and NaCN d. Cresols and aniline

5 Sulphide ores of metals are usually concentrated by froth floatation process. Which one of the following sulphide ores offers an exception and is concentrated by leaching?
a. Sphalerite b. Argentite c. Galena d. Copper pyrites

6 The process used in refining blister copper is
a. Zone refining b. Hoope's c. Poling process d. Cupellation

7 The pair of metals obtained by heating their oxides to higher temperatures
a. ZnO and HgO b. Ag_2O and HgO c. Cu_2O and ZnO d. Cu_2O and Fe_2O_3

8 After partial roasting, the sulphide of copper is reduced by
a. Cyanide process b. Electrolysis
c. Reduction with carbon d. Self-reduction

9 Type metal is an alloy containing
a. Pb, Sb and Sn b. Al, Sb and Pb
c. Pb, Cu and Sn d. Pb, Sb and Si

10 The formulae of the soluble complexes formed during the extraction of Ag and Au by the cyanide process respectively are
a. $Na_2[Ag(S_2O_3)_4]$ and $H[AuCl_4]$ b. $Na[Ag(CN)_2]$ and $Na[Au(CN)_2]$
c. $Na_2[AgCN)_4]$ and $Na_4[Au_6(CN)_5]$ d. $Na_3[Ag(CN)_2]$ and $Na_2[Au(CN)_4]$

One or More than One Options Correct Type

Direction (Q. Nos. 11-20) *This section contains 10 multiple choice questions. Each question has four choices (a), (b), (c) and (d), out of which ONE or MORE THAN ONE are correct.*

11 ZnS and PbS present in a sulphide ore can be separated by using selectively suitable chemical. The incorrect statement in this is
a. Depressant is potassium ethyl xanthate
b. Depressant is NaCN
c. In this ZnS is selectively prevented by NaCN from coming to the froth
d. In this PbS is selectively by NaCN from coming to the froth

Working Space

12 Regarding leaching of bauxite ore which of the following is incorrect?

a. Leaching of bauxite is carried with NaOH
b. In this insoluble $Al(OH)_3$ is formed
c. In this soluble complex of aluminium $Na[Al(OH)_4]$ is formed
d. By leaching the common impurities iron oxide, SiO_2 can be removed

13 In the metal extraction which of the following is subjected to calcinations?

a. Dolomite b. Calamine c. Galena d. Sphalerite

14 Which of the following process is based on the formation of complex compounds?

a. Belgian process b. Mac Arthur Forest process
c. Poling d. Mond's process

15 Lapis lazuli is

a. ferrous sulphate b. copper sulphate
c. sodium alumino silicate d. zinc sulphate

16 $\Delta G°$ *vs* T plot in the Ellingham diagram slopes upwards for the reaction

a. $Mg + \frac{1}{2}O_2 \rightarrow MgO$ b. $2Ag + \frac{1}{2}O_2 \rightarrow Ag_2O$
c. $C + \frac{1}{2}O_2 \rightarrow CO$ d. $CO + \frac{1}{2}O_2 \rightarrow CO_2$

17 Lithopone, a white pigment is a mixture of

a. $ZnSO_4$ b. $BaSO_4$ c. ZnS d. BaS

18 Which is incorrect match?

a. Nessler's reagent $= K_2[HgI_4]$ b. Corrosive sublimate $= Hg_2Cl_2$
c. Litharge $= PbO_2$ d. Lunar caustic $= AgCl$

19 Which of the following is a hydrated oxide mineral?

a. Limonite b. Cassiterite c. Bauxite d. Haematite

20 The Ellingham diagram of a certain metal M is represented in the figure

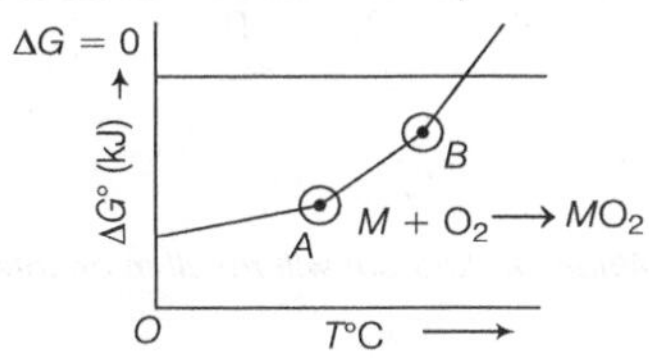

The points A and B respectively corresponds to

a. oxidation and reduction of metal
b. beginning and ending of reduction process
c. minimum and maximum value of ΔG
d. melting and boiling points of metal

Comprehension Type

Direction *(Q. Nos. 21 and 22) This section contains a paragraph, each describing theory, experiments, data, etc. Two questions related to the paragraph have been given. Each question has only one correct answer among the four given options (a), (b), (c) and (d).*

Passage

Metal M is extracted from its ore, which is a double salt. It is a salt of M and potassium. The ore is concentrated and water of hydration present in the double salt is removed. The ore is then reduced to get the metal. The metal so produced used to make Grignard reagent.

21 The ore of the metal in the passage

a. Magnesite b. Epsomite
c. Carnalite d. Spodumene

22 The flame colouration given by the ore mentioned in the passage

a. No colour b. Lilac c. Crimson red d. Brick red

Working Space

Matching List Type

***Direction** (Q. No. 23 and 24) Choices for the correct combination of elements from Column I and Column II are given as options (a), (b), (c) and (d), out of which one is correct.*

23 Match the Column I with Column II and mark the correct option from the codes given below.

	Column I		Column II
i.	Mond's process	p.	$Cr_2O_3 + 2Al \longrightarrow 2Cr + Al_2O_3$
ii.	van Arkel process	q.	$TiCl_4 + 2Mg \longrightarrow Ti + 2MgCl_2$
iii.	Thermite process	r.	$Ni(CO)_4 \xrightarrow{\Delta} Ni + 4CO$
iv.	Kroll's process	s.	$ZrI_4 \xrightarrow{\Delta} Zr + 2I_2$

Codes

	i	ii	iii	iv		i	ii	iii	iv
a.	p	s	r	q	b.	r	s	p	q
c.	q	p	s	r	d.	s	q	p	r

24 Match the Column I with Column II and mark the correct option from the codes given below.

	Column I		Column II
i.	Malachite	p.	Extraction of gold
ii.	Mond process	q.	$CuCO_3 \cdot Cu(OH)_2$
iii.	Cyanide process	r.	Impure metal to volatile complex
iv.	Blistered copper	s.	$2Cu_2O + Cu_2S \longrightarrow 6Cu + SO_2$

Codes

	i	ii	iii	iv		i	ii	iii	iv
a.	p	q	r	s	b.	q	r	p	s
c.	r	p	p	s	d.	q	s	p	p

One Integer Value Correct Type

***Direction** (Q. Nos. 25-29) This section contains 5 questions. When worked out will result in an integer from 0 to 9 (both inclusive).*

25 The value of '*n*' in the molecular formula $Be_nAl_2Si_6O_{18}$ is ....

26 In blue vitriol the number of water molecules directly bonded to Cu^{2+} ion are

27 The number of oxide ores among Rutile, Beryl, Tinstone, Pyrolusite, Chromite, Corundum, Chalcosite, Gypsum, Cryolite, Feldspar and Pitche blende

28 In German silver the percentage of silver is

29 How many of the following metals can be refined by vapour phase refining?
Zr, Zn, Cd, Hg, Ni, Co, Pt, Fe, Ti

Statement Type

***Direction** (Q. No. 30) This section is based on Statement I and Statement II. Select the correct answer from the codes given below.*

a. Both Statement I and Statement II are correct and Statement II is the correct explanation of Statement I
b. Both Statement I and Statement II are correct and Statement II is not the correct explanation of Statement I
c. Statement I is correct but Statement II is incorrect
d. Statement II is correct but Statement I is incorrect

30 **Statement I** F_2 is manufactured by electrolysis of KHF_2 but here anhydrous HF is used as solvent in place of water.

Statement II In the presence of anhydrous HF, more F_2 is produced.

JEE Main & AIEEE Archive

(Compilation of Last 13 Years' Questions)

1 Which series of reactions correctly represent chemical relations related to iron and its compound?

a. $Fe \xrightarrow{\text{Dil. } H_2SO_4} FeSO_4 \xrightarrow{H_2SO_4,\ O_2} Fe_2(SO_4)_3 \xrightarrow{\text{Heat}} Fe$ **(JEE Main 2014)**

b. $Fe \xrightarrow{O_2,\ \text{heat}} FeO \xrightarrow{\text{Dil. } H_2SO_4} FeSO_4 \xrightarrow{\text{Heat}} Fe$

c. $Fe \xrightarrow{Cl_2,\ \text{heat}} FeCl_3 \xrightarrow{\text{Heat, air}} FeCl_2 \xrightarrow{Zn} Fe$

d. $Fe \xrightarrow{O_2,\ \text{heat}} Fe_3O_4 \xrightarrow{CO,\ 600°C} FeO \xrightarrow{CO,\ 700°C} Fe$

2 Which method of purification is represented by the following equation? **(AIEEE 2012)**

$$Ti\,(s) + 2I_2(g) \xrightarrow{523\text{ K}} TiI_4(g) \xrightarrow{1700\text{ K}} Ti\,(g) + 2I_2(g)$$

a. Zone refining
b. Cupellation
c. Polling
d. van Arkel

3 Which of the following factors is of no significance for roasting sulphide ores to the oxides and not subjecting the sulphide ores to carbon reduction directly? **(AIEEE 2008)**

a. Metal sulphides are thermodynamically more stable than CS_2
b. CO_2 is thermodynamically more stable than CS_2
c. Metal sulphides are less stable than the corresponding oxides
d. CO_2 is more volatile than CS_2

4 During the process of electrolytic refining of copper, some metals present as impurity settle as anode mud. These are **(AIEEE 2005)**

a. Fe and Ni
b. Ag and Au
c. Pb and Zn
d. Se and Ag

5 Which one of the following ores is best concentrated by froth floatation method? **(AIEEE 2004)**

a. Magnetite
b. Cassiterite
c. Galena
d. Malachite

6 Aluminium is extracted by the electrolysis of

a. alumina **(AIEEE 2002)**
b. bauxite
c. molten cryolite
d. alumina mixed with molten cryolite

7 Cyanide process is used for the extraction of

a. barium
b. silver **(AIEEE 2002)**
c. boron
d. zinc

8 Refining of impure copper with zinc impurity is to be done by electrolysis using electrodes as **(AIEEE 2002)**

	Cathode	**Anode**
a.	pure copper	pure zinc
b.	pure zinc	pure copper
c.	pure copper	impure copper
d.	pure zinc	impure zinc

JEE Advanced & IIT JEE Archive

(Compilation of Last 10 Years' Questions)

1 Upon heating with Cu_2S, the reagent(s) that give copper metal is/are **(2014 Adv., One or More than One Options Correct Type)**

a. $CuFeS_2$ b. CuO
c. Cu_2O d. $CuSO_4$

2 The carbon-based reduction method is not used for the extraction of **(2013 Adv., One or More than One Options Correct Type)**

a. tin from SnO_2
b. iron from Fe_2O_3
c. aluminium from Al_2O_3
d. magnesium from $MgCO_3$, $CaCO_3$

3 Sulphide ores are common for the metals **(2013 Adv., Only One Option Correct Type)**

a. Ag, Cu and Pb b. Ag, Cu and Sn
c. Ag, Mg and Pb d. Al, Cu and Pb

4 In the cyanide extraction process of silver from argentite ore, the oxidising and reducing agents used are **(2012, Only One Option Correct Type)**

a. O_2 and CO respectively
b. O_2 and Zn dust respectively
c. HNO_3 and Zn dust respectively
d. HNO_3 and CO respectively

5 Oxidation states of the metal in the minerals haematite and magnetite, respectively, are **(2011, Only One Option Correct Type)**

a. II, III in haematite and III in magnetite
b. II, III in haematite and II in magnetite
c. II in haematite and II, III in magnetite
d. III in haematite and II, III in magnetite

6 Extraction of metal from the ore cassiterite involves **(2011, One or More than One Options Correct Type)**

a. carbon reduction of an oxide ore
b. self-reduction of a sulphide ore
c. removal of copper impurity
d. removal of iron impurity

Passage for Q. Nos. (7-9)

Copper is the most noble of the first row transition metals and occurs in small deposits in several countries. Ores of copper include chalcanthite ($CuSO_4 \cdot 5H_2O$), atacamite ($Cu_2Cl(OH)_3$), cuprite (Cu_2O), copper glance (Cu_2S) and malachite ($Cu_2(OH)_2CO_3$). However, 80% of the world copper production comes from the ore chalcopyrite ($CuFeS_2$). The extraction of copper from chalcopyrite involves partial roasting, removal of iron and self-reduction. **(2010, Comprehension Type)**

7 Partial roasting of chalcopyrite produces

a. Cu_2S and FeO b. Cu_2O and FeO
c. CuS and Fe_2O_3 d. Cu_2O and Fe_2O_3

8 Iron is removed from chalcopyrite as

a. FeO b. FeS
c. Fe_2O_3 d. $FeSiO_3$

9 In self-reduction, the reducing species is

a. S b. O^{2-}
c. S^{2-} d. SO_2

10 Match each of the reactions given in Column I with the corresponding product(s) given in Column II and mark the correct code. **(2009, Matching Type)**

	Column I		Column II
i.	Cu + dil. HNO_3	p.	NO
ii.	Cu + conc. HNO_3	q.	NO_2
iii.	Zn + dil. HNO_3	r.	N_2O
iv.	Zn + conc. HNO_3	s.	$Cu(NO_3)_2$
		t.	$Zn(NO_3)_2$

Codes

	i	ii	iii	iv
a.	p,q,r	p	p,s	q,r,t
b.	p	q,s	r,s	r,t
c.	p,s	p	q,t	r,s
d.	p,s	q,s	r,t	q,t

11 Give the number of water molecule (s) directly bonded to the metal centre in $CuSO_4 \cdot 5H_2O$. **(2009, Integer Type)**

12 Give the coordination number of Al in the crystalline state of $AlCl_3$. **(2009, Integer Type)**

13 Match the conversions in Column I with the type(s) of reaction(s) given in Column II. **(2008, Matching Type)**

	Column I		Column II
i.	$PbS \longrightarrow PbO$	p.	Roasting
ii.	$CaCO_3 \longrightarrow CaO$	q.	Calcination
iii.	$ZnS \longrightarrow Zn$	r.	Carbon reduction
iv.	$Cu_2S \longrightarrow Cu$	s.	Self-reduction

Codes

	i	ii	iii	iv
a.	p	q	r,s	p,q
b.	p	r,s	q	r
c.	p	q	r,s	p,s
d.	q,r	p	p	q,s

14 Native silver metal forms a water soluble complex with a dilute aqueous solution of NaCN in the presence of **(2008, Only One Option Correct Type)**

a. nitrogen b. oxygen
c. carbon dioxide d. argon

JEE Archive

15 Extraction of zinc from zinc blende is achieved by **(2007, Only One Option Correct Type)**

a. electrolytic reduction
b. roasting followed by reduction with carbon
c. roasting followed by reduction with another metal
d. roasting followed by self reduction

16 Match the extraction processes listed in Column I with metals listed in Column II and mark the correct options from the codes gives below. **(2006, Matching Type)**

	Column I		Column II
i.	Self reduction	p.	Lead
ii.	Carbon reduction	q.	Silver
iii.	Complex formation and displacement by metal	r.	Copper
iv.	Decomposition of iodide	s.	Boron

Codes

	i	ii	iii	iv
a.	r	p	p,s	q
b.	p,r	p	q	s
c.	p	p,q	q	r,s
d.	q	r,s	q	p

17 Which ore contains both iron and copper? **(2005, Only One Option Correct Type)**

a. Cuprite
b. Chalcocite
c. Chalcopyrite
d. Malachite

18 Write the balanced chemical equation for developing a black and white photographic film. Also give reason, why the solution of sodium thiosulphate on acidification turns milky white and give balance equation of this reaction. **(2005, Subjective Type)**

5. Qualitative Analysis

DPP-1	Principles Involved in Qualitative Analysis of Inorganic Salts
DPP-2	Qualitative Analysis of Acidic Radicals (Anions)
DPP-3	Qualitative Analysis of Basic Radicals (Cations)
DPP-4	Qualitative Analysis of Inorganic Salts (Acidic as and Basic Radicals)
DPP-5	Organic Qualitative Analysis
DPP-6	Quantitative Analysis
DPP-7	Principles Involved in Synthesis, Neutralisation and Kinetic Studies

- **Revisal Problems for JEE Main**
- **Revisal Problems for JEE Advanced**
- **JEE Advanced & IIT JEE Archive**

DPP

DAILY PRACTICE PROBLEMS

DPP-1 Principles Involved in Qualitative Analysis of Inorganic Salts

Subjective Questions

Working Space

Direction *(Q. Nos. 1 and 2) These questions are subjective in nature, need to be solved completely on notebook.*

1 Complete the following sentences by choosing appropriate word(s).

i. ......radical gives apple green colour during flame test but......and......cannot be identified by it.

ii. In sodium carbonate bead test, the colour of bead with MnO is......and with Cr_2O_3 is......

iii. Borax bead test involves formation of a transparent bead of......

iv. The formulae of microcosmic salt and borax are respectively......and.......

v. There is precipitation of solute (PbS) if its ionic product is......than its solubility product.

vi. If solubility product of $M(OH)_3$ is 1×10^{-12} (here M = metal) then 0.001 M, M^{3+} is precipitated at a pH......9.

2 i. Which is better reagents for washing the precipitates of

a. $Ag_2CrO_4(s)$: H_2O or 1M K_2CrO_4 or 1M $AgNO_3$

b. $Ca_3(PO_4)_2(s)$: $CaCl_2$ or Na_3PO_4

c. $PbI_2(s)$: KI or $(CH_3COO)_2Pb$

d. $Hg_2Cl_2(s)$: KCl or $Hg_2(NO_3)_2$

e. $CaF_2(s)$: $CaCl_2$ or NaF

ii. Does the pH increase, decrease or remain the same on addition of each of the following?

a. LiF to HF solution

b. NH_4Cl to NH_3 solution

c. NaCN to HCN solution

d. Na_2CO_3 to $NaHCO_3$ solution

Only One Option Correct Type

Direction *(Q. Nos. 3-14) This section contains 12 multiple choice questions. Each question has four choices (a), (b), (c) and (d), out of which ONLY ONE is correct.*

3 The principle involved in the classification of basic radicals is/are

a. Common ion effect
b. Solubility product
c. Valency of radicals
d. Both (a) and (b)

4 Al^{3+}, Fe^{3+} and Cr^{3+} are grouped together for qualitative analysis as their

a. carbonates are insoluble in $NH_3(aq)$
b. hydroxides are insoluble in $NH_3(aq)$
c. sulphides are soluble in acid
d. None of these

Working Space

5 Which of the following are soluble in excess of sodium hydroxide?

I. CuS II. $AlCl_3$ III. As_2S_3

a. I and II b. II and III c. I and III d. All of these

6 Addition of conc. HCl to saturated $BaCl_2$ solution precipitates $BaCl_2$ because

a. at constant temperature, the product of $[Ba^{2+}]\cdot[Cl^-]^2$ remains constant in saturated solution

b. ionic product $[Ba^{2+}]\cdot[Cl^-]^2$ remains constant in a saturated solution

c. of common ion effect

d. it follows from Le-Chatelier's principle

7 Colourless salt $(x) \xrightarrow{\Delta} y \xrightarrow{Cu^{2+},\,\Delta}$ Coloured bead, x can be

a. borax b. microcosmic salt c. Both (a) and (b) d. CuS

8 Nitrates form a brown ring when treated with HCl/$FeSO_4$ followed by a few drops of conc. H_2SO_4. This test depends upon

a. the reduction of nitrate to nitric oxide

b. the oxidation of nitric oxide to nitrogen dioxide

c. reduction of ferrous sulphate to iron

d. oxidising action of sulphuric acid

9 In which of the following solvents is silver chloride most soluble?

a. 0.1 mol dm^{-3} $AgNO_3$ b. 0.1 mol dm^{-3} HCl

c. H_2O d. $aq.NH_3$

10 In III group of basic radical analysis, NH_4Cl is added before group reagent NH_4OH. This is done to

a. increase the solubility product of III group salts

b. decrease the solubility product of III group salts

c. increase the ionic product of OH^- and III group radicals

d. decrease the degree of ionisation of group reagent

11 To increase significantly the concentration of free Zn^{2+} ion in a solution of the complex ion $[Zn(NH_3)_4]^{2+}$

$$Zn^{2+}(aq) + 4NH_3(aq) \rightleftharpoons [Zn(NH_3)_4]^{2+}(aq)$$

add to the solution some

a. H_2O b. HCl (aq) c. $NH_3(aq)$ d. $NH_4Cl(aq)$

12 X is a colourless salt in cold but becomes yellow on heating. It is soluble in NaOH as well as in dilute HCl. X can be

a. FeO b. PbO c. ZnO d. CdO

13 A solution contains following metal ion : I. Hg^{2+}, II. Zn^{2+}, III. Mn^{2+}, IV Fe^{2+}. Concentration of each metal ion in the solution is 10^{-3} M. When the solution is treated with 10^{-16}M $[S^{2-}]$, in case of which precipitation takes place first?

$[(K_{sp})_{MnS} = 10^{-15}, (K_{sp})_{FeS} = 10^{-23}, (K_{sp})_{ZnS} = 10^{-20}, (K_{sp})_{HgS} = 10^{-54}]$

a. I b. II c. III d. IV

14 H_2S gas is passed into test tubes A and B containing aqueous solution of

A. $ZnCl_2$ B. $Zn(CH_3COO)_2$

The true statement about A and B is

a. Precipitate of ZnS is formed in A but not in B

b. Precipitate of ZnS is formed in A as well as in B

c. No precipitation takes place in either of the test tube

d. Precipitate of ZnS is formed in B but not in A

Working Space

One or More than One Options Correct Type

Direction (Q. Nos. 15 and 16) This section contains 2 multiple choice questions. Each question has four choices (a), (b), (c) and (d), out of which ONE or MORE THAN ONE are correct.

15 Which of the following pair(s) will show common ion effect?

a. $AgCl + NaCl$ b. $NH_4OH + NH_4Cl$
c. $NaCl + HCl$ d. $AgCN + KCN$

16 In the second group of qualitative analysis, H_2S is passed through a solution acidified with HCl. The purpose of adding HCl here is

a. to increase the $[H^+]$ b. limit the $[S^{2-}]$
c. add the Cl^- ion d. decrease the ionisation of H_2S

Comprehension Type

Direction (Q. Nos. 17 and 18) This section contains a paragraph, each describing theory, experiments, data, etc. Two questions related to the paragraph have been given. Each question has only one correct answer among the four given options (a), (b), (c) and (d).

Passage

Common ion effect plays a key role in the qualitative analysis of basic radicals. In the presence of a common ion, the reaction of ionisation of weak electrolyte like AcOH shifts in backward direction. The degree of ionisation of 0.05 M AcOH is x and pK_a of AcOH is 4.74.

17 How is its degree of ionisation gets affected in the presence of 0.01M HCl?

a. Becomes double b. Decreases to half $\left(\frac{x}{2}\right)$
c. Decreases to $\left(\frac{x}{10}\right)$ d. Decreases to $\left(\frac{x}{100}\right)$

18 What is its degree of ionisation if 0.1M AcONa is added to it?

a. 1.8×10^{-2} b. 1.8×10^{-3} c. 1.8×10^{-4} d. 1.8×10^{-5}

Statement Type

Direction (Q. Nos. 19 and 20) This section is based on Statement I and Statement II. Select the correct answer from the codes given below.

a. Both Statement I and Statement II are correct and Statement II is the correct explanation of Statement I
b. Both Statement I and Statement II are correct but Statement II is not the correct explanation of Statement I
c. Statement I is correct but Statement II is incorrect
d. Statement II is correct but Statement I is incorrect

19 **Statement I** Basic radicals have been divided into different groups depending upon their solubility product.

Statement II Salts of most of the alkali metals are miscible with water.

20 **Statement I** Borax bead test is a wet test for the identification of coloured cations.

Statement II A transparent bead of $NaBO_2 + B_2O_3$ is formed which gives a coloured bead with coloured cation.

DPP-2 Qualitative Analysis of Acidic Radicals (Anions)

Subjective Questions

Direction *(Q. Nos. 1-3) These questions are subjective in nature, need to be solved completely on notebook.*

Working Space

1 i. Explain the following.

a. SO_4^{2-} also gives white ppt with lead acetate but ppt dissolves in ammonium acetate solution.

b. In case of $PbSO_4$ or Ag_2SO_4, $Ba(NO_3)_2$ (not $BaCl_2$) is used to test the presence of acidic radical.

ii. How will you identify

a. CO_3^{2-}, if mixture also contains SO_3^{2-},

b. SO_3^{2-}, if mixture also contains SO_4^{2-},

c. Br^- (by layer test) if mixture also contains I^-?

2 i. How can you distinguish between

a. CO_3^{2-} and SO_3^{2-} ion,

b. I^- and Cl^- ion,

c. NO_3^- and NO_2^- ion?

ii. Solution containing Mg^{2+} and NH_4^+ ions is called magnesia mixture and can be used to identify PO_4^{3-} ions. What can be other anion (acidic radical) which can be identified by this magnesia mixture?

iii. Take sodium oxalate in a test tube and heat it strongly. Take a burning matchstick at the mouth of the test tube. You observe that a gas is burning at the mouth of the test tube but on further heating gas is put-off itself.
Which gas burns and which gas puts it off?

3 i. Consider the following activity with salt X.

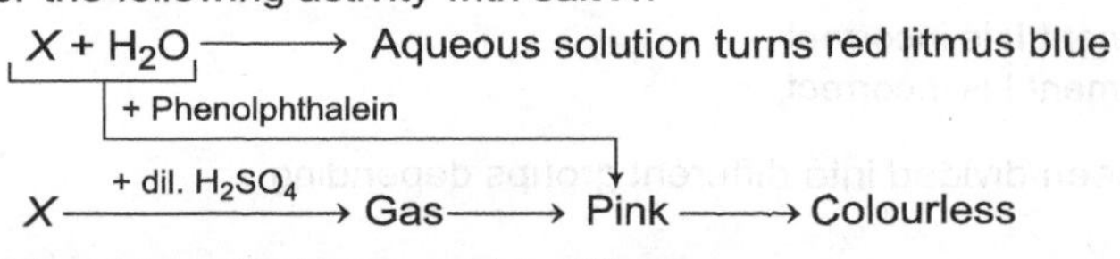

a. Which acidic radical is present in X?

b. What is the nature of solution of X?

c. What type of nature is suggested by the formation of pink colour with phenolphthalein?

d. Why a colourless solution is obtained at the end?

Working Space

ii. Identify *A*, *B*, *C* and *D* and explain reactions.

a. A (green coloured salt) $+ K_2Cr_2O_7 +$ conc. $H_2SO_4 \xrightarrow{\Delta} B$

b. B (reddish brown gas) $+ NaOH \longrightarrow C$ (yellow coloured solution)

c. $C + (CH_3COO)_2Pb \xrightarrow{CH_3COOH} D$ (yellow ppt)

d. $A + NaOH + Br_2$ water $\xrightarrow{\Delta} C$

e. $C + (CH_3COO)_2Pb \xrightarrow{CH_3COOH} D$

Only One Option Correct Type

Direction (Q. Nos. 4-12) *This section contains 9 multiple choice questions. Each question has four choices (a), (b), (c) and (d), out of which ONLY ONE is correct.*

4 Chloride salt gives colourless fumes with conc. H_2SO_4 while iodic salt gives violet fumes in the same test. This is because

a. H_2SO_4 reduces HI to I_2

b. the colour of HI is violet

c. HI gets oxidised to I_2

d. HI gets oxidised to HIO_3

5 *A* releases a colourless pungent gas when treated with dil. H_2SO_4. The released gas turns $Cr_2O_7^{2-}/H^+$ green. The green colour is due to the formation of

a. CrO_4^{2-} b. Cr^{3+}

c. CrO_3 d. CrO_2Cl_2

6 $X \xrightarrow{\text{dil. } H_2SO_4}$ Brown gas

$X \xrightarrow[\text{1-2 drop conc. } H_2SO_4]{FeSO_4}$ Brown ring

The composition of brown ring can be represented as

a. $N_2O \cdot FeSO_4$

b. $[Fe(H_2O)_5NO]^{2+}$

c. NO_2

d. $[Fe(SO_4)_5NO]^{2+}$

7 Mixture *A* is an aqueous solution of two colourless substances *X* and *Y*. On passing Cl_2 (*g*) through the solution, a deep brown colour is developed, while treatment of the solution with $BaCl_2$ or NaOH result in the formation of a white ppt. *X* and *Y* may be

a. $BaSO_4$ and $PbCl_2$

b. NaBr and KCl

c. $NaNO_3$ and $AlCl_3$

d. MgI_2 and $MgSO_4$

8 A solution is a mixture of bromide and iodide salt. When Cl_2 gas is passed followed by the addition of $CHCl_3$, the colour of organic layer will become

a. violet

b. reddish brown

c. colourless

d. blue

Working Space

9 Salt $X \xrightarrow[\text{(ii) } C_2H_5OH]{\text{(i) Conc. } H_2SO_4} Y \xrightarrow{\text{Ignite}}$ Green edged flame

The volatile compound Y burning with green flame has the formula

a. $(C_2H_5)_3B$
b. $(C_2H_5)_3BO_3$
c. $BaSO_4$
d. $C_2H_5B \cdot C_2H_5OH$

10 State whether the following statements are true or false and choose the correct option.

i. Brown ring test is a characteristic test for both NO_2^- and NO_3^-.
ii. A brown gas is evolved when either NO_2^- or NO_3^- is treated with concentrated solution of sulphuric acid.
iii. When treated with alkali in the presence of copper metal, nitrate gives NH_3.

a. TFT
b. TTT
c. FTF
d. TTF

11 Sulphide ions form a purple coloured Y compound with X. The oxidation state of central metal in the conversion of X into Y

a. changes from + 2 to + 3
b. changes from + 3 to + 2
c. changes from + 2 to + 4
d. remains the same

12 Some facts related to compound X are as follows :

i. It occupies 0.35 L per g at NTP.
ii. It decolourises acidified permanganate solution.
iii. Its aqueous solution produces white turbidity with H_2S.

Compound X may be

a. sulphur dioxide
b. sulphur trioxide
c. carbon dioxide
d. sodium carbonate

Matching List Type

Direction *(Q. Nos. 13-15) Choices for the correct combination of elements from Column I and Column II are given as options (a), (b), (c) and (d), out of which one is correct.*

13 Match the Column I with Column II and mark the correct option from the codes given below.

	Column I (Test for acidic radicals)		Column II (Oxidation state of central metal ion in complex involved)
i.	Chromyl chloride test	p.	+ 1
ii.	Ring test	q.	+ 6
iii.	Sodium nitroprusside test	r.	+ 2
iv.	$AgNO_3$ test for Br^-	s.	+ 3

Codes

	i	ii	iii	iv
a.	p	q	r	s
b.	q	p	r	s
c.	q	p	r	p
d.	q	p	s	p

Working Space

14 Match the Column I with Column II and mark the correct option from the codes given below.

	Column I (Reagents)		Column II (Anions)
i.	Silver nitrate	p.	Cl^-
ii.	Ferric chloride	q.	Br^-
iii.	Barium chloride	r.	I^-
iv.	Lead acetate	s.	CH_3COO^-
v.	Chlorine water	t.	CrO_4^{2-}

Codes

	i	ii	iii	iv	v
a.	p,q,r,t	s	t	r,t	q,r
b.	q,r,s	p	p,s	q,s,t	s,t
c.	p,s	s,t	q,r	p	q
d.	p,q,r,t	s	q,s	q,t	q

15 Match the Column I with Column II and mark the correct option from the codes given below.

	Column I (Reactions)		Column II (Characteristic of evolved gases)
i.	$FeSO_4 \xrightarrow{\Delta}$	p.	A mixed anhydride of HNO_2 and HNO_3
ii.	$Na_2SO_3 \xrightarrow{\text{dil. HCl}}$	q.	Forms oleum with H_2SO_4
iii.	$NaCl \xrightarrow{\text{Conc. } H_2SO_4}$	r.	Turns acidified $K_2Cr_2O_7$ green
iv.	$NaNO_2 \xrightarrow[\text{Air}]{\text{dil. HCl}}$	s.	Shows green colour
		t.	Also obtained by the action of moist air on aluminium chloride

Codes

	i	ii	iii	iv
a.	q,r,s	r,s	t	p
b.	p,r,t	p,t	s	p
c.	p,t	r,s	t	p
d.	p,q,r,s	q,t	q	r,s

Comprehension Type

***Direction** (Q. Nos. 16 and 17) This section contains a paragraph, each describing theory, experiments, data, etc. Two questions related to the paragraph have been given. Each question has only one correct answer among the four given options (a), (b), (c) and (d).*

Passage

Gases *A*, *Y* and *Z* are detected by using following reagents :

$$A \xrightarrow{\text{Pyrogallol}} \text{Soluble and the solution turns black}$$

$$\text{Mixture 'X'} \xrightarrow{+\text{ dil. } H_2SO_4} Y + Z \xrightarrow{\text{Lime water}} \text{Turns milky}$$

$$Y \xrightarrow{Cr_2O_7^{2-}/H^+} \text{Green} \qquad Z \xrightarrow{Cr_2O_7^{2-}/H^+} \text{No reaction}$$

Working Space

16 Gases A, Y and Z respectively, are

a. O_2, SO_2, CO_2
b. O_2, CO_2, SO_2
c. O_2, SO_2, SO_3
d. CO_2, SO_2, CO_2

17 The order of analysis of these gases is

a. Z, Y and A
b. Y, A and Z
c. Y, Z and A
d. Z, A and Y

Statement Type

Direction (Q. Nos. 18 and 19) *This section is based on Statement I and Statement II. Select the correct answer from the codes given below.*

a. Both Statement I and Statement II are correct and Statement II is the correct explanation of Statement I
b. Both Statement I and Statement II are correct but Statement II is not the correct explanation of Statement I
c. Statement I is correct but Statement II is incorrect
d. Statement II is correct but Statement I is incorrect

18 **Statement I** NO_2^- is decomposed by urea from a mixture containing nitrate and nitrite both.

Statement II Nitrate interferes in the ring test of nitrite.

19 **Statement I** A colourless gas is evolved when dil. HCl is added to a solution containing HCO_3^- or CO_3^{2-}.

Statement II Both carbonate and bicarbonate ions form a ppt with radical of group VI (which gives reaction with Na_2HPO_4).

One Integer Value Correct Type

Direction (Q. No. 20) *This section contains a question. When worked out will result in an integer from 0 to 9 (both inclusive).*

20 Chromyl chloride test is an identification test for chlorides. Chlorides of how many of the mentioned metals/non-metal give this test?

I. Hydrogen
II. Silver
III. Lead
IV. Antimony
V. Sodium
VI. Tin
VII. Barium
VIII. Mercury

DPP-3 Qualitative Analysis of Basic Radicals (Cations)

Subjective Questions

Working Space

Direction (*Q. Nos.* 1 *and* 2) *These questions are subjective in nature, need to be solved completely on notebook.*

1 i. Give the name of a suitable reagent which can be used to separate the following pairs of ions:

a. Ag^+ and Cu^{2+}
b. Cd^{2+} and Zn^{2+}
c. Ag^+ and Pb^{2+}
d. Al^{3+} and Mg^{2+}
e. Fe^{3+} and Ni^{2+}

ii. Give the name of a common reagent which can be used to differentiate between the following pairs (on the basis of colour of ppt formed).

a. Ag^+ and Ba^{2+}
b. Co^{2+} and Cu^{2+}
c. Cu^{2+} and Pb^{2+}
d. Fe^{3+} and Cu^{2+}

2 Explain the following

i. Concentrated hydrochloric acid cannot be used as a group reagent for the group I cations.

ii. Excess of dilute HCl should be added for the precipitation of cations of group I.

iii. It is necessary to boil off H_2S from filtrate of group II, before using it for analysis of radicals of group III.

iv. The hydroxides of aluminium and iron are insoluble in water. However, NaOH is used to separate one from the other.

v. Ammonium chloride is acidic in liquid ammonia solvent.

Only One Option Correct Type

Direction (*Q. Nos.* 3-10) *This section contains 8 multiple choice questions. Each question has four choices (a), (b), (c) and (d), out of which ONLY ONE is correct.*

3 Consider the following equilibrium

$$AgCl\downarrow + 2NH_3 \rightleftharpoons \underset{\text{(Soluble)}}{[Ag(NH_3)_2]^+} + Cl^-$$

White ppt of AgCl appears on adding

a. NH_3
b. *aq*. NaCl
c. *aq*. HNO_3
d. *aq*. NH_4Cl

4 Solution of X in dil. HCl + H_2O $\longrightarrow$ White turbidity

$$X \xrightarrow{H_2S/HCl} \text{Black ppt } Y. \ Y \text{ is soluble in}$$

a. H_2SO_4
b. YAS
c. HNO_3
d. HCl

Working Space

5 An aqueous solution is a mixture of acetates of zinc, cadmium and copper. If H_2S gas is passed through the solution, the ions that get precipitated as sulphides are

a. Zn^{2+}, Cd^{2+}
b. Cu^{2+}, Cd^{2+}
c. Zn^{2+}, Cu^{2+}
d. Zn^{2+}, Cu^{2+} and Cd^{2+}

6 The colour of a mixture is orange. However, acidification, turns it to green. The mixture may contain

a. HgI_2, CrO_4^{2-}
b. Fe^{2+}, $Cr_2O_7^{2-}$
c. SO_3^{2-}, MnO_4^-
d. Fe^{2+}, CrO_4^{2-}

7 White ppt of *A* on reaction with *aq*. NH_3 are blackened. Select correct statement about *A*.

a. *A* is also called calomel
b. *A* reacts with *aq*. NH_3 forming $HgNH_2Cl$
c. *A* changes to grey on reaction with $SnCl_2$
d. All of the above are correct statements

8 Black ppt is formed when H_2S is passed in an aqueous solution of white substance *X*. *X* is soluble in hot water and the black ppt in hot nitric acid. When conc. H_2SO_4 is added to this solution, a white ppt appears. This white ppt is of

a. $BaSO_4$
b. $SrSO_4$
c. $PbSO_4$
d. $CdSO_4$

9 Consider the following statements regarding reactions of iron salts.

I. They result in the formation of green ferrous hydroxide when exposed to air.
II. $Fe(OH)_3$ is soluble in cold dil. HCl to give pale yellow solution.
III. Addition of colourless KI to the pale yellow solution, changes its colour to brown.
IV. When extracted with carbon tetrachloride, the brown colour (in III) changes into violet.

The true statements among the above are

a. III and IV
b. I and II
c. II, III and IV
d. All of these

10 A solution of a metal ion when treated with KI gives a red precipitate which dissolves in excess KI to give a colourless solution. Moreover, the solution of metal ion on treatment with a solution of cobalt (III) thiocyanate gives rise to deep blue crystalline precipitate. The metal ion is

a. Pb^{2+}
b. Hg^{2+}
c. Cu^{2+}
d. Co^{2+}

One or More than One Options Correct Type

Direction *(Q. Nos. 11-13) This section contains 3 multiple choice questions. Each question has four choices (a), (b), (c) and (d), out of which ONE or MORE THAN ONE are correct.*

11 When metallic Cu is heated with concentrated H_2SO_4, the product obtained is/are

a. $CuSO_4$
b. SO_2
c. CuS
d. Cu_2S

12 During the analysis of basic radicals of II B group, the sulphides are treated with yellow ammonium sulphide

a. to convert them into thio compound
b. to dissolve them
c. to remove sulphide group from them
d. to remove excess of acid

Working Space

13 Ferrous sulphate gives no blue colouration with potassium ferrocyanide but when 'X' is added to the reaction mixture, blue colour appears. X can be

a. MnO_4^-/H^+

b. $Cr_2O_7^-/H^+$

c. H_2O_2/H^+

d. HNO_3 in the presence of H_2SO_4

Statement Type

Direction (Q. Nos. 14-16) This section is based on Statement I and Statement II. Select the correct answer from the codes given below.

a. Both Statement I and Statement II are correct and Statement II is the correct explanation of Statement I
b. Both Statement I and Statement II are correct but Statement II is not the correct explanation of Statement I
c. Statement I is correct but Statement II is incorrect
d. Statement II is correct but Statement I is incorrect

14 **Statement I** Ammonium nitrate can be used in place of ammonium chloride, for the precipitation of radicals of group III.

Statement II Ammonium nitrate and ammonium chloride both are the salts of strong acids and weak bases.

15 **Statement I** Oxides of Zn (II) and Cu (II) can be separated from a mixture by boiling it with aqueous alkali.

Statement II ZnO is solution in alkali (*aq*) while CuO remains undissolved.

16 **Statement** Cu^{2+} and Cd^{2+} solution result in the formation of complex when KCNS is added.

Statement II H_2S gas has no effect over the stable complex $[Cu(CN)_4]^{3-}$.

Matching List Type

Direction (Q. Nos. 17-19) Choices for the correct combination of elements from Column I and Column II are given as options (a), (b), (c) and (d), out of which one is correct.

17 Match the following using three columns with (*X*) containing radicals, (*Y*) the reagents used and (*Z*) the compound formed when (*X*) and (*Y*) react and choose the correct answer using the codes given below

	X		Y		Z
I.	Fe^{2+}	i.	NH_4SCN	A.	Cherry red ppt
II.	Fe^{3+}	ii.	DMG	B.	Turnbull's blue
III.	Ni^{2+}	iii.	$K_4[Fe(CN)_6]$	C.	Prussian blue
IV.	Co^{2+}	iv.	$K_3[Fe(CN)_6]$	D.	Blue

Codes

	I	II	III	IV
a.	(i, C)	(ii, D)	(iii, A)	(iv, B)
b.	(iv,B)	(ii, C)	(iii, A)	(i, D)
c.	(iv,B)	(iii, C)	(ii, A)	(i, D)
d.	(iv, B)	(ii, C)	(i, A)	(iii, D)

Working Space

18 Match the set of basic radicals (in column I) with the group reagent (in column II) and analytical group (in column III).

	Column I		Column II		Column III
i.	Zn^{2+}, Ni^{2+}	1.	H_2S/H^+	p.	I
ii.	Ag^+, Hg_2^{2+}	2.	$(NH_4)_2CO_3/NH_4OH$	q.	II
iii.	Ca^{2+}, Ba^{2+}	3.	$H_2S/NH_4OH, NH_4Cl$	r.	III
iv.	Cu^{2+}, Cd^{2+}	4.	dil. HCl	s.	IV
v.	Al^{3+}, Cr^{3+}	5.	NH_4Cl/NH_4OH	t.	V

Codes

	i	ii	iii	iv	v
a.	3,s	4,p	2,t	1,q	5,r
b.	2,q	3,s	1,q	4,r	5,p
c.	2,s	3,p	4,q	1,r	5,t
d.	2,s	5,p	4,r	3,s	1,q

19 Match the Column I with Column II and mark the correct option from the codes given below.

	Column I (Reagents)		Column II (Cations identified)
i.	$Cr_2O_7^{2-}/H^+$	p.	Pb^{2+}
ii.	MnO_4^-/H^+	q.	Ag^+
iii.	I^-	r.	Fe^{2+}
iv.	CrO_4^{2-}	s.	Ba^{2+}

Codes

	i	ii	iii	iv
a.	r	r	p,q	p,q,s
b.	r	p,q	q,r,s	s
c.	r	p,r	p	q,s
d.	p,r	p	q,r	r,s

One Integer Value Correct Type

Direction *(Q. No. 20) This section contains a question. When worked out will result in an integer from 0 to 9 (both inclusive).*

20 In how many of the following cases, soluble nature of the salt changes ?

i. KCN is added to AgCl (insoluble)
ii. NH_3 is added to AgCl (insoluble)
iii. KI is added to HgI_2 (insoluble)
iv. H_2O is added to $BiCl_3$ (soluble) in HCl
v. H_2O is added to $SbCl_3$ (soluble) in HCl
vi. NH_3 is added to AgI (insoluble)

DPP-4 Qualitative Analysis of Inorganic Salts (Acidic and Basic Radicals)

Subjective Questions

Working Space

Direction (*Q. Nos.* 1-4) *These questions are subjective in nature, need to be solved completely on notebook.*

1 Suggest name of a reagent (including water) that could be used to separate the following pairs of solids ?

i. Hydroxides of Na^+ and Fe^{3+}

ii. Chloride and iodide of Ag^+

iii. Sulphides of Mn^{2+} and Co^{2+}

iv. Hydroxides of Ni^{2+} and Fe^{3+}

v. Cr_2O_3 and hydroxide of Fe^{3+}

2 *A*, an important laboratory reagent, turns red litmus blue, imparts golden yellow colour in flame and is a good precipitating agent. *A* reacts with Zn or Al forming H_2 gas. *A* gives white ppt with $ZnCl_2$ or $AlCl_3$ but ppt dissolves in excess of *A*. What is *A* and explain reactions?

3 *A* is a colourless salt and produces a gas *B* when treated with alkali (the same gas is obtained by the action of water over Mg_3N_2). After the completion of reaction, the reaction mixture is treated with $FeSO_4$ and conc. H_2SO_4, which results in the formation of a brown coloured ring '*C*' between the two layers. *A* on heating strongly forms *D* and *E*. *E* is a neutral oxide of dibasic acid *F*.

i. Identify *A* to *F* and explain reactions.

ii. Also draw the *cis* and *trans* forms of *F* ?

4 Identify *A* to *H*, and explain the reaction.

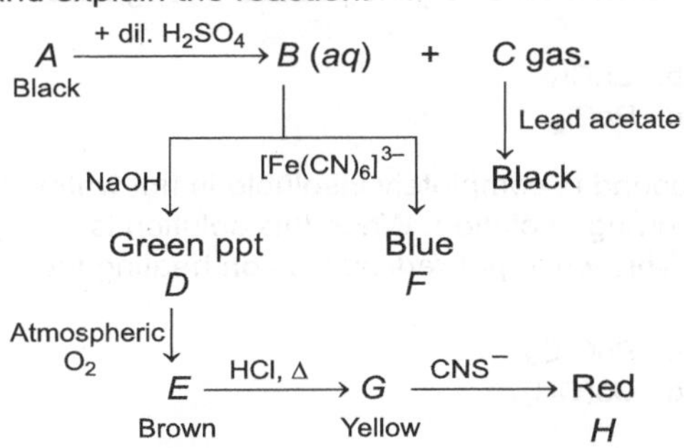

Only One Option Correct Type

Working Space

Direction (Q. Nos. 5-14) *This section contains 10 multiple choice questions. Each question has four choices (a), (b), (c) and (d), out of which ONLY ONE is correct.*

5 Three test tubes contain three unknown colourless liquids. Nearby three label, i.e. I. Alum, II. Hydrated barium chloride and III. KOH are found.

The false statement related to identification of these liquids is

a. Formation of no ppt on adding III into II suggests that II is hydrated barium chloride
b. Formation of white ppt on adding I and II suggests that one of them may be hydrated barium chloride
c. Formation of white ppt on adding I and II suggests that one of them may be alum
d. Formation of white ppt on adding III to I suggests that II is alum

6 $PbCl_2 + KI \longrightarrow \underset{\text{Yellow ppt}}{A} + KCl$

$$A \xrightarrow{KI} \underset{\text{Soluble}}{B}$$

The formulae of compounds *A* and *B* are

a. PbI_4 and K_2PbI_4 respectively
b. $K_2[PbI_4]$ and PbI_4 respectively
c. PbI_2 and $K_2[PbI_4]$ respectively
d. PbI_2 and $K_2[PbI_2]$ respectively

7 Compound *P* is generally used as a laboratory reagent for analysis. Its aqueous solution gives the following reactions:

i. When added with $CuSO_4$ solution, a brown ppt is formed, which turns white when treated with excess of hypo.
ii. When mixed with solution of Ag^+ ion, a yellow curdy ppt is formed, which is insoluble in NH_4OH.

The compound *P* is

a. K_2CrO_4
b. KCNS
c. KI
d. Any of these

8 An aqueous solution of chloride of metal *M*, having excess of KCN, when treated with H_2S will give the precipitate of *M*S only if *M* is

a. Ni
b. Co
c. Cu
d. Cd

9 Salt *X* gives a white ppt when treated with nitric acid followed by silver nitrate. The ppt is soluble in ammonium hydroxide. Further *X* gives white ppt *Y* when H_2S is passed through its solution, saturated with NH_3. *X* also forms white ppt with NaOH which gets dissolved in excess of NaOH. When H_2S is passed through the solution *Y* reappears. *X* is

a. $AlCl_3$
b. $ZnCl_2$
c. $MgCl_2$
d. $BaCl_2$

10 A yellow solid known to be a single compound is completely insoluble in hot water but dissolves in hot dilute HCl to give an orange solution. When this solution is cooled, a white crystalline ppt is formed. This white ppt redissolves on heating the solution. The compound is

a. $Fe(OH)_3$
b. $PbCrO_4$
c. K_2CrO_4
d. $Co(OH)_2$

Working Space

11 $A \text{ (colourless salt)} \xrightarrow{\Delta} \underbrace{B + C}_{\text{gas}} + D;\ D \xrightarrow{H_2O} E$

Get C turns solution E milky. B burns with blue flame. A also decolourises MnO_4^-/H^+. Thus, A, B, C, D and E are

	A	B	C	D	E
a.	CaC_2O_4	CO	CO_2	CaO	$Ca(OH)_2$
b.	CaC_2O_4	CO_2	CO	CaO	$Ca(OH)_2$
c.	$CaCO_3$	CaO	CO	CO_2	$Ca(OH)_2$
d.	$CaOCl_2$	Cl_2	O_2	CaO	$Ca(OH)_2$

12 A salt gives white colour ppt with $AgNO_3$ and deep blue colour in the non-luminous flame of borax bead test. The nature of aqueous solution of this salt and its cause are respectively.

a. acidic, due to hydrolysis of cation
b. acidic, due to hydrolysis of anion
c. basic, due to hydrolysis of cation
d. basic, due to hydrolysis of anion

13 Consider the following statements about salt 'X'.

X reduces $HgCl_2$ solution to white ppt changing to grey.
X turns $FeCl_3$ yellow coloured solution to green.
X gives white ppt with NaOH soluble in excess of NaOH.
X gives yellow dirty ppt on passing H_2S gas, soluble in yellow ammonium sulphide (YAS).
X gives white ppt with HNO_3 / $AgNO_3$.
X can be

a. $SnCl_4$
b. $SnCl_2$
c. $CuCl_2$
d. $AsCl_3$

14 The composition of a salt is $M_2X \cdot 7\,H_2O$. It contains water and M_2X (where, M = monovalent alkaline cation and X = divalent anion) in 1 : 1 ratio by weight. The salt gives a gas with dil. H_2SO_4 which turns $Cr_2O_7^{2-}/H^+$ solution green. The salt can be

a. $K_2SO_4 \cdot 7H_2O$
b. $Na_2SO_3 \cdot 7H_2O$
c. $Na_2S \cdot 7H_2O$
d. $K_2SO_3 \cdot 7H_2O$

One or More than One Options Correct Type

Direction *(Q. Nos. 15-17) This section contains 3 multiple choice questions. Each question has four choices (a), (b), (c) and (d), out of which ONE or MORE THAN ONE are correct.*

15 Which of the following reactions lead to complex formation ?

a. $CuSO_4 + KCN \longrightarrow$
b. $KI + HgCl_2 \longrightarrow$
c. $NiCl_2 + KCN \longrightarrow$
d. $As_2S_3 + (NH_4)_2S \longrightarrow$

16 $$\text{A metal sulphide} \xrightarrow[H_2S]{HCl} \underset{'X'}{\text{Black ppt}} \xrightarrow{YAS} \text{Residue} \xrightarrow[H_2SO_4]{\text{dil.}} \text{Insoluble}$$
$$\text{Residue} \xrightarrow{\text{Aqua-regia}} \text{Soluble}$$

The true statement(s) about the ppt X is/are

a. Nitrate ion of aqua-regia in the presence of H^+ converts S^{2-} ion to free S
b. Chloride ion of aqua-regia in the presence of H^+ converts S^{2-} ion to free S
c. Metal forms a salt with NO_3^-
d. Metal forms a soluble complex with chloride ion

Working Space

17 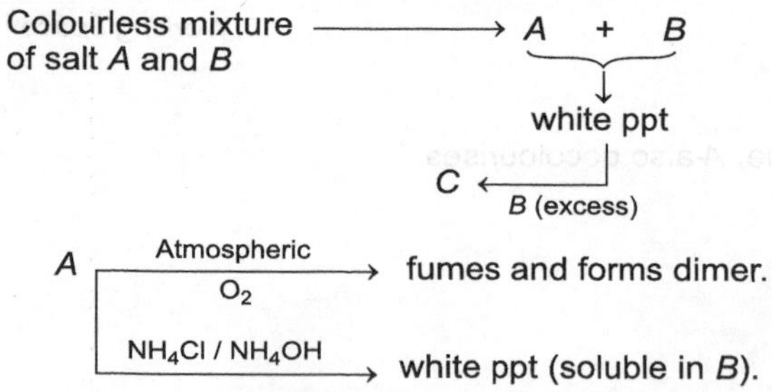

The true statement(s) about the above reaction sequence is

a. *B* turns red litmus blue
b. In excess of *B*, ZnO dissolves to form salt
c. The white fumes are due to aluminium chloride
d. *A* acts as a catalyst in Friedel-Crafts' reaction

Comprehension Type

***Direction** (Q. Nos. 18-20) This section contains a paragraph, describing theory, experiments, data, etc. Three questions related to the paragraph have been given. Each question has only one correct answer among the four given options (a), (b), (c) and (d).*

Passage

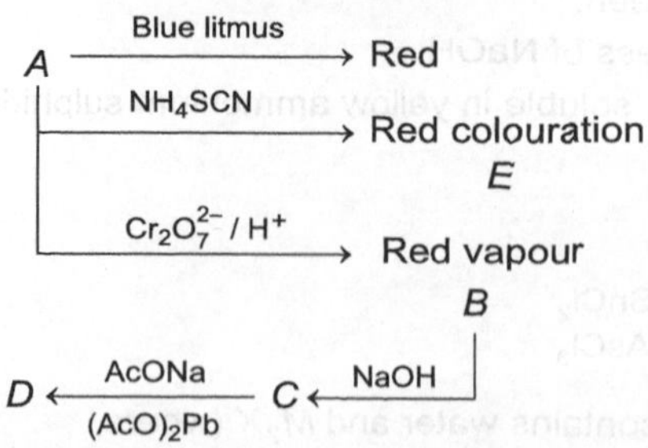

18 What is the hybridisation of central atom in compound '*D*' ?

a. sp^3d
b. sp^3
c. dsp^2
d. sp^2

19 The true statement(s) about the acidic radical of *A* are

I. When treated with MnO_2 a yellow green gas *X* is evolved.
II. The evolved gas *X* oxidises Fe^{2+} to Fe^{3+}
III. The excess of evolved gas *X* forms an explosive with ammonia.

a. I and II
b. II and III
c. III and I
d. I, II and III

20 The formula of *E* is

a. $Fe(SCN)_2$
b. $Fe(SCN)_3$
c. $FeCl_3$
d. $NH_4[Fe(CN)_4]$

DPP-5 Organic Qualitative Analysis

Working Space

Subjective Questions

Direction *(Q. Nos. 1-3) These questions are subjective in nature, need to be solved completely on notebook.*

1 Identify the missing species and answer the related questions

$$\text{Phenol} \xrightarrow[\text{aq. NaOH, }\Delta \text{ (ii) } H^+]{\text{(i) } CCl_4} X \xrightarrow{CH_3COCl} Y \xrightarrow{H_3O^+} X + Q$$

i. What common reagent can be used to identify X and Q ?

ii. Also give other chemical tests for Y, X and Q.

iii. By what common name Y is known ?

2 The molecular formula of an organic compound is $C_3H_6O_2$. This compound on boiling with aqueous acidic solution produces compounds A and B. A gives positive test with Fehling solution as well as with Tollen's reagent. B on boiling with I_2 and *aq*. NaOH produces sodium formate and compound 'C'. C when heated with Ag powder gives acetylene. Identify compound A to C and explain the reactions involved.

3 i. Identify all the isomers having molecular formula C_7H_8O and make distinction between them by chemical tests.

ii. Benzaldehyde gives a positive test with Tollen's reagent but not with Benedict's and Fehling solution. Why?

Only One Option Correct Type

Direction *(Q. Nos. 4-7) This section contains 4 multiple choice questions. Each question has four choices (a), (b), (c) and (d), out of which ONLY ONE is correct.*

4 In which case, compound and its identification test are correctly matched?

a. Phenol	Phthalein test
b. Benzyl amine	Azo dye test
c. Phthalic acid	Fluorescence test
d. Acetamide	Carbylamine test

5 Glucose contains —CHO group and fructose contains $\rangle C{=}O$ (keto group).These two can be distinguished by using

a. Fehling's solution

b. Tollen's reagent

c. Benedict's solution

d. Optical rotation

Working Space

6 Compound X, C_5H_8O does not react appreciably with Lucas reagent at room temperature but gives a ppt with ammoniacal $AgNO_3$. With excess of MeMgBr 0.42 g of (X) gives 224 mL CH_4 at STP. Treatment of 'X' with H_2 in the presence of Pt catalyst followed by boiling with excess of HI gives n-pentane.

The possible structure of X is

a. $HOCH_2C\equiv CCH_2CH_3$ b. $HOCH_2CH_2C\equiv CCH_3$

c. $HOCH_2CH_2CH_2C\equiv CH$ d. $CH_3CH_2\underset{\underset{OH}{|}}{CH}C\equiv CH$

7 A compound liberates CO_2 with $NaHCO_3$ and also gives colour with neutral $FeCl_3$. The possible structure of the compound is

a. 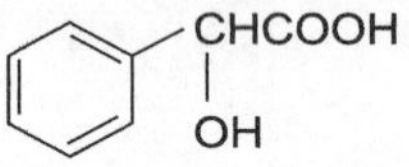

b. 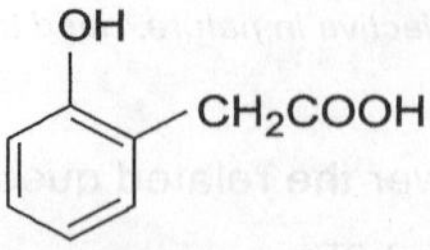

c. 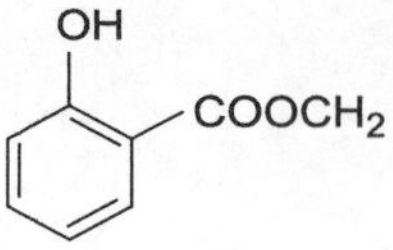

d. 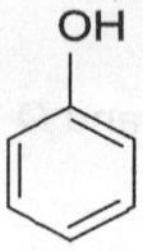

Matching List Type

Direction *(Q. Nos. 8-10)* ***Choices for the correct combination of elements from Column I and Column II are given as options (a), (b), (c) and (d), out of which one is correct.***

8 Match the Column I with Column II and mark the correct answer using the codes given below :

	Column I (Compounds)		Column II (Distinguishing reagent)
i.	Ethene and ethyne	p.	$CuSO_4$ solution
ii.	Formic acid and acetic acid	q.	Ammoniacal Cu_2Cl_2
iii.	Glycine and biuret	r.	Neutral $FeCl_3$
iv.	Benzylamine and o-toluidine	s.	$NaNO_2$/HCl, phenol

Codes

	i	ii	iii	iv		i	ii	iii	iv
a.	p	r	q	s	b.	r	p	q	s
c.	q	r	p	s	d.	q	r	s	p

9 Match the Column I with Column II and mark the correct answer using the codes given below :

	Column I (Functional group)		Column II (Reagent)
i.	Carboxylic acid	p.	$FeCl_3$
ii.	Amide	q.	NaOH
iii.	Amine	r.	HNO_2
iv.	—OH(enolic)	s.	$NaHCO_3$

Codes

	i	ii	iii	iv		i	ii	iii	iv
a.	s	q,r	r	p	b.	q,s	p	s	p
c.	s	q,r	r,s	s	d.	p,q,r	p,q	s	q,s

Working Space

10 Match the Column I with Column II mark the correct option using the codes given below.

	Column I (Compounds)		Column II (Confirmatory test)
i.	Fructose	p.	Dye test
ii.	Benzaldehyde	q.	Silver-mirror test
iii.	Phenol	r.	Liebermann's nitroso test
iv.	Aniline	s.	Malachite green test

Codes

	i	ii	iii	iv		i	ii	iii	iv
a.	q,r	r,s	p,s	q	b.	q	q,s	p,r	p
c.	q	q,s	p,r,s	p	c.	p,q	r	s	p

One or More than One Options Correct Type

Direction (Q. Nos. 11 and 12) This section contains 2 multiple choice questions. Each question has four choices (a), (b), (c) and (d), out of which ONE or MORE THAN ONE are correct.

11 Generally halogens of organic compound are identified by converting them into sodium extract. This fact is not true in case of

a. benzyl chloride
b. carbon tetrachloride
c. 2, 4, 6-trinitro chlorobenzene
d. allyl chloride

12 During the synthesis of butanol-2 from an aldehyde and CH_3MgBr in the presence of NH_4Cl/H_2O, the student was unaware that the flask used for the reaction was contaminated with copper metal. The obtained product(s) in such case

a. gives a yellow ppt with $NaOH / I_2$
b. forms oxime with NH_2OH
c. decolourises bromine water in acidified aqueous solution
d. gives colour with neutral $FeCl_3$ solution

Statement Type

Direction (Q. Nos. 13-15) This section is based on Statement I and Statement II. Select the correct answer from the codes given below.

a. Both Statement I and Statement II are correct and Statement II is the correct explanation of Statement I
b. Both Statement I and Statement II are correct but Statement II is not the correct explanation of Statement I
c. Statement I is correct but Statement II is incorrect
d. Statement II is correct but Statement I is incorrect

13 **Statement I** A white ppt of AgCl is formed when carbon tetrachloride is treated with aqueous solution of $AgNO_3$.

Statement II Carbon tetrachloride is an organic compound.

14 **Statement I** Benedict's solution contains Cu^{2+} ions complexed with citrate ion.

Statement II Due to complex formation, Cu^{2+} does not precipitated as $Cu(OH)_2$.

15 **Statement I** CH_3COCH_2OH reduces Tollen's reagent.

Statement II All carbonyl compounds reduce Tollen's reagent.

Working Space

Comprehension Type

Direction *(Q. Nos. 16-18) This section contains a paragraph, describing theory, experiments, data, etc. Three questions related to the paragraph have been given. Each question has only one correct answer among the four given options (a), (b), (c) and (d).*

Passage

Organic compounds can be separated from a mixture on the basis of their solubility in different reagents, e.g. NaOH is used for extracting an acidic compound, while extraction of basic compound is done with the help of hydrochloric acid.

Suppose an organic mixture contains following compounds.

I. N-methyl aniline II. Benzyl amine
III. 4-methyl aniline IV. Phenol
V. Benzoic acid VI. Methanol
VII. 3-methylaniline VIII. Propane-2-ol
IX. 2-methyl benzoic acid X. 4-methyl benzoic acid

16 HCl can be used to separate a mixture of

a. I and II b. III and IV c. V and IX d. VIII and X

17 A mixture of two compounds from the above is dissolved in $CHCl_3$ and then treated with NaOH and aqueous layer separated. $CHCl_3$ layer containing the compound on heating with KOH gives unpleasent smell of $Z(C_8H_7N)$. Compound in chloroform layer is oxidised to the other compound *A* which on heating with sodalime gives a hydrocarbon *B*.

Compound in NaOH layer on acidification followed by reaction with neutral $FeCl_3$ gives buff coloured ppt. This compound is also a good food preserver.

From the above given compounds the suitable match with the given reactions is

	$CHCl_3$ layer	NaOH layer
a.	I	IX
b.	II	V
c.	III	X
d.	VII	V

18 VI and VIII can be differentiated by

a. iodoform test b. carbylamine
c. phthalein test d. ceric ammonium nitrate test

One Integer Value Correct Type

Direction *(Q. No. 19 and 20) This section contains 2 questions. When worked out will result in an integer from 0 to 9 (both inclusive).*

19 While detecting nitrogen in organic compounds, a prussian blue colour complex is formed. The sum of oxidation number of iron in this complex is

20 An organic compound of the formula $C_3H_8O_3$ yields on acetylation with acetic anhydride, a compound of formula $C_9H_{14}O_6$. How many hydroxy groups are present in the compound?

DPP-6 Quantitative Analysis

Subjective Questions

Working Space

Direction *(Q. Nos. 1 and 2) These questions are subjective in nature, need to be solved completely on notebook.*

1 A litre of a mixture contains 20 g of caustic soda, 20 g of sodium bicarbonate and 20 g of sodium carbonate. What values of 50 mL of the mixture will be used for titration against N—HCl if

i. first titrated with phenolphthalein
ii. methyl orange is added after the first end point
iii. methyl orange is added from the very beginning

2 The solution of $KMnO_4$ was titrated against $N/10\, C_2O_4^{2-}$ solution and it was found that

$$25 \text{ mL of oxalic acid} \equiv 37.5 \text{mL } KMnO_4$$

25 mL of iron solution required 24 mL of standard $KMnO_4$ solution. 100 mL of this solution were reduced by Zn\H_2SO_4 and the obtained solution is diluted to 250 mL. For the 50 mL of this diluted solution, $KMnO_4$ required is 36.75 mL. Calculate the amount of ferrous and ferric ions present in the solution in g/L.

Only One Option Correct Type

Direction *(Q. Nos. 3-12) This section contains 10 multiple choice questions. Each question has four choices (a), (b), (c) and (d), out of which ONLY ONE is correct.*

3 In the titration of oxalic acid *vs* $KMnO_4$, the solution is heated at 60°-70° C to

a. increase the rate of reaction
b. to prepare a homogeneous solution
c. to dissolve the brown ppt formed
d. to reduce the $KMnO_4$ into $MnSO_4$ by H_2SO_4

4 In the reaction of $KMnO_4$ with oxalic acid in acidic medium, nitric acid cannot be used for acidification in place of sulphuric acid. This is because

a. it reacts chemically with $KMnO_4$ solution
b. it interfere with the oxidising action of $KMnO_4$
c. it reduces $KMnO_4$ before its reaction with oxalic acid
d. it gets oxidised by $KMnO_4$

5 To determine the amount of $NaHCO_3$ in a tablet of antacid, it is dissolved with water and the obtained solution is titrated with HCl. The most appropriate indicator for this titration is
(Given, $H_2CO_3\ k_{a_1} = 2.5 \times 10^{-4}$ and $k_{a_2} = 2.4 \times 10^{-8}$)

a. methyl orange
b. phenolphthalein
c. Both (a) and (b)
d. None of these

Working Space

6 For preparing 250 mL of $N/20$ solution of Mohr's salt, the amount of Mohr's salt required is

a. 9.8 g b. 4.9 g c. 19.6 g d. 3.2 g

7 3.92 g of ferrous ammonium sulphate (Mohr's salt) reacts completely with 50 mL $\frac{N}{10}$ $KMnO_4$ solution. The percentage purity of the sample is

a. 50 b. 78.4
c. 80 d. 39.2

8 When $KMnO_4$ solution is added to oxalic acid solution, the decolourisation is slow in the beginning but becomes instantaneous after sometime. This is because

a. CO_2 obtained as a by product makes the medium acidic
b. reaction is exothermic
c. the MnO_4^- acts as self catalyst and catalyses the reaction
d. Mn^{2+} acts as self catalyst

9 When the equal volume of two different acids (I and II) were titrated with NaOH solution of same concentration, the following titration curves were obtained

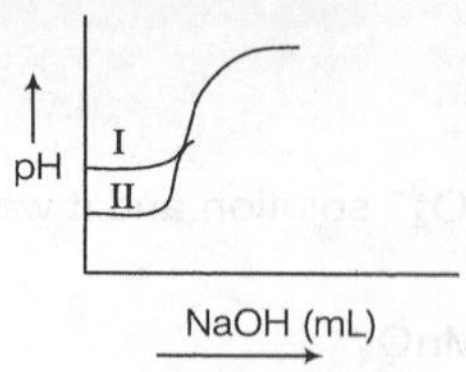

These curves show that

a. I is weak acid but more concentrated than II
b. II is weak acid but more concentrated than I
c. I is weak acid and both are equally concentrated
d. II is a weak acid and both are equally concentrated

10 In the reaction of $KMnO_4$ with an oxalate in acidic medium, MnO_4^- is reduced to Mn^{2+} and $C_2O_4^{2-}$ is oxidised to CO_2. Hence, 50 mL of 0.04 M $KMnO_4$ is equivalent to

a. 100 mL of 0.1 M $H_2C_2O_4$
b. 50 mL of 0.2 M $H_2C_2O_4$
c. 50 mL of 0.1 M $H_2C_2O_4$
d. 25 mL of 0.1M $H_2C_2O_4$

11 0.5 g mixture of potassium dichromate and permanganate was treated with excess of KI in acidic medium. The liberated I_2 require 100 cm^3 of 0.15 N $Na_2S_2O_3$ solution for titration. The percentage amount of dichromate in the mixture is

a. 85.36% b. 14.64%
c. 58.36% d. 26.14%

12 When the pH is below 3, the colour of methyl orange is red and when the pH is greater than 6.3, its colour is yellow and at the intermediate values of pH, it is orange.
What will be the colour of methyl orange in a 0.1M solution of NH_4Br?
[$pK_b(NH_4OH) = 4.74$]

a. Red b. Yellow
c. Orange d. Colourless

Working Space

One or More than One Options Correct Type

Direction (Q. Nos. 13 and 14) *This section contains 2 multiple choice questions. Each question has four choices (a), (b), (c) and (d), out of which ONE or MORE THAN ONE are correct.*

13 Which of the following is the examples of redox titration?

a. Na_2CO_3 *vs* HCl
b. As_2O_3 *vs* I_2
c. Oxalic acid *vs* $KMnO_4$
d. Mohr's salt *vs* $KMnO_4$

14 Which of the following will oxidise 126 g of oxalic acid ($H_2C_2O_4 \cdot 2H_2O$) in the presence of hydronium ion?

a. $\frac{1}{3}$ mol $Cr_2O_7^{2-}$
b. $\frac{2}{5}$ mol MnO_4^-
c. 2 mol $Cr_2O_7^{2-}$
d. $\frac{1}{3}$ mol MnO_4^-

Comprehension Type

Direction (Q. Nos. 15 and 16) *This section contains a paragraph, describing theory, experiments, data, etc. Two questions related to the paragraph have been given. Each question has only one correct answer among the four given options (a), (b), (c) and (d).*

Passage

Potassium dichromate and potassium permanganate both are good oxidising agents in acidic medium and are used in the redox titrations, e.g. Mohr's salt *vs* $KMnO_4$ titration, Mohr's salt *vs* $K_2Cr_2O_7$ titration. The concentration of each solution. i.e. Mohr's salt, $KMnO_4$, $K_2Cr_2O_7$ is 0.1M, and the volume of dichromate required is X and of permanganate required is Y for the oxidation of Mohr's salt.

15 The given Mohr's salt is equivalent to

a. $\frac{1}{5}$ mol MnO_4^-
b. $\frac{1}{6}$ mol $Cr_2O_7^{2-}$
c. 1 equivalent each of MnO_4^- and $Cr_2O_7^{2-}$
d. All of these

16 If concentration of Mohr's salt is 1M, V_1 and V_2 are related as

a. $X > Y$
b. $X = Y$
c. $Y = X$
d. Data insufficient

Statement Type

Direction (Q. No. 17) *This section is based on Statement I and Statement II. Select the correct answer from the codes given below.*

a. Both Statement I and Statement II are correct and Statement II is the correct explanation of Statement I
b. Both Statement I and Statement II are correct and Statement II is not the correct explanation of Statement I
c. Statement I is correct but Statement II is incorrect
d. Statement II is correct but Statement I is incorrect

17 **Statement I** When pH is 3.4, phenolphthalein and methyl orange exist in their quinonoid form.

Statement II The colour of quinonoid form is usually deeper than the benzenoid form.

Working Space

Matching List Type

***Direction** (Q. No. 18) Choices for the correct combination of elements from Column I and Column II are given as options (a), (b), (c) and (d), out of which one is correct.*

18 Match the Column I with Column II and mark the correct option using the codes given below.

	Column I		Column II
i.	$KMnO_4 + H_2SO_4 + Fe^{2+} \longrightarrow$	p.	Equivalent weight = M (of underlined species)
ii.	$CuSO_4 + KI \longrightarrow$	q.	Equivalent weight = $\frac{M}{4}$
iii.	$As_2O_3 + I_2 \xrightarrow{H_2O}$	r.	Redox titration
iv.	Mohr salt + $K_2Cr_2O_7 \xrightarrow{H_2SO_4}$	s.	Iodimetry

Codes

	i	ii	iii	iv
a.	p,r	p,r	q,r,s	p,r
b.	p,r	p,r	q,s	p,r
c.	p	p,r	q,r,s	p,s
d.	p,r	s	q,r,s	p,r

One Integer Value Correct Type

***Direction** (Q. No. 19 and 20) This section contains 2 questions. When worked out will result in an integer from 0 to 9 (both inclusive).*

19 How many of the following are the examples of external indicators?
Starch, potassium ferricyanide, methyl orange, Euriochrome black-T, uranyl acetate, potassium permanganate, uronyl nitrate, phenosaframine

20 How many electrons are involved in the reaction of Mohr's salt with $KMnO_4$?

DPP-7 Principles Involved in Synthesis, Neutralisation and Kinetic Studies

Only One Option Correct Type

Working Space

Direction (Q. Nos. 1-13) *This section contains 13 multiple choice questions. Each question has four choices (a), (b), (c) and (d), out of which ONLY ONE is correct.*

1 Which metal sol cannot be prepared by Bredig's arc method?

a. K b. Cu
c. Au d. Pt

2 Chrome alum and Mohr's salt both are the double salts, i.e. show all the properties of their constitutents when dissolve in water. The ratio of water of crystallisation in alum and Mohr's salt is

a. 1 : 6 b. 6 : 1
c. 1 : 4 d. 4 : 1

3 The conversion of aniline to *p*-nitroaniline is not achieved by using nitrating mixture (a mixture of conc. H_2SO_4 and conc. HNO_3) because

a. it gives *ortho* as a major product
b. it results in the formation of a mixture of *o* and *p*-nitroaniline
c. it causes oxidation of amino group
d. it results in the formation of a mixture of *o* and *p*-nitro acetanilide

4 If the enthalpy of neutralisation of acetic acid with strong base is $-50.6\ kJ\ mol^{-1}$ and normal value of enthalpy of neutralisation is $-55.9\ kJ\ mol^{-1}$, the enthalpy of dissociation of acetic acid is

a. $-5.3\ kJ\ mol^{-1}$ b. $-8.9\ kJ\ mol^{-1}$
c. $+5.3\ kJ\ mol^{-1}$ d. None of these

5 Which reaction gives colloidal solution?

a. $Cu + HgCl_2 \longrightarrow CuCl_2 + Hg$
b. $2HNO_3 + 3H_2S \longrightarrow 3S + 4H_2O + 2NO$
c. $2Mg + CO_2 \longrightarrow 2MgO + C$
d. $Cu + Cu_2Cl_2 \longrightarrow Cu_2Cl_2$

6 When dilute aqueous solution of $AgNO_3$ (excess) is added to KI solution, positively charged sol particles of AgI are formed due to adsorption of

a. K^+ ion b. Ag^+ ion
c. I^- ion d. NO_3^- ion

Working Space

7 The following conversion can be achieved by treatment with

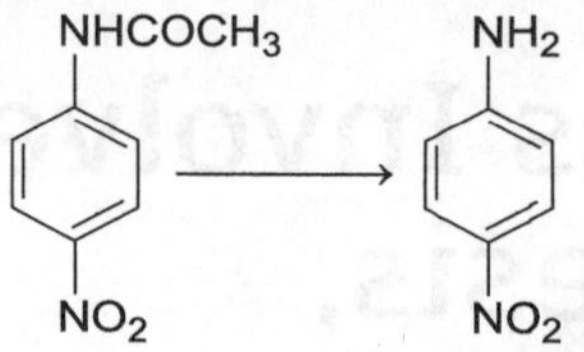

a. dilute acid
b. dilute alkali
c. Both (a) and (b)
d. None of these

8 When diazoamino benzene is treated with hydrochloric acid at 40° C, it results in the formation of

a. hydrazo compound
b. azobenzene
c. aniline yellow
d. No reaction

9 During the preparation of Mohr's salt, a small amount of dilute H_2SO_4 is also added to prevent the following conversion

a. Fe^{2+} to Fe^{3+}
b. $FeSO_4$ to $Fe(OH)_2$
c. $FeSO_4$ to $Fe(OH)_3$
d. Fe^{3+} to Fe^{2+}

10 Leveling bulb is used during experiment to study kinetics of the dissociation of H_2O_2 to ensure

a. uniform pressure difference between the room and the gases in the system
b. pressure within the reaction vessel is same as that in the room
c. same temperature as that of the room
d. None of the above

11 The heats of neutralisation of four acids I, II, III and IV are – 13.7, – 9.4, – 11.2 and – 12.4 kcal, respectively when they are neutralised by a common base. The order of acidic character is

a. I > II > III > IV
b. I > IV > III > II
c. IV > III > II > I
d. IV > II > III > I

12 A solution of 200 mL of 1M KOH is added to 200 mL of 1M HCl and the mixture is well shaken. This rise in temperature T_1 is noted. The same experiment is repeated with 100 mL of each solution and increase in temperature T_2 is again noted. The T_1 and T_2 are related as

a. $T_1 = T_2$
b. T_2 is twice as large as T_1
c. T_1 is twice as large as T_2
d. T_1 is four times as large as T_2

13 Sodium salt of sulphanilic acid

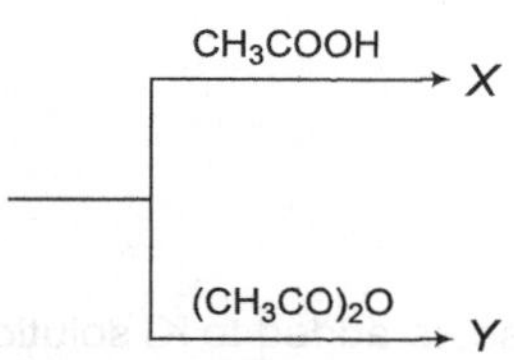

The correct statement about X and Y is/are

a. X is the acetylation product but Y is not
b. Y is an acetanilide but X does not
c. X and Y both are the acetanilide derivatives
d. AcOH is not a acetylation reagent

Working Space

One or More than One Options Correct Type

Direction (Q. Nos. 14 and 15) This section contains 2 multiple choice questions. Each question has four choices (a), (b), (c) and (d), out of which ONE or MORE THAN ONE are correct.

14 The true statements about aniline yellow is/are

a. Its nature is carcinogenic
b. It is also known by the name 4-aminoazobenzene
c. It is an acid dye
d. It is also known as 4-phenylazoaniline

15 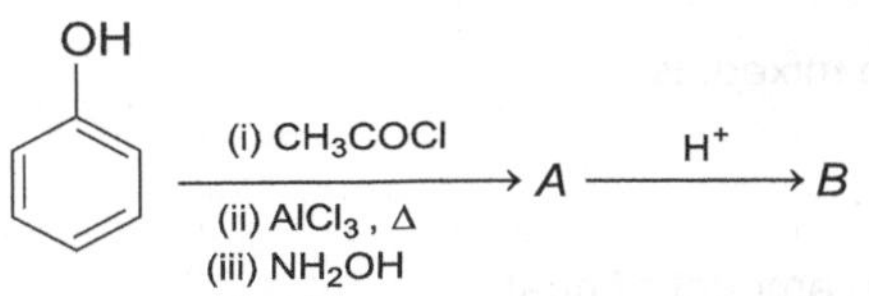

The true statements about the above reaction sequence is/are

a. *o*-hydroxy acetanilide is the only product
b. It involves Fries as well as Beckmann rearrangement
c. *p*-hydroxy acetanilide is obtained from the *E*-isomer of the oxime
d. In first step acetylation takes place at O, not at C

Matching List Type

Direction (Q. Nos. 16) Choices for the correct combination of elements from Column I and Column II are given as options (a), (b), (c) and (d), out of which one is correct.

16 Match the column and choose the correct answer using the codes given below:

	Column I		Column II (Enthalpy of neutralisation)
i.	$HCl\,(aq) + NaOH\,(aq) \longrightarrow NaCl + H_2O$	p.	– 57.3
ii.	$HF + NaOH \longrightarrow NaF + H_2O$	q.	No
iii.	$H^+ + OH^- \longrightarrow H_2O$	r.	> 57.3
iv.	$HCl + NaOH \longrightarrow NaCl + H_2O$ (in non-polar protic solvent)	s.	< 57.3

Codes

	i	ii	iii	iv		i	ii	iii	iv
a.	p	q	r	s	b.	p	r	p	s
c.	q	r	p	s	d.	q	r	q	s

Statement Type

Direction (Q. Nos. 17) This section is based on Statement I and Statement II. Select the correct answer from the codes given below.

a. Both Statement I and Statement II are correct and Statement II is the correct explanation of Statement I.
b. Both Statement I and Statement II are correct and Statement II is not the correct explanation of Statement I.
c. Statement I is correct but Statement II is incorrect.
d. Statement II is correct but Statement I is incorrect.

17 **Statement I** Dissolution of a substance in water may be exothermic or endothermic.
Statement II If $E_h > E_l$, the dissolution will be exothermic and *vice-versa*.

Comprehension Type

Working Space

***Direction** (Q. Nos. 18 and 19) This section contains a paragraph, describing theory, experiments, data, etc. Two questions related to the paragraph have been given. Each question has only one correct answer among the four given options (a), (b), (c) and (d).*

Passage

A student added 50 mL of 1.0 M HCl to 50 mL of 1.0 M NaOH in a coffee cup calorimeter and observed that temperature raises from 21.0°C to 27.5°C. The heat absorbed by the calorimeter is negligible. The total volume of the solution is 100 mL, its density is 1.0 g mL^{-1} and its specific heat is 4.18 J/g.

18 The heat change when 50 mL of each, i.e. HCl and NaOH are mixed, is

a. – 2.86 kJ
b. + 2.86 kJ
c. – 2.72 kJ
d. + 2.72 kJ

19 If 1000 m-mole of HCl are mixed with 1000 m-mole NaOH, the amount of heat change is

a. – 57.26 kJ
b. + 57.26 kJ
c. – 54.4 kJ
d. + 54.5 kJ

One Integer Value Correct Type

***Direction** (Q. No. 20) This section contains a question. When worked out will result in an integer from 0 to 9 (both inclusive).*

20 How many of the following compounds can be used to prepare triiodomethane by reaction with NaOI.

I. Acetone
II. Propan-2-ol
III. Ethanol
IV. Butanol
V. Ethanal
VI. Butanal
VII. Acetophenone
VIII. Benzophenone
IX. Formic acid

Revisal Problems for JEE Main

SET - 1

Only One Option Correct Type

Working Space

Direction *(Q. Nos. 1-29) This section contains 29 multiple choice questions. Each question has four choices (a), (b), (c) and (d), out of which ONLY ONE is correct.*

1 Sodium fusion extract, obtained from aniline, on treatment with iron (II) sulphate and H_2SO_4, in the presence of air gives a Prussian blue precipitate. The blue colour is due to the formation of

a. $Fe_4[Fe(CN)_6]_3$ b. $Fe_3[Fe(CN)_6]_2$ c. $Fe_4[Fe(CN)_6]_2$ d. $Fe[Fe(CN)_6]_3$

2 In the redox titrations, the substance, which is generally taken as a standard is

a. oxidising agent b. reducing agent
c. Any of the two d. The coloured solution

3 Which of the following reagents is used to remove sulphate ions from a solution containing sulphate as well as chloride ions ?

a. Sodium hydroxide b. Lead nitrate
c. Barium sulphate d. Potassium hydroxide

4 During the preparation of acetanilide from aniline, a small amount of zinc is added to the reaction mixture because

a. zinc induces the precipitation
b. zinc prevents the reduction of aniline during the reaction
c. zinc reduces the colour impurities in the aniline and also prevent it from being oxidised
d. zinc forms a white crystalline complex with aniline

5 In alum, the ratio of hydrated monovalent cation $[M(H_2O)]^+$, trivalent cation $[M'(H_2O)_6]^{3+}$ and sulphate ion is

a. 1 : 1 : 1 b. 1 : 2 : 3 c. 1 : 3 : 2 d. 1 : 1 : 2

6 A dark violet colour mixture in the presence of dil HCl changes to pale yellow solution. Mixture may contain

a. MnO_4^-, Fe^{2+} b. $MnO_4^-, C_2O_4^{2-}$
c. $I^-, Cr_2O_7^{2-}$ d. None of the these are correct

7 Colloidal solution of arsenious sulphide can be prepared by

a. electrodispersion method b. peptisation
c. double decomposition d. hydrolysis

8 Schiff's reagent is

a. magenta solution decolourised with sulphurous acid
b. magenta solution decolourised with chlorine
c. ammoniacal cobalt chloride solution
d. ammoniacal manganese sulphate solution

9 Cationic part of chromyl chloride is

a. Cr^{3+} b. CrO_2^+ c. CrO_2^{2+} d. CrO^{2+}

10 CoS (black) obtained in group IV of salt analysis is dissolved in aqua-regia, treated with an excess of $NaHCO_3$ and then Br_2 water is added. An apple green coloured stable complex is formed. It is

a. sodium cobaltocarbonate b. sodium cobaltibromide
c. sodium cobalticarbonate d. sodium cobaltobromide

11 Consider the following reaction,

Nitrite + acetic acid + thiourea $\longrightarrow N_2 + SCN^- + 2H_2O$

Formation of the product in the above reaction can be identified by

Working Space

a. $FeCl_3$/dil. HCl when blood red colour appears
b. $FeCl_3$/dil. HCl when blue colour appears
c. $K_2Cr_2O_7$/HCl when green colour appears
d. $KMnO_4$/HCl when colourless solution is formed

12 Consider the following properties/tests,
I. Silver mirror test II. Libermann's nitroso reaction III. Fehling's test IV. Baeyer's test
Sucrose does not give
a. II and IV b. I and III c. I, II and III d. I to IV

13 In the kinetic study of reaction of iodide ion with hydrogen peroxide, a known volume of sodium thiosulphate is added to
a. oxidise I^- to I_2 b. reduce I_2 to I^-
c. form a blue coloured soluble complex d. induce the reaction rate

14 Which of the following salt does not give positive test for nitrate ion ?
a. KNO_3 b. $NaNO_3$ c. $Pb(NO_3)_2$ d. $Mg(NO_3)_2$

15 In lake test of Al^{3+} ion, there is the formation of coloured floating due to
a. adsorption of litmus by H_2O b. adsorption by litmus of $Al(OH)_3$
c. adsorption of litmus $Al(OH)_4^-$ d. None of these

16 A + tap water $\longrightarrow$ White turbidity (soluble in aqueous NH_3)
(Gives ring test)
$A \xrightarrow{\Delta}$ Residue B + C (oxides of N) + O_2
C ↓ Paramagnetic (VD = 46)

A is
a. $Pb(NO_3)_2$ b. $AgNO_3$ c. $AgNO_2$ d. $NaNO_3$

17 Sulphuric acid is not used for the preparation of original solution in analysis of basic radicals because
a. it acts as a reductant
b. it results in the formation of insoluble sulphate with certain basic radicals
c. it is viscous in nature
d. it results in the formation of soluble complex

18 The titration of Mohr's salt *vs* $KMnO_4$ is a redox titration where permanganate only oxidises Fe^{2+} to Fe^{3+} but we can't use ferrous sulphate or any other salt of Fe^{2+} in place of Mohr's salt because
a. it is less stable than Mohr's slalt b. in air it is oxidised to ferric sulphate
c. in air it loses water of crystallisation d. All of these

19 In which case of mixing of a strong acid and a base each of 1N concentration, temperature increase is highest ?
a. 20 mL acid + 20 mL alkali b. 10 mL acid + 40 mL alkali
c. 25 mL acid + 25 mL alkali d. 35 mL acid + 15 mL alkali

20 The methods of preparation of lyophilic and lyophobic sols respectively are
a. oxidation and reduction b. dissolution and peptisation
c. peptisation and oxidation d. All of these

21 The standard enthalpy of formation of HCl (*g*), H^+ (*aq*) and Cl^- (*aq*) are −92.30, 0.00 and −167.44 kJ mol^{-1} respectively, then the standard enthalpy of solution of HCl (*g*) is
a. −53.7kJ b. +75.14kJ c. −75.14 kJ d. +53.7kJ

22 Why is HCl not used to make the medium acidic in oxidation reaction of $KMnO_4$ in acidic medium?
a. Both HCl and $KMnO_4$ are good oxidants
b. $KMnO_4$ oxidises HCl to Cl_2, which is also a good oxidant
c. $KMnO_4$is a weaker oxidising agent than HCl
d. $KMnO_4$ acts as a reductant in the presence of HCl

Working Space

23 The best way to ensure complete precipitation from saturated $H_2S(aq)$ of a metal ion M^{2+}, as its sulphide, $MS(s)$, is to

a. add an acid
b. increase $[H_2S]$ in solution
c. raise the pH
d. heat the solution

24 X is a white sodium salt. It is neutral to litmus and form a white precipitate insoluble in dilute nitric acid, when treated with silver nitrate. The anion of the salt could be

a. carbonate
b. chloride
c. sulphate
d. sulphide

25 The acid-base indicator methyl red has K_a of 1×10^{-4}. Its acidic form is red while its alkaline form is yellow. If methyl red is added to a colourless solution with pH = 7, what will be the colour of the indicator?

a. Yellow
b. Red
c. Orange
d. None of these

26 When one mole anhydrous $CuSO_4$ is dissolved in excess of water, – 66.4 kJ heat is evolved. When one mole of $CuSO_4 \cdot 5H_2O$ is dissolved in water, the heat change is 11.7 kJ. The enthalpy of hydration of $CuSO_4$ (anhy.) is

a. −78.1 kJ
b. −54.7kJ
c. +54.7kJ
d. +78.1 kJ

27 A mixture contains Cu^{2+}, Al^{3+} and Ni^{2+}. Following steps have been adopted but written in disorder.

I. Filter, boil off H_2S gas and add NH_4Cl, heat and add NH_4OH.
II. Filter, add NH_4OH and pass H_2S gas.
III. Pass H_2S gas into acidified solution of mixture.

Steps will be used in the following order

a. I, II and III
b. III, I and II
c. III, II and I
d. I, III and II

28 An organic compound '*A*' burns with a sooty flame. It gives negative Tollen's test but positive Brady's test. The compound '*A*' is

a. acetophenone
b. acetone
c. salicylic acid
d. benzaldehyde

29 The mass of potassium dichromate crystals required to oxidise 750 cm^3 of 0.6 M Mohr's salt solution is [Molar mass of potassium dichromate = 294 and of Mohr's salt is 392]

a. 0.49 g
b. 0.45 g
c. 22.05 g
d. 2.2 g

Statement Type

Direction (Q. Nos. 30-33) *This section is based on Statement I and Statement II. Select the correct answer from the codes given below.*

a. Both Statement I and Statement II are correct and Statement II is the correct explanation of Statement I
b. Both Statement I and Statement II are correct but Statement II is not the correct explanation of Statement I
c. Statement I is correct but Statement II is incorrect
d. Statement II is correct but Statement I is incorrect

30 **Statement I** When $H_2S(g)$ is passed into aqueous $Zn(CH_3COO)_2$ solution, zinc is precipitated as ZnS.

Statement II CH_3COOH is produced in the reaction which cannot dissolve ZnS formed.

31 **Statement I** $PbCl_2$ and Hg_2Cl_2 precipitates can be separated by hot water.

Statement II Hg_2Cl_2 is blackened by aqueous NH_3.

32 **Statement I** Bromide ion does not interfere in the chromyl chloride test.

Statement II When treated with acidified dichromate, bromide is oxidised to bromine, which dissolves in alkali to give a colourless solution.

33 **Statement I** Benzaldehyde and acetaldehyde can be distinguished by Fehling's solution.

Statement II All carbonyl compounds form bisulphite complex with $NaHSO_3$.

SET - 2

Revisal Problems for JEE Advanced

Only One Option Correct Type

Working Space

Direction (Q. Nos. 1-15) *This section contains 15 multiple choice questions. Each question has four choices (a), (b), (c) and (d), out of which ONLY ONE is correct.*

1 When chlorine is passed to a solution containing potassium halide in the presence of trichloromethane, a violet colour appears, which disappears to give a clear solution in the excess of the gas.

This test shows the presence of

a. I^-
b. Br^-
c. Cl^-
d. Both (a) and (b)

2 Mixture is initially orange in colour. When solution is prepared in dilute acid, it changes to dark brown colour. Mixture contains

a. $HgI_2, Cr_2O_7^{2-}$
b. $I^-, Cr_2O_7^{2-}$
c. I^-, Cu^{2+}
d. I^-, SO_3^{2-}

3 A large excess of $MgF_2(s)$ is maintained in contact with 1.00 L of pure water to produce a saturated solution. When an additional 1.00 L of pure water is added to the solid-liquid mixture and equilibrium re-established, compared to its value in the original saturated solution $[Mg^{2+}]$ will be

a. the same
b. twice as large
c. three as large
d. half as large

4 Aqueous solution of *A* can dissolve AgBr forming a soluble complex *B*. *A* also reacts with aqueous $AgNO_3$ solution giving white ppt *C* changing to black ppt *D*. *A*, *B*, *C* and *D* are

	A	*B*	*C*	*D*
a.	$Na_2S_2O_3$	$Na_3[Ag(S_2O_3)_2]$	$Ag_2S_2O_3$	Ag_2S
b.	$Na_2S_2O_3$	$Ag_2S_2O_3$	Ag_2S	$Na_3[Ag(S_2O_3)_2]$
c.	NH_3	$[Ag(NH_3)_2]Br$	$Ag(OH)$	Ag_2O
d.	NH_3	$Na[Ag(OH)_2]$	Ag_2O	$[Ag(NH_3)_2]Br$

5 Aqueous solution of *A* gives yellow ppt with aqueous K_2CrO_4. *A* may contain

a. Pb^{2+} or Ag^+ or Ba^{2+}
b. Pb^{2+} or Ba^{2+}
c. Ag^+ or Ba^{2+}
d. Pb^{2+} or Ag^+

6 A solution when diluted with H_2O and boiled, gives a white precipitate. On addition of excess NH_4Cl/NH_4OH, the volume of precipitate decreases leaving behind a white gelatinous precipitate. Identify the precipitate which dissolves in NH_4OH/NH_4Cl.

a. $Zn(OH)_2$
b. $Al(OH)_3$
c. $Mg(OH)_2$
d. $Ca(OH)_2$

7 Equal volumes of 1 M HCl and 1 M H_2SO_4 are neutralised by dilute NaOH and *x* and *y* kcal of heat are liberated respectively. *x* and *y* are related as

a. $x = y$
b. $x = 0.5\,y$
c. $x = 0.4\,y$
d. None of these

8 Turnbull's blue and Prussian's blue respectively are

I. $Fe^{II}[Fe^{II}(CN)_6]^{2-}$ II. $Fe^{III}[Fe^{III}(CN)_6]$ III, $Fe^{II}[Fe^{III}(CN)_6]^-$ IV. $Fe^{III}[Fe^{II}(CN)_6]^-$

a. I, II
b. I, III
c. III, IV
d. IV, III

Working Space

9 Nitrite (NO_2^-) interferes in the 'ring-test' of nitrate (NO_3^-). Some of the following reagents can be used for the removal of nitrite.

I. NH_4Cl
II. $(NH_2)_2CS$ (thiourea)
III. NH_2SO_3H (sulphamic acid)
IV. Sulphanilic acid

Correct choice is

a. I, II b. I, II, IV c. I, II, III d. II, III, IV

10 Three test tubes *A*, *B* and *C* contain Pb^{2+}, Hg_2^{2+} and Ag^+ (but unknown). To each aqueous solution NaOH is added in excess. Following changes occur.

A : Black ppt
B : Brown ppt
C : White ppt but dissolves in excess of NaOH

A, *B* and *C* contain respectively

a. Pb^{2+}, Hg_2^{2+} and Ag^+ b. Hg_2^{2+}, Ag^+, Pb^{2+}
c. Ag^+, Pb^{2+}, Hg_2^{2+} d. Ag^+, Hg_2^{2+}, Pb^{2+}

11 Some white colourless crystals are heated. A cracking sound is heard and brown fumes are given off. The residue is yellow-brown in colour. When a glowing splinter is held in the fumes, it is relighted. The fumes consists of

a. O_2 b. NO_2
c. Cl_2 d. NO_2 and O_2

12 When 80 mL of 1.0×10^{-5}M $Ba(OH)_2$ is added to 20 mL of 1.0×10^{-5} M $Fe_2(SO_4)_3$, then

$[K_{sp}(Fe(OH)_3) = 2.6 \times 10^{-39}, K_{sp}(BaSO_4) = 1.1 \times 10^{-10}]$

a. $BaSO_4$ is precipitated but $Fe(OH)_3$ does not
b. $Fe(OH)_3$ is precipitated but $BaSO_4$ does not
c. Both $Fe(OH)_3$ and $BaSO_4$ get precipitated
d. Neither $BaSO_4$ nor $Fe(OH)_3$ get precipitated

13 An indicator is a weak acid and its pH range is 3.1 to 4.5. If the neutral point of the indicator lies in the centre of $[H^+]$ corresponding to given pH range, the ionisation constant of the indicator is

a. 7.9×10^{-4} b. 4.13×10^{-4} c. 3.16×10^{-5} d. 4.00×10^{-3}

14 Four samples of acids and bases are taken for an experiment

I. 100 mL of 1 M NaOH and 100 mL of 1 M HCl
II. 100 mL of 2 M KOH and 100 mL of 1 M H_2SO_4
III. 100 mL of 1 M CH_3COOH and 100 mL of 1 M NaOH
IV. 100 mL of 1.0 M KOH and 100 mL of 0.5 M HNO_3

In each case the enthalpy of neutralisation (ΔH_{neu}) is calculated. The results indicate that

a. ΔH_{neu} is same in all the cases b. $(\Delta H_{neu})_1 = (\Delta H_{neu})_4$
c. $(\Delta H_{neu})_1 = (\Delta H_{neu})_2 = (\Delta H_{neu})_4$ d. $(\Delta H_{neu})_1 \neq (\Delta H_{neu})_2 \neq (\Delta H_{neu})_3 \neq (\Delta H_{neu})_4$

15 Consider the following statements about the following conversion:

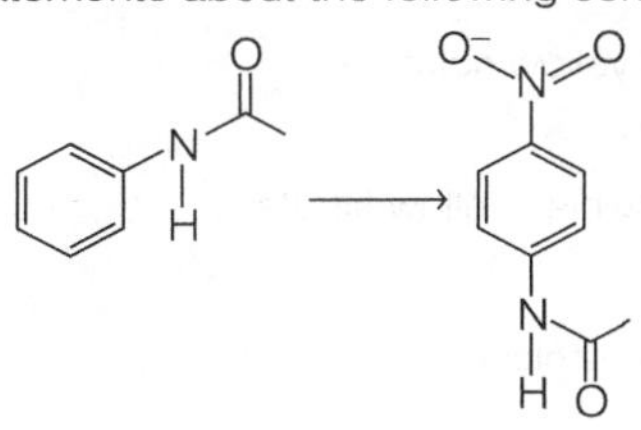

I. The attacking species here is nitronium ion.
II. The temperature of the reaction mixture is maintained below 0° C.
III. *o*-isomer is the major product due to H-bonding.
IV. It involves formation of cyclopentadienyl cation.

The correct statement(s) are

a. I and II b. I, II and IV c. I, III and IV d. All of these

One or More than One Options Correct Type

Working Space

Direction *(Q. Nos. 16-21) This section contains 6 multiple choice questions. Each question has four choices (a), (b), (c) and (d), out of which ONE or MORE THAN ONE are correct.*

16 In an aqueous solution, sulphide ion is present. It will form a

a. white ppt when treated with suspension of $CaCO_3$
b. purple colouration when treated with sodium of thiosulphate solution
c. black ppt when treated with aqueous solution lead acetate
d. yellow ppt when treated with a suspension of $CdCO_3$ in water

17 Removal of phosphate is done in the presence of acetic acid and sodium acetate. It is due to

a. common ion effect b. buffer solution c. solubility product d. pH of system

18 $X \xrightarrow[\text{HCl}]{\text{dil.}} \underset{\text{suffocating smell}}{Y} \xrightarrow[K_2Cr_2O_7]{\text{acidic}} \text{Green}$

$Y \xrightarrow{O_2} Z$

The correct statement about Z is/are

a. it is a decomposition product of ferric sulphate
b. it is used for drying gases
c. it is an anhydride of H_2SO_3
d. it is used to prepare Marshall's acid

19

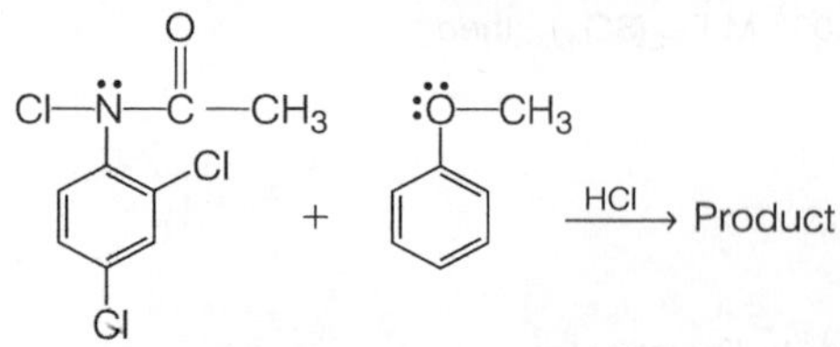

The product(s) of this reaction is/are

a. 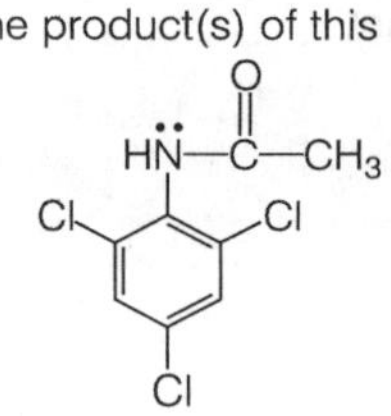

b. 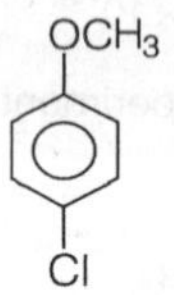

c. 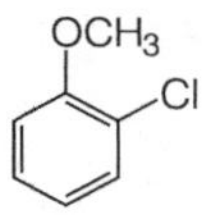

d. No reaction at all

20 For the given aqueous reaction, which of the statement(s) is/are true?

$$KI + K_3[Fe(CN)_6] \xrightarrow{\text{dil. } H_2SO_4} \text{Brownish yellow solution} \xrightarrow{ZnSO_4} \text{White ppt} + \text{Brownish yellow filtrate}$$

$$\text{White ppt} \longrightarrow \text{Colourless solution}$$

a. The first reaction is a redox reaction
b. White ppt is $Zn_3[Fe(CN)_6]_2$
c. Addition of filtrate to $Na_2S_2O_3$ starch solution gives blue colour
d. White ppt is soluble in NaOH solution

21 Isomers of C_7H_9N giving positive azo dye test $\xrightarrow{KMnO_4/OH^-} \xrightarrow[\Delta]{\text{Sodalime}} X$

X can

a. give azo dye test
b. give carbylamine reaction
c. be oxidised to benzoquinone by $Cr_2O_7^{2-}/H^+$
d. be converted into benzyl alcohol on reaction with HNO_2

Working Space

Comprehension Type

***Direction** (Q. Nos. 22-26) This section contains 2 paragraphs, each describing theory, experiments, data, etc. Five questions related to the paragraphs have been given. Each question has only one correct answer among the four given options (a), (b), (c) and (d).*

Passage I

p-amino-N,N-dimethylaniline is added to a strongly acidic solution of *X*. The resulting solution is treated with a few drops of aqueous solution of *Y* to give blue colouration due to formation of an intense blue ppt. The precipitate dissolves in excess of the reagent. Similarly, treatment of solution of *Y* with the solution of potassium hexacyanoferrate (III) leads to a brown colouration due to the formation of *Z*.

22 The compound *X* is

a. $NaNO_3$ b. NaCl
c. Na_2SO_4 d. Na_2S

23 The compound *Y* is

a. $MgCl_2$ b. $FeCl_2$
c. $FeCl_3$ d. $ZnCl_2$

24 The compound *Z* is

a. $Mg_2[Fe(CN)_6]$ b. $Fe[Fe(CN)_6]$
c. $Fe_4[Fe(CN)_6]_3$ d. $K_2Zn_3[Fe(CN)_6]_2$

Passage II

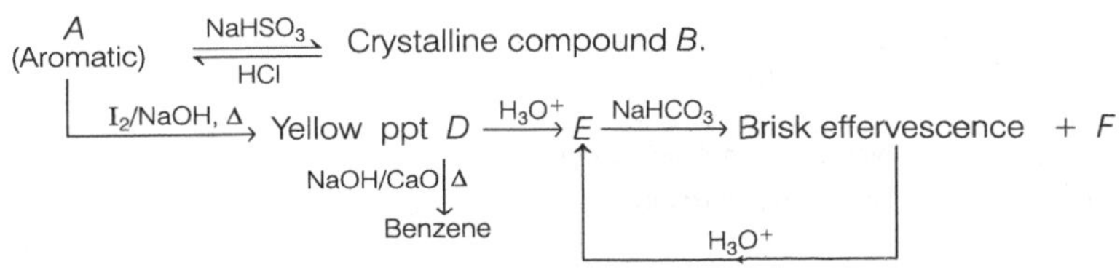

25 If *A* and *E* are present in a mixture, the reagent used for their separation is

a. $NaHCO_3$ b. $NaHSO_3$ c. Both (a) and (b) d. None of these

26 *E* can be confirmed by

a. neutral $FeCl_3$ b. $NaHCO_3$
c. Tollen's reagent d. Benedict's solution

Statement Type

***Direction** (Q. Nos. 27-29) This section is based on Statement I and Statement II. Select the correct answer from the codes given below.*

a. Both Statement I and Statement II are correct and Statement II is the correct explanation of Statement I.
b. Both Statement I and Statement II are correct and Statement II is not the correct explanation of Statement I.
c. Statement I is correct but Statement II is incorrect.
d. Statement II is correct but Statement I is incorrect.

27 **Statement I** $PbCl_2$ and AgCl precipitates can be separated by aqueous NH_3.

Statement II $PbCl_2$ precipitate change to soluble $[Pb(NH_3)_4]Cl_2$.

28 **Statement I** A chocolate colour ppt is formed when potassium ferrocyanide is added to a solution containing Cu^{2+}.

Statement II Here iron is oxidised from its + 2 to + 3 state.

29 **Statement I** Ascorbic acid decompose $NaHCO_3$ liberating CO_2.

Statement II A —COOH group is present in the ascorbic acid.

Matching List Type

Working Space

Direction (Q. Nos. 30-33) *Choices for the correct combination of elements from Column I and Column II are given as options (a), (b), (c) and (d), out of which one is correct.*

30 Match the Column I with Column II mark the correct option from the codes given below.

	Column I (Compound)		Column II (Functional group)
i.	Resorcinol	p.	—OH (enol)
ii.	Picric acid	q.	$—NO_2$
iii.	Salol	r.	—COOH
iv.	Ascorbic acid	s.	—OH (alcohol)

Codes

	i	ii	iii	iv
a.	p	p,q	p,r	s
b.	p	p,q	p,r	p,s
c.	p,q	p,r	r	p
d.	s	p	q	p

31 Match the Column I with Column II mark the correct option from the codes given below.

	Column I (Reaction with dil. HCl)		Column II (Confirmatory test)
i.	Sulphide	p.	Evolved gas turns lime water milky
ii.	Carbonate	q.	Evolved gas turns lead acetate paper black
iii.	Sulphite	r.	Evolved gas turns acidified dichromate solution green
iv.	Thiosulphate	s.	Gives white turbidity

Codes

	i	ii	iii	iv
a.	q	q	p,r	p,r,s
b.	q,r	p	p,r	p,r,s
c.	p	q	r	s
d.	q	s	p	r

32 Match the Column I with Column II mark the correct option from the codes given below.

	Column I (Titration)		Column II (Indicator)
i.	HCl *vs* NaOH	p.	Phenol red
ii.	NaOH *vs* CH_3COOH	q.	Phenolphthalein
iii.	H_2SO_4 *vs* NH_4OH	r.	Methyl orange
iv.	CH_3COOH *vs* NH_4OH	s.	Thymolphthalein

Codes

	i	ii	iii	iv
a.	q,r	s	r	p
b.	r,s	q	p	r
c.	p,q,r,s	q,r,s	p,r	s
d.	r,q,s	q,s	r	p

33 Match the Column I with Column II mark the correct option from the codes given below.

Working Space

	Column I		Column II
i.	$Fe^{3+} + [Fe(CN)_6]^{4-}$	p.	Cherry red
ii.	$Fe^{3+} + CNS^-$	q.	Pink
iii.	Ni^{2+} + DMG	r.	Blood red
iv.	$CrO_2Cl_2 + NaOH + Pb^{2+}$	s.	Blue
v.	$Mn^{2+} + PbO_2 + H^+$	t.	Yellow ppt

Codes

	i	ii	iii	iv	v
a.	q,r,s	q,s	r	p	q
b.	r,s	q	p	r	r
c.	s	r	p	t	q
d.	q	s	r	p	t

One Integer Value Correct Type

Direction *(Q. Nos. 34-36) This section contains 3 questions. When worked out will result in an integer from 0 to 9 (both inclusive).*

34 The number of interfering radicals among the given are
Phosphate, chloride, bromide, fluoride, oxalate, borate, sulphate

35 A metal chloride dissolves endothermally when 8.27 g of its anhydrous form are dissolved in excess of water, X kcal heat is absorbed. If the molar mass of chloride is 74.5 g, the enthalpy of solution is AX. Here, A is

36 In calorimeter, 400 mL 1.0 M HX is added with 400 mL of 1.0 M BOH. It raises the temperature of calorimeter by 4.4° C. If water equivalent of calorimeter is 12 g, specific heat is 1cal $deg^{-1} mL^{-1}$, then the enthalpy of neutralisation is

JEE Advanced & IIT JEE Archive

(Compilation of Last 10 Years' Questions)

1 Among PbS, CuS, HgS, MnS, Ag_2S, NiS, CoS, Bi_2S_3 and SnS_2, the total number of black coloured sulphides is **(2014 Adv., Integer Type)**

Passage for Q. Nos. (2-3)

An aqueous solution of metal ion M_1 reacts separately with reagents Q and R in excess to give tetrahedral and square planar complexes, respectively. An aqueous solution of another metal ion M_2 always forms tetrahedral complexes with these reagents. Aqueous solution of M_2 on reaction with reagent S gives white precipitate which dissolves in excess of S. The reactions are summarised in the scheme given below **(2013 Adv., Comprehension Type)**

$$\text{Tetrahedral} \xleftarrow[\text{excess}]{Q} M_1 \xrightarrow[\text{excess}]{R} \text{Square planar}$$

$$\text{Tetrahedral} \xleftarrow[\text{excess}]{Q} M_2 \xrightarrow[\text{excess}]{R} \text{Tetrahedral}$$

$$\downarrow S\text{, stoichiometric amount}$$

$$\text{White precipitate} \xrightarrow[\text{excess}]{S} \text{Precipitate dissolves}$$

2 M_1, Q and R, respectively are

a. Zn^{2+}, KCN and HCl
b. Ni^{2+}, HCl and KCN
c. Cd^{2+}, KCN and HCl
d. Co^{2+}, HCl and KCN

3 Reagent S is

a. $K_4[Fe(CN)_6]$
b. Na_2HPO_4
c. K_2CrO_4
d. KOH

Passage for Q. Nos. (4-5)

An aqueous solution of a mixture of two inorganic salts,when treated with dilute HCl, gave a precipitate (P) and filtrate (Q). The precipitate P was found to dissolve in hot water. The filtrate (Q) remained unchanged, when treated with H_2S in a dilute mineral acid medium. However, it gave a precipitate (R) with H_2S in an ammoniacal medium. The precipitate R gave a coloured solution (S), when treated with H_2O_2 in an aqueous NaOH medium.

4 The precipitate P contains **(2013 Adv., Comprehension Type)**

a. Pb^{2+} b. Hg_2^{2+} c. Ag^+ d. Hg^{2+}

5 The coloured solution S contains

a. $Fe_2(SO_4)_3$
b. $CuSO_4$
c. $ZnSO_4$
d. Na_2CrO_4

6 Upon treatment with ammoniacal H_2S, the metal ion that precipitates as a sulphide is **(2013 Adv., Only One Option Correct Type)**

a. Fe (III)
b. Al (III)
c. Mg (II)
d. Zn (II)

7 For the given aqueous reaction which of the statement(s) is(are) true?

$$\text{Excess KI} + K_3[Fe(CN)_6] \xrightarrow{\text{Dilute } H_2SO_4} \text{Brownish-yellow solution}$$

$$\downarrow ZnSO_4$$

$$(\text{White precipitate} + \text{Brownish-yellow filtrate})$$

$$\downarrow Na_2S_2O_3$$

$$\text{Colourless solution}$$

(2012, One or More than One Options Correct Type)

a. The first reaction is a redox reaction
b. White precipitate is $Zn_3[Fe(CN)_6]_2$
c. Addition of filtrate to starch solution gives blue colour
d. White precipitate is soluble in NaOH solution

Passage for Q. Nos. (8-10)

When a metal rod M is dipped into an aqueous colourless concentrated solution of compound N, the solution turns light blue. Addition of aqueous NaCl to the blue solution gives a white precipitate O. Addition of aqueous NH_3 dissolves O and gives an intense blue solution. **(2011, Comprehension Type)**

8 The metal rod M is

a. Fe
b. Cu
c. Ni
d. Co

9 The compound N is

a. $AgNO_3$
b. $Zn(NO_3)_2$
c. $Al(NO_3)_3$
d. $Pb(NO_3)_2$

10 The final solution contains

a. $[Pb(NH_3)_4]^{2+}$ and $[CoCl_4]^{2-}$
b. $[Al(NH_3)_4]^{3+}$ and $[Cu(NH_3)_4]^{2+}$
c. $[Ag(NH_3)_2]^+$ and $[Cu(NH_3)_4]^{2+}$
d. $[Ag(NH_3)_2]^+$ and $[Ni(NH_3)_6]^{2+}$

11 Passing H_2S gas into a mixture of Mn^{2+}, Ni^{2+}, Cu^{2+} and Hg^{2+} ions in an acidified aqueous solution precipitates **(2011, Only One Option Correct Type)**

a. CuS and HgS
b. MnS and CuS
c. MnS and NiS
d. NiS and HgS

Passage for Q. Nos. (12-14)

p-amino-N, N-dimethylaniline is added to a strongly acidic solution of *X*. The resulting solution is treated with a few drops of aqueous solution of *Y* to yield blue colouration due to the formation of methylene blue. Treatment of the aqueous solution of *Y* with the reagent potassium hexacyanoferrate (II) leads to the formation of an intense blue precipitate. The precipitate dissolves on excess addition of the reagent. Similarly, treatment of the solution of *Y* with the solution of potassium hexacyanoferrate (III) leads to a brown colouration due to the formation of *Z*.

(2009, Comprehension Type)

12 The compound *X*, is

a. $NaNO_3$ b. $NaCl$
c. Na_2SO_4 d. Na_2S

13 The compound *Y* is

a. $MgCl_2$ b. $FeCl_2$
c. $FeCl_3$ d. $ZnCl_2$

14 The compound *Z* is

a. $Mg_2[Fe(CN)_6]$
b. $Fe[Fe(CN)_6]$
c. $Fe_4[Fe(CN)_6]_3$
d. $K_2Zn_3[Fe(CN)_6]_2$

15 A solution of colourless salt H on boiling with excess NaOH produces a non-flammable gas. The gas evolution ceases after some time. Upon addition of Zn dust to the same solution, the gas evolution restarts. The colourless salt (s) H is (are) **(2008, One or More than One Options Correct Type)**

a. NH_4NO_3 b. NH_4NO_2
c. NH_4Cl d. $(NH_4)_2SO_4$

16 A solution of a metal ion when treated with KI gives a red precipitate which dissolves in excess KI to give a colourless solution. Moreover, the solution of metal ion on treatment with a solution of cobalt (II) thiocyanate gives rise to a deep blue crystalline precipitate. The metal ion is

(2007, Only One Option Correct Type)

a. Pb^{2+} b. Hg^{2+} c. Cu^{2+} d. Co^{2+}

17 $MgSO_4$ on reaction with NH_4OH and Na_2HPO_4 forms a white crystalline precipitate. What is its formula? **(2006, Only One Option Correct Type)**

a. $Mg(NH_4)PO_4$ b. $Mg_3(PO_4)_2$
c. $MgCl_2 \cdot MgSO_4$ d. $MgSO_4$

18 $CuSO_4$ decolourises on addition of KCN, the product is **(2006, Only One Option Correct Type)**

a. $[Cu(CN)_4]^{2-}$
b. Cu^{2+} get reduced to form $[Cu(CN)_4]^{3-}$
c. $Cu(CN)_2$
d. $CuCN$

19 A solution when diluted with H_2O and boiled, it gives a white precipitate. On addition of excess NH_4Cl / NH_4OH, the volume of precipitate decreases leaving behind a white gelatinous precipitate. Identify the precipitate which dissolves in NH_4OH / NH_4Cl.

(2006, Only One Option Correct Type)

a. $Zn(OH)_2$ b. $Al(OH)_3$ c. $Mg(OH)_2$ d. $Ca(OH)_2$

20 A metal nitrate reacts with KI to give a black precipitate which on addition of excess of KI convert into orange colour solution. The cation of metal nitrate is **(2005, Only One Option Correct Type)**

a. Hg^{2+} b. Bi^{3+} c. Sn^{2+} d. Pb^{2+}

Answers *with* Explanations

Answers

1. *p*-Block Elements

DPP-1 General Properties of 15th Group (Nitrogen and Phosphorus)

2. (i) between -3 to $+5$ (ii) (a) -3 and $+5$ (b) -3 and $+3$ (c) -3 and -3 (d) $-\frac{1}{3}$

5. *a* **6.** *b* **7.** *c* **8.** *c* **9.** *b* **10.** *b* **11.** *b* **12.** *c* **13.** *a,c*
14. *a,b,c* **15.** *a,b,d* **16.** *a,d* **17.** *d* **18.** *a* **19.** *d* **20.** *a* **21.** *b* **22.** 4 **23.** 8
24. 4 **25.** *c*

DPP-2 Nitrogen and Phosphorus Compounds

1. (i) Nitrous oxide (N_2O) and nitric oxide (NO)

(ii) In gaseous state they have unpaired electron and because of formation of dimers in liquid and solid state, they are diamagnetic.

(iii) Nitrous oxide

(iv) N_2O_3 and P_2O_3 are acidic, As_2O_3 and Sb_2O_3 are amphoteric and Bi_2O_3 is basic.

2. Normal salts do not undergo neutralisation with bases. Acidic salts are NaH_2PO_3, NaH_2PO_2, Na_2HPO_4 and $Na_2H_2P_2O_7$.

5. *a* **6.** *c* **7.** *a* **8.** *c* **9.** *c* **10.** *a* **11.** *b* **12.** *b* **13.** *a,b,c*
14. *a,c,d* **15.** *b,c,d* **16.** *a,b,c,d* **17.** *b,c,d* **18.** *c* **19.** *b* **20.** *a* **21.** *a* **22.** 5
23. 4 **24.** 2 **25.** *b*

DPP-3 General Properties of 16th Group (Oxygen, Ozone and Sulphur)

2. (i) 2.5 (ii) +6 (iii) -1 (iv) +2 **4.** (iii) $OF_2(+2)$ and $O_2F_2(+1)$

6. *d* **7.** *d* **8.** *b* **9.** *c* **10.** *b* **11.** *c* **12.** *c* **13.** *d* **14.** *d* **15.** *c*
16. *c,d* **17.** *a,b,d* **18.** *b,c,d* **19.** *a,b,c,d* **20.** *b,c,d* **21.** *b* **22.** *d* **23.** *b* **24.** *c*
25. *a* **26.** 6 **27.** 6 **28.** 4 **29.** 2 **30.** *a*

DPP-4 Oxygen and Sulphur Compounds

3. (i) $H_2S_2O_7$ (ii) Sulphur (iii) $H_2S_2O_8$ (iv) SO_3 (v) Conc. H_2SO_4

5. (i) Thiosulphuric acid ($H_2S_2O_3$) and thiosulphurous acid ($H_2S_2O_2$)

(ii) Thiosulphuric acid (iii) Pyrosulphuric acid (v) Thiosulphuric acid

6. *d* **7.** *b* **8.** *d* **9.** *c* **10.** *c* **11.** *a* **12.** *b* **13.** *b* **14.** *a* **15.** *d*
16. *d* **17.** *c,d* **18.** *a,d* **19.** *a,b,c,d* **20.** *b,d* **21.** *d* **22.** *b* **23.** *b* **24.** *c*
25. *a* **26.** 0 **27.** 5 **28.** 4 **29.** 9 **30.** 6

DPP-5 General Properties of 17th Group (Halogens)

1. (i) Bromine (ii) Iodine (iii) Fluorine (iv) Fluorine (v) Chlorine
(vi) Chlorine, bromine and iodine

6. *d* **7.** *b* **8.** *c* **9.** *b* **10.** *b* **11.** *d* **12.** *a* **13.** *d* **14.** *a* **15.** *c*
16. *b,c* **17.** *b,c* **18.** *a,b,d* **19.** *b,c* **20.** *a,c* **21.** *b* **22.** *b* **23.** *c* **24.** *a* **25.** 3
26. 8 **27.** 1 **28.** 5 **29.** 6 **30.** *a*

DPP-6 Compounds of Halogens (F, Cl, Br and I)

2. (i) $HClO < HClO_2 < HClO_3 < HClO_4$ (ii) $HClO > HClO_2 > HClO_3 > HClO_4$
(iii) $HClO < HClO_2 < HClO_3 < HClO_4$ (iv) $I^- > Br^- > Cl > F^-$
(v) $HOF > HOCl > HOBr > HOI$ (vi) $HI > HBr > HCl > HF$

6. *d* **7.** *c* **8.** *a* **9.** *a* **10.** *c* **11.** *c* **12.** *a* **13.** *d* **14.** *b* **15.** *b*
16. *a,c,d* **17.** *a,c,d* **18.** *a* **19.** *a,b,c* **20.** *a,c* **21.** *c* **22.** *b* **23.** *d* **24.** *a* **25.** *c*
26. 8 **27.** 8 **28.** 7 **29.** 3 **30.** *a*

DPP-7 Interhalogens and Pseudo Halogens

1. (i) Fluorine (ii) +1, +3, +5 and +7
2. Bond energy of interhalogen molecule is less than halogen molecules.
5. (i) $K^+ [IF_6]^-$ (ii) KCl + Br (iii) RbCl + ICl (iv) $[I_3]^+ [AlCl_4]^-$

6. *d*	**7.** *a*	**8.** *d*	**9.** *b*	**10.** *c*	**11.** *a*	**12.** *b*	**13.** *d*	**14.** *c*	**15.** *c*
16. *a,c*	**17.** *a,c*	**18.** *a,b,d*	**19.** *a,b,c*	**20.** *c,d*	**21.** *c*	**22.** *a*	**23.** *d*	**24.** *b*	**25.** *b*
26. 4	**27.** 6	**28.** 2	**29.** 4	**30.** 7					

DPP-8 General Properties and Compounds of Xenon

1. (i) Helium (ii) Argon (iii) Xenon (iv) Xenon (v) Xenon (vi) Krypton (vii) Radon
2. (i) $^{226}_{88}Ra \longrightarrow ^{222}_{86}Rn + ^{4}_{2}He$ (ii) + 8 (iii) Fractional distillation
   (iv) Activated coconut charcoal
   (v) KOH absorbs CO_2, pyrogallol solution absorbs O_2 and conc. H_2SO_4 and P_4O_{10} absorb moisture.

6. *a*	**7.** *d*	**8.** *d*	**9.** *b*	**10.** *b*	**11.** *b*	**12.** *a*	**13.** *d*	**14.** *d*	**15.** *c*
16. *a,b,c*	**17.** *a,c,d*	**18.** *b,c,d*	**19.** *b,d*	**20.** *a,b,c,d*		**21.** *c*	**22.** *d*	**23.** *a*	**24.** *d*
25. *a*	**26.** 1	**27.** 3	**28.** 3	**29.** 9	**30.** 8				

Revisal Problems (JEE Main)

1. *c*	**2.** *d*	**3.** *c*	**4.** *a*	**5.** *a*	**6.** *d*	**7.** *a*	**8.** *a*	**9.** *c*	**10.** *a*
11. *a*	**12.** *d*	**13.** *b*	**14.** *c*	**15.** *c*	**16.** *a*	**17.** *c*	**18.** *c*	**19.** *d*	**20.** *a*
21. *d*	**22.** *d*	**23.** *c*	**24.** *c*	**25.** *c*	**26.** *a*	**27.** *c*	**28.** *d*	**29.** *a*	**30.** *b*

Revisal Problems (JEE Advanced)

1. *b*	**2.** *d*	**3.** *a*	**4.** *d*	**5.** *b*	**6.** *a*	**7.** *d*	**8.** *d*	**9.** *d*	**10.** *d*
11. *a,b*	**12.** *b,c*	**13.** *b,d*	**14.** *a,d*	**15.** *a,c,d*	**16.** *a,d*	**17.** *b,c,d*	**18.** *a,b,c,d*	**19.** *a,b,c*	**20.** *a,b,d*
21. *a*	**22.** *d*	**23.** *a*	**24.** *a*	**25.** 4	**26.** 3	**27.** 4	**28.** 4	**29.** 6	**30.** 6

JEE Main & AIEEE Archive

1. *c*	**2.** *a*	**3.** *a,b,c,d*	**4.** *c*	**5.** *a*	**6.** *d*	**7.** *c*	**8.** *a*	**9.** *d*	**10.** *c*
11. *c*	**12.** *c*	**13.** *a*	**14.** *b*	**15.** *c*					

JEE Advanced & IIT JEE Archive

1. *c*	**2.** *a*	**3.** *a,c,d*	**4.** *a*	**5.** *a*	**6.** *d*	**7.** *c*	**8.** *a*	**9.** *b*	**10.** *c*
11. *d*	**12.** *d*	**13.** 5	**14.** 4	**15.** *a*	**16.** *b*	**17.** *a,b,c*	**18.** *b*	**19.** *c*	**20.** *c*
21. *b*	**22.** *d*	**23.** *a*	**24.** *c*	**25.** *a*	**26.** *c*	**27.** *b*	**28.** *b*	**29.** *c*	

2. *d* & *f*-Block Elements

DPP-1 General Properties of *d*-Block Elements

1. (i) Iron (ii) Mercury (iii) Tungsten (iv) Technetium (v) Silver (vi) Scandium and osmium
2. $(n-1)$ (d)-orbitals are complete.
5. (i) Copper iron pyrites ($CuFeS_2$) (ii) Zn, Fe and Co (iii) 87.5% gold

6. *b*	**7.** *a*	**8.** *c*	**9.** *d*	**10.** *a*	**11.** *b*	**12.** *c*	**13.** *d*	**14.** *c*	**15.** *a*
16. *a,b*	**17.** *b,c,d*	**18.** *b,c*	**19.** *b,c*	**20.** *b,d*	**21.** *c*	**22.** *c*	**23.** *b*	**24.** *a*	**25.** 2
26. 3	**27.** 5	**28.** 4	**29.** 4	**30.** *a*					

DPP-2 General Chemical and Magnetic Properties of *d*-Block Elements

3. Cr^{2+} 4. Manganese in 3*d* series, ruthenium in 4*d* series and osmium in 5*d* series.

6. *d*	**7.** *d*	**8.** *a*	**9.** *c*	**10.** *b*	**11.** *c*	**12.** *d*	**13.** *a*	**14.** *d*	**15.** *b*
16. *b,c,d*	**17.** *a,c,d*	**18.** *b*	**19.** *b,c*	**20.** *b,d*	**21.** *d*	**22.** *b*	**23.** *a*	**24.** *d*	**25.** *a*
26. 6	**27.** 7	**28.** 6	**29.** 5	**30.** *a*					

DPP-3 Oxides and Oxo Anions of Cr and Mn

4. (i) ZnO (ii) $ZnO \cdot CoO(CoZnO_2)$ (iii) TiO_2 (iv) ZrO_2 (v) OsO_4 (vi) Hg_2Cl_2 (vii) $HgCl_2$

5. 0.05 M **6.** *a* **7.** *a* **8.** *b* **9.** *d* **10.** *b* **11.** *c* **12.** *c* **13.** *a* **14.** *a*
15. *c* **16.** *a,c,d* **17.** *a,b,c* **18.** *a,b,d* **19.** *b* **20.** *a,b,d* **21.** *d* **22.** *b* **23.** *a* **24.** *b*
25. 10/3 **26.** 8 **27.** 8 **28.** 3 **29.** *d* **30.** *a*

DPP-4 *f*-Block Elements-I (Lanthanides)

2. +4 oxidation state : Cerium, praceodymium, neodymium, terbium and dysprosium
+2 oxidation state : Europium, ytterbium, samarium, thullium, neodymium

6. *d* **7.** *d* **8.** *d* **9.** *a* **10.** *d* **11.** *a* **12.** *b* **13.** *a* **14.** *c* **15.** *c*
16. *a,d* **17.** *a,b,d* **18.** *a,b,c,d* **19.** *a,b,d* **20.** *a,b,c* **21.** *a* **22.** *b* **23.** *a* **24.** 5 **25.** 0
26. 4 **27.** 3 **28.** 8 **29.** a **30.** a

DPP-5 *f*-Block Elements-II (Actinides)

4. (i) Elements after uranium are called transuranium elements.
(ii) Neptunium
(iii) +3 to +7

6. *d* **7.** *b* **8.** *d* **9.** *d* **10.** *a* **11.** *c* **12.** *d* **13.** *b* **14.** *b* **15.** *d*
16. *a,b* **17.** *a,c* **18.** *a,d* **19.** *a,b,c* **20.** *a,b,d* **21.** *a* **22.** *d* **23.** *a* **24.** 9
25. 0 **26.** 4 **27.** 2 **28.** 8 **29.** *b* **30.** *d*

Revisal Problems (JEE Main)

1. *c* **2.** *d* **3.** *a* **4.** *c* **5.** *d* **6.** *c* **7.** *c* **8.** *b* **9.** *c* **10.** *c*
11. *d* **12.** *a* **13.** *b* **14.** *a* **15.** *d* **16.** *b* **17.** *c* **18.** *c* **19.** *c* **20.** *d*
21. *b* **22.** *b* **23.** *a* **24.** *a* **25.** *c* **26.** *d* **27.** *a* **28.** *b* **29.** *c* **30.** *b*

Revisal Problems (JEE Advanced)

1. *a* **2.** *a* **3.** *c* **4.** *a* **5.** *b* **6.** *c* **7.** *a* **8.** *c* **9.** *b* **10.** *c*
11. *a,b,c* **12.** *a,b,d* **13.** *a,d* **14.** *a,b,d* **15.** *b,d* **16.** *a,b,d* **17.** *a,b,c,d* **18.** *c,d* **19.** *a,b,d*
20. *a,b,c* **21.** *b* **22.** *a* **23.** *a* **24.** *d* **25.** 3 **26.** 7 **27.** 6 **28.** 4 **29.** 9
30. 4

JEE Main & AIEEE Archive

1. *a* **2.** *d* **3.** *d* **4.** *d* **5.** *d* **6.** *a* **7.** *c* **8.** *b* **9.** *b* **10.** *d*
11. *d* **12.** *d* **13.** *d* **14.** *d* **15.** *c* **16.** *d* **17.** *a* **18.** *d* **19.** *a* **20.** *b*
21. *b* **22.** *b* **23.** *a* **24.** *a*

JEE Advanced & IIT JEE Archive

1. 7 **2.** *a* **3.** *a,c,d* **4.** *a,c,d* **5.** *b* **6.** *c*

3. Coordination Compounds

DPP-1 Introduction of Coordination Compounds and Werner's Theory

1. 8 ions **6.** *d* **7.** *b* **8.** *a* **9.** *c* **10.** *d*
11. *a* **12.** *c* **13.** *a* **14.** *d* **15.** *a* **16.** *c,d* **17.** *a,b,d* **18.** *b,d* **19.** *a,c,d* **20.** *b,c*
21. *b* **22.** *a* **23.** *a* **24.** 6 **25.** 4 **26.** 0 **27.** 0 **28.** 3 **29.** *b* **30.** *d*

DPP-2 IUPAC Nomenclature of Coordination Compounds

1. (i) $K_3[AgF_4]$ (ii) $Ca[Co(C_2O_4)_2(NH_3)_2]$ (iii) $[Cr(C_6H_6)_2]$ (iv) $Na_3[Ag(S_2O_3)_2]$

2. (i) Tetrammineplatinum (II) tetrachloroplatinate (II)
(ii) Ammonium diamminetetraisothiocyanatochromate (III)
(iii) Potassium trichloro(ethene)platinate (II)
(iv) *Tris* (bipyridine) iron (II) ion

3. (i) Pentamminechloroplatinum (IV) chloride $[Pt(NH_3)_5Cl]Cl_3$
(ii) Potassium hexachloroplatinate (IV) $K_2[PtCl_6]$
(iii) Triamminetrichloroiridium (III) $[Ir(NH_3)_3Cl_3]$
(iv) Tetramminedichlorocobalt (III) chloride $[Co(NH_3)_4Cl_2]Cl$

4. (i) Acetylacetanato (acac)
(ii) Glycinato (gly)
(iii) 2,2′ bipyridine (bipy)
(iv) Diethylenetriamine (dien)

6. *d* **7.** *a* **8.** *a* **9.** *c* **10.** *c*
11. *b* **12.** *c* **13.** *b* **14.** *a* **15.** *c* **16.** *a,b* **17.** *a,b,d* **18.** *a,c* **19.** *a,c* **20.** *c,d*
21. *a* **22.** *c* **23.** *c* **24.** *d* **25.** 4 **26.** 3 **27.** 5 **28.** 1 **29.** 5 **30.** *c*

DPP-3 Isomerism in Coordination Compounds-I

4. (i) Zero (ii) Zero (iii) Two (iv) Two
6. *b* **7.** *c* **8.** *a* **9.** *c* **10.** *c*
11. *b* **12.** *c* **13.** *a* **14.** *d* **15.** *c* **16.** *a,b,c* **17.** *a,c,d* **18.** *a,b* **19.** *a,b,d* **20.** *a,d*
21. *b* **22.** *b* **23.** *c* **24.** *d* **25.** 3 **26.** 3 **27.** 4 **28.** 2 **29.** 3 **30.** *c*

DPP-4 Isomerism in Coordination Compounds-II

1. (i) Zero (ii) Zero
2. $[CoCl_2Br_2]^{2-}$ is tetrahedral while $[PdBr_2Cl_2]^{2-}$ is square planar.
4. *cis*-isomer of $[Pt(NH_3)_2Cl_2]$ **6.** *a* **7.** *b* **8.** *a* **9.** *c* **10.** *c*
11. *d* **12.** *c* **13.** *c* **14.** *c* **15.** *c* **16.** *a,b,c* **17.** *a,b,c* **18.** *b,d* **19.** *b,c* **20.** *b,c*
21. *b* **22.** *d* **23.** *b* **24.** *a* **25.** 9 **26.** 4 **27.** 5 **28.** 0 **29.** 4 **30.** *c*

DPP-5 Valence Bond Theory, Effective Atomic Number and Magnetic Properties

3. $x = 6$ **5.** trithiocyanato iron (III), 5.916 BM
6. *d* **7.** *a* **8.** *b* **9.** *d* **10.** *a*
11. *d* **12.** *d* **13.** *c* **14.** *c* **15.** *a* **16.** *b,d* **17.** *a,c,d* **18.** *a,c,d* **19.** *a,c,d*
20. *a,b,c* **21.** *d* **22.** *a* **23.** *a* **24.** *a* **25.** 4 **26.** 4 **27.** 4 **28.** 4 **29.** 0
30. *b*

DPP-6 Crystal Field Theory (CFT) and Colour of Complexes

2. (i) 492 nm (ii) 20300 cm^{-1} **6.** *d* **7.** *a* **8.** *c* **9.** *a* **10.** *b*
11. *d* **12.** *a* **13.** *d* **14.** *c* **15.** *a* **16.** *a,d* **17.** *a,b,d* **18.** *a,c,d* **19.** *a,b,c,d* **20.** *c,d*
21. *d* **22.** *c* **23.** *c* **24.** *a* **25.** 6 **26.** 4 **27.** 6 **28.** 4 **29.** 6 **30.** *a*

DPP-7 Bonding in Organometallic Compounds, Stability and Applications of Coordination Compounds

6. *b* **7.** *b* **8.** *b* **9.** *c* **10.** *a* **11.** *b* **12.** *a* **13.** *c* **14.** *b* **15.** *d*
16. *a,b* **17.** *a,b* **18.** *a,c,d* **19.** *a,b,c* **20.** *b,d* **21.** *b* **22.** *a* **23.** *a* **24.** *a* **25.** 6
26. 3 **27.** 2 **28.** 6 **29.** 0 **30.** *a*

Revisal Problems (JEE Main)

1. *a* **2.** *a* **3.** *d* **4.** *c* **5.** *d* **6.** *a* **7.** *a* **8.** *c* **9.** *b* **10.** *c*
11. *b* **12.** *a* **13.** *a* **14.** *c* **15.** *c* **16.** *c* **17.** *d* **18.** *b* **19.** *a* **20.** *b*
21. *b* **22.** *c* **23.** *d* **24.** *d* **25.** *a* **26.** *c* **27.** *c* **28.** *a* **29.** *d*

Revisal Problems (JEE Advanced)

1. *b*	2. *c*	3. *c*	4. *c*	5. *c*	6. *b*	7. *d*	8. *c*	9. *a*	10. *b*
11. *a,b*	12. *a,b,d*	13. *a,b,c*	14. *b,c*	15. *a,c*	16. *b,d*	17. *b,d*	18. *b,c,d*	19. *b,c,d*	
20. *a,b,c*	21. *d*	22. *a*	23. *a*	24. *d*	25. *d*	26. 4	27. 6	28. 2	29. 4
30. 2	31. *a*								

JEE Main & AIEEE Archive

1. *b*	2. *c*	3. *b*	4. *c*	5. *c*	6. *b*	7. *c*	8. *b*	9. *d*	10. *b*
11. *d*	12. *a*	13. *d*	14. *b*	15. *d*	16. *b*	17. *d*	18. *d*	19. *b*	20. *c*
21. *a*	22. *b*	23. *a*	24. *d*	25. *a*	26. *d*	27. *b*	28. *a*	29. *b*	30. *d*
31. *c*	32. *c*								

JEE Advanced & IIT JEE Archive

1. *b*	2. 8	3. *b*	4. *a*	5. *d*	6. *c*	7. *b*	8. *c*	9. 6	10. 3
11. *b*	12. *c*	13. *b*	14. *c,d*	15. *a*	16. *b*	17. *c*	18. *c*	19. *b*	20. *a*
21. *b*	22. *b*	23. *c*	24. *a*	25. *c*	26. *a*	27. *a*			

4. Metallurgy

DPP-1 Introduction to Metallurgy and Ores of Some Metals

5. (i) O > Si > Al > Fe > Ca > Na > K > Mg (ii) O > C > H > N > Ca > P > S
(iii) O > H > Cl > Na > Mg > S > Ca > K

6. *b*	7. *c*	8. *d*	9. *c*	10. *a*	11. *d*	12. *b*	13. *c*	14. *b*	15. *d*
16. *a, b, d*	17. *a,b,c,d*		18. *b,c*	19. *b,c,d*	20. *c,d*	21. *a*	22. *c*	23. *a*	24. *a*
25. 5	26. 3	27. 5	28. 5	29. 5	30. *c*				

DPP-2 Metallurgy Process of Concentration of Ores

2. By froth flotation process. **5.** Acid fluxes, Basic fluxes

6. *b*	7. *c*	8. *d*	9. *a*	10. *b*	11. *a*	12. *b*	13. *b*	14. *a*	15. *b*
16. *a,b,c*	17. *a,b*	18. *a,b,c*	19. *a,c*	20. *a,b,d*	21. *a*	22. *b*	23. *a*	24. *c*	25. 3
26. 4	27. 0	28. 3	29. 4	30. *b*					

DPP-3 Extraction of Crude Metals from Concentrated Ores

6. *c*	7. *a*	8. *d*	9. *d*	10. *b*	11. *b*	12. *b*	13. *a*	14. *c*	15. *b*
16. *a,c,d*	17. *a,b,c*	18. *a,b,c*	19. *a,b*	20. *b,c,d*	21. *a*	22. *d*	23. *b*	24. *c*	25. 4
26. 6	27. 8	28. 3	29. 3	30. *c*					

DPP-4 Thermodynamic Principles of Metallurgy, Ellingham's Diagram and Its Applications

6. *b*	7. *c*	8. *a*	9. *d*	10. *a*					
11. *c*	12. *d*	13. *b*	14. *b*	15. *a,d*	16. *c,d*	17. *a,b*	18. *b,c*	19. *c,d*	20. *b*
21. *d*	22. *a*	23. *b*	24. 0	25. 6	26. 6	27. 3	28. 2	29. *d*	30. *d*

DPP-5 Refining of Crude Metals

3. (i) $x = 330 - 350$ K, $y = 450 - 470$ K (ii) $a = 523$ k, $b = 1700$ K

6. *a*	7. *c*	8. *c*	9. *d*	10. *a*	11. *b*	12. *d*	13. *c*	14. *a*	15. *c*
16. *a,b,c*	17. *a,c,d*	18. *a,b,c*	19. *a,b,d*	20. *a,c*	21. *b*	22. *c*	23. *a*	24. *b*	25. 6
26. 0	27. 4	28. 4	29. 4	30. *c*					

DPP-6 Metallurgy of Some Important Metals-I (Cu, Fe, Ag)

6. *a*	**7.** *a*	**8.** *d*	**9.** *b*	**10.** *a*	**11.** *a*	**12.** *c*	**13.** *b*	**14.** *d*	**15.** *b*
16. *b,c,d*	**17.** *a,c*	**18.** *a,c,d*	**19.** *a,c*	**20.** *b,d*	**21.** *b*	**22.** *a*	**23.** *a*	**24.** *b*	**25.** 4
26. 0	**27.** 6	**28.** 0	**29.** 3	**30.** *b*					

DPP-7 Metallurgy of Some Important Metals-II (Al, Zn, Au)

6. *c*	**7.** *a*	**8.** *a*	**9.** *c*	**10.** *c*	**11.** *d*	**12.** *d*	**13.** *a*	**14.** *a*	**15.** *b*
16. *a,b*	**17.** *a,c,d*	**18.** *a,c,d*	**19.** *a,c*	**20.** *a,b*	**21.** *a*	**22.** *c*	**23.** *a*	**24.** *c*	**25.** 3
26. 3	**27.** 4	**28.** 3	**29.** 4	**30.** *c*					

Revisal Problems (JEE Main)

1. *d*	**2.** *d*	**3.** *a*	**4.** *b*	**5.** *b*	**6.** *c*	**7.** *b*	**8.** *c*	**9.** *a*	**10.** *c*
11. *c*	**12.** *a*	**13.** *d*	**14.** *c*	**15.** *c*	**16.** *d*	**17.** *b*	**18.** *c*	**19.** *b*	**20.** *a*
21. *b*	**22.** *c*	**23.** *a*	**24.** *b*	**25.** *b*	**26.** *d*	**27.** *d*	**28.** *a*	**29.** *b*	**30.** *d*

Revisal Problems (JEE Advanced)

1. *d*	**2.** *b*	**3.** *b*	**4.** *d*	**5.** *b*	**6.** *c*	**7.** *b*	**8.** *d*	**9.** *a*	**10.** *b*
11. *b,c*	**12.** *a,c,d*	**13.** *a,b*	**14.** *b,d*	**15.** *c*	**16.** *a,b,d*	**17.** *b,c*	**18.** *b,c,d*	**19.** *a,c*	**20.** *c,d*
21. *c*	**22.** *b*	**23.** *b*	**24.** *b*	**25.** 3	**26.** 4	**27.** 6	**28.** 0	**29.** 3	**30.** *c*

JEE Main & AIEEE Archive

1. *d*	**2.** *d*	**3.** *c*	**4.** *b*	**5.** *c*	**6.** *d*	**7.** *b*	**8.** *c*

JEE Advanced & IIT JEE Archive

1. *b,c,d*	**2.** *c,d*	**3.** *a*	**4.** *b*	**5.** *d*	**6.** *a,d*	**7.** *b*	**8.** *d*	**9.** *c*	**10.** *d*
11. 4	**12.** 6	**13.** *c*	**14.** *b*	**15.** *b*	**16.** *b*	**17.** *c*			

5. Qualitative Analysis

DPP-1 Principles Involved in Qualitative Analysis of Inorganic Salts

1. (i) Barium, beryllium, magnesium (ii) green, yellow (iii) $NaBO_2$ and B_2O_3
(iv) $NaNH_4HPO_4 \cdot 4H_2O, Na_2B_4O_7 \cdot 10H_2O$ (v) greater (vi) greater than

2. (i) (a) H_2O (b) $CaCl_2$ (c) KI (d) KCl (e) NaF
(ii) (a) increases (b) decreases (c) increases (d) increases

3. *b*	**4.** *b*	**5.** *b*	**6.** *c*	**7.** *c*	**8.** *a*	**9.** *d*	**10.** *d*		
11. *b*	**12.** *c*	**13.** *a*	**14.** *d*	**15.** *a,b,d*	**16.** *a,b,d*	**17.** *c*	**18.** *c*	**19.** *b*	**20.** *d*

DPP-2 Qualitative Analysis of Acidic Radicals (Anions)

4. *c*	**5.** *b*	**6.** *b*	**7.** *d*	**8.** *a*	**9.** *b*	**10.** *b*			
11. *d*	**12.** *a*	**13.** *c*	**14.** *a*	**15.** *a*	**16.** *a*	**17.** *c*	**18.** *c*	**19.** *c*	**20.** 3

DPP-3 Qualitative Analysis of Basic Radicals (Cations)

1. (i) (a) Dil. HCl (b) H_2S in the presence of HCl (c) HCl followed by hot water
(d) NH_4OH in NH_4Cl
(e) NH_4OH in NH_4Cl
(ii) (a) Potassium chromate (b) Ammonium thiocyanide
(c) Potassium chromate (d) Ammonium thiocyanide

3. *c*	**4.** *c*	**5.** *d*	**6.** *b*	**7.** *d*	**8.** *c*	**9.** *c*	**10.** *b*		
11. *a,b,c*	**12.** *a,b*	**13.** *a,b,c,d*	**14.** *d*	**15.** *a*	**16.** *d*	**17.** *c*	**18.** *a*	**19.** *a*	**20.** 5

DPP-4 Qualitative Analysis of Inorganic Salts (Acidic and Basic Radicals)

5. *d* 6. *c* 7. *c* 8. *d* 9. *b* 10. *b*
11. *a* 12. *a* 13. *b* 14. *b* 15. *a,b,c,d* 16. *a,d* 17. *a,b,d* 18. *b* 19. *d* 20. *b*

DPP-5 Organic Qualitative Analysis

4. *a* 5. *d* 6. *c* 7. *b* 8. *c* 9. *a* 10. *b*
11. *a,c,d* 12. *a,b,c,d* 13. *d* 14. *a* 15. *c* 16. *b* 17. *b* 18. *a* 19. 5 20. 3

DPP-6 Quantitative Analysis

3. *a* 4. *b* 5. *a* 6. *b* 7. *a* 8. *d* 9. *c* 10. *c*
11. *b* 12. *c* 13. *b,c,d* 14. *a,b* 15. *d* 16. *a* 17. *d* 18. *a* 19. 3 20. 5

DPP-7 Principles Involved in Synthesis, Neutralisation and Kinetic Studies

1. *a* 2. *d* 3. *c* 4. *c* 5. *b* 6. *b* 7. *a* 8. *c* 9. *b* 10. *b*
11. *b* 12. *a* 13. *b* 14. *a,b,d* 15. *b,c,d* 16. *c* 17. *a* 18. *c* 19. *c* 20. 6

Revisal Problems (JEE Main)

1. *a* 2. *a* 3. *b* 4. *c* 5. *d* 6. *a* 7. *c* 8. *a* 9. *c* 10. *c*
11. *a* 12. *a* 13. *b* 14. *c* 15. *b* 16. *b* 17. *b* 18. *d* 19. *c* 20. *b*
21. *c* 22. *b* 23. *c* 24. *b* 25. *a* 26. *a* 27. *b* 28. *a* 29. *c* 30. *a*
31. *b* 32. *a* 33. *b*

Revisal Problems (JEE Advanced)

1. *a* 2. *b* 3. *a* 4. *a* 5. *b* 6. *a* 7. *b* 8. *c* 9. *c* 10. *b*
11. *d* 12. *b* 13. *b* 14. *c* 15. *b* 16. *c,d* 17. *a,b* 18. *a,b* 19. *a,b,c*
20. *a,c,d* 21. *a,b,c* 22. *d* 23. *c* 24. *b* 25. *c* 26. *b* 27. *c* 28. *c* 29. *c*
30. *b* 31. *a* 32. *d* 33. *c* 34. 4 35. 9 36. 9

JEE Advanced & IIT JEE Archive

1. 6 or 7 2. *b* 3. *d* 4. *a* 5. *d* 6. *d* 7. *a,c,d* 8. *b* 9. *a* 10. *c*
11. *a* 12. *d* 13. *c* 14. *b* 15. *a,b* 16. *b* 17. *a* 18. b 19. *a* 20. *b*

Answers

1. *p*-Block Elements

DPP-1 General Properties of 15th Group (Nitrogen and Phosphorus)

1. (i) The single N—N bond is weaker than the single P—P bond because of interelectronic repulsions of the non-bonding electrons, owing to the small bond length in nitrogen. Hence, the catenation tendency is weaker in nitrogen.

(ii) Valence shell configuration for 14th, 15th and 16th group elements is ns^2np^2, ns^2np^3 and ns^2np^4 respectively. Since, ns^2np^3 is relatively more stable than the other two hence, 15th group elements have higher first ionisation energy.

Ionisation energy of C < N > O and Si < P > S

(iii) Valence shell configuration for 15th group elements is ns^2np^3. Due to stable half-filled configuration tendency to gain electron is less for these elements.

Electron gain enthalpy of N < C < O < F
and similarly P < Si < S < Cl

2. (i) N shows wide range of oxidation numbers between − 3 to + 5.

(ii) (a) NH_4NO_3 is taken as $NH_4^+NO_3^-$. Then oxidation state of N is − 3 and + 5.

(b) NH_4NO_2 is taken as $NH_4^+NO_2^-$. Then oxidation state of N is − 3 and + 3.

(c) NH_4CN is taken as $NH_4^+CN^-$. Then oxidation state of N is − 3 and − 3.

(d) HN_3 (hydrazoic acid) In this nitrogen has − 1/3 oxidation state (fractional value).

3. White phosphorus has strained discrete P_4 molecules. Hence, it is highly reactive. Red P has polymeric structure having a chain of P_4 molecules. Hence, it is chemically inert. Black P has layer lattice structure and acts as a conductor of heat and electrictiy.

4. (i) Anomalous properties of nitrogen are due to

(a) small size of nitrogen

(b) more electronegativity

(c) more ionisation energy

(d) non-availability of *d*-orbitals.

(ii) Among the 15th group elements, nitrogen is chemically inert. This is due to the presence of strong triple bond between the two atoms (bond enthalpy =941.4 kJ/mol)

(iii) For nitrogen only four orbitals are available in the valence shell for bonding and it has no *d*-orbitals. Hence, its maximum covalency is 4. Other elements have vacant *d*-orbitals in the valence shell and hence, they can expand their covalency to 6. e.g. PF_6^-.

(iv) In 15th group, arsenic to bismuth inert pair effect, i.e. inertness of ns^2 electrons to take part in bonding increases. Due to this stability of + 5 decreases and + 3 increases.

Thus, stability of $Bi^{3+} > Sb^{3+} > As^{3+}$

and $Bi^{5+} < Sb^{5+} < As^{5+}$

5. (a) Realgar and orpiment have chemical formulae as As_2S_4 and As_2S_3 respectively.

6. (b) The correct order is O > F > N > C.

7. (c) N≡N and N—H bond energies are more for nitrogen while others are weaker compared to phosphorus.

N—N	P—P	N≡N	P≡P	N—H	P—H	N—F	P—F	N—O	P—O
160	209	946	490	391	322	272	490	201	340

8. (c) Boiling point of

He (4 K) < H_2 (20 K) < Ne (27 K) < N_2 (77 K) < Ar (87 K) < O_2 (90 K)

9. (b) Gaseous chemical formed by the oxidation of ammonia with CuO is nitrogen.

$$2NH_3 + 3CuO \longrightarrow 3Cu + N_2 + 3H_2O$$

The same is obtained by reacting excess ammonia with chlorine

$$8NH_3 + 3Cl_2 \longrightarrow 6NH_4Cl + N_2$$

10. (b) $N_2^+ (13) = \sigma 1s^2, \overset{*}{\sigma} 1s^2, \sigma 2s^2, \overset{*}{\sigma} 2s^2, \pi 2p_x^2 = \pi 2p_y^2, \sigma 2p_z^1$

11. (b) The electronegativity decrease down the group with increasing atomic size. However, amongst the heavier elements, the difference is not much pronounced.

Hence, the correct order of electronegativity is as follows:

N(3.0) > P(2.1) > As(2.0) > Sb(1.9) = Bi(1.9)

12. (c) P_4 molecule has 6 bond pairs and 4 lone pairs, i.e. ratio is 3 : 2.

13. (a,c) Being good conductor of heat and electricity, black phosphorus is a most stable allotrope of phosphorus.

14. (a,b,c) Maximum covalency of nitrogen and phosphorus are 4 and 6 respectively. Phosphorus have vacant *d*-orbitals in the outermost shell which can be used for bonding hence, expand their covalency upto 6.

15. (a,b,d) White phosphorus is a translucent white waxy solid. It is poisonous, insoluble in water but soluble in CS_2. Red phosphorus is insoluble in water as well as in CS_2.

16. (a,d) Minimum and maximum oxidation number of phosphorus is −3 and +5 respectively.

The electron gain enthalpy of P < N < S < O.

17. (d) A molecule of N_2 has the $p\pi$-$p\pi$ bonding with each other respectively and the abundance in the earth's crest follows

P > N > As > Sb

18. (a) White P is an allotrope of phosphorus with low ignition temperature.

19. (d) Black P has layer lattice like graphite.

20. (a) (i) → (p), (ii) → (q), (iii) → (r,s), (iv) → (q,t)

21. (b) (i) → (q), (ii) → (r), (iii) → (s), (iv) → (p)

22. (4) 4 chemicals liberate dinitrogen on heating.

$$NH_4NO_2(aq) \xrightarrow{\text{Heat}} N_2(g) + 2H_2O(l)$$

$$(NH_4)_2Cr_2O_7 \xrightarrow{\text{Heat}} N_2 + 4H_2O + Cr_2O_3$$

$$2NaN_3 \xrightarrow{573\text{ K}} 2Na + 3N_2$$

$$2NH_4NO_3 \longrightarrow 2N_2 + O_2 + 2H_2O$$

23. (8) Except in NF_3, N_2O and NO_2^-, in all others N has negative oxidation states.

24. (4) $N_2 = \sigma 1s,^2 \overset{*}{\sigma} 1s,^2 \sigma 2s,^2 \overset{*}{\sigma} 2s,^2 \pi 2p_x^2 = \pi 2p_y^2, \sigma 2p_z^2$

25. (c) P—P single bond in P_4 molecule is much weaker (213 kJ mol^{-1}) than N ≡ N triple bond (941.4 kJ mol^{-1}) in N_2.

DPP-2 **Nitrogen and Phosphorus Compounds**

1. (i) Neutral oxides of nitrogen are nitrous oxide (N_2O) and nitric oxide (NO).

(ii) In the gaseous state, NO and NO_2 have unpaired electron and hence, they are paramagnetic. But in the liquid or solid state, they are diamagnetic due to the formation of dimers.

(iii) Nitrous oxide is the most commonly used inhalation anaesthetic in dentistry.

(iv) N_2O_3, P_2O_3 are acidic, As_2O_3, Sb_2O_3 are amphoteric and Bi_2O_3 is basic.

2. Normal salts do not undergo neutralisation with base.
Acidic salts further undergo neutralisation with a base.
Acidic salts are

NaH_2PO_3, NaH_2PO_2, Na_2HPO_4 and $Na_2H_2P_2O_7$

3. (i) NH_3 has intermolecular hydrogen bonding while PH_3 has no such bonding.

(ii) Stability of hydride depends on *M*—H bond energy, Since *M*—H bond energy decreases down the group hence, NH_3 is more stable. Least stable hydride acts as strong reducing agent, i.e. BiH_2 is a strong reducing agent.

(iii) In NH_3, nitrogen is sp^3-hybridised and expected bond angle is 109°28′. But due to high electron density around the N-atom causes greater repulsions between lone pair and N—H bond pairs. Hence, angle decreases to 107°.

The H—P—H bond angle is 93.5°. This angle indicates that the phosphorus atom is almost unhybridised. The 3 bonds from phosphorus to hydrogen roughly involve the three 3*p*-orbitals of phosphorus and 1s-orbitals of three hydrogen atoms.

4. (i) A mixture having conc. HNO_3 and conc. HCl in 1 : 3 ratio is called aqua-regia. It contains nitrosyl chloride and nascent chlorine.

$$HNO_3 + 3HCl \longrightarrow 2H_2O + 2Cl + NOCl$$

A mixture of conc. HNO_3 and conc. H_2SO_4 in 1 : 1 ratio is called nitration mixture.

(ii) $Au + 3Cl \longrightarrow AuCl_3$ and

$AuCl_3 + HCl \longrightarrow HAuCl_4$ (aurochloric acid)

$Pt + 4Cl \longrightarrow PtCl_4$ and

$PtCl_4 + 2HCl \longrightarrow H_2PtCl_6$ (chloroplatinic acid)

5. (a) In solid state, N_2O_5 exists as nitronium nitrate ($NO_2^+ \; NO_3^-$).
In NO_2^+ nitrogen is *sp* and in NO_3^-, nitrogen is sp^2.

6. (c) Pure phoshine is not inflammable but due to contamination of P_2H_4 and P_4 vapours, it becomes inflammable.

7. (a) Reducing property of P oxyacids is due to P—H bonds in the acid. Correct reducing property order is
$H_3PO_2 > H_3PO_3 > H_3PO_4$

8. (c) $H_4P_2O_6$ is oxyacid of phosphorus which has P—P bond.

9. (c)

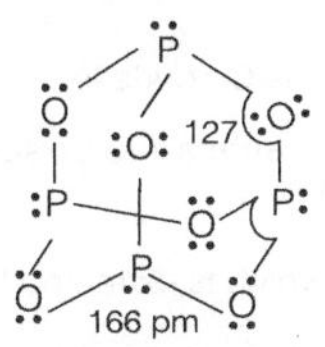

P—O bonds are 12 and lone pairs are for each P one pair and for each oxygen two pairs.

10. (a) NH_2OH and NH_2—NH_2 may be considered as NH_3 derivatives in which H is replaced but —OH and NH_2 respectively. Due to their electron withdrawing nature, these groups decreases electron density over nitrogen making them less basic. The effect of —OH group is stronger than —NH_2.

11. (b) $2PCl_5 \rightleftharpoons [PCl_4^+] \; [PCl_6^-]$
Tetrahedral Octahedral

$PBr_5 \rightleftharpoons [PBr_4]^+ + Br^-$

PBr does not split in the same fashion. The anion $[PBr_6^-]$ is not possible due to large size of Br-atoms. Hence, it splits differently.

12. (b) H_3PO_2, H_3PO_3 and H_3PO_4 are mono, di and tribasic acids respectively.

13. (a,b,c) P_4O_{10}, $POCl_3$ and conc. H_2SO_4 are act as dehydrating agent.

14. (a,c,d) Sodium and potassium nitrate on heating gives nitrite and oxygen but not NO_2.

$$2KNO_3 \longrightarrow 2KNO_2 + O_2$$

15. (b,c,d) $P_4 + 3NaOH + 3H_2O \longrightarrow PH_3 + 3NaH_2PO_2$

$$Ca_3P_2 + 6H_2O \longrightarrow 2PH_3 + 3Ca(OH)_2$$

$$4H_3PO_3 \longrightarrow 3H_3PO_4 + PH_3$$

16. (a,b,c,d) Metal nitride upon hydrolysis gives ammonia.
e.g. $Mg_3N_2 + 6H_2O \longrightarrow 2NH_3 + 3Mg(OH)_2$
As we know Haber's process involves the direct combination of nitrogen and hydrogen with the removal of some amount of heat, i.e. this is exothermic in nature.

$N_2(g) + 3H_2(g) \rightleftharpoons 2NH_3(g)$; $\Delta_f H° = -46.1$ ms/mol

$AgCl(s) + 2NH_3(aq) \longrightarrow [Ag(NH_3)_2]Cl(aq)$
Colourless

Nitrogen atom of ammonia can donate a lone pair of e^- to any species that is why, Lewis base as well as Bronsted-Lowry base.

17. (b,c,d)

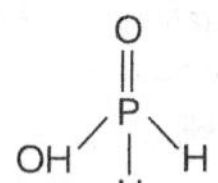

Since, here 1-hydroxyl group is present so, it is monobasic acid.
Since, metaphosphoric, when reacts in the presence of heat orthophosphoric acid is form which is in the form of glacial phosphoric acid.

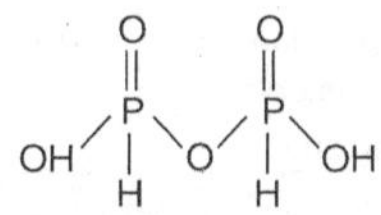

1, P—O—P bond

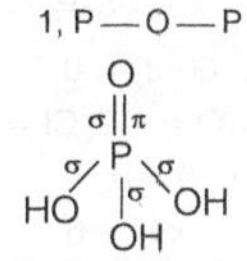

7-sigma, 1-π bond

∴ Option (a) is not correctly matched.

18. (c)

$(HPO_3)_3 =$ 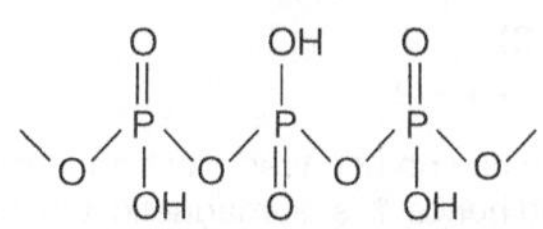

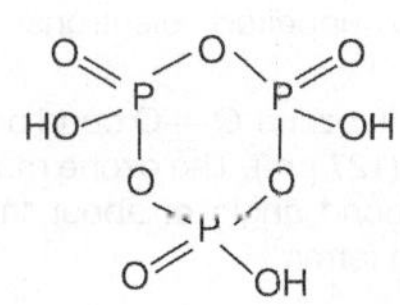

(Cyclotrimetaphosphoric acid)

19. (b) Due to the presence of H in *A* and *B*, they show reducing properties.

20. (a) (i) → (p), (ii) → (s), (iii) → (q), (iv) → (r)

21. (a) (i) → (p,r), (ii) → (p,q,r), (iii) → (q,s), (iv) → (r,t)

22. (5) N_2O_4 and NH_4NO_3 have no bond between nitrogen atoms while others have.

23. (4) Cr, Fe, Ni and Al are passive towards conc. HNO_3.

24. (2) $\underset{(+1)}{2H_3PO_2} \longrightarrow \underset{(-3)(+5)}{H_3\ PO_4}$

Equivalent weight, E_1(oxidation) = $M/4$ and

Equivalent weight, E_2 (reduction)

$= M/4$

Equivalent wight of H_3PO_2 in the disproportionation reaction

$= E_1$ (oxidation) + E_2 (reduction)

$= M/4 + M/4 = M/2$

Molecular nitrogen is less reactive than that of oxygen because nitrogen has high dissociation energy in comparison to oxygen so, its reactivity is less. Also bond length of nitrogen is shorter than oxygen because of the presence of triple bond between nitrogen atoms. So, both are true but not correct explanation.

25. (b) Due to e^- withdrawing capacity of fluoride ion, it withdraws e^- from nitrogen in NF_3 make it weaker ligand while presence of e^- donating methyl group makes the nitrogen in $N(CH_3)_3$ a strong ligand. In aqueous medium, NF_3 furnishes fluoride ion.

$$NF_3 \xrightarrow{(aq)} NF_2 + F^-$$

DPP-3 General Properties of 16th Group (Oxygen, Ozone and Sulphur)

1. (i) Among the chalcogens, electron affinity order is S > Se > Te > Po > O

Reason : Oxygen is very small in size among chalcogens . The repulsion between the electron being added to the atom and the electrons already present in the atom is significant in oxygen due to small size than the others. So, oxygen has negative electron gain enthalpy than sulphur.

(ii) Oxygen molecule is diatomic involving *pπ-pπ* bonding. Its small size produce weak intermolecular forces and is thus gas at room temperature. Sulphur do not form π- bonds like oxygen. Its molecule has a puckered ring structure of 8 sulphur atoms joined by single covalent bonds. The van der Waals' forces acting between S_8 molecules are stronger and hence, sulphur is a solid.

(iii)

Nitrogen configuration $1s^2 2s^2 2p^3$ (stable) (half-filled shell)	1st ionisation energy of nitrogen (1402 kJ/mol)
Oxygen configuration is $1s^2 2s^2 2p^4$ (less stable)	1st ionisation energy of oxygen (1314 kJ/mol)
N^+ configuration $1s^2 2s^2 2p^2$ (less stable)	2nd ionisation energy of nitrogen (2855 kJ/mol)
O^+ configuration $1s^2 2s^2 2p^3$ (stable) (half-filled shell)	2nd ionisation energy of oxygen (3388 kJ/mol)

2. (i) $Na_2S_4O_6$: Here, Na = + 1, S = x, O = −2

Then $2(+1) + 4(x) + 6(-2) = 0 \quad \therefore \quad x = 2.5$

(ii) SO_2Cl_2 : Here, S = x, O = −2, Cl = −1

Then $x + 2(-2) + 2(-1) = 0$

$\therefore \quad x = +6$

(iii) H_2S_2 : Here, H = + 1, S = x Then $2(+1) + 2(x) = 0$

$\therefore \quad x = -1$

(iv) $S_2O_3^{2-}$: Here, S = x, O = −2

Then $2(x) + 3(-2) = -2$

$\therefore \quad x = +2$

3. (i) Ozone molecule has no unpaired electron (3 bond pairs and 6 lone pairs) and hence, it is diamagnetic. On the other hand, oxygen has two unpaired electrons and hence, it is paramagnetic.

(ii) Due to resonance in ozone, O—O bond order is 1.5 and bond lengths are same (127 pm). The ozone molecule is angular as expected with a bond angle of about 117°. It is resonance hybrid of two main forms

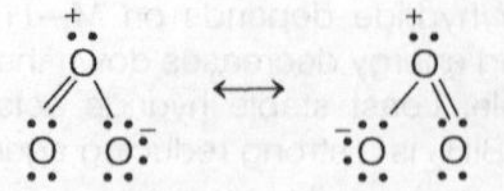

(iii) Bond order in oxygen is 2 while in ozone is 1.5. Hence, bond length in ozone (127 pm) is higher than oxygen (121 pm).

$$\text{Bond order} \propto \frac{1}{\text{bond length}}$$

(iv) The amount of ozone in a gas mixture can be determined by passing the gas into KI solution. The iodine that is liberated is titrated with standard hypo solution.

$$O_3 + 2I^- + H_2O \longrightarrow I_2 + 2OH^- + O_2$$

$$I_2 + 2S_2O_3^{2-} \longrightarrow 2I^- + S_4O_6^{2-}$$

4. (i) Oxygen shows anomalous properties due to

(a) small size of oxygen (b) high electronegativity

(c) absence of *d*-orbitals in the valence shell

(ii) Oxygen has no *d*-orbitals in the valence shell. By using the two unpaired electrons, it commonly shows divalency . On the other hand, S, Se, Te show divalency in ground state and tetra, hexa valency in the excited state using the vacant *d*-orbitals.

(iii) In $OF_2(+2)$ and in $O_2F_2(+1)$.

(iv) In superoxides, oxygen possess −1/2, oxidation state e.g. KO_2.

5. (i) Ozone as oxidising agent and oxidises mercury into mercurous oxide. Due to this, mercury loses its mobility and sticks to the glass. This is called tailing of mercury.

$$2Hg + O_3 \longrightarrow Hg_2O + O_2$$

(ii) Both oxygen and sulphur have 6 valence electrons. To get complete octet, they gain a pair of electrons and act as Lewis acids.

$$O + 2e^- \longrightarrow O^{2-}$$

and $\quad S + 2e^- \longrightarrow S^{2-}$

(iii) The bleaching action of ozone is due to its oxidising action on the organic matter .

Coloured substance + [O] ⟶ Colourless

6. (d) In tellurium, differentiating electron enters into 5*p* subshell. Its valence shell configuration is $5s^2 5p^4$.

7. (d) Sulphur exists in elemental form which is extracted by Frasch process.

Sulphide ores are zinc blende (ZnS), galena (PbS), copper pyrites ($CuFeS_2$), cinnabar (HgS).

Sulphate ores are gypsum ($CaSO_4 \cdot 2H_2O$), epsom ($MgSO_4 \cdot 7H_2O$).

8. (b) Se is non-conductor in dark but acts as conductor when exposed to light.

9. (c) 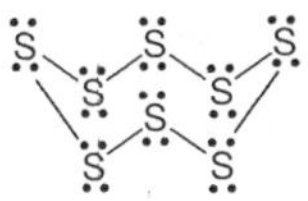

10. (b) Presence of lone pairs on bonded atoms generally lowers the bond energy.

Bond energy order is

O—O(146) < (S—S) (266) < S—H(366) < O—H(464) kJ/mol

11. (c) Stratosphere layer strongly absorb harmful UV-radiations (λ 255 nm).

12. (c) It is 369 K or 96°C. The stable form at room temperature is rhombic sulphur which transforms to monoclinic sulphur when heated about 369 K.

13. (d) Thermal stability $\propto$ bond dissociation energy

$$\Delta_{diss}(H-E)/\text{kJ mol}^{-1} \propto \frac{1}{\text{size of central atom}}$$

(where, E = group 16 elements)

On moving down the group, bond dissociation energy decreases due to increase in bond length. Thus, the order of bond dissociation energy or thermal stability is

$H_2O > H_2S > H_2Se > H_2Te > H_2Po$

14. (d) O_2 molecule, $\sigma 1s^2, \sigma^* 1s^2, \sigma 2s^2, \sigma^* 2s^2$

$$\sigma 2p_z^2 \begin{bmatrix} \pi 2p_y^2 \\ \pi 2p_x^2 \end{bmatrix}, \begin{bmatrix} \pi^* 2p_y^1 \\ \pi^* 2p_x^1 \end{bmatrix}$$

In O_2, bond order $= \frac{1}{2}(N_b - N_a) = \frac{1}{2}(10-6) = 2$

In O_2^+, bond order $= \frac{1}{2}(10-5) = 2.5$

15. (c) $O_2F_2 \Rightarrow 2x + 2 \times -1 = 0 \Rightarrow x = +1$

$O_3 \Rightarrow 3x = 0 \Rightarrow x = 0$

$KO_2 \Rightarrow +1 + 2(x) = 0 \Rightarrow x = -\frac{1}{2}$

$BaO_2 \Rightarrow 2 + 2x = 0 \Rightarrow x = -1$

16. (c,d) Electron affinity and ionisation energy decreases down the group due to increase in size.

17. (a,b,d) In S_8 molecule, sulphur has oxidation state is +2, +4, +6.

18. (b,c,d) Ozone is thermodynamically unstable w.r.t oxygen since, its decomposition into oxygen results in liberation of heat. (ΔH is negative). Hence, it is exothermic conversion.

$$2O_3(g) \xrightleftharpoons[\text{or Ultraviolet radiation}]{\text{Silent electric discharge}} 2O_2(g),\ \Delta H = -284\ \text{kJ mol}^{-1}$$

19. (a,b,c,d) All the given statements are correct.

20. (b,c,d) Liquid oxygen is attracted to and actually sticks to the poles of the magnet. It is due to electron distribution as unpaired electrons are present while other do not stick to the poles of magnet.

21. (b) $H_2S_2O_7$ has no peroxy bond while $H_2S_2O_8$ has one peroxy bond.

$H_2S_2O_7 \Rightarrow 2 \times 1 + 2 \times x + 7(-2) = 0$

or, $2 + 2x - 14 = 0,\ x = +6$

$H_2S_2O_8 \Rightarrow 2 \times 1 + 2x + 6(-2) + 2(-1) = 0$

or, $2 + 2x - 12 - 2 = 0$,

$x = +6$

22. (d) Oxygen does not show +4 and +6 oxidation states because oxygen has no d-orbitals in its valence shell.

23. (b) (i) $\rightarrow$ (p), (ii) $\rightarrow$ (q), (iii) $\rightarrow$ (r), (iv) $\rightarrow$ (s)

24. (c) (i) $\rightarrow$ (t), (ii) $\rightarrow$ (p), (iii) $\rightarrow$ (q), (iv) $\rightarrow$ (r), (v) $\rightarrow$ (s)

25. (a) (i) $\rightarrow$ (p), (ii) $\rightarrow$ (r), (iii) $\rightarrow$ (q,r,s), (iv) $\rightarrow$ (t)

26. (6) $(NH_4)_2Cr_2O_7$, NH_4NO_2 gives dinitrogen while NH_4NO_3 gives N_2O.

27. (6) Te in ground state

$5s^2$ $5p^4$

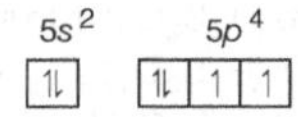

Te in fully excited state

$5s$ $5p$ $5d$

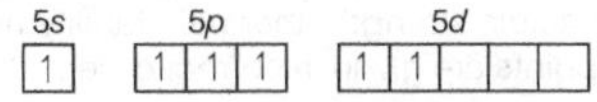

28. (4) O_3, NO_2, SO_2 and H_2S show bent shape.

29. (2) PbS $\Rightarrow$ +2

$SO_2 \Rightarrow +4$

$SF_6 \Rightarrow +6$

$Na_2S_2O_3 \Rightarrow +2$

$H_2SO_3 \Rightarrow +4$

30. (a) H—S bond is weaker than H—O bond hence, H_2S is more acidic than H_2O.

DPP-4 Oxygen and Sulphur Compounds

1. (i) H—S bond is weaker than H—O bond. Moreover, the conjugate base HS^- is more stable than OH^-. Hence, H_2S is stronger acid than H_2O with higher K_a value.

(ii) Due to the presence of lone pair of electrons on sulphur, it acts as Lewis base and due to the presence of vacant d-orbitals, it can also act as Lewis acid.

(iii) In SO_3, sulphur atom gain an electron pair due to polarisation of S$=$O in reactions.

2. (i) Conc. H_2SO_4 reacts with CaF_2 liberating colourless HF vapours.

Conc. H_2SO_4 reacts with $CaBr_2$ liberating first colourless HBr vapours. These are further oxidised by H_2SO_4 into reddish colour Br_2 vapours.

$$CaF_2 + H_2SO_4 \longrightarrow CaSO_4 + 2HF$$

$$CaBr_2 + H_2SO_4 \longrightarrow CaSO_4 + 2HBr$$

$$2HBr + H_2SO_4 \longrightarrow SO_2 + Br_2 + 2H_2O$$

(ii) Conc. H_2SO_4 acts as oxidising agent while H_2S as a reducing agent. Hence, both of them involve in a redox reaction.

$$H_2S + H_2SO_4 \longrightarrow S + SO_2 + 2H_2O$$

(iii) Conc. H_2SO_4 has great affinity for water molecules. During addition of water, large amount of heat is released. Temperature rises suddenly and bumping takes place. To avoid any accident, acid should be added to water with constant stirring.

(iv) SO_2 reacts with NaOH solution forming sodium sulphite. This then reacts with more SO_2 to form sodium bisulphite.

$$2NaOH + SO_2 \longrightarrow Na_2SO_3 + H_2O$$

$$Na_2SO_3 + SO_2 + H_2O \longrightarrow 2NaHSO_3$$

3. (i) $H_2S_2O_7$ (ii) S_8 (Sulphur) (iii) $H_2S_2O_8$

(iv) SO_3 (v) Conc. H_2SO_4

4. (i) Acidified $K_2Cr_2O_7$ acts as an oxidising agent and oxidises SO_2 and it is reduced to chromium compound which is green in colour.

$$K_2Cr_2O_7 + H_2SO_4 + 3SO_2 \longrightarrow K_2SO_4 + \underset{\text{Green}}{Cr_2(SO_4)_3} + H_2O$$

(ii) In H_2S, sulphur has its lowest oxidation state −2. In chemical reactions, it can only increase and cannot decrease its oxidation state. Hence, it acts as reducing agent only. In SO_2 the oxidation state of S is +4, i.e. it can be increased or decreased. Hence, SO_2 can act as both oxidant and reductant.

$$\underset{\text{Reductant}}{2H_2S} + \underset{\text{Oxidant}}{SO_2} \longrightarrow 3S + 2H_2O$$

$$K_2Cr_2O_7 + H_2SO_4 + \underset{\text{Reductant}}{3SO_2} \longrightarrow K_2SO_4 + Cr_2(SO_4)_3 + H_2O$$

(iii) Conc. H_2SO_4 acts as a strong dehydrating agent. When in contact with sugar, all water molecules are taken up and carbon is left out.

$$C_{12}H_{22}O_{11} \xrightarrow{\text{Conc. } H_2SO_4} 12C + 11H_2O$$

(iv) With conc. H_2SO_4, Cu metal liberates SO_2 gas and with dil. acid no reaction.

$$Cu + 2H_2SO_4\,(\text{conc.}) \longrightarrow CuSO_4 + SO_2 + H_2O$$

$$Cu + 2H_2SO_4\,(\text{dil.}) \longrightarrow \text{No reaction}$$

5. (i) Thiosulphuric acid ($H_2S_2O_3$) and thiosulphurous acid ($H_2S_2O_2$).

(ii) Thiosulphuric acid ($H_2S_2O_3$) in this S has +2 and +6.

(iii) Pyrosulphuric acid ($H_2S_2O_7$)

(iv) Thiosulphuric acid ($H_2S_2O_3$) forms hypo ($Na_2S_2O_3 \cdot 5H_2O$).

6. (d) Order of acidic strength, thermal stability and melting points and boiling points are given in correct order.

7. (b) Mixed oxides are made up of two simple oxides like Pb_3O_4 ($2PbO + PbO_2$), Fe_3O_4 ($FeO + Fe_2O_3$), Mn_3O_4 ($2MnO + MnO_2$).

8. (d) Sulphur having valency 6 do not possess lone pair while the sulphur with valency 4 possess one lone pair.

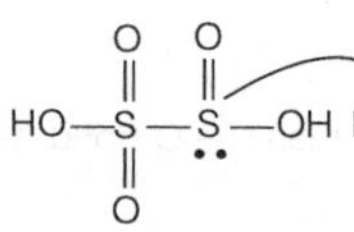

9. (c) Conc. H_2SO_4 is a strong dehydrating agent. Many wet gases can be dried by passing them through sulphuric acid. H_2SO_4 removes water from organic compounds. It is evident by its charring action on carbohydrates.

$$C_{12}H_{22}O_{11} \xrightarrow{H_2SO_4} 12C + 11H_2O$$

10. (c) Disulphur dichloride

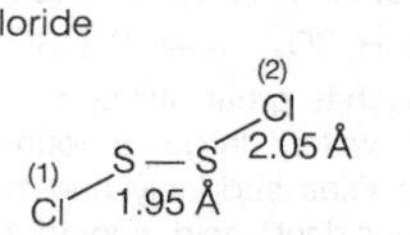

The angle between the $\overset{(1)}{Cl}$—S—S and S—S—$\overset{(2)}{Cl}$ planes is 90°. The structure is referred to as *gauche*.

11. (a) $5SO_2 + 2MnO_4^- + 2H_2O \longrightarrow 5SO_4^{2-} + 4H^+ + 2Mn^{2+}$

12. (b) The chemical reaction of H_2SO_4 as a result of its low volatility, strong acidic character, strong affinity for water and ability to act as an oxidising agent.

13. (b) In thiosulphuric acid, S has both −2 and + 6 oxidation states.

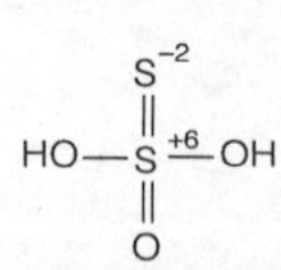

14. (a) In gas phase, SO_3 molecule is triangular planar having sp^2-hybridisation of the S-atom. There are three sigma bonds formed by sp^2-p overlap and three pi-bonds are by $p\pi$-$p\pi$ overlap and two $p\pi$-$d\pi$ overlap.

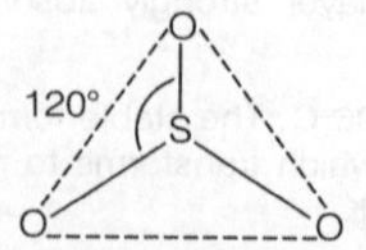

15. (d) H_2SO_5 is known as Caro's acid or peroxysulphuric acid.

16. (d) The oxidation state of sulphur in caro's acid is + 6. The structure of the acid is

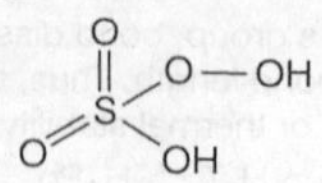

Caro's acid (H_2SO_5)

The two of the attached O-atoms impart +2 oxidation state each to the sulphur atom and rest two O-atoms (single bonded) imparts +1 oxidation state each to the S-atom. Hence, S-atom has a total of +6 oxidation state

17. (c,d) In SO_2, the oxidation state of S is +4, i.e. it can be increased or decreased. Hence, SO_2 can act as both oxidant and reductant.

18. (a,d) $SOCl_2$ and SO_3 are sp^2-hybridised with triangular planar geometry.

19. (a,b,c,d) Sodium thiosulphate is an inorganic compound also used to treat fungal infection of skin diseases.

20. (b,d) (i) CO_2 and OCS both are sp-hybridised and have linear geometry.

$$\underset{CO_2}{O{=}C{=}O}, \quad \underset{OCS}{O{=}C{=}S}$$

(ii) SO_4^{2-} and PO_4^{3-} both are sp^3-hybridised and have tetrahedral geometry.

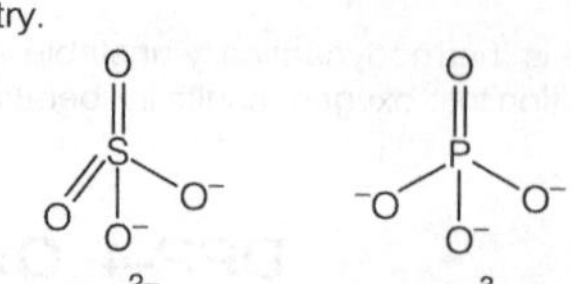

(SO_4^{2-}) (PO_4^{3-})

21. (d) In $S_2O_3^{2-}$, central atom sulphur is sp^3-hybridised and it has tetrahedral shape.

22. (b) The structure of $Na_2S_4O_6$

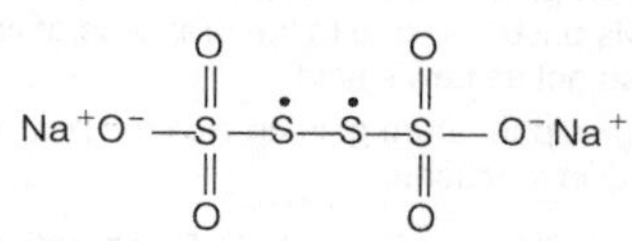

In this structure, two central sulphur atoms have zero oxidation number because electron pair forming the S—S bond remain in the centre. Let the oxidation number of (remaining atoms) S atom be x.

$$2(+1) + 6(-2) + 2x + 2(0) = 0$$

For Na for O

$2-12+2x=0$ or $x=\frac{+10}{2}=+5$

$\therefore$ Two central S-atoms have zero oxidation state and two terminal S- atoms have +5 oxidation state.

23. (b) The oxidation states of sulphur in H_2SO_3, H_2SO_4 and H_2SO_5 are respectively +4, +6 and +6.

(i) $\rightarrow$ (p), (ii) $\rightarrow$ (r), (iii) $\rightarrow$ (r)

24. (c) (i) $\rightarrow$ (p), (ii) $\rightarrow$ (q), (iii) $\rightarrow$ (s)

25. (a) (i) $\rightarrow$ (p,q,r,s), (ii) $\rightarrow$ (p,r), (iii) $\rightarrow$ (q,t), (iv) $\rightarrow$ (q,r,s)

26. (0)

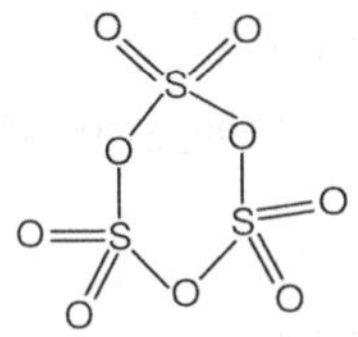

27. (5) II group radicals Pb^{2+}, Cu^{2+}, Cd^{2+}, Hg^{2+}, Sn^{2+} are precipitated with H_2S in acidic medium.

IV group cations Zn^{2+}, Mn^{2+}, Co^{2+}, Ni^{2+} are precipitated with H_2S in alkaline medium.

28. (4) The four chemical species among the given molecules having sp^3-hybridisation of central atom are SO_2Cl_2, SF_2, SO_3^{2-} and H_2SO_4.

29. (9)

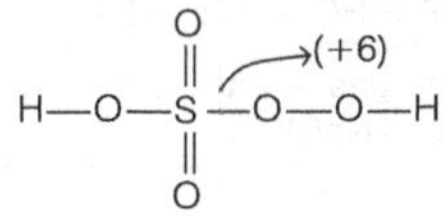

H_2SO_5 (peroxomonosulphuric acid)

It has nine covalent bonds.

30. (6)

DPP-5 **General Properties of 17th Group** (Halogens)

1. (i) Bromine
(ii) Iodine
(iii) Fluorine
(iv) Fluorine
(v) Chlorine
(vi) Chlorine, bromine and iodine

2. (i) Due to small size of F atoms, the interelectronic repulsions between non-bonding electrons on the fluorine atoms is very high. Such repulsions are less in chlorine molecule. Hence, bond energy of F_2 (158.8 kJ/mol) is less than Cl_2 (242.6 kJ/mol).

(ii) Fluorine is very small in size among halogens. The repulsion between the electrons being added to the atom and the electrons already present on the atom is significant in fluorine due to small size than in chlorine. Hence, electron affinity of fluorine (–333 kJ/mol) is less than chlorine (–349 kJ/mol).

(iii) In oxyacids the central atom always shows positive oxidation state. Fluorine being most electronegative never shows positive oxidation state. Hence, F does not form oxyacids . However, it has one oxyacid HOF in which central atom is oxygen.

(iv) An element that shows disproportionation must show atleast 3 different oxidation states. F_2 has only 0 and – 1 oxidation states. But Cl_2, Br_2 and I_2 show variable oxidation states, i.e. – 1 to + 7. Hence, F_2 does not undergo disproportionation but other halogens do.

(v) Cl_2 bleaches by oxidation while SO_2 does it by reduction. The reduced product gets oxidised again and the colour returns.

3. (i) I_2 is a covalent molecule and it is less soluble in polar solvent like water. But I_2 combines with KI and forms an ionic compound KI_3 which is more soluble in water.

(ii) Iodine is bigger in size and it has low ionisation energy, electronegativity among the halogens. Hence, it shows +ve oxidation states more commonly than others.

(iii) Halogens absorb part of the light in the visible region which cause excitation of outer electrons to higher energy levels. The excitation energy depends on the size of the halogen atom, the smaller the atom, the greater is the excitation energy. As different colours are absorbed by halogens, they display different complimentary colours.

e.g. F_2 absorbs violet light and thus appears yellow while iodine absorbs yellow and green portions of light and thus appears violet.

(iv) It is due to

(a) low enthalpy of dissociation of F—F bond.

(b) high hydration enthalpy of F^- ion.

(c) high SRP value, i.e. + 2.87 V.

4. (i) $2KMnO_4 + 16HCl \longrightarrow 2KCl + 2MnCl_2 + 5Cl_2 + 8H_2O$

(ii) $2NaOH + Cl_2 \longrightarrow NaCl + NaOCl + H_2O$

(iii) $2F_2 + 2H_2O \longrightarrow 4HF + O_2$

(iv) $3I_2 + 6NaOH \longrightarrow 5NaI + NaIO_3 + 3H_2O$

5. (i) Reasons for anomalous properties

(a) Small size

(b) Highest electronegativity

(c) Low F—F bond dissociation enthalpy than expected

(d) Non-availability of *d*-orbitals

(e) More hydration energy of fluoride ion

(ii) (a) Fluorine shows –1 but others show –1 to +7 oxidation states.

(b) F does not form polyhalide ions while others form.

(c) F forms only one oxyacid while others form different oxyacids.

(d) AgF soluble in water but other silver halides do not.

(e) HF is a low boiling liquid due to hydrogen bonding while others are gases.

(iii) Fluorine is the most electronegative element and it has no *d*-orbitals in the valence shell. Hence, it always show – 1 oxidation state and monovalency in its compounds.

Cl, Br, I having vacant *d*-orbitals, in the excited states commonly show a valency of 1, 3, 5, 7 and their oxidation states vary from – 1 to + 7.

6. (d) Due to increase in atomic size, ionisation enthalpy decreases down the group.

7. (b) Melting point increases with halogens from fluorine to iodine because van der Waals' forces are stronger down the groups.

8. (c) Atomic and ionic radii increase down the group due to increasing number of quantum shells.

9. (b) As the electronegativity of halogen decreases, the dipole moment of the hydrogen halide and hence, the ionic character decreases in the same order, i.e. HI < HBr < HCl < HF.

10. (b) The correct order of bond dissociation enthalpy is

$$\underset{(242.6)}{Cl_2} > \underset{(192.8)}{Br_2} > \underset{(158.8)}{F_2} > \underset{(151.1)}{I_2}$$

11. (d) Thermal stability decreases as the strength of H-X bond decreases which in turn decreases as the size of the halogen increases, i.e. HF > HCl > HBr > HI.

12. (a) When chlorine gas is passed through hot NaOH solution, it produces NaCl and $NaClO_3$.

$$6NaOH + 3\overset{0}{Cl_2} \longrightarrow 5Na\overset{-1}{Cl} + Na\overset{+5}{ClO_3}$$

Oxidation state varies from 0 to –1 and 0 to +5.

13. (d) Fluorine has no *d*-orbitals in the valence shell. Hence, it always show –1 oxidation state and monovalency in its compounds.

14. (a) Cl_2 bleaches by oxidation while SO_2 does it by reduction.

15. (c) Solid iodine is a molecular solid.

16. (b,c) The cryolite (Na_3AlF_6) and fluorspar (CaF_2) are two ores of fluoride.

17. (b,c) The negative electron gain enthalpy of fluorine is less than that of chlorine. It is due to small size of fluorine atom.

18. (a,b,d) Fluorides have abnormal solubilities than other halides.

19. (b,c) Certain deep seaweeds contain 0.5% of iodine in their ashes. Caliche or crude chile saltpetre contain about 0.2% of $NaIO_3$.

20. (a,c) 6NaOH (hot and conc.) + $3Cl_2 \longrightarrow 5NaCl + NaClO_3 + 3H_2O$

21. (b) All the halogens are highly reactive. They react with metals and non-metals to form halides due to readily acceptance of an electron.

22. (b) In calcium oxychloride (CaOCl)Cl) has different oxidation states. Hence, chlorine has +1 and –1 oxidation states.

23. (c) (i) → (p), (ii) → (q,r,t), (iii) → (q), (iv) → (q,s)

24. (a) (i) → (q,r), (ii) → (s,p), (iii) → (u), (iv) → (t)

25. (3) Three lone pairs are present on the central iodine in triiodide ion.

$$I_2 + I^- \longrightarrow [I—I \leftarrow I]^-$$

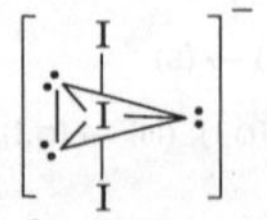

sp^3*d*-hybridisation (linear)

26. (8) Fluorine atom have 2+7 electrons so an F_2 molecule contain 18 electrons.

$$F_2(18)\ \sigma 1s^2, \overset{*}{\sigma}1s^2, \sigma 2s^2, \overset{*}{\sigma}2s^2, \sigma 2p_x^2, \begin{bmatrix} \pi 2p_y^2 & \overset{*}{\pi}2p_y^2 \\ \pi 2p_z^2 & \overset{*}{\pi}p_z^2 \end{bmatrix}$$

Hence, 8 electrons occupy the antibonding orbitals.

27. (1) Fluorine show only one positive oxidation state of +1 in HOF while all of the halogens show positive oxidation state of +1, +3, +5 and +7.

28. (5) Carnallite is a double salt. It will dissociate into simple substances or ions completely when dissolved in water.

$$KCl \cdot MgCl_2 \cdot 6H_2O \longrightarrow K^+ + Mg^{2+} + 3Cl^- + 6H_2O$$

Hence, 5 ions are produced.

29. (6) AgCl, AgBr, AgI, CaF_2, $PbCl_2$, $HgCl_2$, are water insoluble.

30. (a) Halogen molecule absorbs light in the visible region as a result of which their electrons are excited to higher energy level while the remaining light is transmitted.

The colour of halogens is actually the colour of this transmitted light, i.e. halogens have complementary colours.

DPP-6 Compounds of Halogens (F, Cl, Br and I)

1. (i) In anhydrous state, HCl is a covalent compound whereas in aqueous solution, HCl combines with water to form H_3O^+ and Cl^- ions.

Therefore, it is a good conductor of electricity.

$$HCl + H_2O \longrightarrow H_3O^+ + Cl^- \text{ ions}$$

(ii) The boiling point of covalent compounds (HF, HCl, HBr, HI) increases with increase in their molecular weight due to increase in van der Waals' forces of attraction among molecules. Furthermore, HF showing H-bonding possess higher boiling point or less volatile nature. Thus, volatile nature of halogen acids may be represented as

Volatile nature HCl > HBr > HI > HF

or boiling point order HCl < HBr < HI < HF

(iii) HF attacks glass bottles. HF reacts with silicates of the glass forming fluorosilicates.

$$Na_2SiO_3 + 6HF \longrightarrow Na_2SiF_6 + 3H_2O$$

2. (i) $HClO < HClO_2 < HClO_3 < HClO_4$

(ii) $HClO > HClO_2 > HClO_3 > HClO_4$

(iii) $HClO < HClO_2 < HClO_3 < HClO_4$

(iv) $I^- > Br^- > Cl^- > F^-$

(v) HOF > HOCl > HOBr > HOI

(vi) HI > HBr > HCl > HF

3. (i) When $KClO_3$ is heated with conc. HCl, a mixture of Cl_2 and ClO_2 is formed. This mixture is called euchlorine.

$$2KClO_3 + 4HCl \longrightarrow 2KCl + 2ClO_2 + Cl_2 + 2H_2O$$

(ii) H_2SO_4 acts as strong oxidising agent. H_2SO_4 oxidises HI form KI into iodine. But H_3PO_4 does not oxidise HI.

$$2KI + H_2SO_4 \longrightarrow K_2SO_4 + 2HI$$

$$2HI + H_2SO_4 \longrightarrow I_2 + SO_2 + 2H_2O$$

(iii) % of ionic nature depends on bond length and extent of polarity. The formula for calculating % of ionic nature

$$= \frac{\text{dipole moment observed}}{\text{dipole moment for 100\% ionic}} \times 100$$

This gives nearly 40% ionic nature for HF.

4. (i) Cl_2O has even number of electrons while ClO_2 has odd number of electrons.

(ii) In Cl_2O due to larger size of chlorine atoms repulsion between non-bonding electrons on the chlorine atoms overcome the lone pair-lone pair repulsions.

Hence, bond angle is 111°. In OF_2, due to lone pair-lone pair repulsions, bond angle decreases to 103° from the tetrahedral angle.

(iii) HOCl is unstable and liberate nascent oxygen on decomposition. The nascent oxygen acts as powerful oxidant and hence, HOCl acts as bleaching agent and oxidising agent.

$$HOCl \longrightarrow HCl + [O]$$

5. (i) Bleaching powder, formula is $CaOCl_2$. It is a mixed salt. It can be considered as Ca(OCl)Cl. Hence, chlorine has +1 and −1 oxidation states.

(ii) The amount of chlorine obtained from a sample of bleaching powder by treatment with excess of dil. acid or CO_2 is called available chlorine. A good sample contains 35-38% of available chlorine (Theoretical value 49% from $CaOCl_2 \cdot H_2O$).

(iii) Bleaching action of bleaching powder is due to release of Cl_2 at the time of application. But on long standing, it decomposes into $CaCl_2$ and $Ca(ClO_3)_2$ which do not liberate chlorine.

$$6CaOCl_2 \longrightarrow 5CaCl_2 + Ca(ClO_3)_2$$

(iv) Available chlorine in bleaching powder can be determined by treating aqueous solution of bleaching powder with KI and treating the liberated I_2 using standard $Na_2S_2O_3$ solution.

$$CaOCl_2 + H_2O \longrightarrow Cl_2 + Ca(OH)_2$$
$$Cl_2 + 2KI \longrightarrow 2KCl + I_2$$
$$I_2 + 2Na_2S_2O_3 \longrightarrow 2NaI + Na_2S_4O_6$$

Thus, $CaOCl_2 \equiv Cl_2 \equiv I_2 \longrightarrow 2Na_2S_2O_3$

6. (d) The correct order of % of ionic character is

$$\underset{43\%}{HF} > \underset{18\%}{HCl} > \underset{12\%}{HBr} > \underset{15\%}{HI}$$

7. (c) The reaction in which a species is simultaneously reduced and oxidised to form two different products is known as disproportionation reaction.

8. (a) As we move down the group, stability of oxides decreases because the size increases.

9. (a) $Au + 4H^+ + NO_3^- + 4Cl^- \longrightarrow [AuCl_4]^- + NO + 2H_2O$

$3Pt + 16H^+ + 4NO_3^- + 18Cl^- \longrightarrow 3[PtCl_6]^{2-} + 4NO + 8H_2O$

10. (c) Both are sp^3-hybridised.

11. (c) The structure of perchloric acid ($HClO_4$) is

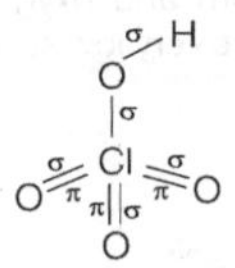

5σ and 3π-bonds

12. (a) The correct order of acidic strength is $Cl_2O_7 > SO_2 > P_4O_{10}$.

13. (d) The correct order is $\underset{-1}{HI} < \underset{0}{I_2} < \underset{+1}{ICl} < \underset{+7}{HIO_4}$

14. (b) ClO_3^- has pyramidal while NO_3^- has trigonal planar shape.

15. (b) In ClO_2, due to larger size of chlorine atom, repulsion between non-bonding electrons on the chlorine atoms overcome the lone pair-lone pair repulsion. Hence, the order of bond angle $ClO_2 > Cl_2O > H_2O > OF_2$

16. (a,c,d) Paramagnetic oxide has unpaired electrons or odd number of electrons.

17. (a,c,d) $Cl_2 + 2OH^- \longrightarrow Cl + OCl^- + H_2O$ is not a disproportionation reaction.

18. (a) $NaCl + H_2SO_4 \xrightarrow{420\,K} NaHSO_4 + HCl(g)$

$H_2SO_4 + 2H^+ + 2Br^- \longrightarrow Br_2 + SO_2 + H_2O$

$H_2SO_4 + 8H^+ + 8I^- \longrightarrow 4I_2 + H_2S + 4H_2O$

19. (a,b,c) Iodine cannot displace Cl^- and Br^- from their salt and bromine cannot displace Cl^- from their salt.

20. (a,c) $CaOCl_2 + CO_2 \longrightarrow CaCO_3 + Cl_2$

$$2CaOCl_2 \xrightarrow[CoCl_2]{\Delta} 2CaCl_2 + O_2$$
$$CaOCl_2 + H_2SO_4(\text{excess}) \longrightarrow CaSO_4 + H_2O + Cl_2$$
$$2CaOCl_2 + H_2S \longrightarrow CaCl_2 + H_2O + S$$

21. (c) Greenish yellow gas is chlorine.

$$4NaCl + MnO_2 + H_2SO_4 \downarrow MnCl_2 + 4NaHSO_4 + 2H_2O + Cl_2$$

22. (b) Chloride salt produce colourless HCl vapours with conc. H_2SO_4 and these become dense due to NH_4Cl on exposure to ammonia.

23. (d) (i) → (s), (ii) → (p), (iii) → (p), (iv) → (p,q,r).

24. (a) (i) → (q), (ii) → (r), (iii) → (r)

25. (c) (i) → (q,r), (ii) → (p,s,t), (iii) → (r,s), (iv) → (p,s,t).

26. (8)

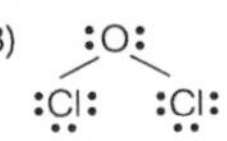

Two lone pairs on oxygen and three lone pairs on each chlorine.

27. (8)

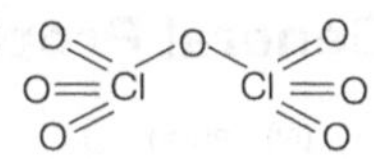

28. (7) Except NO_3^-, all others have sp^3 hybridised atoms.

29. (3) Aqua-regia is a mixture of one mole of conc. HNO_3 and 3 moles of conc. HCl.

$$3HCl + HNO_3 \longrightarrow 2H_2O + NOCl + 2Cl$$

30. (a) On dehydration with P_2O_5 at −10°C, $HClO_4$ gives Cl_2O_7 which is the anhydride of perchloric acid.

$$2HClO_4 + P_2O_5 \longrightarrow Cl_2O_7 + 2HPO_3$$

This reaction has been used for the preparation of Cl_2O_7.

DPP-7 Interhalogens and Pseudo Halogens

1. (i) Since, fluorine (F) is small and most electronegative halogen, all the other halogens combine with it and give fluorides, hence, fluorides are maximum.

(ii) +1, + 3, + 5 and +7.

2. The bond energy of interhalogen molecule (X — X′) is less than the halogen molecule (X_2).

3. The ions (cations or anions) composed of similar halogen atoms (Cl_3^+, Br_3^+, I_3^+) or dissimilar halogen atoms (ICl_2^-, $IBrCl^-$) are called polyhalide ions. Ionic compounds having either polyhalide cations, or polyhalide anions are called polyhalides.

e.g. $K + CrBr_3$, $K[ICl_2]$, $Cr[ClF_4]$

4. (i) Both form insoluble salts with Ag^+, Pb^{2+}, Hg^+, etc.

(ii) Both act as ligands and form complexes with transition metal ions.
e.g. $[Cu(CN)_4]^{2-}, [CuCl_4]^{2-}$

5. (i) $K^+ [IF_6]^-$ (ii) KCl + Br (iii) RbCl + ICl
(iv) $[I_3]^+ [AlCl_4]^-$.

6. (d) Pseudo halides are uni-negative charged similar to halide ions.

7. (a) Electropositive character of halogens is in order I > Br > Cl. Central atom is bigger in size, more electropositive nature form stable interhalogen compound.

8. (d) $[I_3]^-$ = Linear; BrF_5 = square pyramidal; ClF_3 = 'T' shaped.

9. (b) It has square pyramidal shape and it has 5 bond pairs and one lone pair.

10. (c) Both have sp^3d^2-hybridisation with square pyramidal shape.

11. (a) $[BrF_2]^+ = sp^3$, angular;
$[ICl_4]^- = sp^3d^2$, square planar;
$IF_5 = sp^3d^2$, square pyramidal;
$BrF_3 = sp^3d$, 'T' shaped

12. (b) $ClF + H_2O \longrightarrow HOCl + HF$

13. (d) $CsBr_3$ contains Cs^+, Br^- and lattice Br_2 molecules respectively.

14. (c) In IF_7, iodine is the least electronegative halogen, so its higher oxidation number (+7) is more stable than those of the lighter member of the group.

15. (c) Due to pentagonal bipyramidal geometry, IF_7 has zero dipole moment.

16. (a,c) Cl ionises as $2ICl \longrightarrow I^+ + ICl_2^-$
At cathode, $2I^+ + 2e^- \longrightarrow I_2$
At anode, $2ICl_2^- \longrightarrow I_2 + 2Cl_2 + 2e^-$

17. (a,c) When a polyhalide containing more than one type of halogen atom undergo thermal dissociation, the metal monohalide is always formed, which has highest lattice energy. In this, CsCl has more lattice energy than CsI.

18. (a,b,d) $(CN)_2$, $(OCN)_2$ and $(SeCN)_2$ are pseudo halogens.
$(CN)_2$-cyanogen, $(OCN)_2$ = oxy-cyanogen
$(SeCN)_2$ = seleno cyanogen. BrF_5 is an interhalogen.

19. (a,b,c) KCN (K—C ≡ N);
KI_3(I—K · I—I); $CsBr_3$(Br—Cs · Br—Br) all three have both ionic as well as covalent bonds But K_2O (K—O—K) has only ionic bonds.

20. (c,d) ICl_4^- and BrF_4^- both have sp^3d^2-hybridisation.

21. (c) Thermal stability of diatomic interhalogen increases with increase in difference between the electronegativities of the halogen atoms.
Hence, increasing order of the stability of given interhalogens (in options) will be
BrCl < ClF < BrF < IF

22. (a) $BrCl + H_2O \longrightarrow HOBr + HCl$

23. (d) Interhalogens are covalent compounds.

24. (b) Pseudo halogens have similar chemical properties as that of halogens like complex formation.

25. (b) (i) → (s,t), (ii) → (q,t), (iii) → (r,t), (iv) → (s)

26. (4) CN^-, SCN^-, $SeCN^-$, N_3^-

27. (6) In this geometry, each lone pair is 90° with the bond pairs. Hence, lone pair-bond pair repulsions are six.

28. (2) ICl_4^-, BrF_4^-.

29. (4) 3 lone pairs and 1 bond pair.

30. (7) N≡C—C≡N, It has 3σ and 4π-bonds.

DPP-8 General Properties and Compounds of Xenon

1. (i) Helium (ii) Argon (iii) Xenon (iv) Xenon
(v) Xenon (vi) Krypton (Lazy = Ar and Stranger = Xe)
(vii) Radon (discovered by Dorn).

2. (i) ${}^{226}_{88}Ra \longrightarrow {}^{222}_{86}Rn + {}^{4}_{2}He$
(ii) $x - 12 = -4 \quad \therefore \; x = +8$
(iii) Fractional distillation which is based on difference in the boiling points of the gases in air.
(iv) Activated coconut charcoal.
(v) Caustic potash (KOH) solution (absorbs CO_2), pyrogallol solution (absorbs O_2) and conc. H_2SO_4 and P_4O_{10} absorb moisture.

3. (i) Helium and neon do not form compounds due to their very high IP values and absence of empty *d*-orbitals in their valence shells.
(ii) Among the noble gases, only krypton and xenon form compounds. They form compounds mainly with fluorine and oxygen. The reasons are
(a) Krypton and xenon have relatively lower IP values.
(b) They have empty *d*-orbitals in their valence shells.
(iii) Bartlett realised that ionisation enthalpy of O_2 (1175 kJ/mol) and xenon (1170 kJ/mol) are almost same. Bartlett then prepared the first noble gas compound $Xe^+ PtF_6^-$ by reacting Xe with PtF_6.
(iv) A mixture of helium and oxygen is used in modern diving apparatus due to its very low solubility in blood under pressure unlike N_2.

4. (i)

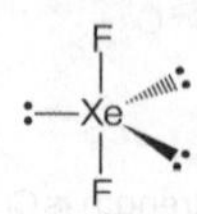

(a)

In this *lp-lp* repulsions = 0
bp-bp repulsions = 0
bp-lp repulsions = 6

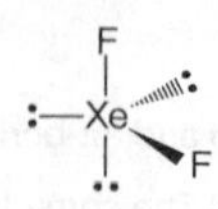

(b)

In this *lp-lp* repulsions = 2
bp-bp repulsions = 1
bp-lp repulsions = 3

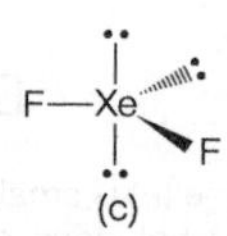

(c)

In this *lp-lp* repulsions = 2
bp-bp repulsions = 0
bp-lp repulsions = 4

Among these, first one is correct and other two are wrong.

(ii) Among these, first one is correct.

(a)

lp-lp repulsions = 0
lp-bp repulsions = 8
bp-bp repulsions = 4

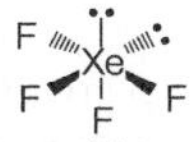

(b)

lp-lp repulsions = 1
lp-bp repulsions = 6
bp-bp repulsions = 5

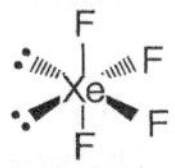

(c)

lp-lp repulsions = 1
lp-bp repulsions = 4
bp-bp repulsions = 4

(iii) $XeF_4 + SbF_5 \longrightarrow [XeF_3]^+ + [SbF_6]^-$

(iv) Xenon atom has 8 valence electrons (octet) when it forms compounds its valence electrons become more than octet.
e.g. In XeF_2, XeF_4, XeF_6 xenon has 10, 12 and 14 valence electrons respectively.

5. (i) Noble gases form a number of compounds in which the gases are trapped within cavities of crystal lattices of certain organic and inorganic substances.These are called clathrates or cage compounds.
e.g. when quinol is allowed to crystallise from its aqueous solution in the presence of Ar, Kr or Xe at 10-40 atm pressure, noble gas atoms get trapped within the lattice quinol.

(ii) He and Ne do not form clathrates as they are too small and they escape easily from the cavities of the quinol.

(iii) (a) Separation of He and Ne from other noble gases.
(b) Clathrates of Ar, Kr, Xe provide convenient means of storing and transportation of these gases.

6. (a) $\frac{C_p}{C_V}$ value for noble gases is 1.66.

7. (d) These three gases were discovered by Travers and Ramsay

8. (d) For hydrogen like atom

$$\text{IP of He}^+ = \text{IP of H} \times \frac{Zt^2}{n^2}$$

For He, $Z = 2, n = 1$

$$\therefore \text{ IP of He}^+ = 13.6 \times \frac{2^2}{1^2} = 54.4 \text{ eV}$$

9. (b) Hybridisation of the compounds are

$$XeF_2 \longrightarrow sp^3d, \quad XeF_4 \longrightarrow sp^3d^2$$
$$XeO_4 \longrightarrow sp^3, \quad XeF_6 \longrightarrow sp^3d^3$$

10. (b) The structure of O_2PtF_6 is $O_2^+[PtF_6]^-$.

Oxidation state of $O_2 = 2x = +1$

$x = +\frac{1}{2}$

11. (b)

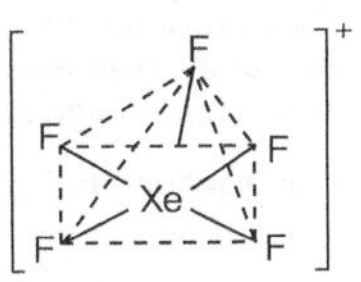

12. (a)

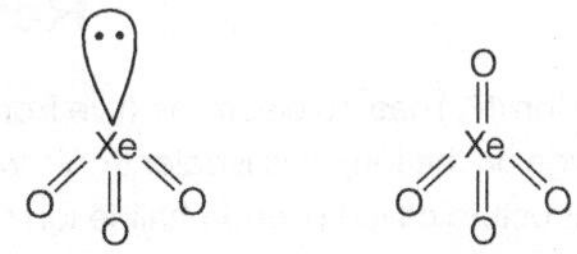

XeO_3 $\quad$ XeO_4

13. (d) XeF_6 do not produce any result reacting with CsF.

14. (d) All the cases are possible reactions.

15. (c) He is used in atomic reactor for cooling.

16. (a,b,c) Radon formed on α-decay of radium.

17. (a,c,d) In dewar's method, coconut charcoal is used, which adsorb different gases at different temperatures.

18. (b,c,d) $2XeF_2 + 2H_2O \longrightarrow 2Xe + 4HF + O_2$

19. (b,d) In XeF_2, *lp* are at equatorial position and XeF_4 is square planar while SF_4 has *see-saw* structure.

20. (a,b,c,d)

21. (c)

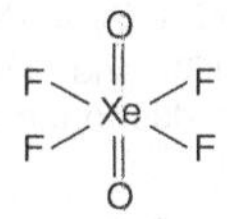

22. (d) Rest of properties increases from He to Xe.

23. (a)

24. (d) Structure of XeF_4 is square planar.

25. (a) (i) → (q), (ii) → (r), (iii) → (s), (iv) → p

26. (1)

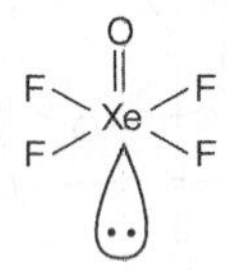

$XeOF_4$-1 *lp*

27. (3) XeF_2, XeF_4, XeO_4 are non-polar, i.e. have zero dipole moment due to symmetrical structure.

28. (3) Electron dot structure of XeF_2 is

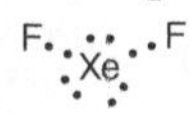

3-*lp*

29. (9) Electronic configuration of Ar is

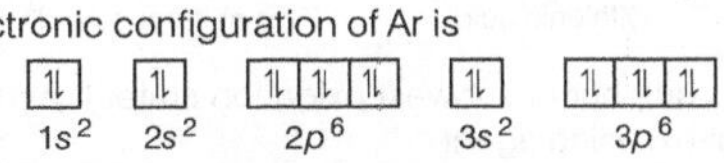

Here, 9 electrons each has $+\frac{1}{2}$ and $-\frac{1}{2}$ value of magnetic moment spin.

30. (8) The electronic configuration of Xe is $[Kr]4d^{10}5s^25p^6$
depending on the energy provided 8 electrons (from 5 *s* and 5*p*) can go into excited state is 6*s*, 4*f*, 5*d*, 6*p*, 7*s*, 5*f*, 6*d*, 7*p*, 8*s*, etc.

Revisal Problems (JEE Main)

1. (c) The azide ion (N_3^-) has 16 electrons (five from each N and one from the charge on the ion). It is isoelectronic with N_2O.
It has linear structure called pseudo halide ion not pseudo halogen.

2. (d) H_3PO_2, H_3PO_3 and H_3PO_4 are monobasic, dibasic and tribasic acids respectively.

3. (c) $2Pb(NO_3)_2 \longrightarrow 2PbO + 4NO_2 + O_2$

4. (a) NF_3 is sp^3-hybridised and have trigonal pyramidal shape.

5. (a) It readily dissolves in cold water forming metaphosphoric acid.

$$P_4O_{10} + 2H_2O\ (\text{cold}) \longrightarrow 4HPO_3$$

With hot water, it gives phosphoric acid.

$$P_4O_{10} + 6H_2O\ (\text{hot}) \longrightarrow \underset{\text{Phosphoric acid}}{4H_3PO_4}$$

6. (d) $NH_4NO_3 \longrightarrow 2H_2O + N_2O\,(A)$

$4NH_3 + 5O_2 \longrightarrow 6H_2O + 4NO\,(B)$

$2Pb(NO_3)_2 \xrightarrow{\Delta} 2PbO + O_2 + 4NO_2\,(C)$

Both nitric oxide (NO) and nitrogen dioxide (NO_2) are paramagnetic in nature due to the presence of odd number of electron(s) in outer shell.

7. (a) $\underset{\text{Ammonium dichromate}}{(NH_4)_2Cr_2O_7(s)} \longrightarrow Cr_2O_3(s) + N_2(g) + 4H_2O(g)$

$NH_4NO_2 \longrightarrow N_2 + 2H_2O$

The X gas is nitrogen.

8. (a)

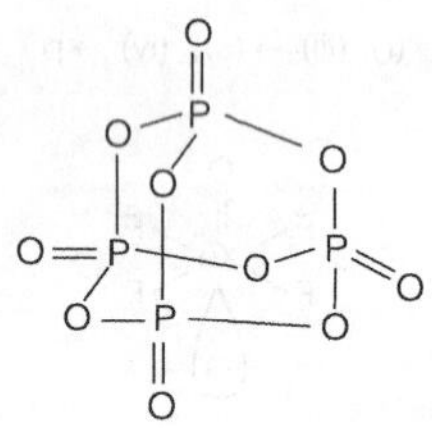

(P_4O_{10})

9. (c) $HO—SO_2—OH + PCl_5 \longrightarrow ClSO_2Cl + POCl_3 + 2HCl$

10. (a)

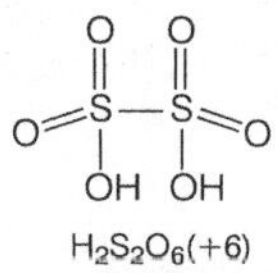

$H_2S_2O_6(+6)$
Dithionic acid

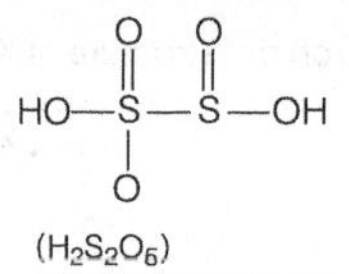

($H_2S_2O_5$)
Pyro sulfurous acid

11. (a) In this sulphur has lowest oxidation state. It acts as reducing agent not oxidising agent.

12. (d)

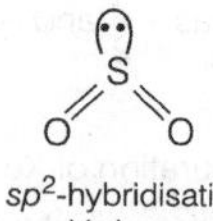

sp^2-hybridisation
V-shape

13. (b) XeF_4, BrF_4^- and ICl_4^- are sp^3d^2-hybridised with square planar shape.

14. (c) In O_2 molecule, the antibonding $\overset{*}{\pi}2p_x$ and $\overset{*}{\pi}2p_y$ orbitals are singly occupied. Hence, it is paramagnetic in nature.

15. (c) Effective nuclear charge increases in the same order hence, it becomes difficult to remove electron.

16. (a) Milliequivalents of hypo = milliequivalents of iodine

$= 100 \times 0.25$

Gram equivalents of hypo = 0.025

Number of moles of hypo = number of milliequivalents of hypo

$= 0.025$

Weight of hypo = $0.025 \times 248 = 6.2$ g

17. (c) SO_3 has one $p\pi$-$p\pi$ and two $p\pi$-$d\pi$ bonds. Among chlorine oxyacids, the anion that has two $p\pi$-$d\pi$ bonds is chlorate ion i.e. ClO_3^- ion. Oxidation state of Cl in ClO_3^- is + 5.

18. (c) $2Cl_2 + 2Ca(OH)_2 \longrightarrow Ca(OCl)_2 + CaCl_2 + H_2O$

19. (d) CO_3^{2-}, SO_3 and NO_3^- are sp^2-hybridised.

20. (a) BrF_3 has 'T' shape while others have symmetrical shapes.

21. (d) $3\overset{0}{Cl_2} + 6NaOH \longrightarrow 5Na\overset{-1}{Cl} + Na\overset{+5}{Cl}O_3 + 3H_2O$

22. (d) ClF_4^+ is sp^3d-hybridised with 4 bond pair and one lone pair.

23. (c) $BF_3 = 120°$, $CF_4 = 109°\text{-}28'$, $NF_3 = 103°$ and $XeF_2 = 180°$.

24. (c) Fluorine is prepared by electrolysis of fused KHF_2, made as a mixture of $KF(s)$ and $HF(g)$

$$2KHF_2 \xrightarrow[70°-100°C]{\text{Electrolysis}} H_2 + F_2 + 2KF$$

25. (c) $XeF_6 + 3H_2O \longrightarrow XeO_3 + 6HF$

XeO_3 is a colourless explosive solid and has a pyramidal molecular structure.

26. (a) $E = \frac{3}{2}RT$

Thus, $p\Delta V = p(V_2 - V_1) = (pV_2 - pV_1)$

Since, $pV = RT$ for 1 mole of gas. The above equation becomes

$$p\Delta V = R\Delta T$$

or $R = p\Delta V/\Delta T$

$\therefore$ $C_p = C_V + R = \frac{3}{2}R + R = \frac{5}{2}R$

Thus, $\frac{C_p}{C_V} = \frac{5}{3} = 1.67$

For monoatomic gases,

$$\frac{C_p}{C_V} = 1.67$$

27. (c) Both have linear shape.

28. (d) $XeF_4 + SbF_4 \longrightarrow [XeF_3]^+\ [SbF_6]^-$

29. (a) The mixture of 80% helium and 20% oxygen is used to arrest breathing in asthma and other respiratory diseases because helium is light and is less soluble in blood under high pressure.

30. (b) XeO_3 is pyramidal with sp^3-hybridisation and others are correct key.

Revisal Problems (JEE Advanced)

1. (b) In this, iron has +1 and NO has +1 oxidation state. Iron has 3 unpaired electrons with magnetic moment $\sqrt{15}$.

$$2NaNO_3 + H_2SO_4 \longrightarrow 2NaHSO_4 + 2HNO_3$$
$$2HNO_3 + 6FeSO_4 + 3H_2SO_4 \longrightarrow 3Fe_2(SO_4)_3 + 2NO + 4H_2O$$
$$(Fe(H_2O)_6)SO_4 + NO \longrightarrow \underset{\text{(Brown ring)}}{[Fe(H_2O)_5NO]^{2+}} + SO_4^{2-} + H_2O$$

2. (d) Both have tetrahedral geometry.

3. (a) $O_2 + PtF_6 \longrightarrow O_2^+[PtF_6]^-$. PtF_6 is one of the few oxidising agents sufficiently powerful to oxidise O_2.

4. (d) The pair of Ne and Na has the highest difference in their first ionisation energy.

5. (b) NF_3 does not react with water at ambient temperature however, slowly hydrolysed by basic solution at 100°C.

6. (a) Both N_3^- and XeF_2 are linear.

7. (d) 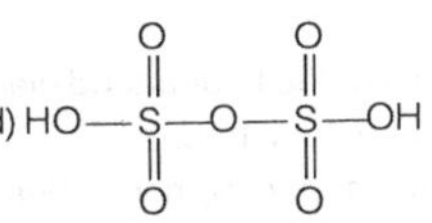

$(H_2S_2O_7)$

8. (d) SO_2, SO_3, Cl_2O, Cl_2O_7, N_2O_3, P_4O_{10} are acid anhydrides.

9. (d) $3Cu(s) + 2NO_3^-(aq) + 8H^+(aq) \longrightarrow 3Cu^{2+}(aq) + 2NO(g) + 4H_2O(l)$

10. (d) Here, oxidation state of hydrogen (+1) remains constant and oxygen (−2) remains constant. But S in SO_2 reduces from S(+4) to S(0) and sulphur in H_2S oxidises from S(−2) to S (0).

11. (a,b) NO and O_2^+ have 15 electrons each. According to MOT, the bond order in them is 2.5.

12. (b,c) Two of the sulphur atoms present in the tetrathionate anion $(S_4O_6^{2-})$ are in oxidation state 0 and two are in oxidation state +5.

13. (b,d) Ozone is sp^2-hybridised, has bent structure and is diamagnetic in nature.

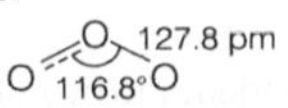

14. (a,d) CuS and Ag_2S are black while CdS is yellow and MnS is flesh coloured.

15. (a,c,d) In HF, there is strong intermolecular H-bonding between F and H. Hence, HF is lower volatile hydride.

16. (a,d) XeF_4, ClF_3 and ICl_4^- have two lone pairs each.

17. (b,c,d) Ozone does not oxidise $KMnO_4$ as Mn is already in maximum +7 oxidation state.

18. (a,b,c,d)

(a) In NO_2^-, N—O bond order is 1.5 while in NO_3^-, bond order is 1.33.

(b) NH_3 is stronger Bronsted base than PH_3. Reason is N being smaller has more electron density on nitrogen.

(c) SF_4 is hydrolysed to sulphurous acid easily while SF_6 do not.

(d) Sulphur in vapour state has S_2 molecules exhibit paramagnetism like O_2.

19. (a,b,c)

(a) The HI formed initially is oxidised by H_2SO_4 into iodine.

(b) Both have pyramidal geometry.

(c) Due to inert pair effect, Bi^{5+} gains two electrons and is converted into Bi^{3+}.

(d) NH_3 is not linear while N_3^- is linear.

20. (a,b,d) S_8 molecule has non-planar cyclic structure.

21. (a) $Cl_2 + 2NaOH(\text{cold and dil.}) \longrightarrow \underset{(P)}{NaOCl} + NaCl + H_2O$

NaOCl is sodium hypochlorite.

$3Cl_2 + 6NaOH(\text{hot and conc.}) \longrightarrow \underset{(Q)}{NaClO_3} + 5NaCl + 3H_2O$

$NaClO_3$ is sodium chlorate.

22. (d) $Cl_2 + SO_2 \xrightarrow{\text{Charcoal}} SO_2Cl_2(R)$

$$10SO_2Cl_2 + P_4 \longrightarrow 10SO_2 + 4PCl_5(S)$$
$$PCl_5 + 4H_2O \longrightarrow 5HCl + H_3PO_4(T)$$

23. (a)

24. (a) (i) → (p,s), (ii) → (q,s), (iii) → (r,t), (iv) → (q,t)

$$(3Cu + 8HNO_3(\text{dil.}) \longrightarrow 3Cu(NO_3)_2 + 4H_2O + 2NO$$
$$Cu + 4HNO_3(\text{conc.}) \longrightarrow Cu(NO_3)_2 + H_2O$$
$$4Zn + 10HNO_3 \longrightarrow 4Zn(NO_3)_2 + 5H_2O + N_2O$$
$$Zn + 4HNO_3(\text{conc.}) \longrightarrow Zn(NO_3)_2 + 2H_2O + \underset{\text{Nitrogen dioxide}}{2NO_2}$$

25. (4) $PCl_5 + 4H_2O \longrightarrow H_3PO_4 + 5HCl$

26. (3)

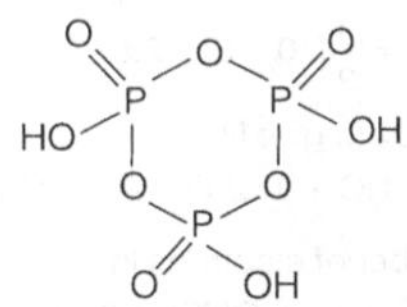

27. (4) H_3O^+, OSF_2, SO_3^{2-} and NF_3 have pyramidal shape.

28. (4) Al, Fe, Cr, Ni are rendered passive by conc. HNO_3.

29. (6) SO_3, XeO_3, H_3PO_4, ClO_4^-, SO_4^{2-}, $XeOF_2$ have $d\pi$-$p\pi$ bonding.

30. (6) sp^3d^2-hybridisation.

JEE Main & AIEEE Archive

1. (c) Acidic strength of oxoacids depends upon the stability of conjugate base. More stable the conjugate base of an acid, stronger is the acid.

$$HClO_4 \rightleftharpoons ClO_4^- + H^+$$
$$HClO_3 \rightleftharpoons ClO_3^- + H^+$$
$$HClO_2 \rightleftharpoons ClO_2^- + H^+$$
$$HOCl \rightleftharpoons ClO^- + H^+$$

Conjugate bases stabilises by resonance.

$ClO_4^- \longrightarrow$

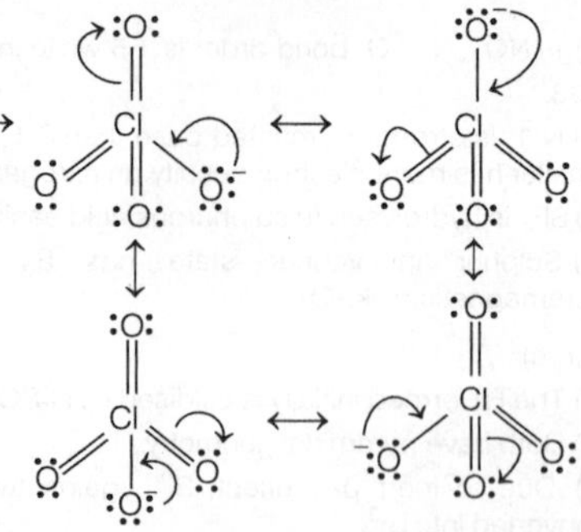

$ClO_3^- \longrightarrow$

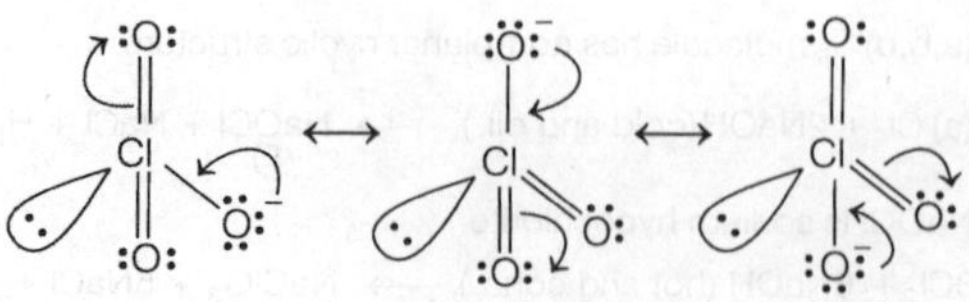

$ClO_2^- \longrightarrow$

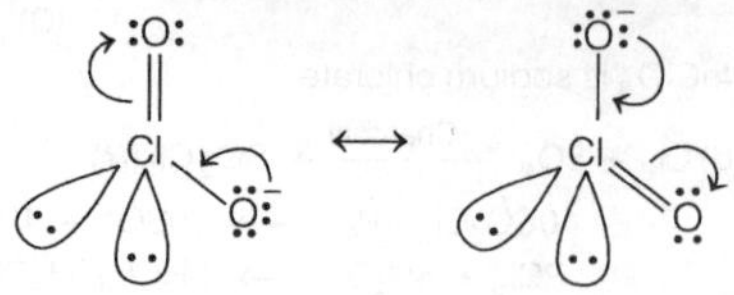

ClO^- does not show resonance. Thus, the stability order of conjugate base due to resonance is

$$ClO_4^- > ClO_3^- > ClO_2^- > ClO^-$$

Hence, the acidic strength of given oxoacids is as follows:

$HClO_4 > HClO_3 > HClO_2 > HClO$

2. (a) The electronic configuration of NO[$15e^-$] is given by

$\sigma 1s^2, \overset{*}{\sigma} 1s^2, \sigma 2s^2, \overset{*}{\sigma} 2s^2, \sigma 2p_z^2, \pi 2p_x^2 = \pi 2p_y^2, \overset{*}{\pi} 2p_z^1$

Thus, NO possess one unpaired electron, hence it is paramagnetic in nature. However, it dimerises at low temperature to become diamagnetic.

$$2NO \rightleftharpoons N_2O_2$$

Bond order of NO $= \frac{1}{2}[10-5] = 2.5$

It combines with O_2 to give NO_2.

$$NO + 1/2\, O_2 \longrightarrow NO_2$$

3. (a,b,c,d) (a) Number of electrons in

$ONCl = 32\, e^-$

$ONO^- = 24\, e^-$

Thus, ONCl and ONO^- are not isoelectronic.

(b) O_3 has bent shape structure.

Central atom O is sp^2-hybridised with one lone pair of electrons.

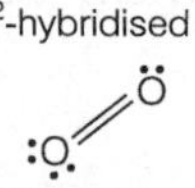

(c) Ozone is violet black in solid state.

(d) O_3 has no unpaired electrons, thus it is diamagnetic.

Hence, all the given statements are correct.

4. (c) (a) $2NaNO_3 \xrightarrow{\Delta} 2NaNO_2 + O_2$

(b) $2KClO_3 \xrightarrow{\Delta} 2KCl + 3O_2$

(c) $CaCO_3 \xrightarrow{\Delta} \underset{\text{Basic}}{CaO} + \underset{\text{Acidic}}{CO_2}$

(d) $NH_4NO_3 \xrightarrow{\Delta} N_2O + H_2O$

5. (a), (a) Thermal stability of the hydrides decreases as we go down the group in periodic table for group 15 (N-family).

$$BiH_3 < SbH_3 < AsH_3 < PH_3 < NH_3$$

Least stable — Most stable

M—H	—	255	247	322	391
Bond—energy (kJ mol^{-1})					

(b) Due to absence of d-orbital, nitrogen cannot form $d\pi$-$p\pi$ bond, thus it is correct.

(c) The N—N bond (BE 160 kJ mol^{-1}) is weaker than P—P bond (BE 209 kJ mol^{-1}). Thus, it is correct.

(d) N_2O_4 can form two resonance structures.

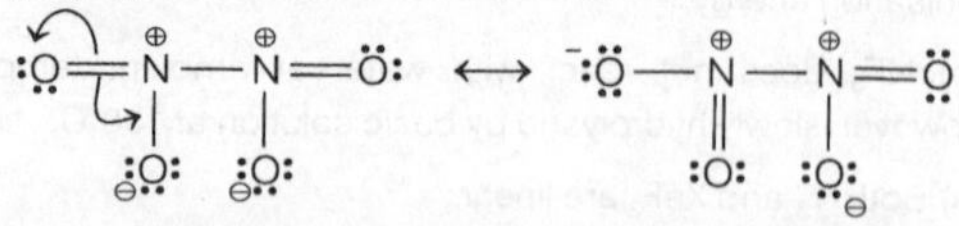

Thus, it is correct.

6. (d) (a) S_2 molecule is paramagnetic due to unpaired electrons in MO and is blue coloured compound, thus true.

(b) The vapour at 200°C consists mostly of S_8 rings, thus correct.

(c) At 600°C, the gas mainly consists of S_2 molecules, thus correct.

(d) Oxidation states of sulphur are

−2	in	H_2S	less than +4
0	in	S_8	
+2	in	$S_2O_3^{2-}$	
+4	in	SO_2	
+6	in	SO_3	

7. (c) The correct increasing basic strength

$$SbH_3 < AsH_3 < PH_3 < NH_3$$

NH_3 is most basic because of its small size, the electron density of electron pair is concentrated over small region.

As the size increases, the electron density gets diffused over a large surface area and hence, the ability to donate the electron pair (basicity) decreases. Hence, basic strength decreases down the group.

8. (a) XeF_6 has much tendency to hydrolyse. The reverse reaction is more spontaneous.

$$XeF_6 + 3H_2O \longrightarrow XeO_3 + 6HF$$

9. (d) $(NH_4)_2SO_4 \longrightarrow \underset{\text{Acidic}}{2NH_4^+} + SO_4^{2-}$

10. (c) Hypophosphorous acid (H_3PO_2) is a monobasic acid and has only one ionisable H because two H-atoms are directly attached to phosphorus.

The structure of H_3PO_2 is

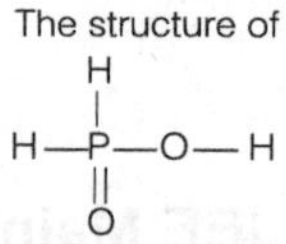

ionisable as directly attached to O.

11. (c) In a group, ΔG_f° (HX) changes from negative to positive downwards.

$$HF(g),\ \Delta G = -273.20 \text{ kJ mol}^{-1}$$

$$HI(g),\ \Delta G = +1.72 \text{ kJ mol}^{-1}$$

Thus, HF is thermally stable and HI not.

Thus, HF > HCl > HBr > HI

12. (c) Helium is not used to produce and sustain powerful superconducting magnets. All others are the uses of helium.

13. (a) In Holmes signals of the ship, mixture of CaC_2 and Ca_3P_2 is used.

14. (b) $2HCl + \frac{1}{2}O_2 \longrightarrow H_2O + Cl_2$

15. (c) Phosphorus (3rd period element) can raise covalency facilitating vacant d-orbitals.

P (ground): [Ne] $3s^2\ 3p^3\ 3d^0$

P (excited): [Ne] $3s^1\ 3p^3\ 3d^1$

JEE Advanced & IIT JEE Archive

1. (c) **Plan** This problem can be solved by using concept involved in chemical properties of xenon oxide and xenon fluoride.

XeF_6 on complete hydrolysis produces XeO_3.

XeO_3 on reaction with OH^- produces $HXeO_4^-$ which on further treatment with OH^- undergo slow disproportionation reaction and produces XeO_6^{4-} along with $Xe(g)$, $H_2O(l)$ and $O_2(g)$ as a by-product.

Oxidation half-cell in basic aqueous solution

$$HXeO_4^- + 5OH^- \longrightarrow XeO_6^{4-} + 3H_2O + 2e^-$$

Reduction half-cell in basic aqueous solution

$$HXeO_4^- + 3H_2O + 6e^- \longrightarrow Xe + 7OH^-$$

Balanced overall disproportionation reaction is

$$4HXeO_4^- + 8OH^- \longrightarrow \underbrace{3XeO_6^{4-} + Xe}_{\text{2 gaseous products}} + 6H_2O$$

Complete sequence of reaction can be shown as

$$XeF_6 + 3H_2O \longrightarrow XeO_3 + 3H_2F_2$$

$$\downarrow OH^-$$

$$HXeO_4^-$$

$$\downarrow OH^-/H_2O \text{ (disproportionation)}$$

$$XeO_6^{4-}(s) + Xe(g) + H_2O(l) + O_2(g)$$

2. (a) **Plan** This problem is based on chemical properties of phosphorus.

White phosphorus on reaction with thionyl chloride ($SOCl_2$) produces phosphorus trichloride.

$$P_4(s) + 8SOCl_2(l) \longrightarrow 4PCl_3(l) + 4SO_2(g) + 2S_2Cl_2(g)$$

But if amount of thionyl chloride ($SOCl_2$) is in excess then it produces phosphorus pentachloride.

$$P_4 + 10SOCl_2(l) \longrightarrow 4PCl_5 + 10SO_2$$

3. (a,c,d) **Plan** Due to resonance, bond lengths between two atoms are equal. Species is said to be diamagnetic if all electrons are paired.

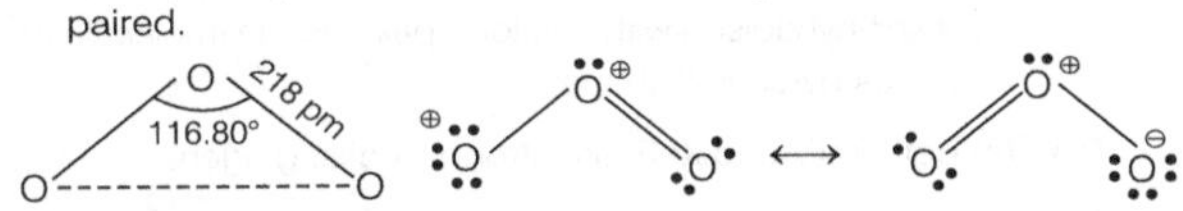

O_3 is bent molecule. All electrons paired thus, it is diamagnetic.

$$2O_3 \longrightarrow 3O_2 \quad \Delta H^\circ = -142 \text{ kJ mol}^{-1}$$

Thermal decomposition of O_3 is exothermic.

Explanation for Q. Nos. (4-5)

$$2NaOH + Cl_2 \xrightarrow{\text{Cold}} NaCl + \underset{P}{NaOCl} + H_2O$$

$$6NaOH + 3Cl_2 \xrightarrow{\text{Hot}} 5NaCl + \underset{Q}{NaClO_3} + 3H_2O$$

$$\underset{\text{hypochlorous acid}}{HOCl} \xrightarrow{NaOH} \underset{P}{NaOCl}$$

$$\underset{\text{Chloric acid}}{HClO_3} \xrightarrow{NaOH} \underset{Q}{NaClO_3}$$

$$Cl_2 + SO_2 \longrightarrow \underset{R}{SO_2Cl_2}$$

$$10SO_2Cl_2 + P_4 \longrightarrow \underset{S}{4PCl_5} + 10SO_2$$

$$PCl_5 + 4H_2O \longrightarrow \underset{T}{H_3PO_4} + 5HCl$$

4. (a) **5.** (a)

6. (d) (i) $2PbO_2 + 2H_2SO_4 \xrightarrow{\text{Warm (r)}} 2PbSO_4 + O_2 + 2H_2O$

(ii) $Na_2S_2O_3 + H_2O \xrightarrow{Cl_2(s)} NaHSO_4 + HCl$

(iii) $N_2H_4 \xrightarrow{I_2(q)} N_2 + HI$

(iv) $XeF_2 \xrightarrow{NO(p)} Xe + NOF$

Thus, (i) → (r), (ii) → (s), (iii) → (q), (iv) → (p)

7. (c) The involved redox reactions are

$$2H^+ + OCl^- + 2I^- \longrightarrow Cl^- + I_2 + H_2O \quad \text{...(i)}$$

$$I_2 + 2S_2O_3^{2-} \longrightarrow 2I^- + S_4O_6^{2-} \quad \text{...(ii)}$$

Also the n-factor of $S_2O_3^{2-}$ is one as

$$2S_2O_3^{2-} \longrightarrow S_4O_6^{2-} + 2e^-$$

[one 'e^-' is produced per unit of $S_2O_3^{2-}$]

Molarity of $Na_2S_2O_3$ = 0.25 N × 1 = 0.25 M

mmol of $Na_2S_2O_3$ used up = 0.25 × 48 = 12

Now, from stoichiometry of reaction (ii)

12 mmol of $S_2O_3^{2-}$ would have reduced 6 mmol of I_2.

From stoichiometry of reaction (i)

mmol of OCl^- reduced = mmol in I_2 produced = 6

Molarity of household bleach solution = $\frac{6}{25}$ = 0.24 M

Alternate Method

Milliequivalent of $Na_2S_2O_3$ = milliequivalent of OCl^-

= 0.25 × 48 = 12

Also n-factor of OCl^- = 2 [$Cl^+ \longrightarrow Cl^-$, gain of $2e^-$]

mmol of OCl^- = $\frac{12}{2}$ = 6 mmol. Remaining part is solved in the same manner.

8. (a) Bleaching powder is Ca(OCl) Cl. Therefore, the oxoacid whose salt is present in bleaching powder is HOCl. Anhydride of HOCl is Cl_2O as

$$2HOCl \longrightarrow Cl_2O + H_2O$$

Note *The oxidation number of element in anhydride and oxoacid remains the same.*

9. (b) Let oxidation number of N be x.

In HNO_3, $+1 + x + 3(-2) = 0 \Rightarrow x = +5$

In NO, $x - 2 = 0 \Rightarrow x = +2$

In N_2, $x = 0$

In NH_4Cl, $x + 4 - 1 = 0$

$\Rightarrow$ $x = -3$

10. (c) The reaction of white phosphorus with aqueous alkali is

$$P_4 + 3NaOH + 3H_2O \longrightarrow PH_3 + NaH_2PO_2$$

In the above reaction, phosphorus is simultaneously oxidised $[P_4(0) \longrightarrow NaH_2\overset{+1}{P}O_2]$ as well as reduced $[P_4(0) \longrightarrow \overset{-3}{P}H_3]$.

Therefore, this is an example of disproportionation reaction. Oxidation number of phosphorus in PH_3 is –3 and in NaH_2PO_2 is +1. However, +1 oxidation number is not given in any option, one might think that NaH_2PO_2 has gone to further decomposition on heating.

$$2NaH_2PO_2 \xrightarrow{\Delta} Na_2H\overset{+5}{P}O_4 + \overset{-3}{P}H_3$$

11. (d) In XeO_2F_2, the bonding arrangement around the central atom Xe is

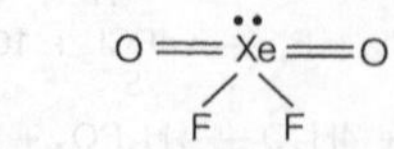

4σ-bonds + 1. $lp = 5$

Hybridisation of Xe = sp^3d

sp^3d-hybridisation corresponds to trigonal bipyramidal geometry. Also, in trigonal bipyramidal geometry, lone pairs remain present on equatorial positions in order to give less electronic repulsion.

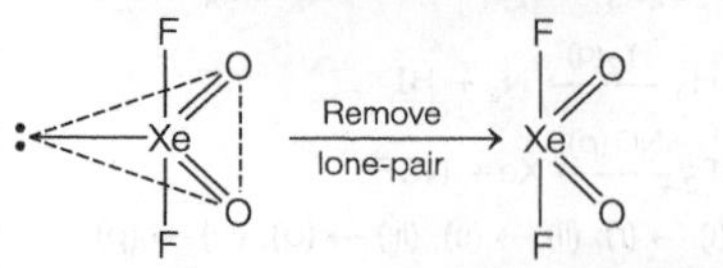

Note *According to Bent's rule, the more electronegative atoms must be present on axial position. Hence, F are kept on axial positions.*

12. (d) $Ba(N_3)_2 \xrightarrow{Heat} Ba(s) + 3N_2(g)$

Azide salt of barium can be obtained in purest form as well as the decomposition product contain solid Ba as by product alongwith gaseous nitrogen, hence no additional step of separation is required.

Other reactions are

$$NH_4NO_3 \xrightarrow{Heat} N_2O + 2H_2O$$

$$2NH_3 + 3CuO \xrightarrow{Heat} 3Cu + 3H_2O + N_2$$

$$(NH_4)_2Cr_2O_7 \xrightarrow{Heat} Cr_2O_3 + 4H_2O + N_2$$

13. (5) Br_2 is disproportionated in basic medium as

$$3Br_2 + 3Na_2CO_3 \longrightarrow 5NaBr + NaBrO_3 + 3CO_2$$

14. (4) PCl_5 produces $POCl_3$ with the following reagents

$$PCl_5 + SO_2 \longrightarrow POCl_3 + SOCl_2$$
$$PCl_5 + H_2O \longrightarrow POCl_3 + 2HCl$$
$$PCl_5 + H_2SO_4 \longrightarrow SO_2Cl_2 + 2POCl_3 + 2HCl$$
$$6PCl_5 + P_4O_{10} \longrightarrow 10\ POCl_3$$

15. (a) (i) $(CH_3)_2SiCl_2 + H_2O \longrightarrow (CH_3)_2Si(OH)_2 + 2HCl$

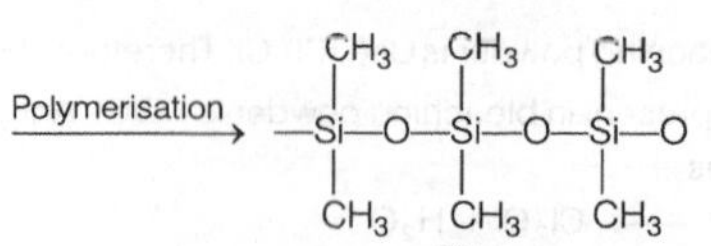

(ii) $3XeF_4 + 6H_2O \longrightarrow XeO_3 + 2Xe + 12HF + \frac{3}{2}O_2$

(iii) $Cl_2 + H_2O \longrightarrow HCl + HOCl \longrightarrow HCl + \frac{1}{2}O_2$

(iv) $VCl_5 + H_2O \longrightarrow VOCl_3 + 2HCl$

16. (b) In limited supply of oxygen, phosphorus is oxidised to its lower oxide P_4O_6 while excess of oxygen gives P_4O_{10}. A mixture of O_2 and N_2 is used for controlled oxidation of phosphorus into P_4O_6.

17. (a,b,c) The structures of these oxides are

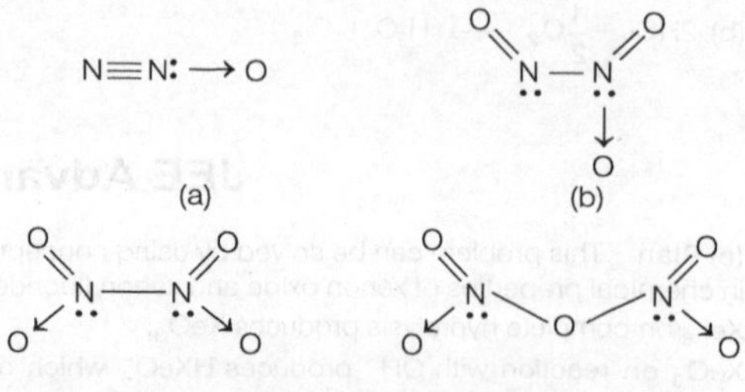

Hence, options (a), (b) and (c) have N—N bonds.

18. (b) Sodium thiosulphate, $Na_2S_2O_3$ gets oxidised by chlorine water as

$$Na_2S_2O_3 + 4Cl_2 + 5H_2O \longrightarrow 2NaHSO_4 + 8HCl$$

$FeCl_3$ oxidises $Na_2S_2O_3$ to $Na_2S_4O_6$.

19. (c) Due to greater solubility in water and prone to microbial attack, nitrates are less abundant in earth's crust.

20. (c) NH_3 is stronger Lewis base than PH_3. In a group of hydrides, basic strength decreases down the group.

21. (b) White phosphorus undergo disproportionation in alkaline medium.

$$P_4 + NaOH \longrightarrow \overset{-3}{P}H_3 + NaH_2\overset{+1}{P}O_2$$

22. (d) In P_4, all phosphorus are sp^3-hybridised and has 75% p-character.

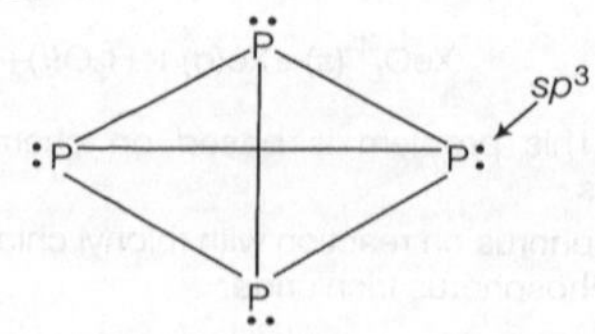

23. (a) Ar, being inert, provide inert atmosphere in arc welding and prevent from undesired oxidation.

24. (c)

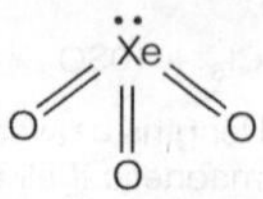

Xe is sp^3-hybridised with one lone pair. Hence, molecule of XeO_3 has pyramidal shape.

25. (a) Both XeF_4 and XeF_6 are strong oxidising agent.

26. (c) In $KMnO_4$, Mn is already in its highest oxidation state (+7), cannot be oxidised by any oxidising agent.

27. (b) $PbO_2 + HNO_3 \longrightarrow Pb(NO_3)_2 + H_2O + O_2$

28. (b) Equimolar amounts of NO and NO_2 at –30°C gives $N_2O_3(l)$ which is a blue liquid.

$$NO(g) + NO_2(g) \xrightarrow{-30°C} \underset{(Blue)}{N_2O_3(l)}$$

29. (c) Black phosphorus is thermodynamically most stable allotrope of phosphorus. It is due to three dimensional network structure of polymeric black phosphorus.

2. *d* & *f*-Block Elements

DPP-1 General Properties of *d*-Block Elements

1. (i) Iron (ii) Mercury (iii) Tungsten (iv) Technetium
(v) Silver (vi) Scandium and osmium

2. In these elements, the $(n-1)-d$-orbitals are complete and covalent bond formation in d-orbitals is not possible.

3. Due to lanthanide contraction, the elements of 5*d* series have virtually the same size as the corresponding elements just above them in 4*d* series. As a result, 4*d* and 5*d* series of elements have similar properties.

4. (i) $Fe^{3+} = [Ar]\,3d^5 4s^0$ is more stable configuration than $Fe^{2+} = [Ar]\,3d^6 4s^0$

(ii) Silver has $3d^{10}4s^1$ but Ag^{2+} ion has $3d^9$ configuration with incomplete d-orbitals. But Zn or Zn^{2+} has $3d^{10}$ configuration.

5. (i) Copper iron pyrites, $CuFeS_2$

(ii) Zn, Fe and Co

(iii) 24 carats = 100% gold. Hence, 21 carat = 87.5% gold

6. (b) Since,1st transition series belongs to 3*d* orbital and 2nd transition series belongs to 4*d*-orbital. According to the given option, atomic number 39 to 48 ($5s^2 4d^{1\text{-}10}$) belongs to second transition series.

7. (a) Mn^{2+}/Mn have -1.18 SRP value.

8. (c) For d-block elements, group number is equal to the number of electrons in $(n-1)\,d$ subshell + number of electrons in valence shell (*n*th shell).

9. (d) $[Kr]\,4d^{10}5s^0$

10. (a) $Sc(3d^1 4s^2)$ has 631 kJ mol^{-1} and Zn $(3d^{10}4s^2)$ has 906 kJ mol^{-1}

11. (b) Fe $(Z=26)$ in $+3$ oxidation state show $Fe^{3+}([Ar]\,3d^5)$ configuration. Ni(28) in +2 oxidation state show $Ni^{2+}([Ar]3d^8)$ configuration.

12. (c) The element lying above $Z = 42$ has Z value 24, i.e. chromium. Hence, Cr configuration is $[Ar]3d^5 4s^1$.

13. (d) Zinc has lowest melting point than all.

14. (c) Due to lanthanide contraction, Ag and Au have nearly same size.

15. (a) $V^{2+}/V(-1.18) < Fe^{2+}/Fe\ (-0.44)$

$< Ni^{2+}/Ni < (-0.25) < Cu^{2+}/Cu\,)(+0.34)$

16. (a,b) Due to lanthanide contraction.

17. (b,c,d) Only reactive metal zinc liberates hydrogen on reaction with acid.

$$Zn + 2HCl \longrightarrow ZnCl_2 + H_2$$

18. (b,c) An element with pseudo noble gas configuration has 18 electrons in the highest energy level instead of 8.

$$Zn^{2+}(28) = 3s^2, 3p^6, 3d^{10}$$
$$Ag^+\ (46) = 4s^2 4p^6 4d^{10}$$

19. (b,c) V, Nb, Ta have bcc structure and Fe, Ru, Os have hcp structure

20. (b,d) Both Fe^{2+} and Co^{3+} have $[Ar]\,3d^6$ configuration while V^{2+} and Cr^{3+} have $[Ar]\,3d^3$.

21. (c) Due to lanthanide contraction.

22. (c) The high melting point of Cr metal is attributed to the involvement of greater number of electrons from $(n-1)\,d$ in addition to *ns* electrons in the interatomic metallic bonding.

23. (b) (i) → (s), (ii) → (r), (iii) → (q), (iv) → (p)

24. (a) (i) → (r), (ii) → (s), (iii) → (p), (iv) → (q)

25. (2) Ni has $3d^8 4s^2$ and Pt has $5d^9 6s^1$.

26. (3) Chromium has 6 unpaired electrons and hence, its spin value is –3 or +3. Cr,

1	1	1	1	1		1
		$3d^5$				$4s$

There are 6 unpaired electrons in chromium.

$$S = \frac{1}{2} + \frac{1}{2} + \frac{1}{2} + \frac{1}{2} + \frac{1}{2} + \frac{1}{2} = \frac{6}{2} = 3$$

27. (5) $l = 2$ means d-subshell. Zinc has 10 electrons in d-subshell in which 5 electrons are + 1/2 spin and 5 electrons are – 1/2.

28. (4) Zn, Cd, Sc and La do not show variable valencies.

29. (4) Ytterium, zirconium, technetium and cadmium only have ns^2 while all other have ns^1.

30. (a) Osmium has the electronic configuration $5d^6 6s^2$. As 5*d* and 6s are close in energy, all the 8 electron can participate in bonding.

DPP-2 General Chemical and Magnetic Properties of *d*-Block Elements

1. Cr^{2+} is reducing as its configuration changes from d^4 to d^3 (by losing one electron) and the latter has half-filled t_{2g} level.
Mn^{3+} (with d^4) gains an electron and become Mn^{2+} which has half-filled stable configuration having extra stability.

2. Copper has high enthalpy of atomisation $(\Delta_a H°)$ and low enthalpy of hydration $(\Delta_{hyd} H°)$.
Hence, formation of $Cu^{2+} \longrightarrow$ Cu has
$E° = +0.34$ V while $Cu \rightarrow Cu^{2+}$ has $E° = -0.34$ V (less favourable).

3. Cr^{2+} is stronger reducing agent than Fe^{2+}.
Reason is that case of $Cr^{2+} \rightarrow Cr^{3+}$, Cr changes from $d^4 \rightarrow d^3$.
In case of $Fe^{2+} \rightarrow Fe^{3+}$, Fe changes from $d^6 \rightarrow d^5$.
In aqueous medium, d^3 is more stable than d^5.

4. In 3d series, manganese, 4d series ruthenium and in 5d series, osmium show maximum oxidation state.

5. Mn^{2+} is more stable than Mn^{3+} state. Hence, $Mn^{3+} (3d^4 4s^0) \longrightarrow Mn^{2+} (3d^5 4s^0)$ is spontaneous but Fe^{3+} is more stable than Fe^{2+}. Hence, $Fe^{3+} (3d^5 4s^0) \longrightarrow Fe^{2+} (3d^6 4s^0)$ is non-spontaneous.

6. (d) $(n-1)d^5 ns^2$ attains the maximum oxidation state of + 7.

7. (d) Ti^{2+}, V^{3+}, Cr^{4+} and Mn^{5+} show $1s^2 2s^2 2p^6 3s^2 3p^6 4s^0 3d^2$ electronic configuration.

8. (a) $Mn^{2+} (d^5) > Cr^{2+} (d^4) > Fe^{2+} (d^6) > Co^{2+} (d^7)$

Mn^{2+} is more stable due to half-filled configuration, i.e. d^5.

9. (c) The variability of oxidation state, a characteristic of transition element arises due to incomplete filling of d-orbitals.

10. (b) Cu (I) undergoes spontaneous disproportionation.

$$Cu^+ \rightleftharpoons Cu^{2+} + e^-, \; E^\circ = -0.153 \text{ V}$$
$$Cu^+ + e^- \rightleftharpoons Cu, \; E^\circ = +0.521 \text{ V}$$
$$2Cu^+ \rightleftharpoons Cu + Cu^{2+}, \; E^\circ_{cell} = +0.368 \text{ V}$$

Thus, ΔG is negative (spontaneous) for disproportionation.

11. (c) Gun metal is an alloy of Cu, Zn and Sn. It contains 88% Cu, 10% Sn and 2% Zn.

12. (d) Ni^{2+} and Ti^{3+} ions are coloured in aqueous solution because they contain unpaired electrons.

13. (a) We know that white light is composed of many different colours. When white light is allowed to fall on $[Ti(H_2O)_6]^{3+}$ ion, this ion absorbs the light of green and yellow colour and transmits is almost purple colour. The purple colour of the transmitted light is called complimentary colour of that of absorbed light.

14. (d) Metal ion is Fe^{3+} which have 5 unpaired electrons.

15. (b) $Mn^{2+} (d^5)$ is more stable than $Mn^{3+} (d^4)$ because Mn^{2+} have half-filled configuration.

16. (b,c,d) Due to the presence of unpaired electrons, transition elements are paramagnetic in nature. Mn^{2+} contains 5 unpaired electrons while Fe^{2+} contains 4 unpaired electrons.

Manganese exhibits all the oxidation states from + 2 to + 7.

17. (a,c,d) It is a copper alloy with nickel and often zinc. The usual formation is 60% copper, 20% nickel and 20% zinc.

18. (b) Magnetic moment is 3.87 BM means that the ion has 3 unpaired electrons.

19. (b,c) +8 oxidation state shown by Ru and Os.

20. (b,d) Ni^{2+} and Fe^{2+} have green colour and V^{4+} and Cr^{2+} have blue colour.

21. (d) In transition metal complexes, a change in electron distribution between the metal and a ligand give rise to charge transfer bonds.

Here, charge transfer may occur from the ligand molecular orbitals to the empty or partially filled metal d-orbitals.

22. (b) $Ag^+ (d^{10})$ and $Au^+ (d^{10})$ are diamagnetic due to absence of unpaired number of electrons.

23. (a) (i) → (s), (ii) → (p), (iii) → (q) (iv) → (r)

24. (d) (i) → (s), (ii) → (s), (iii) → (p), (iv) → (q)

25. (a) Fenton's reagent ($FeSO_4 + H_2O_2$) is used for oxidising alcohol to aldehyde. *Zeigler-Natta* catalyst $[Al(C_2H_5)_3 + TiCl_4]$ is used for manufacture of polythene.

(i) → (q), (ii) → (r), (iii) → (s), (iv) → (p)

26. (6) German silver (Cu + Zn + Ni), nichrome (Ni 60% + Cr 20% + Fe), monel metal (Cu + Ni 66%), constanton (Cu + Ni), invar (Fe + Ni), alnico (Al + Ni + Co + Cu).

27. (7) In both, x and y value is 7.

28. (6)

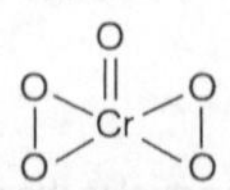

4 oxygen atoms attached by peroxide linkage. So, their oxidation states are taken as –1.

Now, let the oxidation state of Cr be x.

then $\quad x + 4 \times (-1) - 2 = 0$

$\quad x = 6$

29. (5) In metal carbonyls, metal oxidation state is zero.

30. (a) Colours of $Cr_2O_7^{2-}$, CrO_4^{2-}, MnO_4^-, MnO_4^{2-}, Cu_2O and Ni-DMG complex are due to charge transfer transition.

(iv) Sodium sulphite is a reducing agent. It reduces acidified $K_2Cr_2O_7$ to chromic sulphate which is green in colour.

$$Cr_2O_7^{2-} + 3SO_3^{2-} + 8H^+ \longrightarrow 2Cr^{3+} + 3SO_4^{2-} + 4H_2O$$

DPP-3 Oxides and Oxo Anions of Cr and Mn

1. (i) $2Mn^{3+} + 2H_2O \longrightarrow Mn^{2+} + MnO_2 + 4H^+$

(ii) $4MnO_4^- + 4OH^- \longrightarrow 4MnO_4^{2-} + O_2 + 2H_2O$

(iii) $3MnO_4^{2-} + 2H_2O \longrightarrow MnO_2 + 2MnO_4^- + 4OH^-$

(alkaline medium)

$3MnO_4^{2-} + 4H^+ \longrightarrow MnO_2 + 2MnO_4^- + 2H_2O$

(acidic medium)

2. Equation for the formation of A is

$2MnO_2$ (Pyrolusite) $+ 4KOH + O_2 \longrightarrow 2K_2MnO_4$ ((A) Potassium manganate) $+ 2H_2O$

Formation of B

$3K_2MnO_4$ (A) $+ 2H_2SO_4 \longrightarrow 2KMnO_4$ (B) $+ MnO_2 + 2K_2SO_4 + 2H_2O$

Equations for the reactions of (B)

(i) $2KMnO_4 + H_2O + KI \longrightarrow 2MnO_2 + 2KOH + KIO_3$ (C)

(ii) $2KMnO_4 + H_2SO_4 \longrightarrow Mn_2O_7$ (D) $+ K_2SO_4 + H_2O$

$2Mn_2O_7 \longrightarrow 4MnO_2 + 3O_2$ (E)

3. (i) In the solution, the following equilibrium exists

$$Cr_2O_7^{2-} + H_2O \rightleftharpoons 2CrO_4^{2-} + 2H^+$$

In acidic medium (pH < 7), it exists as $Cr_2O_7^{2-}$ ions and has orange colour while in basic medium (pH > 7), it exists as CrO_4^{2-} ions and has yellow colour.

(ii) $K_2Cr_2O_7$ is not hygroscopic and can be used as primary standard for the volumetric estimation of ferrous salts, sulphides, etc.

(iii) Acidified solution of dichromate ions forms a deep blue colour with H_2O_2 due to the formation of chromium peroxide CrO_5 $[CrO(O_2)_2]$. This peroxo compound can be extracted into ether, where it is stable, forming the adduct, $Py \cdot CrO\ (O_2)_2$.

$$Cr_2O_7^{2-} + 2H^+ + 4H_2O_2 \longrightarrow 2CrO_5 + 5H_2O$$

4. (i) ZnO (ii) $ZnO \cdot CoO(CoZnO_2)$
(iii) TiO_2 (iv) ZrO_2 (v) OsO_4 (vi) Hg_2Cl_2 (vii) $HgCl_2$

5. $N_1V_1 = N_2V_2$ or $N_1 \times 40 = 0.1 \times 100$

$\therefore\ N_1 = 0.25$ N

Molarity of $KMnO_4 = \dfrac{0.25}{5} = 0.05$ M

6. (a) Oxide having lower oxidation number is basic and as oxidation number increase, acidic nature increases.

7. (a) TiO > VO > CrO > FeO with the decrease in size of metal atom from Ti to Fe, the basic character of their monoxides decreases.

8. (b) $2MnO_4^- + 16H^+ + 5S^{2-} \longrightarrow 2Mn^{2+} + 8H_2O + 5S$

9. (d) $2MnO_4^- + H_2O + I^- \longrightarrow 2MnO_2 + IO_3^- + 2OH^-$

10. (b) One equivalent of $KMnO_4$ = One equivalent of $K_2Cr_2O_7$ = One equivalent of Fe^{2+} ion or 1/5 moles of $KMnO_4$ = 1/6 moles of $K_2Cr_2O_7$ = One mole of Fe^{2+} ion.

Hence, one mole of $KMnO_4$ reacts with 5 moles of Fe^{2+} ion and one mole of $K_2Cr_2O_7$ reacts with 6 moles of Fe^{2+} ion.

11. (c) This is called ligand to metal charge transfer phenomenon.

12. (c) The green manganate is paramagnetic with one unpaired electron but the permanganate is diamagnetic.

13. (a)

14. (a) $2MnO_2 \xrightarrow[\text{Oxidised with air}]{\text{Fused with KOH}} MnO_4^{2-}$

and $MnO_4^{2-} \xrightarrow[\text{in alkaline solution}]{\text{Electrolytic oxidation}} MnO_4^-$

15. (c) $CrO_4^{2-} + 4H_2O + 3e^- \longrightarrow Cr(OH)_3 + 5OH^-, E^\circ = -0.13$ V

$Cr_2O_7^{2-} + 14H^+ + 6e^- \longrightarrow 2Cr^{3+} + 7H_2O, E^\circ = 1.33$ V

16. (a,c,d) $K_2Cr_2O_7 + 4NaCl + 6H_2SO_4 \longrightarrow 2KHSO_4 + 4NaHSO_4 + \underset{\substack{\text{Chromyl chloride} \\ \text{(orange red)}}}{2CrO_2Cl_2} + 3H_2O$

17. (a,b,c) Mn_2O_7 is acidic while others are amphoteric.

18. (a,b,d) $2KMnO_4 \xrightarrow{\Delta} K_2MnO_4 + MnO_2 + O_2$

$2KNO_3 \xrightarrow{\Delta} 2KNO_2 + O_2$

$(NH_4)_2Cr_2O_7 \xrightarrow{\Delta} N_2 + Cr_2O_3 + 4H_2O$

$4K_2Cr_2O_7 \xrightarrow{\Delta} 4K_2CrO_4 + 2Cr_2O_3 + 3O_2$

19. (b) Potassium dichromate is used as a primary standard in volumetric analysis.

20. (a,b,d) $2MnO_4^- + 5NO_2^- + 6H^+ \longrightarrow 2Mn^{2+} + 5NO_3^- + 3H_2O$

21. (d) ZnO and V_2O_5 are amphoteric oxides.

22. (b) $3MnO_4^{2-} + 4H^+ \longrightarrow 2MnO_4^- + MnO_2 + 2H_2O$

23. (a) (i) → (q,t) (ii) → (p,r) (iii) → (q,r,t) (iv) → (q,s)

24. (b) (i) → (q) (ii) → (s) (iii) → (r) (iv) → (p)

(i) In acidic medium,

$$MnO_4^- + 8H^+ + 5e^- \longrightarrow Mn^{2+} + 4H_2O$$

Hence, – M/5

(ii) In strongly alkaline medium

$$MnO_4^- + e^- \longrightarrow MnO_4^{2-}$$

Hence, = M/1

(iii) In neutral or weakly basic medium

$$MnO_4^- + 2H_2O + 3e^- \longrightarrow MnO_2 + 4OH^-$$

Hence, = M/3

(iv) $Cr_2O_7^{2-} + 14H^+ + 6e^- \longrightarrow 2Cr^{3+} + 7H_2O$

Hence, equivalent weight is M/6.

25. $\left(\dfrac{10}{3}\right) KMnO_4 + FeC_2O_4 + H_2SO_4 \longrightarrow K_2SO_4 + MnSO_4 + Fe_2(SO_4)_3 + CO_2 + H_2O$

Balanced equation,

$6KMnO_4 + 10FeC_2O_4 + 24H_2SO_4 \longrightarrow 3K_2SO_4 + 6MnSO_4 + 5Fe_2(SO_4)_3 + 20\,CO_2 + 24H_2O$

Thus, 6 moles of $KMnO_4$ react with $10\,FeC_2O_4$.

Hence, 2 moles of $KMnO_4$ react with

$$\frac{10 \times 2}{6} = \frac{10}{3} \text{ mol } FeC_2O_4$$

26. (8) Milliequivalents of $K_2Cr_2O_7$ reacted with $H_2S\ (N_1V_1)$ = milliequivalents of $KMnO_4$ reacted with $H_2S (N_2V_2)$,

i.e. milliequivalent of $KMnO_4$ = milliequivalent of H_2S

Therefore, $50 \times 0.04 \times 6 = V_2 \times 0.03 \times 5$

or $V_2 = 80$ mL

27. (8)

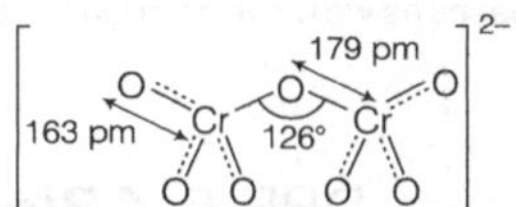

28. (3) Normality = molarity × number of electrons gained per mole of dichromate in a redox reaction. Hence, Normality = Molarity × 6

29. (d) $2CrO_4^{2-} + 2H^+ \rightleftharpoons Cr_2O_7^{2-} + H_2O$

30. (a) $Cr_2O_7^{2-} + 14H^+ + 6I^- \longrightarrow 2Cr^{3+} + 7H_2O + 3I_2$

DPP-4 *f*-Block Elements-I (Lanthanides)

1. (i) Among lanthanides, + 3 oxidation state is most stable.

(ii) Ce^{4+} readily gains one electron to get stable +3 state. Hence, Ce^{4+} is strong oxidising agent in aqueous solution.

$$Ce^{4+} + e^- \longrightarrow Ce^{3+}, E^\circ = +1.74$$

(iii) On the other hand, Sm^{2+} loses one electron to get +3 state and acts as reducing agent.

2. +4 oxidation state of cerium, praceodymium, neodymium, terbium and dysprosium.

+2 oxidation state of europium (Eu), ytterbium (Yb), samarium (Sm), thullium (Tm), neodymium (Nd).

3. They do not absorb light in the visible region but absorb in UV region.

4. Lanthanides originally extracted from oxides which were considered to be rare. Hence, they are called rare-earths. But the term rare-earth is avoided now because many of these elements are no longer rare but are abundant.

5. (i) Monazite mineral (Ce,La) PO_4 which is chiefly worked for thorium and the lanthanides are obtained as by products.

(ii) The steady decrease in the atomic and ionic (Ln^{3+}) radii as the atomic number of the lanthanoid element increases is called lanthanoid contraction.

This is because the additional electron goes to 4f-subshell. This subshell being large and diffuse have poor shielding effect.

The effective nuclear charge increases which causes the decrease in the size of electron charge cloud.

The lanthanoid contraction is extraordinary in nature as all the inner-transition elements are present in the same group as well as in same period. It means that all elements have position in periodic table in spite of there is regular decrease in size with increase in atomic number.

6. (d) Er (68), Tm (69), Yb (70) and Lu (71) elements are present in same group as well as same period.

7. (d) Samarium (Sm) (Z = 62) belongs to III B group and 6th period.

8. (d) Due to lanthanoid contraction.

9. (a) Tb (Z = 65) have [Xe] $4f^9 5d^0 6s^2$ configuration. Tb^{4+} have half-filled 4f subshell, i.e. $4f^7$.

10. (d) Europium (Eu) has [Xe] $4f^7 5d^0 6s^2$ configuration. The oxidation state of + 2 and + 3 are exhibited by these elements, e.g. Eu^{2+} is [Xe] $4f^7$ and Eu^{3+} is [Xe]$4f^6$.

11. (a) Misch metal is lanthanide metal (95%) and iron (5%) and traces of S, C, Ca, Al.

12. (b) The basicity decreases from $Ce(OH)_3$ to $Lu(OH)_3$. Due to lanthanide contraction, the decrease in the size of the cation increases the covalent character between the lanthanide ion and hydroxide ion thereby, reducing the basic character of the lanthanide hydroxide.

13. (a) Gadolinium (Gd) have [Xe] $4f^7\ 5d^1 6s^2$ configuration.

14. (c) Electronegativity increases as we move from left to right due to lanthanide contraction.

15. (c) The hardness increases with increasing atomic number.

16. (a,d) $Ce^{3+}(4f^1)$ and $Yb^{3+}(f^{13})$ are colourless because they do not absorb in the visible region. The $E°$ value for Ce^{4+}/Ce^{3+} is + 1.74 V which suggest it is strong oxidising agent.

17. (a,b,d) Lanthanoids belong to 3rd group and 6th period of the periodic table.

18. (a,b,c,d) La^{3+}, Gd^{3+} and Lu^{3+} have empty ($4f^0$) half-filled ($4f^7$) and full-filled ($4f^{14}$) orbitals. $Ce^{3+}(4f^1)$ is colourless because it does not absorb in the visible region.

19. (a,b,d) Basic strength decreases from $La(OH)_3$ to Lu $(OH)_3$.

20. (a,b,c) The pair of elements Zr - Hf, Nb - Ta, Mo - W possess almost the same size. After the lanthanoids the increase in radii from second to third transition series almost vanishes.

21. (a) More negative the SRP value, stronger is the reducing agent.

22. (b) Ce^{4+} is [Xe] $4f^0$, it tends to revert to more stable oxidation state of +3 by gain of an electron. That is why, in aqueous solution, Ce^{4+} is good oxidising agent.

23. (a) (i) $\rightarrow$ (p,t) (ii) $\rightarrow$ (r) (iii) $\rightarrow$ (q,s) (iv) $\rightarrow$ (q,s)

24. (5) Gd: [Xe] $4f^7 5d^1 6s^2$, Lu: [Xe] $4f^{14} 5d^1 6s^2$,
Eu: [Xe] $4f^7 5d^0 6s^2$, Yb: [Xe] $4f^{14} 5d^0 6s^2$
Eu^{2+}: [Xe] $4f^7 5d^0 6s^0$,
Gd^{3+}: [Xe] $4f^7 5d^0 6s^0$, Eu^{3+}: [Xe] $4f^6 5d^0 6s^0$,
Tb^{4+}: [Xe] $4f^7 5d^0 6s^0$, Ce^{4+}: [Xe] $4f^0 5d^0 6s^0$,

25. (0) Ytterbium has configuration [Xe] $4f^{14}\ 5d^0 6s^2$

26. (4) La^{3+}, Ce^{4+}, Yb^{2+} and Lu^{3+} are diamagnetic in nature.

27. (3) Eu^{2+}, Sm^{2+}, Yb^{2+} are reducing agents.

28. (8) The highest oxidation state shown by the transition element is Ru (+8) in 4d series and Os in 5d series.

29. (a) Both statements I and II are correct. Statement II correctly explains the statement I.

30. (a) Many trivalent lanthanide ions are coloured due to the presence of f-electrons, in solid state as well as in aqueous solutions.

DPP-5 *f*-Block Elements-II (Actinides)

1. (i) f-block elements are called inner transition elements as these elements lie between two transition elements are called elements in which differentiating electron enters f-subshell of anti-penultimate shell, i.e. $(n-2)$ f-subshell.

The general outer electronic configuration is

$(n-2) f^{1-14} (n-1) d^{0-1} ns^2$,

where n = 6 or 7.

(ii) Actinium belong to d-block as the differentiating electron enters into 6d-subshell.

2. When 5f-orbitals are beginning to be occupied, they will penetrate less into inner core of electrons. In other words, 5f-electrons themselves provide poor shielding from element to element in the series.

Hence, actinide contraction is greater from element to element than lanthanide contraction.

3. Actinides show a greater range of oxidation states due to the fact that the 5f, 6d and 7s levels are comparable energies.

These show in general +3 oxidation states. The elements in the first half of the series frequently show higher oxidation states but decreases in succeeding elements.

The maximum oxidation state increases from +4 in Th to +5, + 6 and +7 respectively in Pa, U and Np but decreases in succeeding elements.

4. (i) Elements after uranium, (i.e. $Z > 92$) are called transuranium elements.

(ii) Neptunium

(iii) + 3 to + 7

5. (i) The first actinides to be discovered were thorium and uranium. The other actinides that were discovered in small portions in nature were actinium and protactinium.

(ii) The evidence that shows actinides are very reactive. They have low solubilities as halide compounds; and the lighter, trivalent. Actinide compounds are unstable in aqueous solutions and can be easily precipitated out of acidic and basic environments.

6. (d) Thorium ($Z = 90$) have [Rn] $5f^0 6d^2 7s^2$ configuration.

7. (b) General EC = [Rn] $5f^{1-14} 6d^{0-1} 7s^2$.

8. (d) The wide range of oxidation state of actinoids is due to the fact that the 5*f*, 6*d* and 7*s* energy levels are of comparable energies and all these three subshells can participate in comparison to lanthanoid. The 5 *f* orbitals extend farther from the nucleus than the 4*f* orbitals.

9. (d) 4*f* electrons have greater shielding effect but 5*f* electrons have poor shielding effect.

10. (a) Americium ($Z = 95$) and curium (Cm, $Z = 96$) have [Rn] $5f^7 6d^0 7s^2$ and [Rn] $5f^7 6d^1 7s^2$ configuration, respectively.

11. (c) Actinides are highly electropositive and reactive metals.

12. (d) Actinoids are from thorium ($Z = 90$) to Lawrencium ($Z = 103$) in the Periodic table So, rutherfordium ($Z = 104$) is not an actinoid.

13. (b) UF_6 is used in enrichment of uranium for nuclear power plant.

14. (b) Plutonium ($Z = 94$) have [Rn] $5f^6 6d^0 7s^2$ configuration and show oxidation state upto +7.

15. (d) Cm ($Z = 96$) have [Rn] $5f^7 6d^1 7s^2$,

Lr ($Z = 103$) have [Rn] $5f^{14} 6d^1 7s^2$

Np ($Z = 93$) have [Rn] $5f^4 6d^1 7s^2$

16. (a,b) Uranium is used in the production of nuclear energy by the process of nuclear fission.

Plutonium is fissionable material and is used for fueling atomic reactors.

17. (a,c) Actinoids are radioactive elements and uranium is the heaviest natural element.

18. (a,d) Correct statements are oxides and hydroxides of actinides are more basic than lanthanoids and actinoids have greater tendency to form complexes in comparison to lanthanoids.

19. (a,b,c) Am ($Z = 95$) have [Rn] $5f^7 6d^0 7s^2$

Am^{2+} have [Rn] $5f^7 6d^0$, Cm ($Z = 96$) have [Rn] $5f^7 6d^1 7s^2$

NP ($Z = 93$) have [Rn] $5f^4 6d^1 7s^2$

20. (a,b,d) Alkalis have no action with actinoids metals.

21. (a) Number of alpha particles emitted

$$= \frac{\text{Change in mass number}}{4} = \frac{16}{4} = 4\alpha$$

Number of beta particles liberated

= Double the alpha particles − change in atomic number
$= 8 - 7 = 1$

22. (d) The end product of neptunium series is bismuth-209.

23. (a) (i) → (q), (ii) → (s), (iii) → (r), (iv) → (p)

24. (9) Berkelium ($Z = 97$) : [Rn] $5f^9 6d^0 7s^2$

25. (0) Thorium +4 configuration ($Z = 90$) = [Rn] $5f^0 6d^0 7s^0$

26. (4) Only U, Np, Pu and Am show +6 oxidation state.

27. (2) The element with atomic number 100 is fermium. Its configuration is [Rn] $5f^{12} 6d^0 7s^2$.

28. (8) Number of alpha particles (α) liberated = change in mass number/4 = 238 − 206/4 = 8

29. (b) The energy gap among 4*f*, 5*d* and 6*s* is large where as the energy gap among 5*f*, 6*d* and 7*s* is small.

30. (d) Acitinides form relatively more stable complexes as compared to lanthanides.

Revisal Problems (JEE Main)

1. (c) A metal is soluble in acid if it displaces H_2 from solution, which is determined by the metal's standard reduction potential. Cu, Ag, Pt and Au cannot displace H_2.

2. (d) Only Cu shows +1 oxidation state.

3. (a) In MnO_2, manganese is in Mn^{4+} form. Hence, the configuration is [Ar] $3d^3$.

4. (c) Rutile is a mineral compared primarily of titanium dioxide, TiO_2.

5. (d) The order of increasing atomic radii is Ni < Mn < Ti < La

6. (c) Colour of the compounds of the same transition metal ion depends upon the nature of ligand attached to it, e.g. $[CuCl_4]^{2-}$ is greenish yellow, $[Cu(H_2O)_4]^{2+}$ is blue but $[Cu(NH_3)_4]^{2+}$ is violet or deep blue.

7. (c) Both have 23 electrons each.

8. (b) Metal oxide with higher oxidation number exhibit acidic nature.

9. (c) Smaller the ion, more is the hydration and also hydration energy.

10. (c) Cobalt +2 ion is blue pink in colour.

11. (d) Addition of high proportions of Mn gives hardness to steel which is used for making rails of rail roads.

12. (a) Both Hg^{2+} and Hg_2^{2+} do not have unpaired electron.

13. (b) In (i), (ii) and (iii), same element is undergoing both oxidation and reduction.

14. (a) Mn_2O_7 = acidic, CrO_3 = acidic, Cr_2O_3 = amphoteric,

CrO = basic, V_2O_5 = acidic, V_2O_4 = amphoteric oxides.

15. (d) Interstitial compounds are chemically inert.

16. (b) Consider one mole of the oxide, moles of $M = 0.98$ and moles of $O^{2-} = 1$

Let moles of $M^{3+} = x$ and moles of $M^{2+} = 0.98 - x$

During charge balance $(0.98 - x) \times 2 + 3x - 2 = 0$

On solving, we get $x = 0.04$,

Hence, % of $M^{3+} = \frac{0.04}{0.09} \times 100 = 4.08$

17. (c) All others have sp^3 hybridisation with tetrahedral shape. $[Ni(CN)_4]^{2-}$ have dsp^2-hybridisation.

18. (c) $Mn^{3+} + e^- \longrightarrow Mn^{2+}$, $E° = +1.57$ V

$Co^{3+} + e^- \longrightarrow Co^{2+}$, $E° = +1.97$ V

19. (c) $Cr_2O_7^{2-} + 2OH^- \longrightarrow 2CrO_4^{2-} + H_2O$

20. (d) (i) One equivalent dichromate = one equivalent ferrous oxalate or 1/6 mole dichromate = 1/3 mole ferrous oxalate

Three moles dichromate = six moles ferrous oxalate or one mole ferrous oxalate = 0.5 mole dichromate

(ii) In ferric oxalate, only oxalate ion undergo oxidation
1/6 mole, dichromate = 1/2 mole ferric oxalate
Two mole dichromate = six mole ferrous oxalate or
one mole ferrous oxalate = 0.33 mole dichromate

21. (b) $2MnO_4^- + 5C_2O_4^{2-} + 16H^+ \longrightarrow 2Mn^{2+} + 10CO_2 + 8H_2O$

$2MnO_4^- + 10Cl^- + 16H^+ \longrightarrow 2Mn^{2+} + 5Cl_2 + 8H_2O$

22. (b) $2CrO_4^{2-} + 2H^+ \longrightarrow Cr_2O_7^{2-} + H_2O$

23. (a) In neutral or weakly basic medium, permanganate ion is reduced to MnO_2.

24. (a) $5S^{2-} + 2MnO_4^- + 16H^+ \longrightarrow 2Mn^{2+} + 5S + 8H_2O$

25. (c) $2MnO_2 + 4KOH + O_2 \longrightarrow 2K_2MnO_4 + 2H_2O$

$3MnO_4^{2-} + 4H^+ \longrightarrow 2MnO_4^- + MnO_2 + 2H_2O$

26. (d) Eu =[Xe] $4f^7 5d^0 6s^2$; Eu shows +2 and +3 oxidation states.

27. (a) Promethium is synthetic and radioactive element.

28. (b) Ti^{3+} have one d-electron which is responsible for its colour. d-d transition occurs here.

29. (c) Oxidising agent gains electron and reducing agent loses electrons.

Ce(IV) gain electron to become Ce (III) and hence, strong oxidising agent.

Eu (II) loses one electron to become Eu (III) and hence, act as reducing agent.

30. (b) Eu = [Xe] $4f^7 5d^0 6s^2$ and Yb = [Xe] $4f^{14} 5d^0 6s^2$. Removal of 3rd electron requires more energy from these elements.

Revisal Problems (JEE Advanced)

1. (a) The substance which have unpaired electron shows paramagnetic behaviour. The number of unpaired electrons present in these substances as follows :

$(d^5)\ Mn^{2+}$ = 5 unpaired e^-

$(d^6)\ Fe^{2+}$ = 4 unpaired e^-

$(d^7)\ Co^{2+}$ = 3 unpaired e^-

(d^{10}) Zn = no unpaired

2. (a) Consider one mole of the oxide, moles of $M = 1.8$ and moles of $O^{2-} = 1$

Let moles of $M^{2+} = x$ and moles of $M^+ = 1.8 - x$

During charge balance

$$(1.8 - x) \times 1 + 2x - 2 = 0$$

On solving, we get

$$x = 0.20$$

Hence, percentage of $M^{2+} = \frac{0.20}{1.8} \times 100 = 11$

3. (c) Stability of higher oxidation states increases down the group. WO_3 is more stable in +6. Hence, it does not act as an oxidising agent.

4. (a) The number of unpaired electrons in Mn^{2+}, Cr^{2+} and V^{2+} are 5, 4 and 3, respectively.

The magnetic moment (BM) can be calculated by the spin-only formula

$$\mu_{BM} = \sqrt{n(n+2)}$$

where, n is the number of unpaired electrons.

Hence, the magnetic moment of Mn^{2+}, Cr^{2+} and V^{2+} are 5.91, 4.89 and 3.87 respectively.

5. (b) $0.1\ M\ Cr_2O_7^{2-} = 0.6\ N\ Cr_2O_7^{2-}$

and similarly, $0.1\ M\ MnO_4^- = 0.5\ N\ MnO_4^-$

$Fe^{2+} = Cr_2O_7^{2-} = MnO_4^-$

$= 30$ mL of $0.1\ M\ Cr_2O_7^{2-}$

$= V$ mL of $0.1\ M\ MnO_4^-$

or 30 mL of $0.6\ N\ Cr_2O_7^{2-} = V$ mL of $0.5\ M\ MnO_4^-$

Hence, volume of MnO_4^- solution required

$$= \frac{30 \times 0.6}{0.5} = 36 \text{ mL}$$

6. (c) Americium have +3, +4, +5 and +6 oxidation states.

7. (a) (a,b,c,d) Amalgam shows zero oxidation state, calomel (Hg_2Cl_2) shows +1 oxidation state while corrosive sublimate $(HgCl_2)$ shows +2 oxidation state.

8. (c) The SRP value is as follows :

$Mn^{3+}/Mn^{2+} = +1.57$

$Fe^{3+}/Fe^{2+} = +0.77$

$Cr^{3+}/Cr^{2+} = -0.41$

$V^{3+}/V^{2+} = -0.26$

9. (b) Among the given compound, $[CrO_4]^{2-}$ show +6 oxidation state which only reduces therefore, it is strongest reducing agent.

10. (c) $K_2Cr_2O_7 + 4H_2SO_4 + 3H_2S \longrightarrow K_2SO_4 + Cr_2(SO_4)_3 + 7H_2O + 3S$

11. (a,b,c) AgBr is insoluble in water while AgF, $AgNO_3$ and $AgClO_4$ are soluble in water.

12. (a,b,d) The element which gives the greatest number of oxidation states occur in or near the middle of the series.

13. (a,d)

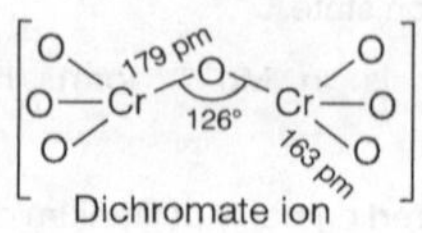

Dichromate ion

14. (a,b,d) Nessler's reagent is a mixture of $2K_2HgI_4 + 3KOH$. It is used to test $NH_3(g)$ and NH_4^+.

15. (b,d) Because second electron will remove from half-filled d-orbital in case of chromium and full-filled d-orbital in case of copper. Both these orbitals are more stable relative to other orbitals.

16. (a,b,d) Metallic radii of zinc is greater than copper.

Zn (137 pm) > Cu (128 pm)

17. (a, b,c,d) MnO_2 used in dry battery cells; TiO_2 is used in pigment industry; *Zeigler-Natta* catalyst is $Al(C_2H_5)_3 + TiCl_4$ and AgBr is used in photographic plates.

18. (c,d) Cr shows highest oxidation state + 6. Mercury exists as Hg^+ and Hg^{2+} ions.

19. (a,b,d) $KMnO_4$ oxidises H_2O_2, NO_2^- and $S_2O_3^{2-}$ in acidic medium.

20. (a,b,c) Eu^{2+}, Gd^{3+} and Tb^{4+} have 7 unpaired electrons. Lu^{3+} have no unpaired electron.

21. (b) $\underset{(A)}{2KMnO_4} \longrightarrow \underset{(B)}{K_2MnO_4} + \underset{(C)}{MnO_2} + \underset{(D)}{O_2}$

$\underset{(C)}{2MnO_2} + 4KOH + O_2 \longrightarrow \underset{(B)}{2K_2MnO_4} + 2H_2O$

$\underset{(C)}{MnO_2} + 4NaCl + 4H_2SO_4 \longrightarrow$

$MnCl_2 + 2NaHSO_4 + \underset{(E)}{Cl_2} + 2H_2O$

22. (a) $\underset{(A)}{2KMnO_4} \rightarrow \underset{(B)}{K_2MnO_4} + \underset{(C)}{MnO_2} + \underset{(D)}{O_2}$

Hence, $A = KMnO_4$ and $C = MnO_2$

23. (a) (i) → (s), (ii) → (p), (iii) → (q), (iv) → (r)

24. (d) (i) → (q), (ii) → (p), (iii) → (s), (iv) → (r)

25. (3) $Hg_2(NO_3)_2 \longrightarrow Hg_2^{2+} + 2NO_3^-$

26. (7) Tb = [Xe] $4f^9 5d^0 6s^2$ and Tb^{4+} = [Xe] $4f^7 5d^0 6s^0$

27. (6) Magnetic moment 4.90 BM indicates that Fe has 4 unpaired electrons. Then iron is Fe^{2+}.

28. (4) In solid $CuSO_4 \cdot 5H_2O$, copper is coordinated to 4 water molecules.

29. (9) $2MnO_4^- + I^- + H_2O \longrightarrow 2MnO_2 + IO_3^- + 2OH^-$

30. (4) La^{3+} – colourless, Pr^{3+} – yellow green
Lu^{3+} – colourless, Ce^{3+} – colourless
Tm^{3+} – green, Yb^{3+} – colourless, Er^{3+} – green

JEE Main & AIEEE Archive

1. (a) (a) V^{2+} = 3 unpaired electrons, Cr^{2+} = 4 unpaired electrons
Mn^{2+} = 5 unpaired electrons, Fe^{2+} = 4 unpaired electrons

Hence, the order of paramagnetic behaviour should be

$$V^{2+} < Cr^{2+} < Fe^{2+} < Mn^{2+}$$

(b) Ionic size decreases from left to right in the same period.

(c) (As per data from NCERT)

$Co^{3+}/Co^{2+} = 1.97$; $Fe^{3+}/Fe^{2+} = 0.77$; $Cr^{3+}/Cr^{2+} = -0.41$,

Sc^{3+} is highly stable (it does not show + 2).

(d) The oxidation state increases as we go from group 3 to group 7 in the same period.

2. (d) SRP value normally increases from left to right in the period of *d*-block elements. Some SRP value are exceptionally higher due to stability of product ion, e.g. $E^\circ_{Mn^{3+}/Mn^{2+}} = +1.57$ V;

$$E^\circ_{Co^{3+}/Co^{2+}} = +1.97 \text{ V}$$

Thus, $E^\circ_{M^{3+}/M^{2+}}$ is highest for Co.

3. (d) (a) $FeO > Fe_2O_3$ (basic). Thus, it is correct statement.

(b) $FeCl_2 > FeCl_3$ (ionic), larger the charge, greater the polarising power. Thus, greater the covalent nature. Thus, the given statement is correct.

(c) Fe^{2+} salts are more ionic thus less volatile than Fe^{3+} salts. Thus, given statement is correct.

(d) Greater the covalent nature, more easily they are hydrolysed. Thus, $FeCl_3$ is more hydrolysed than $FeCl_2$.
Thus, statement (d) is incorrect.

4. (d) (a) There is a gradual decrease in the radii of the lanthanoids with increasing atomic number a case of lanthanide contraction, thus true.

(b) Ionisation potential for the formation of Lu^{3+} is comparatively low, hence +3 state is favourable, thus true.

(c) Due to lanthanoid contraction, Zr and Hf; Nb and Ta, Mo and W have the same size and thus, similar properties and thus separation is not easy, thus true.

(d) Formation of + 4 state requires very high energy, thus incorrect.

5. (d) Gd (64)

$[Xe]_{54}$ $4f^7$: | ↑ | ↑ | ↑ | ↑ | ↑ | ↑ | ↑ |

$5d^1$: | ↑ | | | | | $6s^2$: | ↑↓ |

All the electrons in 4*f*-orbital are unpaired, hence stable. Thus, Gd (64) has EC as $[Xe]_{54}$ $4f^7$, $5d^1$, $6s^2$ instead of $[Xe]_{54}$ $4f^8$, $6s^2$.

6. (a) Usually across the first transition series, the negative values for standard electrode potential decrease except for Mn due to stable d^5-configuration. So, correct order : Mn > Cr > Fe > Co

7. (c) The most common oxidation state of lanthanoid is +3. Lanthanoids in +3 oxidation state usually have unpaired electron in *f*-subshell and impart characteristic colour in solid as well as in solution state due to *f*-*f* transition.
(Except lanthanum and lutetium)

8. (b) As oxidation state increases, electronegativity increases thus acidic character increases not basic.

9. (b) Lanthanoids : [Xe] $f^{1\text{-}14} 5d^{0\text{-}1} 6s^2$

Actinoids : [Rn]$5f^{1\text{-}14} 6d^{0\text{-}1} 7s^2$

Lanthanoids and actinoids use core *d* and *f*-orbitals also to show higher oxidation state. As actinoids have comparatively low energy difference between *f* and *d*-orbitals, thus they show more number of oxidation states.

10. (d) 4*f*-orbital is nearer to nucleus as compared to 5*f*-orbital. Therefore, shielding of 4*f* is more than 5*f*-orbital.

11. (d) The actinoids (5*f*-element) exhibit more number of oxidation states in general than the lanthanoids because 5*f*-orbitals extend farther from the nucleus than the 4*f*-orbitals.

Thus, the attraction on the 5*f*-electrons in actinoid series decreases. Hence, they can easily participate in bonding and therefore, exhibits more number of oxidation states.

12. (d) Imperfect shielding on outer electrons by 4*f*-electrons from the nuclear charge is due to the diffused large shape of 4*f*-orbitals. This imperfect shielding is unable to counterbalance the effect of the increased nuclear charge. Hence, the net result shows contraction in size.

13. (d) Following reaction takes place during Bessemerisation.

$$2Cu_2O + Cu_2S \longrightarrow 6Cu + SO_2$$

14. (d) $Hg_2Cl_2 + 2NH_4OH \longrightarrow \underset{\text{(White)}}{Hg(NH_2)Cl} + \underset{\text{(Black)}}{Hg\downarrow} + NH_4Cl + 2H_2O$

15. (c) +3 and +4 states are shown by Ce in aqueous solution. Thus, statement (c) is incorrect.

16. (d) $(n-1)d^8ns^2$ | ↑↓ | ↑↓ | ↑↓ | ↑ | ↑ | | ↑↓ | 4(as in Ni)

$(n-1)d^8$ ns^2

$(n-1)d^5ns^1$ | 1 | 1 | 1 | 1 | 1 | | 1 | 6(as in Cr)

$(n-1)d^5$ ns^1

$(n-1)d^3ns^2$ | 1 | 1 | 1 | | | | 1⇂ | 5(as in V)

$(n-1)d^3$ ns^2

$(n-1)d^5ns^2$ | 1 | 1 | 1 | 1 | 1 | | 1⇂ | 7(as in Mn)

$(n-1)d^5$ ns^2

17. (a) Cu, Ag, Au are called coinage metals because in the early times, they were used in making coins.

18. (d) Metal oxides lying below Hg in electrochemical series decompose to form metal.

$$ZnO, MgO, CuO \xrightarrow{\Delta} \text{No effect}$$

$$Ag_2O \xrightarrow{\Delta} Ag + O_2$$

$$AgNO_3 \xrightarrow{\Delta} Ag_2O \longrightarrow Ag + O_2$$

19. (a) Mercury (Hg) solidifies after glass, so used mercury for making good quality mirrors.

20. (b) Ammonia can dissolve precipitate of AgCl due to the formation of complex as given below

$$AgCl + 2NH_3 \longrightarrow [Ag(NH_3)_2Cl]$$

21. (b) HgI_2 becomes soluble in water, when KI is added to it due to the formation of K_2HgI_4 complex which is soluble in water.

$$HgI_2 + 2KI \longrightarrow K_2HgI_4 \text{(Soluble)}$$

But when HgI_2 is heated, HgI_2 decomposes to give violet coloured fumes of I_2 that deposits on the cooler parts of the test tube.

$$HgI_2 \xrightarrow{\Delta} Hg + I_2 \uparrow$$

22. (b) $2CrO_4^{2-} + 2H^+ \longrightarrow Cr_2O_7^{2-} + H_2O$

23. (a) Cerium, $Ce_{58} = [Xe]\,4f^1 5d^1 6s^2$. Its most stable oxidation state is +3 but +4 is also existing.

24. (a) Larger the number of unpaired electrons, greater the magnetic moment, $\mu = \sqrt{n(n+2)}$

where, n = Number of unpaired electrons

$Mn_{25} = [Ar]\,4s^2, 3d^5$; $Mn^{2+} = [Ar]\,3d^5 4s^0$

| 1 | 1 | 1 | 1 | 1 |

$3d^5$

Number of unpaired electrons = 5

$Fe_{26} = [Ar]\,4s^2, 3d^6$; $Fe^{2+} = [Ar]\,3d^6, 4s^0$

| 1⇂ | 1 | 1 | 1 | 1 |

$3d^6$

Number of unpaired electrons = 4

$Ti_{22} = [Ar]\,4s^2 3d^2$; $Ti^{2+} = [Ar]\,3d^2 4s^0$

| 1 | 1 | | | |

$3d^2$

Number of unpaired electrons = 2

$Cr_{24} = [Ar]\,3d^5, 4s^1$

$Cr^{2+} = [Ar]\,3d^4, 4s^0$

| 1 | 1 | 1 | 1 | |

$3d^4$

Number of unpaired electrons = 4

Hence, Mn^{2+}: $[Ar]\,3d^5$ with five unpaired electrons have maximum magnetic moment.

JEE Advanced & IIT JEE Archive

1. (7) Acidified $K_2Cr_2O_7$, $CuSO_4$, H_2O_2, Cl_2, O_3, $FeCl_3$ and HNO_3 oxidise aqueous iodide to iodine. Alkaline $KMnO_4$ oxidise aqueous iodide to IO_3^-.

$Na_2S_2O_3$ is a strong reducing agent which on reaction with I_2 produces I^-.

$$Na_2S_2O_3 + I_2 \longrightarrow 2I^- + Na_2S_4O_6$$

Therefore, no reaction takes place between $Na_2S_2O_3$ and iodide ion.

2. (a) The aqueous solution of $CuSO_4$ consists of the complex $[Cu(H_2O)_4]^{2+}$ ion which absorbed in orange-red region and impart deep blue colouration to solution.

3. (a,c,d) Solubilities of silver halides in water decreases from fluoride (AgF) to iodide (AgI). Silver fluoride is readily soluble in water hence, when $AgNO_3$ solution is added to HF solution (HF being weak acid, its solution maintain very low concentration of F^-) no precipitate of AgF is formed.

HCl, HBr and HI being all strong acid, forms precipitates of AgCl, AgBr and AgI when $AgNO_3$ solution is added to their aqueous solution.

$$HCl(aq) + AgNO_3(aq) \longrightarrow \underset{\text{(Curdy white)}}{AgCl\,(s)} + HNO_3(aq)$$

$$HBr\,(aq) + AgNO_3(aq) \longrightarrow \underset{\text{(Pale yellow)}}{AgBr\,(s)} + HNO_3(aq)$$

$$HI\,(aq) + AgNO_3(aq) \longrightarrow \underset{\text{(Yellow)}}{AgI\,(s)} + HNO_3(aq)$$

The solubilities decreases from AgCl to AgI, AgCl dissolves in aqueous ammonia, AgBr dissolves only slightly in concentrated ammonia while AgI does not dissolve in ammonia solution.

$Na_2S_2O_3$ solution dissolve all three, AgCl, AgBr, AgI by forming complex $[Ag(S_2O_3)_2]^{3-}$ as $S_2O_3^{2-}$ is a stronger complexing agent than ammonia.

4. (a,c,d) In neutral medium,

$$MnO_4^- \longrightarrow MnO_2 \; (Mn^{7+} + 3e^- \longrightarrow Mn^{4+})$$

In alkaline medium,

$$MnO_4^- \longrightarrow MnO_2 \; (Mn^{7+} + 3e^- \longrightarrow Mn^{4+})$$

In acidic medium,

$$MnO_4^- \longrightarrow Mn^{2+} \; (Mn^{7+} + 5e^- \longrightarrow Mn^{2+})$$

5. (b) In $CuCl_2$, Cu^{2+} has d^9 configuration, exhibit d-d transition and show colour.

Similarly, in $VOCl_2$, V^{4+} has d^1 configuration, can exhibit d-d transition and show colour.

6. (c) $KMnO_4$ is itself a very strong oxidising agent, O_3 cannot oxidise it.

7. $A = [Ti(H_2O)_6]^{3+}$ and M = Ti, $B = TiO_2$, Ti(IV) has no electron in 3d-orbital, no d-d transition is possible, therefore MCl_4 is colourless. In A, there is one electron in 3d-orbital and its d-d transition is responsible for colour.

3. Coordination Compounds

DPP-1 Introduction of Coordination Compounds and Werner's Theory

1. General formula of an alum is $X_2SO_4 \cdot Y_2(SO_4)_3 \cdot 24H_2O$, where, X = monovalent, Y = trivalent cation.

e.g. potash alum

$$K_2SO_4 \cdot Al_2(SO_4)_3 \cdot 24H_2O \longrightarrow 2K^+ + 2Al^{3+} + 4SO_4^{2-} + 24H_2O$$

An alum on ionisation produce 8 ions in solution.

2. The common coordination number of cobalt is 6.

Among the three chloride ions, only one of them satisfy both primary and secondary valency while the remaining two satisfy only primary valency.

$$[Co(NH_3)_5Cl]Cl_2$$

This satisfy both primary and secondary valency.

In $K_4[Fe(CN)_6]$ both primary and secondary valency are satisfied by CN^- ion.

3. Werner's theory could not answer the following :

(i) Why only certain elements form coordination compounds?

(ii) Why coordination sphere has a definite geometry?

(iii) It could not explain magnetic property and colour of complexes.

4. (i) In this complex, iridium coordination number is 6. In this, chloride ions satisfy both primary and secondary valencies. It is a neutral complex and do not dissociate. Hence, non-conductor.

$$IrCl_3 \cdot 3NH_3 = [Ir(Cl)_3(NH_3)_3]$$

(ii) Homoleptic complex possesses same kind of ligands while heteroleptic complex possesses different kind of ligands.

e.g. $PtCl_4 \cdot 6NH_3 = [Pt(NH_3)_6]Cl_4$ (homoleptic) and

$PtCl_4 \cdot 2NH_3 = [Pt(NH_3)_2Cl_4]$ (heteroleptic)

5. $[Co(H_2O)_6]^{2+}$ complex is pink and is almost colourless when dilute so that writing done with it is practically invisible. Upon the application of heat, water is driven off and the blue complex $Co[CoCl_4]$ is formed. Upon standing, water is slowly taken up from the atmosphere and the original colourless complex is regenerated.

$$\underset{\text{(Pink)}}{2[Co(H_2O)_6]Cl_2} \rightleftharpoons \underset{\text{(Blue)}}{Co[CoCl_4]} + 12H_2O$$

(Colourless when dilute)

6. (d) In $[Pt(en)_2Cl_2]$ the charge on the complex ion is +2 and coordination number is 6. Hence, the primary and secondary valency are +2 and 6 respectively.

7. (b) The complete formula of Zeise's salt $K[Pt(n^2\text{—}C_2H_4)Cl_3]H_2O$. The complex anion has −1 charge. Hence, primary and secondary valency are +1 and 4 respectively.

8. (a) $[Pt(NH_3)_2Cl_4]$ is neutral complex and no ions are formed in aqueous solution.

$$K_2[PtCl_6] \longrightarrow 2K^+ + [PtCl_6]^{2-}$$

It does not have Cl^- ion to give precipitate of AgCl.

9. (c) One third of the chloride is precipitated means 1 mole of AgCl is formed.

$$[Co(NH_3)_4Cl_2]Cl \longrightarrow [Co(NH_3)_4Cl]^+ + Cl^-$$

Hence, 1 Cl^- ion is formed and it will give 1 mole of AgCl with $AgNO_3$.

10. (d) Mohr's salt is a double salt and it will give test for Fe^{2+} ion. It dissociates into simple substances or ions completely when dissolved in water.

$$\underset{\text{Mohr's salt}}{FeSO_4 \cdot (NH_4)_2SO_4 \cdot 6H_2O} \longrightarrow Fe^{2+} + 2NH_4^+ + 2SO_4^{2-} + 6H_2O$$

11. (a) Ligands which can ligate through two different atoms is called ambident ligands, e.g. NO_2^- and SCN^- ion.

In NO_2^-

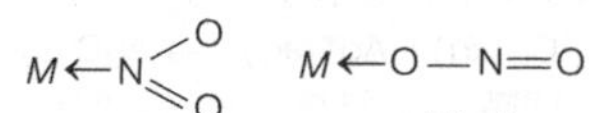

nitrito–N　　　　nitrito–O

It can coordinate either through nitrogen or through oxygen to a central metal atom/ion. Similarly, SCN^- ion can coordinate through the sulphur or nitrogen atom.

12. (c) The acetylacetonate (acac) forms six membered ring as follows:

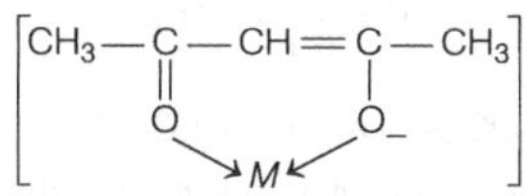

13. (a) In complex $[Fe(H_2O)_5NO]^{2+}$

Let the oxidation state of Fe is x.

$\therefore\ x + 5 \times 0 + 1 = 2$

$x + 1 = 2 \Rightarrow x = 1$

In complex $[Fe(CN)_5NO]^{2-}$

Let the oxidation state of Fe is y.

$\therefore\ y - 5 \times 1 + 1 = -2$

$y - 4 = -2$

$y = 2$

14. (d) In $[Ni(CO)_4]$

Let the oxidation state of Ni is x.

$x + 4 \times (0) = 0$ (Carbonyl (CO) is a neutral ligand and have zero charge.)

$x = 0$

Hence, Ni have zero oxidation state.

15. (a) $CH_3COCH_2COCH_3 \rightleftharpoons CH_3\text{—}\overset{O^-}{\overset{|}{C}}=CH\text{—}\overset{O}{\overset{||}{C}}\text{—}CH_3$

It can ligate through two oxygen atoms.

16. (c,d) Common alum is $K_2SO_4 \cdot Al_2(SO_4)_3 \cdot 24H_2O$ and carnallite is $KCl \cdot MgCl_2 \cdot 6H_2O$.

17. (a,b,d) The complex $K_4[Fe(CN)_6]$ whose formula can be written like that of double salt. $Fe(CN)_2 \cdot 4KCN$, dissociates to give K^+ and $[Fe(CN)_6]^{4-}$ ions in the aqueous solution.

$$K_4[Fe(CN)_6] \longrightarrow 4K^+ + [Fe(CN)_6]^{4-}$$

↓ Does not ionise to give Fe^{2+} and CN^- ion.

Hence, it has ionic, covalent and coordinate covalent bonds.

18. (b,d) In $K_3[Fe(CN)_6]$, the oxidation state of Fe is +3.

Fe (III) contains d^5 electron which is stable configuration. Hence, it need not an extra electron. So, it does not behave like an acid.

In aqueous solution, their complex produce only 4 ions.

$$K_3[Fe(CN)_6] \longrightarrow 3K^+ + [Fe(CN)_6]^{3-}$$

19. (a,c,d) $[Co(en)_3]Cl_3$ produces 4 ions in solution as follows :

$$[Co(en)_3]Cl_3 \longrightarrow [Co(en)_3]^{3+} + 3Cl^-$$

20. (b,c) In the aqueous solution of $Pd(NH_3)_2Cl_2$, the atoms of chlorine are in coordination sphere and the van't Hoff factor of the compound are unite.

21. (b) Since, the oxidation number of Co is +3 hence, the complex compound would be $[CoCl \cdot 5H_2O]Cl_2$.

$$\underset{1\text{ mol}}{[CoCl \cdot 5H_2O]Cl_2} \longrightarrow \underset{1\text{ mol}}{[CoCl \cdot 5H_2O]^{2+}} + \underset{2\text{ mol}}{2Cl^-}$$

Hence, 1 mole of $[CoCl \cdot 5H_2O]Cl_2$ gives 2 moles of Cl^- ion.

Given, moles of $[CoCl \cdot 5H_2O]Cl_2 = 200\text{ mL} \times 0.01\text{ M} = 2$ millimol

$\therefore$ 2 millimol of $[CoCl \cdot 5H_2O]Cl_2$ will give 4 millimol of Cl^-.

$$\underset{1\text{ mol}}{Cl^-(aq)} + \underset{1\text{ mol}}{Ag^+(aq)} \longrightarrow \underset{1\text{ mol}}{AgCl(s)}$$

Now, to neutralise 4 millimol of Cl^-, 4 millimol of $AgNO_3$ is required.

Let the volume of $AgNO_3 = V$ mL

$\therefore$ Millimoles of $AgNO_3 = 0.1 \times V$

$$0.1 \times V = 4 \text{ millimol}$$

$$V = \frac{4}{0.1} = 40\text{ mL}$$

22. (a) $[Co(NH_3)_5Cl]Cl_2 \longrightarrow [Co(NH_3)_5Cl]^{2+} + 2Cl^-$

Hence, 3 ions are present for one mole of the complex in the solution.

23. (a) (i) $\rightarrow$ (r), (ii) $\rightarrow$ (q), (iii) $\rightarrow$ (p), (iv) $\rightarrow$ (s)

24. (6) Formula of Mohr's salt is $FeSO_4 \cdot (NH_4)_2SO_4 \cdot 6H_2O$. Here, x is 6.

25. (4) $AgNO_3$ solution is added in excess of 1 M solution of $CoCl_3 \cdot xNH_3$.

$$\underset{1\text{ mol}}{Cl^-(aq)} + \underset{1\text{ mol}}{Ag^+(aq)} \longrightarrow \underset{1\text{ mol}}{AgCl(s)}$$

The value of x is 4 as $CoCl_3 \cdot 4NH_3$.

26. (0) $AgNO_3 + Cl^- \longrightarrow AgCl + NO_3^-$

$$M = \frac{\text{weight}}{\text{Molecular weight}} \times \frac{1000}{V} \Rightarrow 1 = n \times \frac{1000}{100} \Rightarrow n = 0.01$$

27. (0) In $Co(CO)_6$

Let the oxidation state of Co is x.

$x + 6 \times 0 = 0$ [Carbonyl (CO) is a neutral ligand hence, zero charge]

$x = 0$

Hence, Co has zero oxidation state.

28. (3) Coordination number of platinum is 6.

$\therefore$ Number of chloride ions precipitated by adding AgCl are 3.

29. (b) Statement I is true because CO is a neutral ligand, Statement II is true because

EAN of Fe

= Atomic number of Fe + Electrons gained by coordination

$= 26 + 2 \times 5 = 36$

But Statement II is not the correct explanation. Correct explanation is that oxidation state of Fe in $Fe(CO)_5$ is zero because CO is a neutral ligand.

30. (d) **Statement I** The oxidation number of Pt in Zeise's salt is +2.

Statement II Zeise's salt is ionic complex with the formula $K[PtCl_3(C_2H_4)]$.

Hence, statement II is correct but statment I is incorrect.

DPP-2 IUPAC Nomenclature of Coordination Compounds

1. (i) $K_3[AgF_4]$

(ii) $Ca[Co(C_2O_4)_2(NH_3)_2]$

(iii) $[Cr(C_6H_6)_2]$

(iv) $Na_3[Ag(S_2O_3)_2]$

2. (i) Tetrammineplatinum(II) tetrachloroplatinate(II)

(ii) Ammonium diamminetetraisothiocyanatochromate (III)

(iii) Potassium trichloro(ethene)platinate(II)

(iv) *Tris*(bipyridine)iron (II) ion

3. (i) The complex is $[Pt(NH_3)_5Cl]Cl_3$ pentamminechloroplatinum (IV) chloride.

(ii) The complex is $K_2[PtCl_6]$ potassium hexachloroplatinate (IV).

(iii) The complex is $[Ir(NH_3)_3Cl_3]$ triamminetrichloroiridium(III).

(iv) The complex is $[Co(NH_3)_4Cl_2]Cl$ tetramminedichlorocobalt(III)chloride.

4. (i) Acetylacetonato (acac)

(ii) Glycinato (gly)

(iii) 2, 2′ bipyridine (bipy)

(iv) Diethylenetriamine (dien)

5. (i) Ligand having two or more donor atoms sometimes form 5 or 6 membered ring with metal ion. This is called a chelate complex and ligands are chelating ligands.

e.g. ethylenediamine (en), oxalate ion (ox), EDTA, etc.

(ii) Dimethylglyoxime (dmg) is a bidentate ligand with uninegative charge.

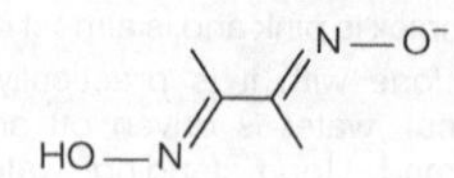

In this ligand, one N and one O-atom are donor atoms.

(iii) A ligand that donates 4 pairs of electrons.

e.g. triethylenetetramine (trien) $\begin{array}{l} CH_2NH(CH_2)_2NH_2 \\ | \\ CH_2NH(CH_2)_2NH_2 \end{array}$

6. (d) Sodium nitroprusside reagent is $Na_2[Fe(CN)_5NO]$. In this compound, NO is present as NO^+ and Fe has +2 charge. On this basis, the actual name should be sodium pentacyanonitrosyliumferrate (II).

7. (a) The formula of complex hexamminecobalt (III) chloride sulphate is $[Co(NH_3)_6]ClSO_4$.

8. (a) The IUPAC name of $[Cr(NH_3)_5(NCS)][ZnCl_4]$ is pentammine isothiocyanatochromium (III) tetrachlorozincate (II).

9. (c) The IUPAC name of complex $[(CO)_5Mn-Mn(CO_5)]$ is decacarbonyldimanganate (0).

10. (c)

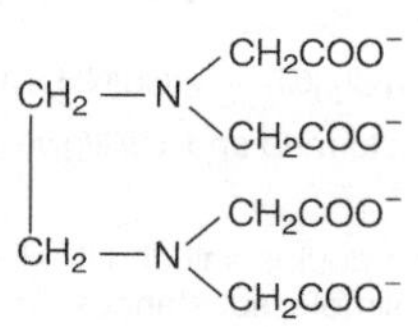

Ethylenediaminetetraacetate ion $(EDTA)^{4-}$ is an important hexadentate ligand. It can bind through two nitrogen and four oxygen atoms to a central metal ion.

11. (b) The IUPAC name of $[Ni(PPh_3)Cl_2]^{2+}$ is dichloro bis (triphenylphosphine) nickel (II) ion.

12. (c) The brown ring test for nitrates depend on the ability of Fe^{2+} to reduce nitrates to nitric oxide, which reacts with Fe^{2+} to form a brown coloured complex.

$$NO_3^- + 3Fe^{2+} + 4H^+ \longrightarrow NO + 3Fe^{3+} + 2H_2O$$

$$[Fe(H_2O)_6]^{2+} + NO \longrightarrow \underset{\text{(Brown)}}{[Fe(H_2O)_5(NO)]^{2+}} + H_2O$$

13. (b) The IUPAC name of the compound $K[SbCl_5Ph]$ is potassium pentachloro(phenyl) antimonate (V).

14. (a) The name of $K_2[OsCl_5N]$ is potassium pentachloroazoosmate (VI).

15. (c) The IUPAC name of $NaBH_4$ is sodium tetrahydridoborate (III).

16. (a,b) Complexes $K[Pt(NH_3)Cl_5]$ and $[Ag(CN)_2]^-$ have been given their respective IUPAC names.

17. (a,b,d) DMG (bidentate and neutral) is not named correctly. acac, bipy and gly have correct names respectively.

18. (a,c) The correct name of ligand in IUPAC name are thiosulphate $(S_2O_3^{2-})$ and cyclopentadienyl $[(C_5H_5)^-]$.

19. (a,c) $K_2[AgF_4]$ and $[Cr(C_6H_6)_2]$ have not correct IUPAC names.

20. (c,d) When a di or polydentate ligand used its two or more donor atoms to bind a single metal ion. It is said to be a chelate ligand. The number of such ligating groups is called the denticity of the ligand. Such complexes are called chelate complex tend to be more stable than similar complexes containing unidentate ligands.

21. (a) The correct name of $[CoBr(NH_3)_4NO_2]$ is tetramminebromonitrocobalt (II).

22. (c) The IUPAC name of $[Ni(NH_3)_4][NiCl_4]$ is tetramminenickel (II) tetrachloronickelate (II).

23. (c) (i) → (q,t), (ii) → (q,s), (iii) → (r,t), (iv) → (p,s)

24. (d) (i) → (q), (ii) → (p), (iii) → (s,t), (iv) → (r)

25. (4)

26. (3) In $[Fe(EDTA)]^-$, Let the oxidation state is x

then $x - 4 = -1 \Rightarrow x = 3$

27. (5) The structure of potassiumdicyanodioxalatonickelate (II) is $K_4[Ni(CN)_2(ox)_2]$.

$$K_4[Ni(CN)_2(ox)_2] \longrightarrow 4K^+ + [Ni(CN)_2(ox)_2]^-$$

This produce 5 ions in solution.

28. (1) EDTA is hexadentate ligand hence, its coordination number is 6.

29. (5)

30. (c) $[Ni(dmg)_2]$: *bis* (dimethylglyoximato) nickel (II)

DPP-3 **Isomerism in Coordination Compounds-I**

1. In $[Pt(gly)_2]$, the ligand glycinate ion is unsymmetrical bidentate and this can have *cis* and *trans*-isomers. But in $[Pt(en)_2]^{2+}$, the ligand is bidentate symmetrical and hence, it does not exhibit geometrical isomerism.

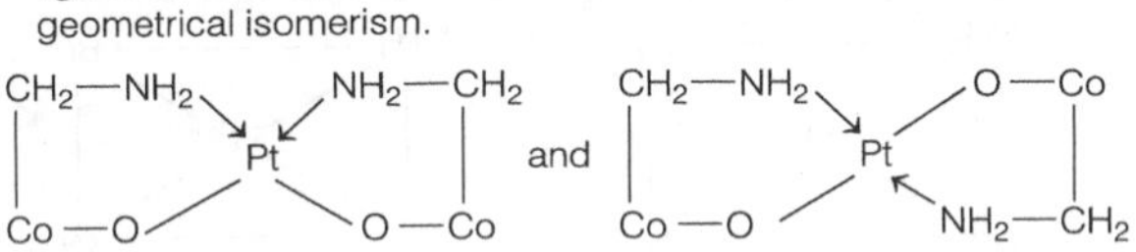

cis and *trans* glycine complex

2. *Fac* and *mer* isomers of an $[Ma_3b_3]$ are geometrical isomers. Due to symmetry in the complex, it failed to show optical isomerism. e.g.

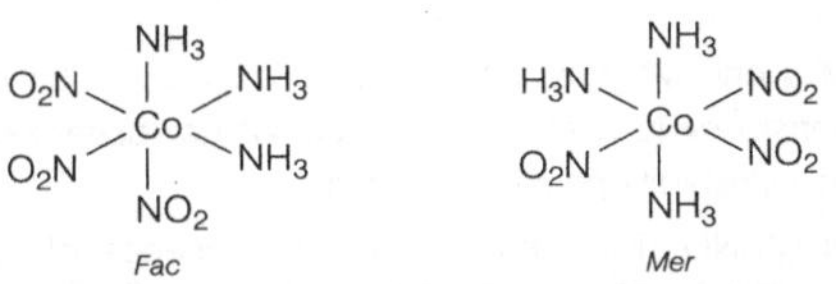

The facial (*fac*) and meridional (*mer*) isomers of $[Co(NH_3)_3(NO_2)_3]$.

3. 15 geometrical isomers are possible and each has its mirror image and hence, total 30 isomers are possible.

4. Platinum(II) forms square planar complexes while platinum(IV) forms octahedral complexes.

(i) Zero (Ma_3b) type do not show

(ii) Zero (Ma_5b) type do not show

(iii) Two isomers (Ma_2bc) type has *cis* and *trans*-isomers.

(iv) Two isomers. This is $[Ma_4bc]$ type and hence, this has *cis* and *trans* isomers.

5. Structure of EDTA

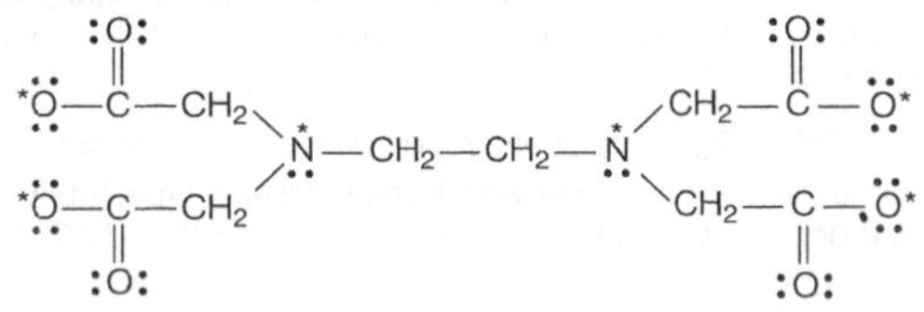

It is hexadentate ligand. It can bind through two nitrogen and four oxygen atoms to a central metal ion shown as (*).

6. (b)

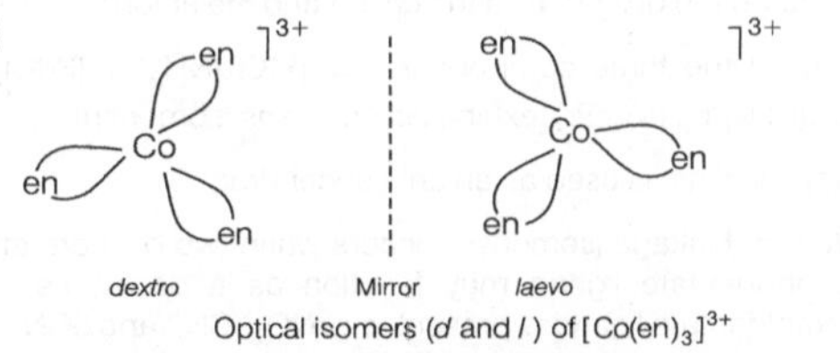

Optical isomers (*d* and *l*) of $[Co(en)_3]^{3+}$

Optical isomers have mirror images that cannot be superimposed on one another. These are called as enantiomers. The molecules or ions that cannot be superimposed are called chiral.

7. (c) Octahedral complexes of formula $[Ma_4b_2]$ in which the two ligands *b* may be oriented *cis* or *trans* to each other.

e.g.

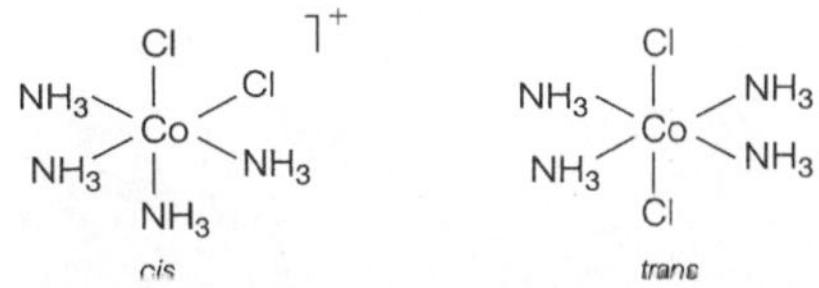

Geometrical isomers (*cis* and *trans*) of $[Co(NH_3)_4Cl_2]^+$.

8. (a) Ligand which can ligate through two different atoms is called ambidentate ligand. NO_2^- ion can coordinate either through nitrogen or through oxygen to a central metal atom/ion.

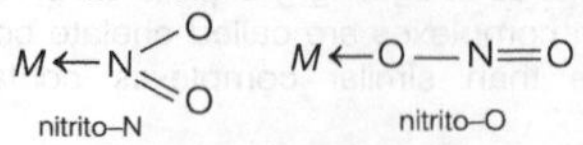

nitrito–N nitrito–O

Hence, NO_2^- ligand show linkage isomerism. e.g. the complex $[Co(NH_3)_5(NO_2)]Cl_2$ which is obtained as the red form, in which the nitrite ligand is bound through oxygen (—ONO) and as the yellow form, in which the nitrite ligand is bound through nitrogen ($—NO_2$).

9. (c) The complex $[Co(NH_3)_4Br_2]Cl$ show geometrical and ionisation isomerism.

Geometrical isomers

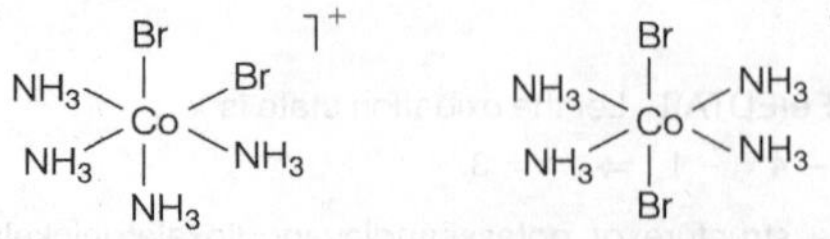

Ionisation isomers

$[Co(NH_3)_4Br_2]Cl$ and $[Co(NH_3)_4BrCl]Br$

10. (c) Due to symmetry in the complex $[Co(NH_3)_3Cl_3]$, it failed to show optical isomerism. The type of coordination entity show *fac* and *mer* isomerism or geometrical isomerism.

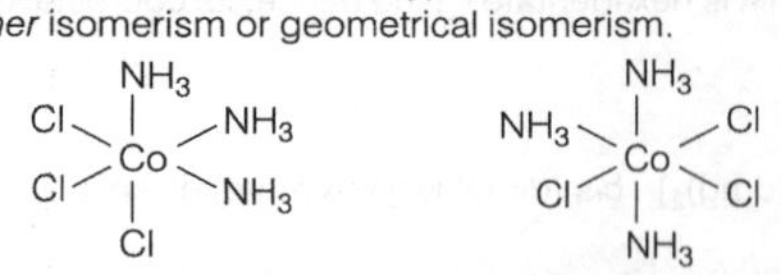

Fac *Mer*

The facial (*fac*) and meridional (*mer*) isomers of $[Co(NH_3)_3Cl_3]$.

11. (b) Ionisation isomerism is due to the exchange of groups between the complex ion and the ions outside it, e.g. $[Cr(H_2O)_4Cl(NO_2)Cl]$ and $[Cr(H_2O)_4Cl_2]NO_2$ show ionisation isomerism.

12. (c) $[CoNH_3Cl_3]$ do not exhibit optical isomerism. Optical isomerism is shown by $M(AA)_2X_2$ type, $M(AA)_3$ type, $[M(AA)B_2X_2]$ type complexes containing hexadentate ligand and $ML_a L_b L_c L_d$ Left type.

13. (a) Coordination isomerism is possible when both positive and negative ions of a salt are complex ions and two isomers differ in distribution of ligands in the cation and the anion.

14. (d) All the three compounds, i.e. $[PtCl_2(NH_3)_2]$, $[PdCl_2BrI]$ and $[Pt(NH_3)(py)(Cl)(Br)]$ exhibit *cis* and *trans* isomerism.

15. (c) *cis* platin is used as an anticancer drug.

16. (a,b,c) Linkage isomerism occurs when two or more atoms in a monodentate ligand may function as a donor, i.e. when an ambidentate ligand is present. e.g. NO_2^-, CN^- and SCN^-.

17. (a,c,d) $K[Ag(CN)_2]$ show linkage isomerism.

$[CrCl(NH_3)_5]I_2$ show ionisation isomerism.

$[PtCl_4I_2]$ show geometrical isomerism.

18. (a,b) This type of isomerism arises from the interchange of ligands between cationic and anionic entities of different metal ions present in a complex.

19. (a,b,d) Due to symmetry in the complex $[Cu(NH_3)_4]Cl_2$, it failed to explain geometrical isomerism.

20. (a,d) Geometrical isomerism is not observed in complexes of coordination number 4 having tetrahedral geometry. The platinumglycinato complex $[Pt(gly)_2]$ show geometrical isomerism (*cis* and *trans* form).

21. (b) The oxidation state and coordination number of cobalt in the complex $[Co(NH_3)_4Br_2]^+$ are +3 and 6.

22. (b) Structure (a) = structure (c) and, structure (b) = structure (d)

23. (c) $[Co(NH_3)_4Cl_2]^+$ exhibits geometrical isomerism.

cis-$[Co(en)_2Cl_2]^+$ because of the absence of symmetry elements exhibit optical isomerism.

$[Co(NH_3)_5(NO_2)]Cl_2$ because of the presence of ambidentate (NO_2) ligand exhibit linkage isomerism. $[Co(NH_3)_6][Cr(CN)_6]$ exhibits coordination isomerism because here, cation and anion both are complex.

(i) → (s), (ii) → (p), (iii) → (t), (iv) → (r)

24. (d) (i) → (r,s), (ii) → (p,q,t), (iii) → (p,q,t), (iv) → (r,s)

25. (3) For a complex of the type *MABCD*, three geometrical isomers are possible.

26. (3) The complex has *cis* and *trans* isomers. The *cis* has non-superimposable mirror images. Hence, total are 3.

27. (4) $[Cu(NH_3)Cl][Pt(NH_3)Cl_3]$, $[Cu(NH_3)Cl_3][Pt(NH_3)_3Cl]$

$[Cu(NH_3)_4][PtCl_4]$, $[Pt(NH_3)_4][CuCl_4]$

28. (2) This is $[Ma_3b_3]$ type of complex and this has two geometrical isomers called facial (*fac* isomer) and meridional (*mer* isomer).

29. (3) [*Mabcd*] type complexes exist in three isomeric forms.

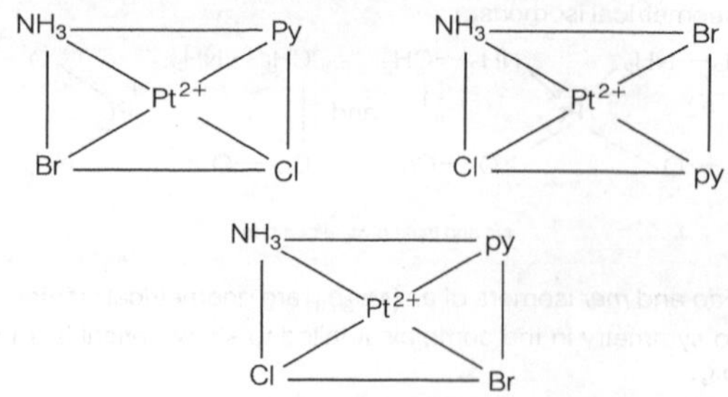

30. (c) MX_6 and MX_5L type complexes do not exhibit geometrical isomerism. *cis* $[Co(en)_2Cl_2]^+$ because of the absence of symmetry elements exhibit optical isomerism.

$[Co(NH_3)_5(NO_2)]Cl_2$ because of the presence of ambidentate (NO_2) ligand exhibit linkage isomerism. $[Co(NH_3)_6][Cr(CN)_6]$ exhibits coordination isomerism because here cation and anion both are complex.

DPP-4 **Isomerism in Coordination Compounds-II**

1. (i) Zero
(ii) Zero, since it has pyramidal geometry.

2. $[CoCl_2Br_2]^{2-}$ is tetrahedral while $[PdBr_2Cl_2]^{2-}$ is square planar.

3. Both geometrical and optical isomerism.

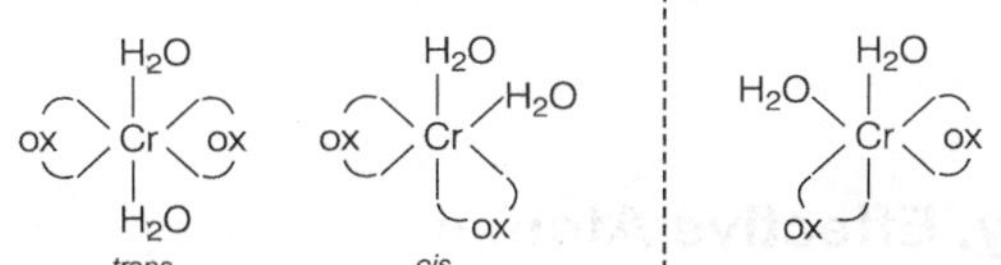

trans *cis* Mirror

4. *cis*-isomer of $[Pt(NH_3)_2Cl_2]$.

5. Propylenediamine (pn), ethylenediamine (en) and oxalate ion = 5, trimethylenediamine (tn) = 6.

6. (a) $[Co(NH_3)_4Br_2]Cl$ can show both geometrical and ionisation isomerism.

Geometrical isomers

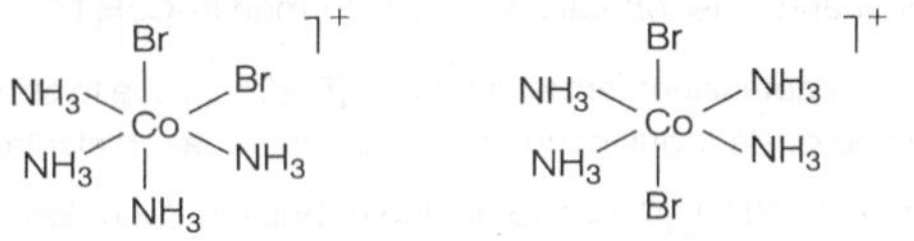

cis *trans*

Ionisation isomers

$[Co(NH_3)_4Br_2]Cl$ or $[Co(NH_3)_4BrCl]Br$

7. (b) $[Co(en)_3]^{3+}$, where, en = $(NH_2CH_2CH_2NH_2)$

The complex show optical isomerism. Optical isomers are mirror image that cannot be superimposed on one another. These are called as enantiomers. The molecules or ions that cannot be superimposed are called chiral.

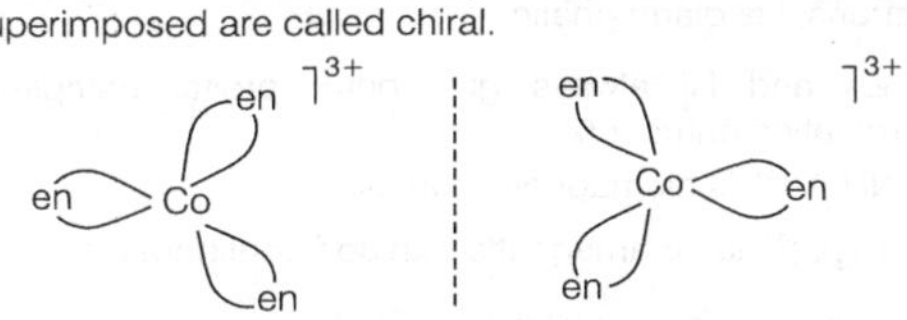

Mirror

Optical isomers (*d* and *l*) of $[Co(en)_3]^{3+}$

8. (a) The complex $[Co(C_2O_4)_2Cl_2]^{3-}$ show both geometrical and optical isomers.

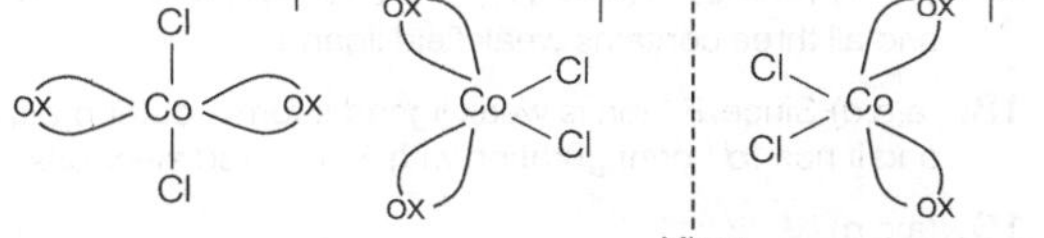

trans-$[CoCl_2(ox)_2]^{3-}$ Mirror *cis*-$[CoCl_2(ox)_2]^{3-}$

9. (c) The complex $[Co(NH_3)_5NO_2]Cl_2$ produces three ions in aqueous solution.

$$[Co(NH_3)_5NO_2]Cl_2 \longrightarrow [Co(NH_3)_5NO_2]^+ + 2Cl^-$$

The aqueous solution on treatment with an excess of $AgNO_3$ gives two moles of AgCl because we have 2 moles of Cl^- ion in aqueous solution.

Linkage isomers

$[Co(NH_3)_5NO_2]Cl_2$ and $[Co(NH_3)_5(ONO)]Cl_2$

The nitrite ligand is bound through oxygen (—ONO) in red form. The nitrite ligand is bound through nitrogen (—NO_2) in yellow form.

10. (c) $[Cr(SCN)_2(NH_3)_4]^+$ shows linkage, geometrical and optical isomerism.

11. (d) $[Co(en)_3]Br$ shows optical isomerism.

12. (c)

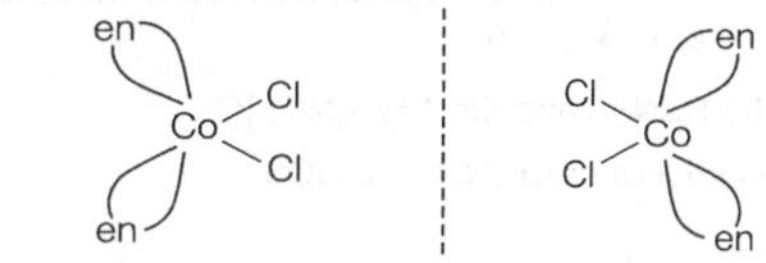

Mirror

cis-$[Co(en)_2Cl_2]$

13. (c) In $[Co(NH_3)_6][Cr(NO_2)_6]$, the $[Co(NH_3)_6]^{3+}$ ion migrates towards the negative electrode where cobalt would be deposited. In $[Cr(NH_3)_6][Co(NO_2)_6]$, the $[Cr(NH_3)_6]^{3+}$ ion migrates towards the negative electrode where chromium would be deposited.

14. (c) The given complexes are the examples of coordination isomers.

15. (c) $[Co(EDTA)]^-$ ion shows both geometrical and optical isomerism.

16. (a,b,c) $[Co(NH_3)_4(NO_2)_2]Cl$ exhibits linkage isomerism, ionisation isomerism and geometrical isomerism.

17. (a,b,c) Both given complexes can be identified by $BaCl_2$, $AgNO_3$ and $PbCl_2$.

18. (b,d) (a) $[Cr(NH_3)_5Cl]^{2+}$ no isomerism and $[Cr(NH_3)_4Cl_2]^+$ show only geometrical.

(b) $[Cr(NH_3)_4Cl_2]^+$ only geometrical and $[Pt(NH_3)_2(H_2O)Cl]^+$ only geometrical.

(c) $[CoBr_2Cl_2]^{2-}$ no isomerism because complex is tetrahedral and $[PtBr_2Cl_2]^{2-}$ show geometrical.

(d) $[Pt(NH_3)_3NO_3]Cl$ only ionisation and $[Pt(NH_3)_3Cl]Br$ only ionisation.

19. (b,c) Tetrahedral complexes do not show geometrical isomerism because the relative position of the unidentate ligands attached to the central metal atoms are the same with respect to each other.

20. (b,c)

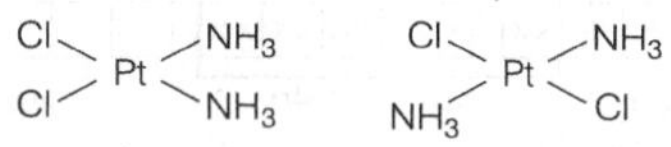

cis *trans*

Geometrical isomers (*cis* and *trans*) of $[Pt(NH_3)_2Cl_2]$

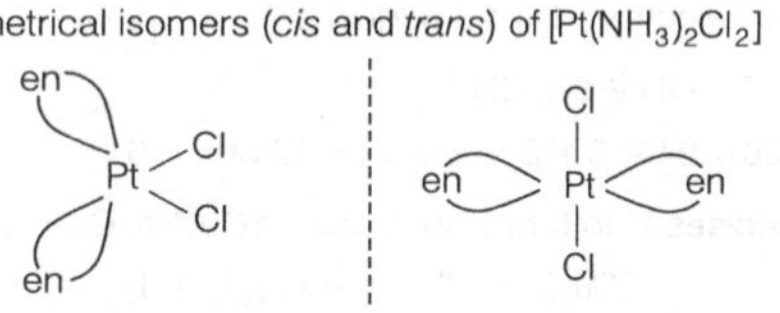

Mirror

cis-$[Pt(en)_2Cl_2]Cl_2$ *trans*-$[Pt(en)_2Cl_2]Cl_2$

21. (b) $\underset{A}{[Cr(NH_3)_4ClBr]Cl} \longrightarrow [Cr(NH_3)_4ClBr]^+ + Cl^-$

Complex (*A*) give white ppt. of AgCl with $AgNO_3$.

22. (d) **Geometrical isomers**

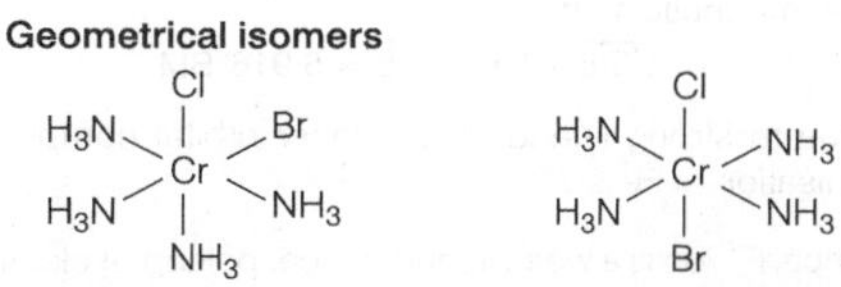

cis-form *trans-form*

Ionisation isomers $[Cr(NH_3)_4ClBr]Cl$ and $[Cr(NH_3)_4Cl_2]Br$

23. (b) (i) → (q), (ii) → (p), (iii) → (s), (iv) → (r)

24. (a) (i) → (r), (ii) → (q,s), (iii) → (p), (iv) → s

25. (9) The complex can be shown in three forms as given below : $[Pt(NCS)_2(en)_2]^{2+}$, $[Pt(SCN)_2(en)_2]^{2+}$ and $[Pt(SCN)(NCS)(en)_2]^{2+}$, Each one has a *trans*-isomer and dextro *cis* and laevo *cis*. Hence, total isomers are 3 × 3 = 9.

26. (4) $[Co(NH_3)_5SCN]Cl$ and $[Co(NH_3)_5NCS]Cl$
$[Co(NH_3)_5Cl]SCN$ and $[Co(NH_3)_5Cl]NCS$

27. (5) When $[Fe(EDTA)]^-$ exhibits optical isomerism, the complex will form five rings.

28. (0) This is a tetrahedral complex.

29. (4) The complex compound $[Cu''(NH_3)_4][Pt''Cl_4]$ will form four isomers.

30. (c) Correct Statement II is the donor atom linking to the central metal atom is different in different linkage isomers.

DPP-5 Valence Bond Theory, Effective Atomic Number and Magnetic Properties

1. (i) In 1927, Sidgwick suggested that a metal atom or ion formed a coordinate bond accepting both the electrons from donor (ligands), i.e. ligands were considered to be Lewis bases and metal ions as Lewis acids. He also suggested that transition metal ions will accept electron pairs from donors until the metal ion in the complex acquired an effective atomic number (EAN) of the nearest noble gas. That is the sum of the electron on the metal plus the electron donated by the ligands was called the effective atomic number.

$EAN = Z - ON + 2 \times CN$

where, Z = atomic number, ON = oxidation number of metal ion, CN = coordination number

(ii) $Mn_2(CO)_{10}$

$2Mn(2 \times 25) = 50$

$10CO(2 \times 10) = 20$

One Mn— Mn bond = 2

EAN for $Mn_2(CO)_{10} = 72$

EAN per Mn atom $= 72/2 = 36e^-$ (Kr)

2. Fe(Z = 26). Iron configuration ($3d^6$, $4s^2$)

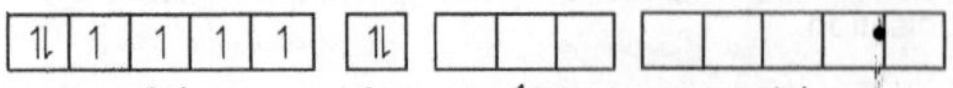

3d 4s 4p 4d

$[Fe(CO)_5]$

dsp^3

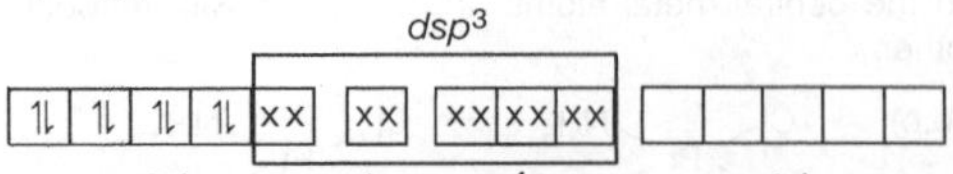

3d 4s 4p 4d

Hence, $[Fe(CO)_5]$ has trigonal bipyramidal geometry and has no unpaired electron. It is diamagnetic.

3. EAN = Z − ON + 2 × CN

Here, 36 = 24 − 0 + 2 × x or 2x = 12 or x = 6.

4. Cu^{2+} oxidises I^- to I_2 or I^- ions reduce Cu^{2+} to Cu^+.

$$2Cu^{2+} + 4I^- \longrightarrow Cu_2I_2 + I_2$$

5. $A = Fe(SCN)_3$ and $B = [FeF_6]^{3-}$

IUPAC name of A is trithiocyanato iron (III).

Electronic configuration of Fe(III) ion $-3d^5$. In this, unpaired electrons = 5

Hence, magnetic moment

$= \sqrt{5(5+2)} = \sqrt{35} = 5.916$ BM

6. (d) Due to strong ligand, this is inner orbital complex and the hybridisation of Fe is d^2sp^3.

7. (a) Since, F^- ion is a weak ligand hence, pairing of electrons does not occur and outer orbital complex is formed in $[CoF_6]^{3-}$.

8. (b) $[Ni(CN)_4]^{2-}$ in this, Ni has dsp^2 hybridisation with square planar geometry and it is diamagnetic.

9. (d) A magnetic moment of 1.73 BM corresponds to one unpaired electron. Among the given, (a) and (b) are diamagnetic and (c) is paramagnetic with 3 unpaired electrons ($3d^7$ configuration).

10. (a) F^- ion is weak ligand hence, pairing of electrons does not occur and outer orbital complex is formed in $[CoF_6]^{3-}$.

11. (d) The potassium ferrocyanide, $K_4[Fe(CN)_6]$ due to strong ligand pairing of electrons occur and it has no unpaired electrons.

12. (d) In $[Co(NH_3)_6]^{3+}$ due to pairing cobalt configuration $t_{2g}^6 e_g^0$ and it is diamagnetic.

In $[CoF_6]^{3-}$, F^- being weak ligand pairing does not occur and Co^{3+} ion has configuration $t_{2g}^4 e_g^2$. It has 4 unpaired electrons.

13. (c) In $[NiCl_4]^{2-}$, Ni has sp^3 hybridisation with 2 unpaired electrons, i.e. paramagnetic.

In $[PdCl_4]^{2-}$, Pd has dsp^2 hybridisation and it has no unpaired electrons, i.e. diamagnetic.

14. (c) Zn and Ni always give outer orbital complexes with coordination number 6.

$[Co(NH_3)_6]^{3+}$ is diamagnetic complex.

$[Cr(NH_3)_6]^{3+}$ is paramagnetic with $3d^3$ configuration.

15. (a) $[Cu(NH_3)_4]^{2+}$ and $[Ni(NH_3)_4]^{2+}$ the central atoms have dsp^2 hybridisation.

16. (b,d) The complex which involves d-orbitals of outer shell are called high spin complex as well as outer orbital complex.

17. (a,c,d) $[ZnCl_4]^{2-}$, $[CuCl_4]^{2-}$ and $[Ni(CO)_4]$ have sp^3 hybridisation and all three contains weak field ligands.

18. (a,c,d) Since, F^- ion is weak ligand it forms outer orbital complex and it has $3d^7$ configuration with 3 unpaired electrons.

19. (a,c,d)

$K_2[HgI_4]$ In this EAN = 80 − 2 + 8 = 86 (Equal to noble gas, radon).

$[Pd(NH_3)Cl_2]$ In this EAN = 46 − 2 + 8 = 52 (Not equal to noble gas, xenon)

$[CdI_4]^{2-}$ In this EAN = 48 − 2 + 8 = 54 (Equal to noble gas, xenon)

$Co_2(CO)_8$ In this EAN = Electrons from Co-atom (27) + electrons from 4 CO molecules (8) + one electron from Co—Co bond (1) = 36 (Equal to noble gas krypton)

20. (a,b,c) $[Fe(H_2O)_6]^{2+}$ and $[Fe(CN)_6]^{4-}$ have different hybridisation, magnetic moment and colour also.

21. (d) Cu(Z = 29) copper configuration ($3d^{10}4s^1$)

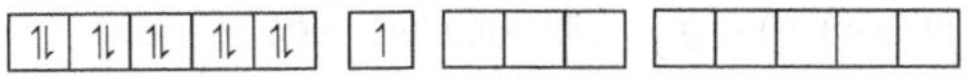

$[Cu(NH_3)_4]^{2+}$

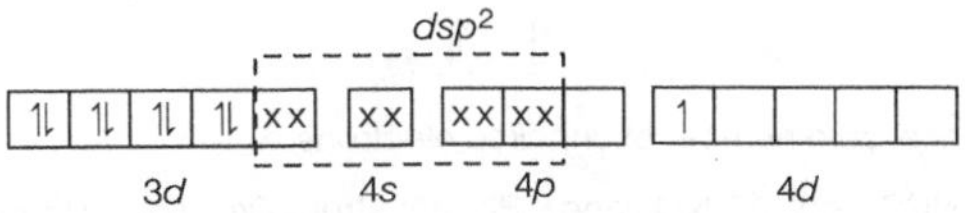

Hence, $[Cu(NH_3)_4]^{2+}$ has square planar geometry and 1 unpaired electron in $4d$ orbital which is transferred from $3d$ orbital with magnetic moment 1.73 BM.

22. (a) It is inner orbital complex, diamagnetic.

23. (a) (i) → (p,s), (ii) → (p,s), (iii) → (q,t), (iv) → (q,r)

24. (a) A → (iii) (b) B → (i) (c) C → (iv) (d) D → (ii)

25. (4)

Complex	EAN (Z − ON + 2 × CN)
$K_3[Fe(CN)_6]$	$26 - 3 + 12 = 35$
$[Ru(CO)_5]$	$44 - 0 + 10 = 54$ (noble gas)
$[Cr(NH_3)_6]^{3+}$	$24 - 3 + 12 = 35$

Complex	EAN (Z − ON + 2 × CN)
$[Co(NH_3)_6]^{3+}$	$27 - 3 + 12 = 36$ (noble gas)
$[Ni(NH_3)_6]^{2+}$	$28 - 2 + 12 = 38$
$[Fe(CO)_5]$	$26 - 0 + 10 = 36$ (noble gas)
$[W(CO)_6]$	$74 - 0 + 12 = 86$ (noble gas)

26. (4) Diamagnetic are
$[Zn(OH)_4]^{2-}$, $K_4[Fe(CN)_6]$, $[PdBr_4]^{2-}$, $[Ni(CO)_4]$

27. (4) In this, Mn has +3 oxidation state. Mn^{3+} configuration $= [Ar]3d^4$. This forms outer complex. Hence, unpaired electrons are 4.

28. (4) $Ru^{2+} = [Kr]\ 4d^6$. This forms outer complex. Hence, unpaired electrons are 4.

29. (0) In $[Fe(dipy)_3]^{2+}$ complex ion, Fe has $3d^6$ configuration. Due to pairing, this has no unpaired electrons.

30. (b) Statement I $[Ni(CN)_4]^{2-}$ is square planar and diamagnetic whereas $[NiCl_4]^{2-}$ is tetrahedral and paramagnetic.

Statement II $[Ni(CN)_4]^{2-}$ is square planar while $[NiCl_4]^{2-}$ is tetrahedral.

DPP-6 **Crystal Field Theory** (CFT) **and Colour of Complexes**

1. NH_3 is a stronger ligand than H_2O. So, $[Ni(NH_3)_6]^{2+}$ absorbs higher energy light than $[Ni(H_2O)_6]^{2+}$. Being green $[Ni(H_2O)_6]^{2+}$ probably absorbs red light while $[Ni(NH_3)_6]^{2+}$ absorbs higher energy yellow-green light and exhibit violet colour.

2. (i) $\Delta = h\nu = \dfrac{hc}{\lambda}$

Hence,

$$\lambda = \frac{hc}{\Delta} = \frac{6.625 \times 10^{-34}\ \text{Js} \times 3 \times 10^8\ \text{ms}^{-1} \times 6.022 \times 10^{23}}{243 \times 10^3\ \text{J}}$$

$$= 492 \times 10^{-9}\ \text{m} = 492\ \text{nm}$$

(ii) Wave number $(\bar{\nu}) = \dfrac{1}{\lambda} = \dfrac{1}{492 \times 10^{-7}\ \text{cm}}$

$$= 0.00203 \times 10^7$$

$$= 2.03 \times 10^4 = 20300\ \text{cm}^{-1}$$

3. In all three complexes, central metal ion is same, hence the increasing ligand field strength are in the order $H_2O < NH_3 < NO_2^-$

Therefore, the energies absorbed for excitation will be in the order. $[Ni(H_2O)_6]^{2+} < [Ni(NH_3)_6]^{2+} < [Ni(NO_2)_6]^{4-}$ and wavelengths will be in opposite order.

4. In tetrahedral geometry, none of the orbitals point directly at the ligands. Moreover, there are only 4 ligands instead of 6 as in octahedral geometry. As a result, the crystal field splitting in tetrahedral complexes is only about half of that in octahedral complexes. Consequently, Δ_t is almost smaller than spin-pairing energy, P and nearly all tetrahedral complexes are high spin.

5. For any given metal cation, the magnitude of CFSE depends on the nature of ligands .
   (i) The ligands which affect only a small degree of crystal field splitting are termed as weak field ligands.
   (ii) The ligands which affect only a large degree of crystal field splitting are termed as strong field ligands.
   e.g. spectrochemical series
   $I^- < Br^- < Cl^- < F^- < OH^- < ox^{2-} < H_2O < py$

← Weak ligands→
$\approx NH_3 < en < dipy < NO_2^- < CN^- < CO$
← Strong ligands→

6. (d) In $[Fe(H_2O)_6]^{3+}$, iron has $3d^5$ configuration. Since, H_2O is weak ligand the distribution is $t_{2g}^3 e_g^2$.

Hence, CFSE $= -3 \times 0.4 + 2 \times 0.6 = 0.0$

7. (a) As in complex $[Co(NH_3)_5Cl]^{2+}$, unpaired electrons are present, it requires less energy for excitation. Hence, wavelength of light absorbed is highest $\left(E \propto \dfrac{1}{\lambda}\right)$.

8. (c) CFSE for $[CoCl_4]^{2-}$ will be $\Delta_t = \dfrac{4}{9} \times \Delta_o$

i.e. $\Delta_t = \dfrac{4}{9} \times 18000 = 8000\ \text{cm}^{-1}$

9. (a) Crystal field splitting energy depends on strength of ligand. Strong ligands have more value of CFSE (Δ_o).
Hence, $Br^- < F^- < NCS < CN^-$

10. (b) Splitting of orbitals in octahedral complex for d^4 weak field.
CFSE $= -0.4 \times 3 + (+0.6 \times 1)\ \Delta_o = -0.6\ \Delta_o$

11. (d)

12. (a) For high spin d^6, CFSE $= -4 \times 0.4 + 2 \times 0.6 = -0.4$

For low spin d^6, CFSE $= -6 \times 0.4 = -2.4$

13. (d) The order of CFSE for tetrahedral complexes
$[CoCl_4]^{2-} < [CoBr_4]^{2-} < [Co(NCS)_4]^{2-}$

14. (c) Increasing order of ligand field strength
$I^- < Br^- < SCN^- < Cl^- < S^{2-} < F^- < OH^-$
$< C_2O_4^{2-} < H_2O < NCS^- < edta^{4-} < NH_3 < en < CN^- < CO$

15. (a) Weak field ligand have minimum wavelength of absorption.

16. (a,d) Ti^{3+} has $3d^1$ configuration and Cu^{2+} has $3d^9$ configuration.

17. (a,b,d) If $\Delta_o < P$ high spin complexes are favoured.

18. (a,c,d) For d^5 in weak field,
$CFSE = -3 \times 0.4 + 2 \times 0.6 = 0$
For d^{10} in weak and strong ligand field,
$CFSE = -6 \times 0.4 + 4 \times 0.6 = 0$

19. (a,b,c,d) Geometry of complex, charge on metal ion, nature of ligands and series of d orbital are the factors on which CFSE value depends.

20. (c,d) Cr (III) ion have $3d^3$ configuration.
Cr (III) ion

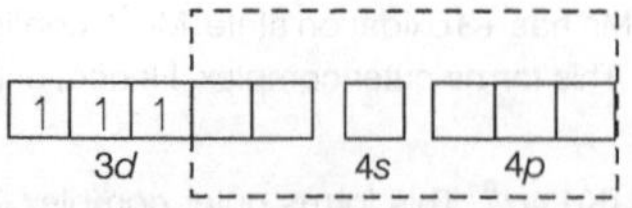

Hence, it will form always d^2sp^3 hybridisation and show colour due to the presence of 3 unpaired electrons.

21. (d) The total crystal field stabilisation energy is given by
$CFSE_{(octahedral)} = -0.4n_{t_{2g}} + 0.6n_{e_g}$ where $n_{t_{2g}}$ and n_{e_g} are the number of electron occupying the t_{2g} and e_g orbitals respectively.
For d^7 in strong field,

$\uparrow$ — e_g

$\uparrow\downarrow$ $\uparrow\downarrow$ $\uparrow\downarrow$ t_{2g}

$CFSE = -0.4 \times 6 + 0.6 \times 1 = -2.4 + 0.6 = -1.8\Delta_o$

22. (c) In $[Mn(H_2O)_6]^{2+}$, manganese have d^5 configuration. H_2O is a weak field ligand. Hence, it will form high spin complex.

$\uparrow$ $\uparrow$ e_g

$\uparrow$ $\uparrow$ $\uparrow$ t_{2g}

i.e. $t_{2g}^3 e_g^2$

In $[Mn(CN)_6]^{4-}$, manganese have d^5 configuration. CN^- ion is a strong field ligand. It will form low spin complex by pairing of electron in t_{2g} orbitals.

— — e_g

$\uparrow\downarrow$ $\uparrow\downarrow$ $\uparrow$ t_{2g}

i.e. $t_{2g}^5 e_g$

23. (c) (i) → (r), (ii) → (s), (iii) → (q), (iv) → (p)

24. (a) (i) → (q), (ii) → (p,s), (iii) → (r,t), (iv) → (s)

25. (6) In d^6 low spin octahedral complex,

— — e_g

$\uparrow\downarrow$ $\uparrow\downarrow$ $\uparrow\downarrow$ t_{2g}

Hence, zero number of unpaired electrons.

26. (4) NCS^-, edta^{4-}, NO_2^- and PPh_3 are strong field ligand than H_2O.

27. (6) In $[Cr(H_2O)_6]^{2+}$ chromium have d^4 configuration and H_2O is a weak field ligand.
d^4 (weak field)

$\uparrow$ — e_g

$\uparrow$ $\uparrow$ $\uparrow$ t_{2g}

In $[Cr(CN)_6]^{4-}$ chromium have d^4 configuration and CN^- ion is a strong field ligand. Hence, CN^- ion causing pairing of electron.
d^4 (strong field)

— — e_g

$\uparrow\downarrow$ $\uparrow$ $\uparrow$ t_{2g}

Hence, total number of unpaired electrons in both the complex are 6.

28. (4) d^7 (weak octahedral field)

$\uparrow$ $\uparrow$ e_g

$\uparrow\downarrow$ $\uparrow\downarrow$ $\uparrow$ t_{2g}

Hence, 3 unpaired electrons.

29. (6) The pairing starts with six electrons in the tetrahedral complexes which have weak field ligand.

30. (a) In $[Co(NH_3)_6]^{3+}$ Co, have d^6 configuration and NH_3 is strong field ligand.
d^6 — — e_g

$\uparrow\downarrow$ $\uparrow\downarrow$ $\uparrow\downarrow$ t_{2g}

$CFSE_{(oct)} = -0.4 \times 6 + 0.6 \times 0 = -2.4\Delta_o$

DPP-7 Bonding in Organometallic Compounds, Stability and Applications of Coordination Compounds

1.

σ-bonded complex	π-bonded complex	σ and π-bonded complex
In this metal and carbon atom of the ligand are joined together with a sigma bond. e.g. $(CH_3)_4Sn$, $(CH_3)_3Al$	These use π-bonds of organic ligands to form complexes. e.g. Zeise's salt $K[PtCl_3(C_2H_4)] \cdot H_2O$ 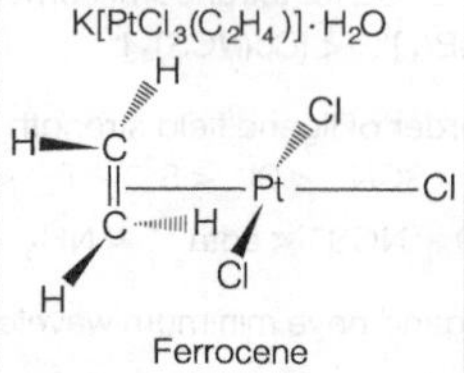 Ferrocene $[Fe(\eta^5-C_5H_5)_2]$	Complexes having CO, NO ligands (π acid). e.g. $[Ni(CO)_4]$, $[Fe(CO)_5]$

2. Gold complex formed is $[AuCl_4]^-$, tetrachloroaurate (III) ion
Platinum complex formed is $[PtCl_6]^{2-}$, hexachloroplatinate (IV) ion

3. In metal carbonyls, metal-carbon bond possess both sigma and pi character. Here, CO acts as σ donor and π acceptor. The effect of π bond strengthens the σ bond and *vice versa*. This is called synergic effect (working together).

4. When a didentate or polydentate ligand coordinate with the central metal atom or ion, a five or six membered ring is formed. This effect is called chelate effect. Due to this, stability of the complex increases.
e.g. $[Ni(en)_3]^{2+}$ is more stable than $[Ni(NH_3)_6]^{2+}$.
$[PtCl_2(en)]$ is more stable than $[PtCl_2(NH_3)_2]$.

5. (i) $K_3[Fe(CN)_6]$ is more stable because in this, Fe is in +3 oxidation state while in $K_4[Fe(CN)_6]$, Fe is in +2 state. Fe in +2 gets oxidised easily into Fe (+3).

(ii) $[Cu(en)_2]^{2+}$ has only two rings in the structure. On the other hand, $[Fe(EDTA)]^-$ has 5 rings in the structure.

6. (b) $Na_2S + Na_2[Fe(CN)_5NO] \longrightarrow \underset{\text{(Violet)}}{Na_4[Fe(CN)_5NOS]}$

7. (b) Wilkinson's catalyst is a rhodium complex $[(Ph_3P)_3RhCl]$ and it is used as a catalyst for hydrogenation of alkenes reactions.

8. (b) Tollen's reagent is ammoniacal silver nitrate solution. It is used to distinguish aldehydes and ketones, reducing and non-reducing sugars.

9. (c) $Au + HNO_3 + HCl \longrightarrow H_3O^+ [AuCl_4]^- \cdot 3H_2O$

$$\downarrow$$

$$AuCl_4$$

$$Pt \xrightarrow{\text{Aqua-regia}} H_2[PtCl_6] \xrightarrow{\text{Heat}} PtCl_4 + 2HCl$$

10. (a) $(CH_3)_4Sn$ is a σ-bonded organometallic compound.

11. (b) The pigment responsible for photosynthesis, i.e. chlorophyll is a coordination compound of magnesium.

12. (a) $2Na_2S_2O_3 + AgBr \longrightarrow Na_3[Ag(S_2O_3)_2] + NaBr$

13. (c) The quantity of Ca, Mg salts in water is estimated by using standard EDTA solution using erichrome black T as an indicator.

14. (b) CO as ligand is sigma donor and pi acceptor ligand.This is called pi-acid.

15. (d) $Fe^{3+} + 3CNS^- \longrightarrow \underset{\text{(Blood red)}}{[Fe(CNS)_3]}$

$Fe^{3+} + 3[Fe(CN)_6]^{4-} \longrightarrow \underset{\text{(Prussian blue)}}{Fe_4[Fe(CN)_6]}$

16. (a,b) Stable complex is obtained when sum of metal *d*-electrons and electrons donated by ligands and the overall charge of the complex equals to 18. Chromium and tungsten obey 18-electron rule or effective atomic number rule.

In $[Cr(CO)_6]$

Atomic number of chromium = 24

$\therefore$ Cr contains electrons = $24\,e^-$

6 CO ligand donate (6×2 electrons) = $12\,e^-$

EAN of $[Cr(CO)_6] = 36e^-$

Thus, it may be stated that the complex $[Cr(CO)_6]$ obeys EAN rule and it will be stable complex. $[W(CO)_6]$ can be solved same as above.

17. (a,b) The porphyrin ring contains pyrrote unit are linked by four (— CH ═) bridges. Porphyrine are derived from porphin, varying according to the nature of substituents. Porphyrin ring act as tetradentate ligands with four pyrole like nitrogens surrounding a central site of metals. Iron complex of the substituted porphin is here. When magnesium lie at the centre of substituted porphin ring; the resulting complex is called chlorophyll.

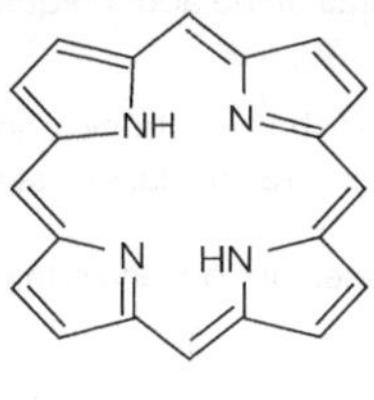

18. (a,c,d) η^6–arene is 6-electron ligand, η^2—C_2H_4 is 2-electron ligand η^5-cyclopentadienyl is 5-electron donor. All these 3 ligands creates π-bonded complex. CO as a ligand is sigma donor and pi-acceptor ligand.

19. (a,b,c) $[Fe(NCS)_6]^{3-}$, $[Fe(H_2O)_6]^{2+}$, $[Cu(H_2O)_4]^{2+}$ are correct.

20. (b,d) $Fe^{3+} + 3[Fe(CN)_6]^{4-} \longrightarrow \underset{\text{(Prussian blue)}}{Fe_4[Fe(CN)_6]}$

$[Fe(NCS)_3]$ = Blood red

$CoCl_2 + 4NH_4CNS \longrightarrow \underset{\text{(Blue)}}{(NH_4)_2[Co(CNS)_4]} + 2NH_4Cl$

21. (b)

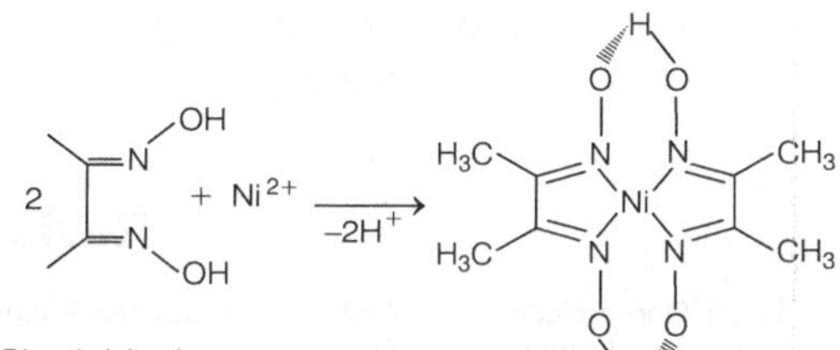

Dimethylglyoxime

22. (a) *cis*-diamminedichloroplatinum (II), $[cis[Pt(NH_3)_2Cl_2]$ is used in cancer chemotherapy.

The mechanism of *cis*-platin is based on Pt-DNA interaction. Platinum binds with N_7 of two intrastrand adjacent guanine bases of DNA.

Their binding changes the active conformation of DNA leading to inhibiting of DNA replication or cell division.

cis - $Pt(NH_3)_2Cl_2 + H_2O \cdots\cdots$ *cis* - $[Pt(NH_3)_2(OH)_2Cl]^+ + Cl^-$

The hydrolysis reaction of *cis*-platin is an important aspect of the biological activity.

DNA coordinated *trans* complex begin to decrease after six hour whereas the *cis*-isomer is then accumulated in the cell nucleus.

23. (b) (i) $\rightarrow$ (r), (ii) $\rightarrow$ (q), (iii) $\rightarrow$ (s), (iv) $\rightarrow$ (p)

24. (a) (i) $\rightarrow$ (p,r), (ii) $\rightarrow$ (s), (iii) $\rightarrow$ (q), (iv) $\rightarrow$ (t)

25. (6)

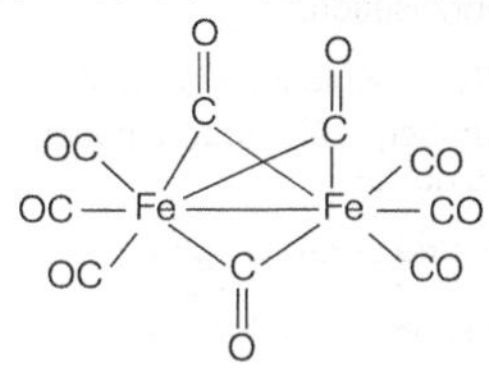

$Fe_2(CO)_9$

$Fe_2(CO)_9$ contains 3 bridging CO groups placed symmetrically around the (Fe—Fe) bond and 6 terminal carbonyl groups.

26. (3) $Fe_2(CO)_9$ has three bridging CO groups.

27. (2)

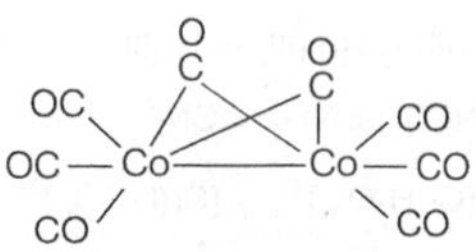

$Co_2(CO)_8$ (Solid)

Each cobalt atom is bonded to three terminal CO group and share two CO ligands as bridging groups.

28. (6) Each Ir bonded to three iridium atoms and three CO molecules. The number of Ir—Ir bonds are 6.

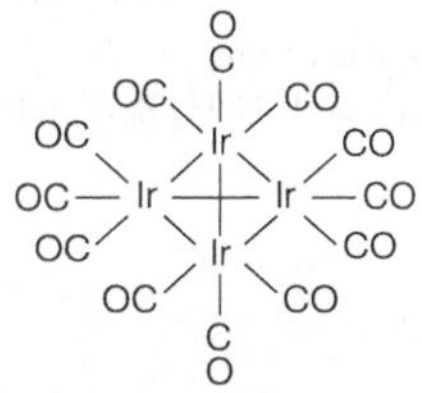

29. (0) $Mn(CO)_5$ is radical with 17-electron configuration. Therefore, it undergoes dimerisation through the formation of *M*— *M* single bonds and complex their valence shells to avoid ligand-ligand repulsion the $Mn(CO)_5$ units acquire staggered configuration.

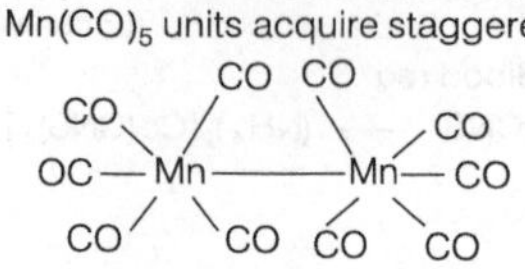

$Mn_2(CO)_{10}$

30. (a) In $Mn(\pi - C_3H_5)(CO)_4$

$$Mn = 25$$
$$\pi\text{-}C_3H_5 = 3$$
$$4(CO) = 8$$

Hence, EAN = 36

Hence, π-C_3H_5 ligand acts as a 3-electron donor and complex obey effective atomic number rule.

Revisal Problems (JEE Main)

1. (a) One molecule of potash alum produces 8 ions in solution ($K_2SO_4 \cdot Al_2(SO_4)_3 \cdot 24H_2O$)

One molecule of Mohr's salt produces 5 ions in solution

$(FeSO_4) \cdot (NH_4)_2SO_4 \cdot 6H_2O)$

One molecule of potassium ferrocyanide produces 5 ions in solution $[K_4[Fe(CN)_6]$.

One molecule of carnalite produces 5 ions in solution $(KCl \cdot MgCl_2 \cdot 6H_2O)$

2. (a) Thiosulphate is not a chelating agent.

3. (d) $[Co(NH_3)_3(NO_2)_3]$ complex is poor electrolytic conductor.

4. (c) EDTA is a hexadentate ligand. Its coordination number is six.

5. (d) Nickel complexes with coordination number six are always outer complexes with Ni (sp^3d^2) hybridisation.

6. (a) In the complex $[Cu(NH_3)_4](ClO_4)$, copper has dsp^2 (square planar) while in complex $[Cu(NH_3)_4]ClO_4$, copper has sp^3 (tetrahedral) hybridisation.

7. (a) When EAN calculation is done for $[Ni(H_2O)_6]^{2+}$, i.e. [Z–oxidation number) + 2 × (coordination number)], we get 38. For all others we get 36.

8. (c) $[MnBr_4]^{2-}$ is tetrahedral complex.

9. (b) 100 mL 0.1M complex = 0.01 moles

$$\underset{0.01\ \text{moles}}{[Co(NH_3)_5Cl]Cl_2} \xrightarrow{\text{Excess } AgNO_3} \underset{0.02\ \text{moles}}{2AgCl}$$

10. (c) Tetramminenickel (II) tetrachloronickelate (II)

11. (b) $[Co(en)_3]Cl_3$ shows only optical isomerism.

12. (a) Fe (III) < Cr (III) < Rh (III) < Ir (III)

13. (a) $Cl^- < F^- < NCS^- < NH_3 < CN^-$

14. (c) $[CrCl_6]^{3-} > [Cr(H_2O)_6]^{3+} > [Cr(NH_3)_6]^{3+} > [Cr(CN)_6]^{3-}$

15. (c) Chlorotetracyano superoxoferrate (II) ion.

16. (c) Magnetic moment 3.87 BM means it has three unpaired electrons. This is possible in d^3 (in weak as well as in strong field ligand).

17. (d) Diamagnetic complexes are $[Ni(CN)_4]^{2-}$ and $Ni(CO)_4$.

Paramagnetic complexes are

$[NiCl_4]^{2-}$ and $[Ni(H_2O)_6]^{2+}$ (attracted)

18. (b) Complex A is $[Co(NH_3)_5SO_4]Cl$
Complex B is $[Co(NH_3)_5Cl]SO_4$ — Ionisation isomerism

19. (a) $[Co(en)_2Cl_2]^+$ this has *cis* and *trans* isomers.

The *cis* isomer exists as enantiomeric pair .Hence, total isomers are 3.

20. (b) CFSE $= -(0.4 \times 6\Delta_o + 3P) = -\frac{12.5}{5}\Delta_o + 3P$

($\because$ P means pairing energy which is the energy required to pair two electrons against electron-electron repulsion in the same orbital)

21. (b) In $K_3[Fe(CN)_6]$, iron is in +3 oxidation state.

$\because$ In strong ligand field, configuration is $t_{2g}^5 e_g^0$. Hence, CFSE is $-5 \times 0.4 = -2.0\Delta_o$.

Since, it has one unpaired electron, magnetic moment is $\sqrt{3}$ BM.

22. (c) Due to synergic effect, CO bond order decreases and bond length increases.

23. (d) The covalent character of the bond between metal and ligand.

24. (d) $Cl < H_2O < NH_3 < CO$

25. (a) The number of unpaired electrons in the square planar $[Pt(CN)_4]^{2-}$ is zero.

26. (c) In decacarbonyl dimanganese (0), the number of CO group bonded in between the Mn atoms are shown in the following figure

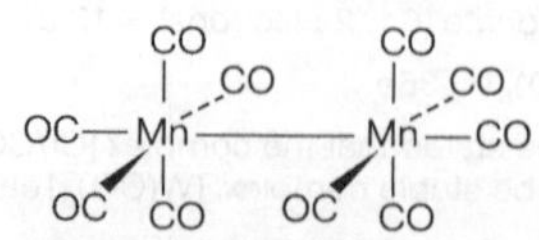

27. (c) CO is a ligand that forms sigma bond with metal by donating electrons and pi bond by accepting electrons from the central metal ion.

This is called synergic effect.

28. (a) When an equation is obtained by adding two or more equations the new equilibrium constant is the product of the equilibrium constants of the added equations.

Therefore, $K = K_1 \times K_2 \times K_3$.

29. (d) Iron atom ($Z = 26$) has electronic configuration $3d^64s^2$. In this complex iron is in +3 oxidation state and has electronic configuration $3d^5$.

It has been experimentally observed that the complex has one unpaired electron.

Revisal Problems (JEE Advanced)

1. (b) In $[FeF_6]^{3-}$, iron is Fe^{3+} with $3d^5$ configuration.
This has 5 unpaired electrons (*P*).
In $[V(H_2O)_6]^{2+}$, vanadium has $3d^3$ configuration.
This has 3 unpaired electrons (*Q*).
In $[Fe(H_2O)_6]^{2+}$, iron is Fe^{2+} with $3d^6$ configuration. This has 4 unpaired electrons (*R*).

2. (c) Complexes $[Zn(NH_3)_4]^{2+}$, $[NiCl_2(PPh_3)_2]$ and $[Cu(CN)_4]^{3-}$ are tetrahedral (sp^3).
Complex $[Cu(NH_3)_4]^{2+}$ is square planar with dsp^2.

3. (c) $K_4[Mn(CN)_6]$

4. (c) Nickel complexes with CN 6 are outer orbital complexes with two unpaired electrons. Hence, it is paramagnetic.

5. (c) In $[Mn(CN)_6]^{4-}$ Mn is +2 with $3d^5$ configuration. Due to strong ligand, this has one unpaired electron with magnetic moment 1.73 BM.
In $[MnBr_4]^{2-}$, Mn is +2 with $3d^5$ configuration. Due to weak ligand this has five unpaired electrons with magnetic moment 5.92 BM.

6. (b) $Cu^{2+} + K_4[Fe(CN)_6] \longrightarrow \underset{\text{(Chocolate brown)}}{Cu_2[Fe(CN)_6]}$.

7. (d) Chelate complexes $(Ni(en)_2(H_2O)_2]^{2+}$ are more stable than similar complexes formed with monodentate complexes.

8. (c) In the brown ring complex, $[Fe(H_2O)_5NO]^{2+}$ iron oxidation number is (+1). Here, Fe has $3d^7$.
Hence, it has 3 unpaired electrons with $\mu = \sqrt{15}$ BM. This forms outer orbital complex.

9. (a) CN = 6,ON = +2

10. (b) $\Delta_t = \frac{4}{9}\Delta_o$

11. (a,b) $[AuCl_4]^-$ is square planar and diamagnetic.
$[PdCl_2(PPh_3)_2]^-$ square planar and diamagnetic.

12. (a,b,d) $Co_2(CO)_8$, $Mn_2(CO)_{10}$, $Ir_4(CO)_{12}$

13. (a,b,c) $[Ag(NH_3)_2]^+$ is linear. Oxidation state of sulphur in sodium nitroprusside is +2 . $Fe(CO)_3$ is triangle pyramidal.

14. (b,c) Tetrahedral complexes have smaller crystal field splitting than octahedral complexes. $[Co(H_2O)_6]^{2+}$ is transformed into $[CoCl_4]^{2-}$.

15. (a,c) The *trans* effect causes the introduction of the ligand *trans* to chloride rather than *trans* to ammonia in both the cases.

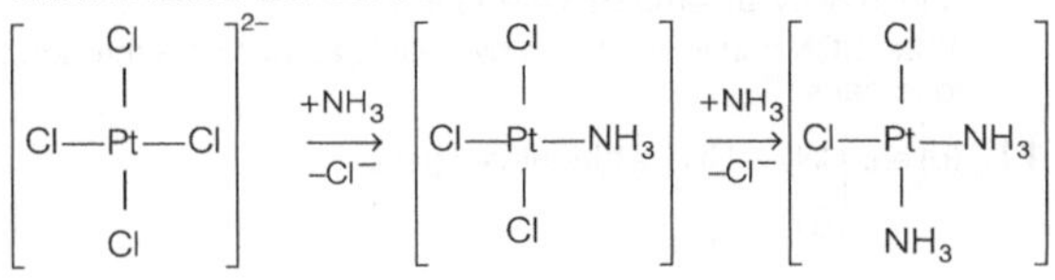

X

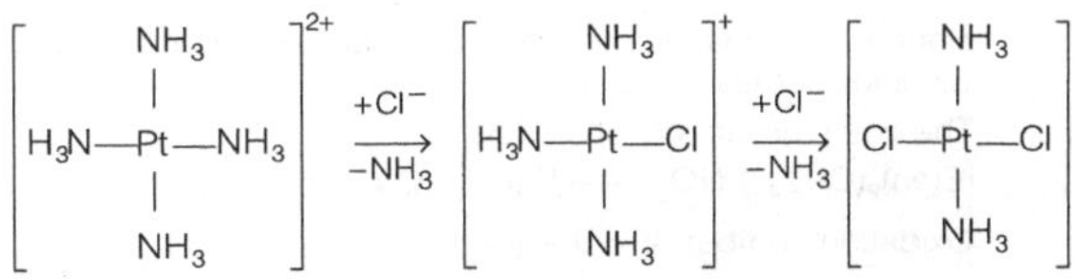

Y

16. (b,d) EAN of $[Ag(NH_3)_2]^+ = (47 - 1 + 4)$
$= 50$
EAN of $[Ni(en)_3]^{2+} = (28 - 2 + 12)$
$= 38$
EAN of $[Mn_2(CO)_{10}] = (25 - 0 + 10 + 1)$
$= 36$
(Here, Mn is 25, from 5 ligands 10 and from Mn—Mn bond 1)
EAN of Ir in $[IrCl_6]^{3-} = (77 - 3 + 12) = 86$

17. (b,d) $[Co(en)_3]Cl_3$ and *cis* $[Co(en)_2Cl_2]^+$, $[Co(en)_2C_2O_4]^+$ and $[Co(EDTA)]^-$

18. (b,c,d) Oxalate ion, ethane-1,2-diamine, 2-2′-dipyridine

19. (b,c,d) d^5 in strong ligand field $= t_{2g}^5 e_g^0$ and
$$CFSE = -0.4 \times 5\Delta_o$$
d^5 in weak ligand field $= t_{2g}^3 e_g^2$ and
$$CFSE = -0.4 \times 3 + 0.6 \times 2 = 0$$
d^{10} in strong or weak ligand field $t_{2g}^6 e_g^4$ and
$$CFSE = -0.4 \times 6 + 0.6 \times 4 = 0$$

20. (a,b,c) OH^-, NH_2^-, CO

21. (d) $[CoBrCl(en)_2]\ ClH_2O$

22. (a) $2[CoBrCl(en)_2]ClH_2O + Pb(NO_3)_2 \longrightarrow PbCl_2\downarrow$
For one mole complex, 0.5 moles of $PbCl_2$ is formed.

23. (a) (i) → (p), (ii) → (q), (iii) → (s), (iv) → (t)
If all ligands is $[CoCl(en)_2H_2O]BrCl$ are replaced by SCN^- we get $[Co(SCN)_6]^{3-}$.
In this, cobalt has $[Ar]3d^6$ configuration with 4 unpaired electrons. Hence, the magnetic moment is 4.9 BM.

24. (d) (i) → (q), (ii) → (r), (iii) → (p), (iv) → (t)

25. (d) (i) → (t), (ii) → (s), (iii) → (p), (iv) → (q)

26. (4) $Fe^{3+} = 3d^5$ and CN^- is strong ligand, paramagnetic.
* $Co^{3+} = 3d^6$ and NH_3 is strong ligand, diamagnetic.
* $Co^{3+} = 3d^6$ and C_2O_4 is a strong ligand, diamagnetic.
* $Ni^{2+} = 3d^8$ and Ni octahedral complexes are always paramagnetic.
* $Pt^{2+} = 5d^8$ and CN^- is square planar complex, diamagnetic.
* $Zn^{2+} = 3d^{10}$ and it is diamagnetic.

27. (6) CFSE $= -(0.4 \times 3 + 1 \times 0.6\Delta_o) = -0.6\Delta_o = -6$ Dq

28. (2) In this $[Co(H_2O)_3F_3]$ cobalt has $3d^6$ configuration. In weak field, the distribution of electrons is $t_{2g}^4 e_g^2$. This has two unpaired electrons in t_{2g} orbitals.

29. (4) Two enantiomeric pairs are possible.

30. (2) $[Cu(en)_2]SO_4 \longrightarrow [Cu(en)_2]^{2+} + SO_4^{2-}$ (two ions)

31. (a) The chelate complexes containing 5 and 6 membered rings are comparatively more stable. In general, chelating ligands which do not contain unsaturated groups.

JEE Main & AIEEE Archive

1. (b)

Ligand	L_1	L_2	L_3	L_4
Coloured region of absorbed wavelength	Red	Green	Yellow	Blue

Coloured regions of visible light (VIBGYOR) are arranged in the increasing order of wavelength or decreasing order of energy [$\because E \propto 1/\lambda$]. Therefore,

$$\underset{(L_1)}{E_{Red}} < \underset{(L_2)}{E_{yellow}} < \underset{(L_3)}{E_{green}} < \underset{(L_4)}{E_{blue}}$$

This suggests that L_4 will absorbs higher energy wavelength followed by L_2, L_3 and L_1.

Since, higher absorbed energy corresponds to greater extent of crystal field splitting and thus, higher field strength of the ligand. Hence, the increasing order of ligand strength of the four ligands is

$$L_1 < L_3 < L_2 < L_4$$

2. (c) Optical isomerism is exhibited by only those complexes which lacks elements of symmetry.

$[Co(NH_3)_3Cl_3]$ shows facial as well as meridional isomerism. But both the forms contain plane of symmetry.

Thus, this complex does not exhibit optical isomerism.

3. (b) Two Br, two (en) and one Cr are parts of complex.

Charge on the complex is

$$\left.\begin{aligned} 2(Br) &= -2 \\ 2(en) &= 0 \\ 1(Cr) &= +3 \end{aligned}\right\} = +1$$

Thus, complex ion is $[Cr(en)_2Br_2]^+$.

Since, anion is bromide thus, complex is $[Cr(en)_2Br_2]Br$.

4. (c) $\underbrace{[Cr(NH_3)_6]}_{\text{Coordinate sphere}} \underbrace{Cl_3}_{\text{Ionisable}} \rightleftharpoons$

$$[Cr(NH_3)_6]^{3+} + 3Cl^- \xrightarrow{AgNO_3} \underset{\text{(White precipitate)}}{AgCl \downarrow}$$

In $[Cr(NH_3)_6]^{3+}$, Cr exists as Cr^{3+} ion.

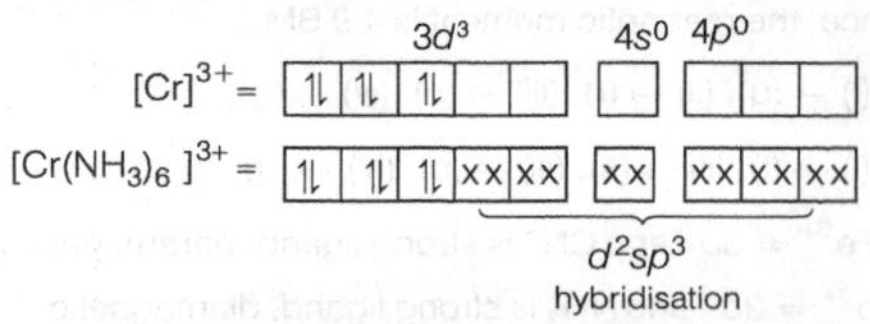

The outer orbital electronic configuration of Cr is $3d^3\ 4s^0$.

The orbital diagram is represented in the figure

Here, [xx] indicates lone-pair of NH_3 donated to Cr.

Hence,

(a) d^2sp^3-hybridisation, octahedral. Thus, correct.

(b) There are three unpaired electrons, hence paramagnetic Thus, correct.

(c) d^2sp^3-inner orbital complex, thus incorrect.

(d) Due to ionisable Cl^- ions, white precipitate with $AgNO_3$, thus correct.

5. (c) $[NiCl_4]^{2-}$; oxidation number of Ni, $x - 4 = -2$

$\therefore\ x = +2$, $Ni_{(28)}$ [Ar] $3d^8$, $4s^2$

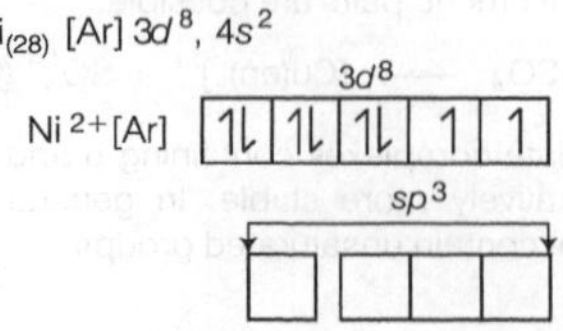

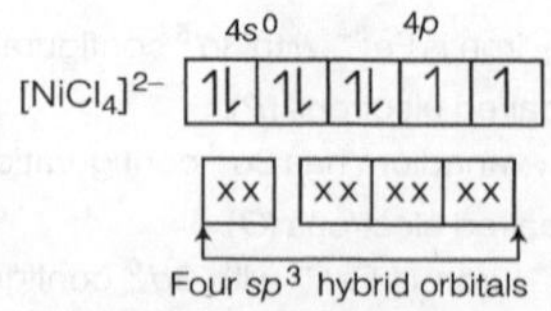

sp^3-hybrid orbitals, tetrahedral. Since, Cl^- is a weak ligand, thus unpaired electrons are not paired. Lone pairs from 4 Cl^- are accommodated in four sp^3 hybrid orbitals.

$$n = \text{unpaired electron} = 2, \text{paramagnetic}$$

Magnetic moment (spin only)

$$= \sqrt{n\,(n+2)}\ \text{BM} = \sqrt{8} = 2.828\ \text{BM}$$

6. (b) Based on spectrochemical series, ligands arranged in increasing order of crystal field strength are as

$$NH_3 < en < CN^- < CO$$

7. (c)

	Complex	Isomerism shown
(a)	$[Co(en)_3]^{3+}$	Optical only
(b)	$[Ni(NH_3)_5Br]^+$	No geometrical isomer
(c)	$[Co(NH_3)_2(en)_2]^{3+}$	*cis* and *trans*
(d)	$[Cr(NH_3)_4(en)]^{3+}$	No geometrical isomer

8. (b) Complex $[Co(en)_3]^{3+}$ has no plane of symmetry and centre of symmetry. That's why, it is optically active.

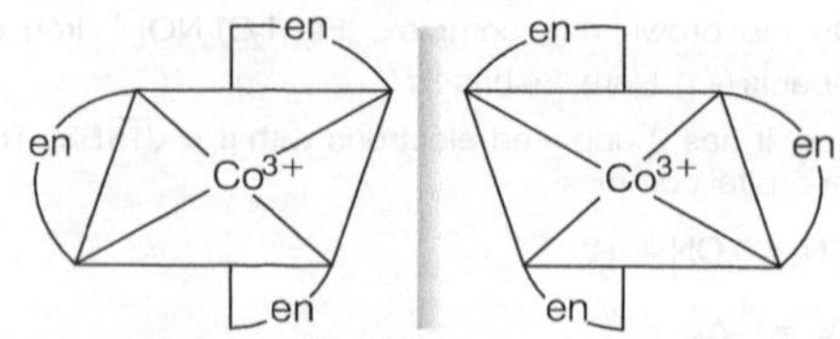

9. (d)

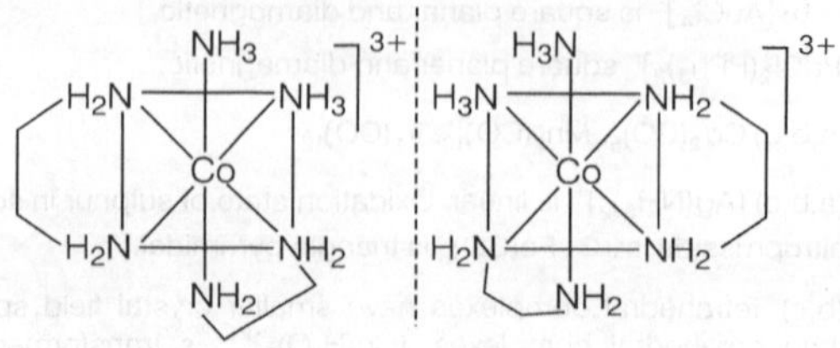

10. (b) Linkage isomers are caused due to the presence of ambidentate ligands.

$[Pd(PPh_3)_2(NCS)_2]$ and $[Pd(PPh_3)_2(SCN)_2]$ are linkage isomers due to SCN, an ambidentate ligand.

Thus, SCN can linked to Pd by both N as well as S through their lone pairs.

11. (d) en, $\begin{matrix} CH_2{-}NH_2 \\ | \\ CH_2{-}NH_2 \end{matrix}$ is bidentate ligand.

$C_2O_4^{2-}$ is also bidentate ligand.

Hence, coordination number of the element E in the given complex comes out to be 6.

The complex can be ionised as

$[E(en)_2(C_2O_4)]\ NO_2 \longrightarrow [E(en)_2\ (C_2O_4)]^+\ NO_2^-$

Oxidation number, $x + 0 + (-2) = +1$

$$x = 3$$

12. (a) CFSE (crystal field splitting energy) for octahedral complex, Δ_o depends on the strength of negative ligand. Spectrochemically, it has been found that the strength of splitting is as follows :

$$CO > \underline{CN^-} > NO_2^- > en > \underline{NH_3} > py >$$

$$NCS^- > H_2O > O^{2-} > ox^{2-} > OH^- > F^- >$$

$$Cl^- > SCN^- > S^{2-} > Br^- > I^-$$

13. (d) Cl^- is a weak ligand but Cl^- cause the pairing of electrons with large Pt^{2+} and consequently give dsp^2 hybridisation and square planar geometry.

14. (b) $Fe(CO)_5$ is an organometallic coordinated compound in which electrons from *d*-orbital of Fe are partially transferred to antibonding molecular orbitals ($\overset{*}{\pi}$) of CO. This type of bonding is called **π-back bonding.**

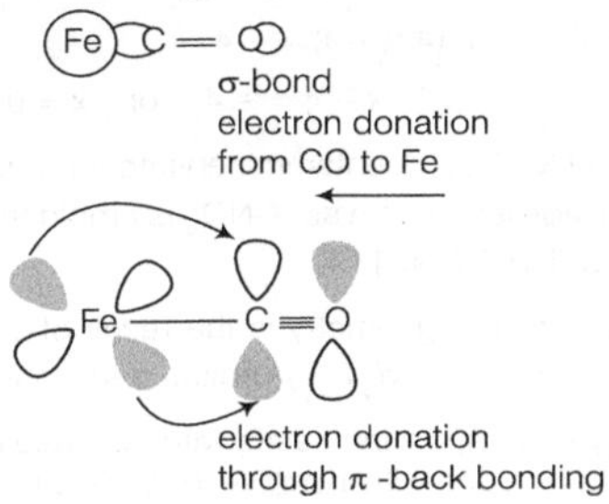

Thus, $Fe(CO)_5$ possesses both σ and π-character.

15. (d) The IUPAC name for the complex

$[Co(NO_2)(NH_3)_5]Cl_2$

Pentammine nitrito-N-cobalt (III) chloride

16. (b) $_{28}Ni = [Ar]\,4s^2, 3d^8$

$Ni^{2+} = [Ar]\,3d^8$

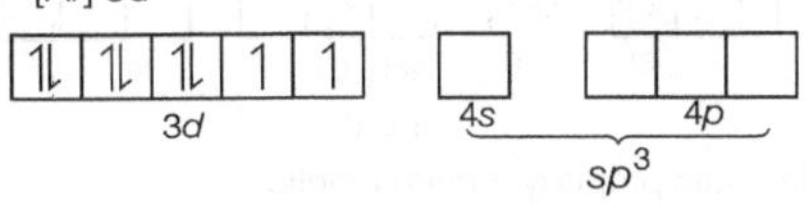

Nickel has two unpaired electrons and geometry is tetrahedral due to sp^3 hybridisation.

17. (d) (a) d^5 in strong field

n = unpaired electron = 1

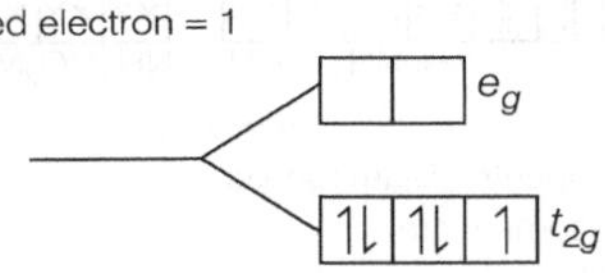

Magnetic moment $= \sqrt{n(n+2)} = \sqrt{3}$ BM = 1.73 BM

(b) d^3 in strong/weak field

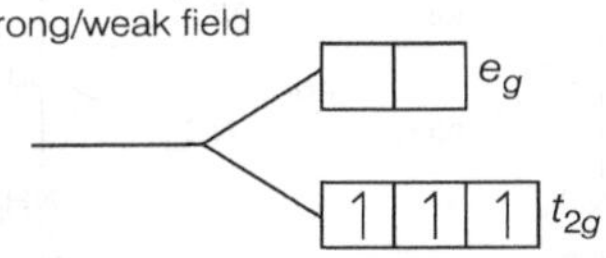

$n = 3$

Magnetic moment $= \sqrt{15} = 3.87$ BM

(c) d^4 in weak field

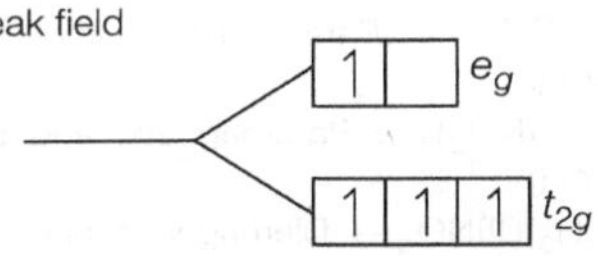

$n = 4$

Magnetic moment $= \sqrt{24} = 4.90$ BM

(d) d^4 in strong field

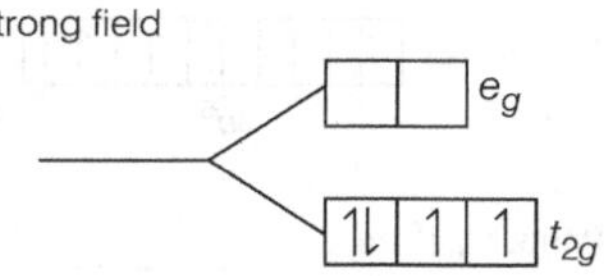

$n = 2$

Magnetic moment $= \sqrt{8} = 2.83$ BM

18. (d) Lanthanide contraction is due to poor shielding of one of 4*f* electron by another in the subshell.

19. (b) Due to lanthanoid contraction, the atomic radii of the elements belonging to the same group of the 2nd and 3rd transition series become same. Thus, $_{40}Zr$ (160 pm) and $_{72}Hf$ (159 pm) have almost same radii.

20. (c) $K_3[Fe(CN)_6]$

Cation Anion

Oxidation state of Fe in anion = + 3

Thus, it is potassium hexacyanoferrate (III).

21. (a)

		Hybridisation/ Unpaired electrons	Magnetic moment
(a)	$[Co(CN)_6]^{3-}$	$d^2sp^3(0)$	0
(b)	$[Fe(CN)_6]^{3-}$	$d^2sp^3(1)$	$\sqrt{3}$ BM
(c)	$[Mn(CN)_6]^{3-}$	$d^2sp^3(2)$	$\sqrt{8}$ BM
(d)	$[Cr(CN)_6]^{3-}$	$d^2sp^3(3)$	$\sqrt{15}$ BM

Thus, least paramagnetism is (a).

22. (b)

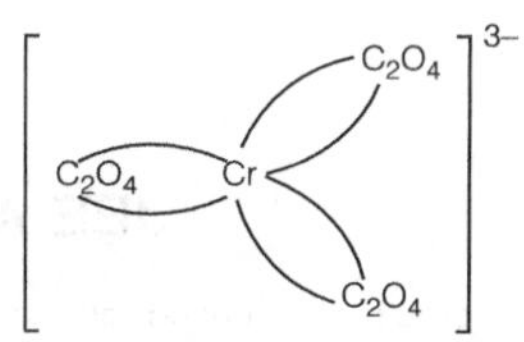

Mirror image is not superimposable, hence optical isomerism is possible.

23. (a) Coordination number is the maximum covalency shown by a metal or metal ion. It is the maximum number of ligands attached to metal by sigma bonds or coordinate bonds.

24. (d) All the given compounds exhibit geometrical (*cis* and *trans* isomers) but only *cis* isomer of $[Co\,(en)_2Cl_2]^+$ possess optical isomers (*d* and *l*) due to the presence of symmetrical didentate, en [ethylenediamine].

25. (a) Number of unpaired electrons in $[Fe(CN)_6]^{4-}$ is zero.

Thus, magnetic moment $= \sqrt{n\,(n+2)} = 0$ BM

(*n* = unpaired electrons)

n in $[MnCl_4]^{2-} = 5, \sqrt{35}$ BM

n in $[CoCl_4]^{2-} = 3, \sqrt{15}$ BM

26. (d)

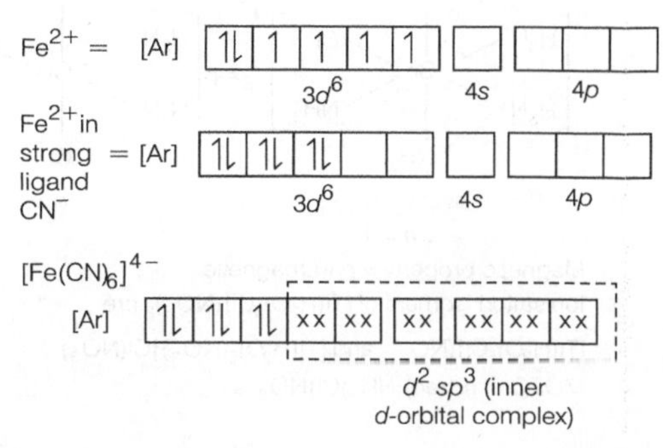

(b) Mn^{2+} = [Ar]

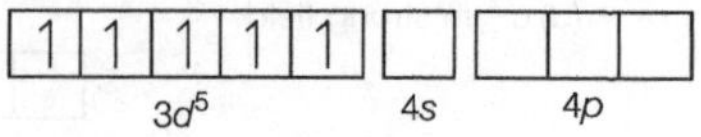

$[Mn(CN)_6]^{4-}$ =

[Ar]

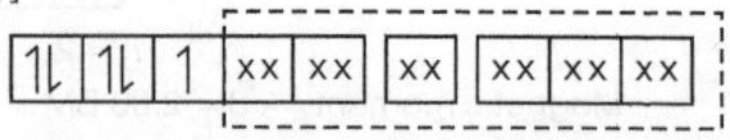

d^2sp^3 (inner d-orbital complex)

(c) Co^{3+} =

[Ar]

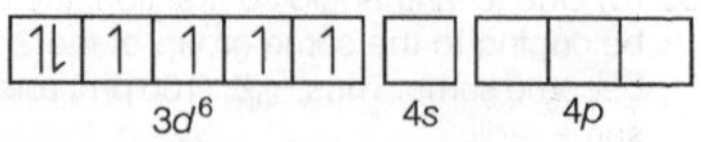

$[Co(NH_3)_6]^{3+}$ =

[Ar]

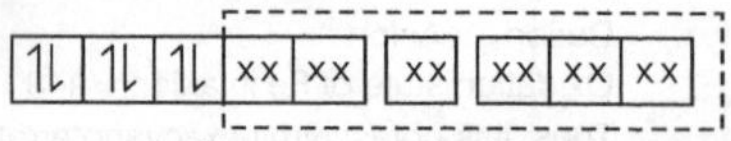

NH_3 is a strong ligand

d^2sp^3 (inner d-orbital complex)

(d) Ni^{2+} =

[Ar]

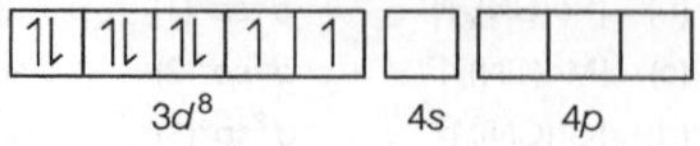

In this case also, NH_3 is a strong ligand but electrons remain unpaired, since only one orbital is left vacant in $3d$. Thus, $[Ni(NH_3)_6]^{2+}$

[Ar]

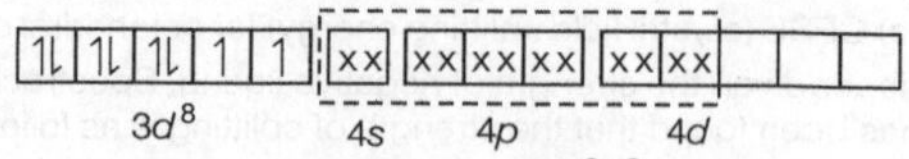

sp^3d^2

(outer d-orbital complex)

27. (b) In acidic solutions, N-atoms of NH_3 donates their lone pair electrons to protons forming NH_4^+ ions.

Therefore, lone pair electrons of NH_3 are not available for forming the complex with copper ions.

28. (a) The oxidation state of nickel is

$[Co(NH_3)_5Cl]\,Cl_2 \rightleftharpoons \underbrace{[Co(NH_3)_5Cl]^{2+} + 2Cl^-}_{\text{Three ions}}$

29. (b) The oxidation state of nickel is

$K_4\,[Ni(CN)_4] \longrightarrow 4K^+ + [Ni(CN)_4]^{4-}$

$[Ni(CN)_4]^{4-}$, $x + (4 \times -1) = -4$

or $x - 4 = -4$ or $x = 0$

30. (d) $[Cr(NH_3)_5NO_2]\,Cl_2$ (pentamminenitritochromium (III) chloride) exhibits linkage isomerism as $—NO_2$ is ambidentate ligand. Isomer is $[Cr(NH_3)_5ONO]\,Cl_2$.

31. (c) A square planar geometry is the result of dsp^2 hybridisation where inner d subshell ($d_{x^2-y^2}$ orbital) participates.

32. (c) Stability of complex increases with increase in charge on the central metal ion and with increase in basic strength of ligand. In $[Fe(CN)_6]^{3-}$, oxidation state of Fe is +3 and basicity of CN^- is higher than OH^-, Cl^- and H_2O.

JEE Advanced & IIT JEE Archive

1. (b) **Plan** This problem is based on concept of VBT and magnetic properties of coordination compound.

Draw VBT for each coordination compounds.

If unpaired electron is present then coordination compound will be paramagnetic otherwise diamagnetic.

Coordination compounds of $[MA_4B_2]$ type show geometrical isomerism.

Molecular orbital electronic configuration (MOEC) for various coordination compounds can be drawn using VBT as

MOEC for $[Cr(NH_3)_4Cl_2]Cl$ is

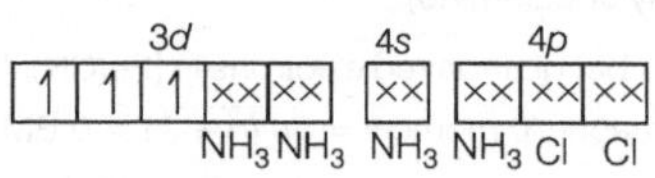

Number of unpaired electrons (n) = 3

Magnetic property = paramagnetic

Geometrical isomers of $[Cr(NH_3)_4Cl_2]^+$ are

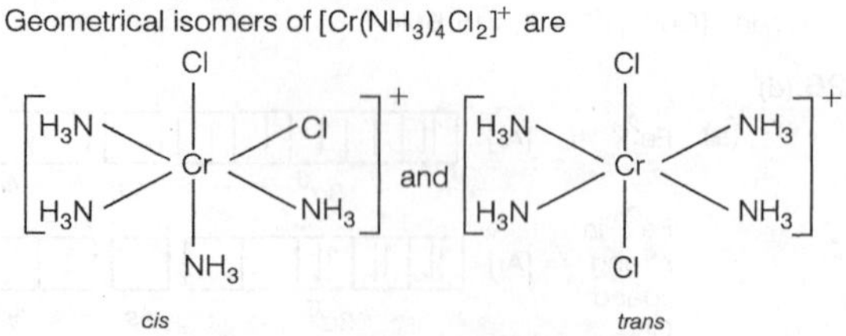

$n = 1$

Magnetic property = paramagnetic

Ionisation isomers of $[Ti(H_2O)_5Cl](NO_3)_2$ are

$[Ti(H_2O)_5Cl](NO_3)_2$ and $[Ti(H_2O)_5(NO_3)]Cl(NO_3)$

MOEC of $[Pt(en)(NH_3)Cl]NO_3$ is

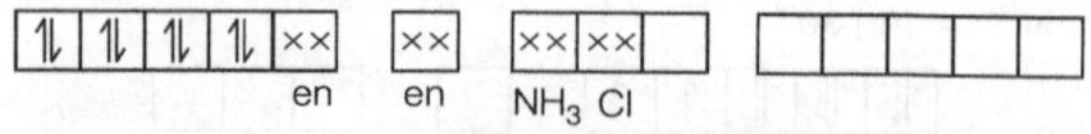

$n = 0$

Magnetic property = diamagnetic

Ionisation isomers are $[Pt(en)(NH_3)Cl]NO_3$ and $[Pt(en)NH_3(NO_3)]Cl$

MOEC of $[Co(NH_3)_4(NO_3)_2]NO_3$ is

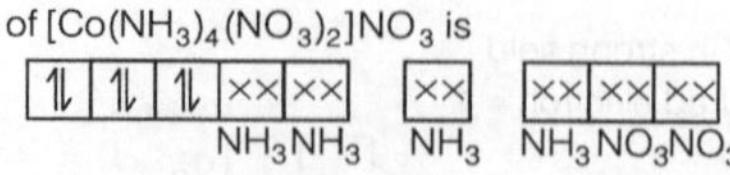

$n = 0$

Magnetic property = Diamagnetic

Geometrical isomers are

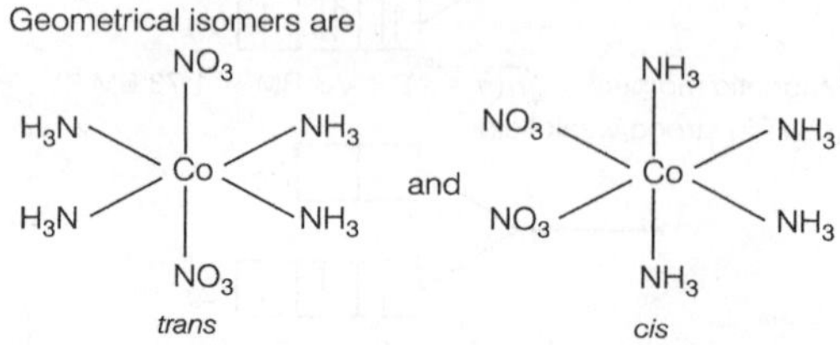

Thus, magnetic property and isomerism in given coordination compounds can be summarised as

(i) $[Cr(NH_3)_4Cl_2]Cl \rightarrow$ Paramagnetic and exhibits *cis-trans* isomerism (r)

(ii) $[Ti(H_2O)_5Cl](NO_3)_2 \rightarrow$ Paramagnetic and exhibits ionisation isomerism (p)

(iii) $[Pt(en)(NH_3)Cl]NO_3 \rightarrow$ Diamagnetic and exhibits ionisation isomerism (s)

(iv) $[Co(NH_3)_4(NO_3)_2]NO_3 \rightarrow$ Diamagnetic and exhibits *cis-trans* isomerism (q)

(i) → (r), (ii) → (p), (iii) → (s), (iv) → (q)

2. (8) **Plan** EDTA is a multidentate ligand as it can donate six pairs of electrons – two pair from the two nitrogen atoms and four pair from the four terminal oxygens of the —COO$^-$ groups.

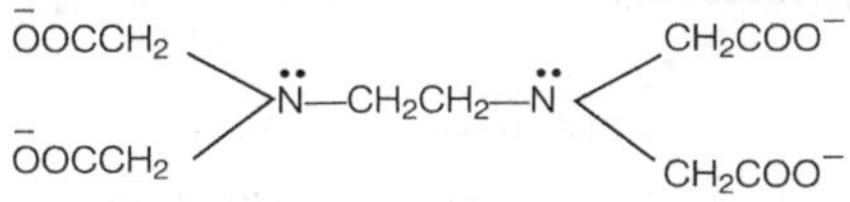

The structure of a chelate of a divalent Co^{2+} with EDTA is shown as

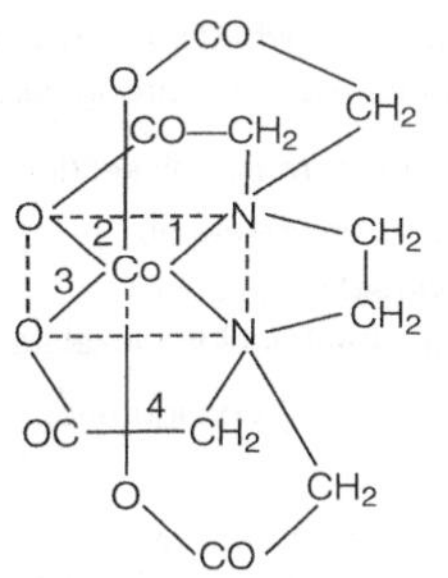

Each N has four N—Co—O bonds thus, total eight N—Co—O bonds.

3. (b) **Plan** Spin only magnetic moment have the formula $\sqrt{n(n+2)}$ BM, where n is the number of unpaired electrons. In presence of weak ligand (as H_2O, Cl^-, F^-), there is no pairing of electrons and electrons donated by ligands are filled in outer vacant orbitals.

In presence of strong ligand (as CN^-, CO, NH_3, en), electrons are paired and electrons from ligands are filled in available inner orbitals.

Complex	Atomic number of central atom	ON	EC	Unpaired electrons	Magnetic moment
P : $[FeF_6]^{3-}$ weak ligand	26	+3	$[Ar]3d^5$	5	$\sqrt{35}$ BM
Q : $[V(H_2O)_6]^{2+}$ weak ligand	23	+2	[Ar]	3	$\sqrt{15}$ BM
R : $[Fe(H_2O)_6]^{2+}$	26	+2	$[Ar]3d^6$	4	$\sqrt{24}$ BM

Thus, order of spin only magnetic moment = $Q < R < P$

4. (a) Ionisation isomers.

5. (d) First of all, the compound has complex positive part $[Co(H_2O)_4(NH_3)_2]^{3+}$ therefore, according to IUPAC conventions, positive part will be named first. Secondly, in writing name of complex, ligands are named first in alphabetical order irrespective of its charge, hence "ammine" will be written prior to "aqua".

Note *In alphabetical order, original name of ligands are considered not the initials of prefixes. Also, special precaution should be taken in spelling name of* NH_3 *ligand as it is ammine. Therefore, name of the complex* $[Co(H_2O)_4(NH_3)_2]Cl_3$ *is*

Diamminetetraaquacobalt (III) chloride.

6. (c) In the given complex, $NiCl_2\{P(C_2H_5)_2(C_6H_5)\}_2$ nickel is in + 2 oxidation state and the ground state electronic configuration of Ni^{2+} ion in free gaseous state is

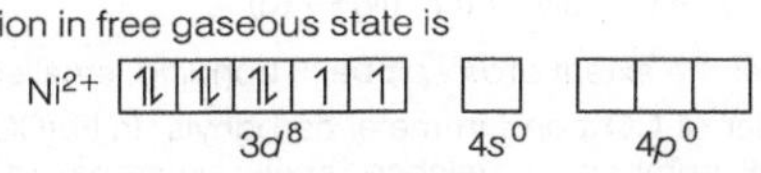

For the given four coordinated complex to be paramagnetic, it must possess unpaired electrons in the valence shell. To satisfy this condition, four lone pairs from the four ligands occupies the four sp^3 hybrid orbitals as

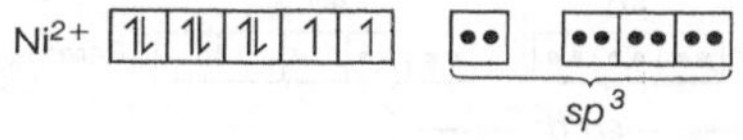

Therefore, geometry of paramagnetic complex must be tetrahedral. On the other hand, for complex to be diamagnetic, there should not be any unpaired electrons in the valence shell. This condition can be fullfilled by pairing electrons of $3d$ orbitals against Hund's rule as

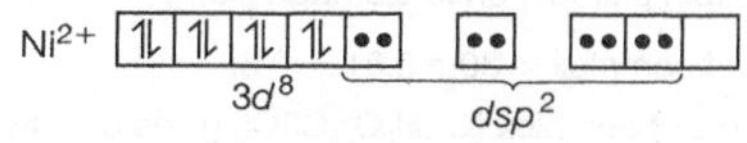

The above electronic arrangement gives dsp^2 hybridisation and therefore, square planar geometry of the complex.

7. (b) $Ni^{2+} + 4CN^- \longrightarrow [Ni(CN)_4]^{2-}$

Here, Ni^{2+} has d^8 configuration with CN^- as strong ligand.

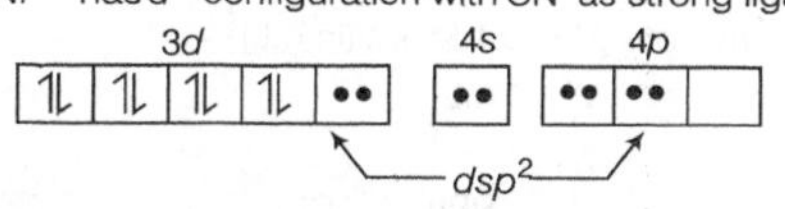

d^8 configuration in strong ligand field gives dsp^2 hybridisation, hence square planar geometry.

$$Ni^{2+} + 4Cl^- \longrightarrow [NiCl_4]^{2-}$$

Here, Ni^{2+} has d^8 configuration with Cl^- as weak ligand.

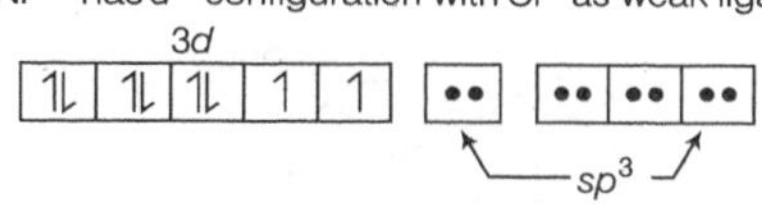

d^8 configuration in weak ligand field gives sp^3 hydridisation, hence tetrahedral geometry.

Ni^{2+} with H_2O forms $[Ni(H_2O)_6]^{2+}$ complex and H_2O is a weak ligand.

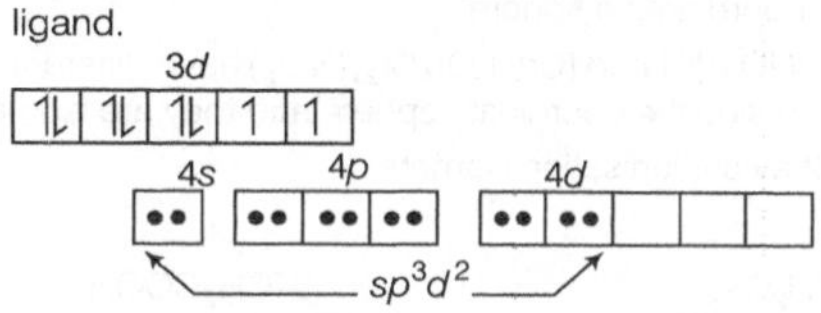

Therefore, $[Ni(H_2O)_6]^{2+}$ has octahedral geometry.

8. (c) For a diamagnetic complex, there should not be any unpaired electron in the valence shell of central metal.

In $K_3[Fe(CN)_6]$, Fe (III) has d^5-configuration (odd electrons), hence it is paramagnetic.

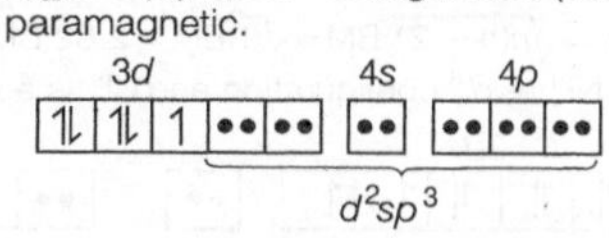

In $[Co(NH_3)_6]Cl_3$, Co (III) has d^6-configuration in a strong ligand field, hence all the electrons are paired and the complex is diamagnetic.

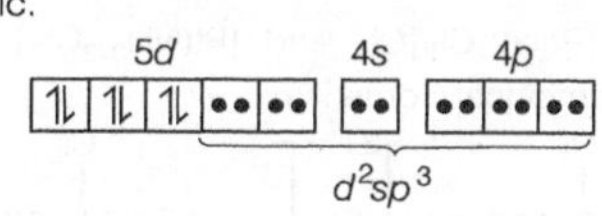

In $Na_3[Co(ox)_3]$, Co (III) has d^6-configuration and oxalate being a chelating ligand, very strong ligand and all the six electrons remains paired in lower t_{2g} level, diamagnetic.

In $[Ni(H_2O)_6]Cl_2$, Ni (II) has $3d^8$-configuration and H_2O is a weak ligand hence

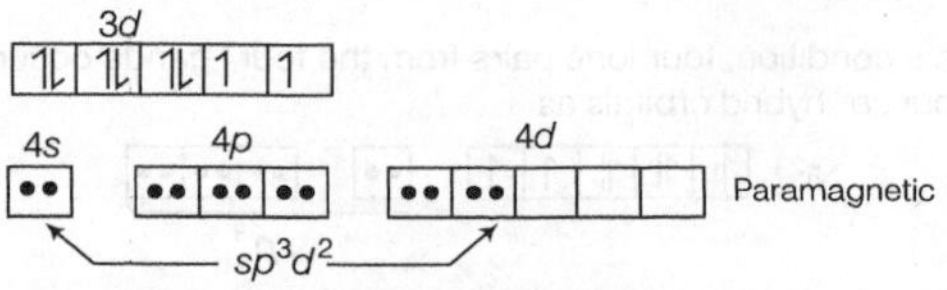

In $K_2[Pt(CN)_4]$, Pt(II) has d^8 configuration and CN^- is a strong ligand, hence all the eight electrons are spin-paired. Therefore, complex is diamagnetic.

In $[Zn(H_2O)_6](NO_3)_2$, Zn (II) has $3d^{10}$ configuration with all the ten electrons spin paired, hence diamagnetic.

9. (6) mmol of complex = 30 × 0.01 = 0.3

Also, 1 mole of complex $[Cr(H_2O)_5Cl]Cl_2$ gives only two moles of chloride ion when dissolved in solution.

$$[Cr(H_2O)_5Cl]Cl_2 \longrightarrow [Cr(H_2O)_5Cl]^{2+} + 2Cl^-$$

mmol of Cl^- ion produced from its 0.3 mmol = 0.6

Hence, 0.6 mmol of Ag^+ would be required for precipitation.

0.60 mmol of Ag^+ = 0.1 M × V(in mL)

V = 6 mL

10. (3)

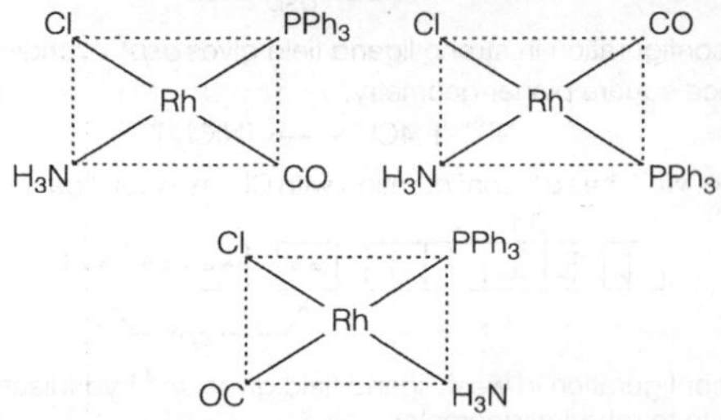

3 isomers.

11. (b) Ionisation isomers are the complexes that produces different ions in solution, i.e. they have ions interchanged inside and outside the coordination sphere.

$[Cr(H_2O)_4Cl(NO_2)]Cl$ and $[Cr(H_2O)_4Cl_2](NO_2)$ have different ions inside and outside the coordinate sphere and they are isomers. Therefore, they are ionisation isomers.

12. (c)

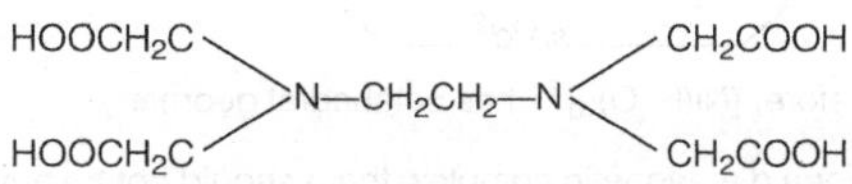

13. (b) Magnetic moment = 2.83 BM indicates that there is two unpaired electrons.

$$u = \sqrt{n(n+2)}\ \text{BM} = \sqrt{8}\ \text{BM} = 2.82\ \text{BM}$$

In $[NiCl_4]^{2-}$, Ni has d^8 configuration and Cl^- is a weak ligand :

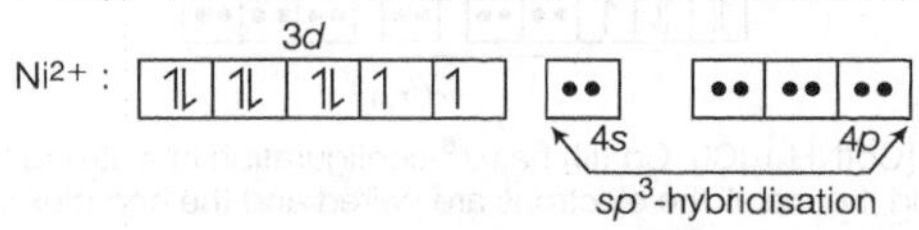

14. (c,d) Both $[Pt(en)_2Cl_2]Cl_2$ and $[Pt(NH_3)_2Cl_2]$ are capable of showing geometrical isomerism.

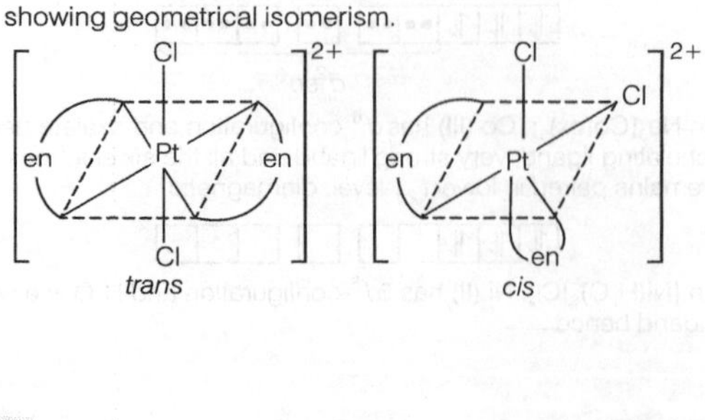

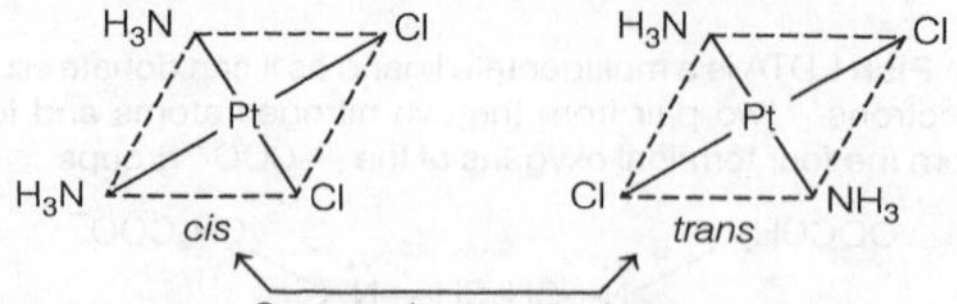

Square planar complex

15. (a) In $Cr(CO)_6$, $3d^6$, has no unpaired electrons, zero magnetic moment because CO is strong field ligand.

16. (b) In $Ni(CO)_4$, Ni is sp^3 hybridised while in $[Ni(CN)_4]^{2-}$, Ni^{2+} is dsp^2 hybridised because CO is strong field ligand.

17. (c) $[Ni(NH_3)_4]^{2+}$ = tetraamminenickel (II)

$[NiCl_4]^{2-}$ = tetrachloronickelate (II)

Cationic part is named first, hence tetraamminenickel (II)-tetrachloronickelate(II)

18. (c) CuF_2 : Cu^{2+} has $3d^9$ configuration, allowed d-d transition, hence, coloured.

19. (b) Both statements are true. However, axis of symmetry is not a criteria of optical isomerism. Optical inactivity of the two geometrical isomers of $[M(NH_3)_4Cl_2]$ is due to the presence of plane of symmetry.

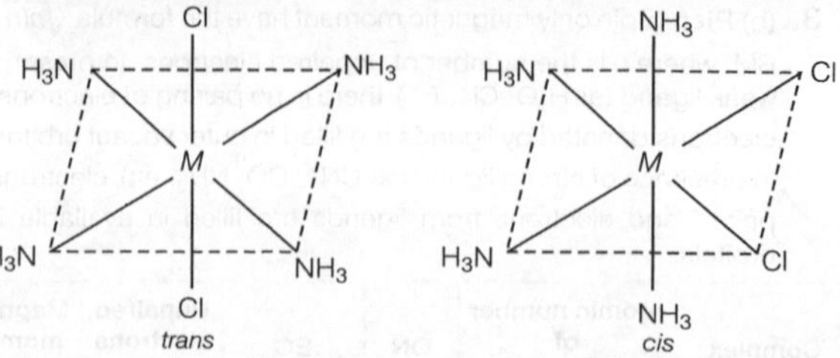

trans (has several planes of symmetry) cis (has several planes of symmetry)

20. (a) In the complex $[Fe(H_2O)_5NO]SO_4$, Fe is in + 1 oxidation state because NO is in +1 state. Also, NO is a strong ligand, complex has $3d^7$ configuration at Fe(I) as

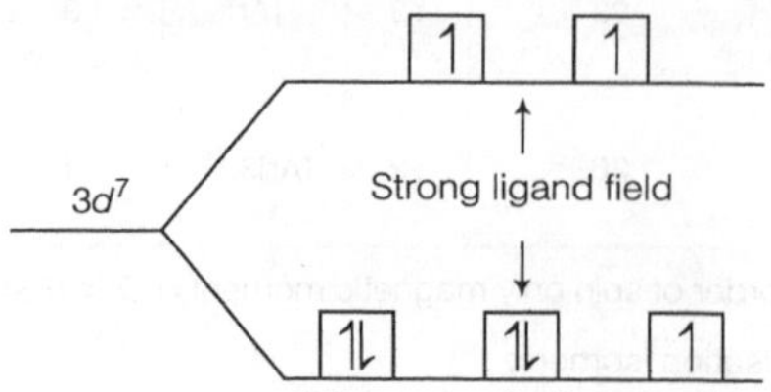

(Three unpaired electrons)

21. (b) (i) $[Co(NH_3)_4(H_2O)_2]Cl_2$: Co^{2+}, $3d^7$ show geometrical isomerism, paramagnetic.

(ii) $Pt(NH_3)_2Cl_2$: Pt^{2+} has d^8 configuration with all paired electrons.

Show geometrical isomerism, diamagnetic.

(iii) $[Co(H_2O)_5Cl]Cl$: Co^{2+}, $3d^7$

cannot show geometrical isomerism, paramagnetic.

(iv) $[Ni(H_2O)_6]Cl_2$: Ni^{2+}, $3d^8$, weak ligand, has two unpaired electrons. Paramagnetic but cannot show geometrical isomerism.

(i) → (r), (ii) → (p), (iii) → (s), (iv) → (q)

22. (b) Greater the extent of $d\pi$ - $p\pi$ back bonding, smaller will be the bond order of CO bond in metal carbonyls. In $Fe(CO)_5$, there is maximum number of valence shell electrons (d-electrons), greatest chances of $p\pi$-$d\pi$ back bonding, lowest bond order of CO bond.

23. (c) In CO, bond order = 3. In metal carbonyls like $Fe(CO)_5$, due to $d\pi - p\pi$ back bonding, bond order of CO decreases slightly therefore, bond length increases slightly.

24. (a) $A = K_2[Ni(CN)_4]$; $B = K_2[NiCl_4]$

A : Potassium tetracyanonickelate (II)

B : Potassium tetrachloronickelate (II)

25. (c) A is diamagnetic, square planar complex because of strong ligand field of CN^-.

$Ni(CN)_4^{2-}(Ni^{2+})$:

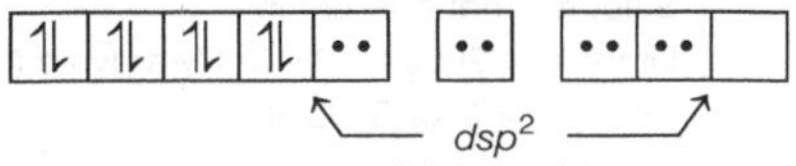

(Diamagnetic)

B is paramagnetic, tetrahedral complex because of weak ligand field of Cl^-.

$NiCl_4^{2-}(Ni^{2+})$:

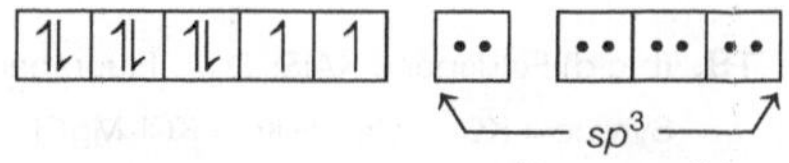

(Paramagnetic)

26. (a) Described in sol.25, A has dsp^2 hybridisation while B has sp^3 hybridisation of Ni.

27. (a) $[Co(NH_3)_4Br_2]Cl$ and $[Co(NH_3)_4BrCl]Br$ are ionisation isomers.

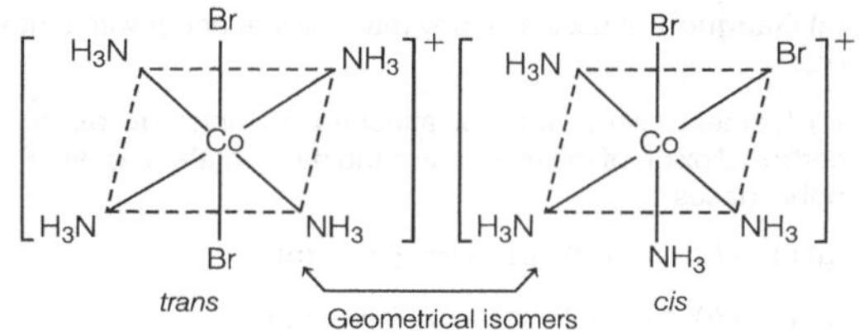

Geometrical isomers

4. Metallurgy

DPP-1 Introduction to Metallurgy and Ores of Some Metals

1. Minerals are naturally occurring chemical compounds of a metal, which may be associated with more impurities. But ore is a chosen mineral of a metal, from which metal is extracted profiteably on a large scale, in pure form. Thus, all ores are minerals but all minerals are not ores.

e.g. the ore of Fe is haematite, while the minerals having Fe may be magnetite, iron pyrites, siderite, limonite etc.

2. The cation of these metals i.e. Cu^+, Zn^{2+}, Ag^+ and Hg^{2+} have pseudo inert gas configuration and Pb^{2+} has inert gas configuration. These have high polarising power and hence can easily polarise the bigger sulphide ion than the smaller oxide ion. As a result sulphides of these metals are even more stable than the oxides.

3. **Ruby** It is the red gemstone. It is aluminium oxide (Al_2O_3) containing a small percentage of chromium oxide (Cr_2O_3).

Sapphire It is a typically blue gemstone variety of the mineral corundum, an aluminium oxide (α-Al_2O_3) containing trace amounts of elements such as iron, titanium.

Emerald It is a gemstone and a variety of the mineral beryl ($Be_3Al_2(SiO_3)_6$) coloured green by trace amounts of chromium and sometimes vanadium. It is the main ore of beryllium.

4. The metals which are below hydrogen in electrochemical series i.e. which are less electropositive metals can occur in native state. These are not readily attacked by oxygen, moisture, CO_2 etc.

e.g. Au, Pt, Pd and Ag.

5. Abundance of elements in

(i) Earth crust : O > Si > Al > Fe > Ca > Na > K > Mg

(ii) Human body : O >C >H >N > Ca > P >S

(iii) Sea water : O > H > Cl > Na > Mg >S > Ca > K

6. (b) Oxygen is highest abundant non-metal, while aluminium is the highest abundant non-metal in the earth crust.

7. (c) Chalcopyrite is $CuFeS_2$.

8. (d) Cryolite (Na_3AlF_6) is used in the manufacture of aluminium. Electrolysis of (Al_2O_3) dissolved in molten Na_3AlF_6.

9. (c) In chromite ore is ($FeO \cdot Cr_2O_3$)

Let the oxidation state of Fe be x.

$$\therefore \quad x - 2 + 2 \times 3 + 3 \times (-2) = 0$$

$$x - 2 + 6 - 6 = 0 \Rightarrow x = +2$$

Let the oxidation state of Cr be y

$$\therefore \quad 2y + 2 - 2 + 3 \times (-2) = 0$$

$$2y - 6 = 0$$

$$y = 3$$

Hence, oxidation state of Fe and Cr are +2 and +3 respectively.

10. (a) The sulphide ore are common for the metals Ag, Cu and Pb due to pseudo inert gas and inert gas configuration. The common ores are Ag_2S (silver glance), Cu_2S (chalcocite) and PbS (lead glance).

11. (d) In haematite (Fe_2O_3). Let the oxidation state of Fe be x.

$$\therefore \quad 2x - 6 = 0 \Rightarrow x = +3$$

It is one of the number of iron oxides the others being iron (II) oxide (FeO) which is rare and iron (III) oxide (Fe_2O_3) also known as haematite. It occurs in nature as the magnetite. It contains both Fe^{2+} and Fe^{3+} ions and formulated as $FeO \cdot Fe_2O_3$.

12. (b) Argentite (Ag_2S) is sulphide and cuprite (Cu_2O) is oxide.

13. (c) A mineral is called an ore if a metal can be profiteably extracted from it. e.g. Aluminium occur in earth's crust in the form of two minerals i.e. bauxite ($Al_2O_3 \cdot 2H_2O$) and clay ($Al_2O_3 \cdot 2SiO_2 \cdot 2H_2O$). Out of these two minerals aluminium can be conveniently and economically extracted from bauxite while no easy and cheap method is so far available for the extraction of aluminium from clay. Therefore, the ore of aluminium is bauxite.

14. (b) Siderite is $FeCO_3$.

15. (d) Crystalline sodium sulphate $Na_2SO_4 \cdot 10H_2O$.

16. (a,b,d) Anglesite = $PbSO_4$.

17. (a,b,c,d) Calamine ($ZnCO_3$), Siderite ($FeCO_3$), Malachite ($CuCO_3 \cdot Cu(OH)_2$), Calcite ($CaCO_3$), Azurite ($2CuCO_3 \cdot Cu(OH)_2$), Witherite ($BaCO_3$).

18. (b,c) Plaster of Paris ($CaSO_4 \cdot \frac{1}{2}H_2O$) is obtained by heating gypsum ($CuSO_4 \cdot 2H_2O$) litharge is lead (II) oxide (PbO). Both of these are not mineral.

19. (b,c,d) Feldspar = $KAlSi_3O_8$, Fluorspar = CaF_2
Sylvine = KCl, Carnallite = $KCl \cdot MgCl_2 \cdot 6H_2O$

20. (c,d) Ruby silver = Pyrargyrite ($3Ag_2S \cdot Sb_2S_3$ or Ag_3SbS_3),
Horn silver = AgCl
German silver = An alloy of Cu + Zn + Ni and
Quick silver = Mercury.

21. (a) **Gangue** The useless stony material that occur with a metallic ore.

22. (c) Elements which are not attached by moisture, oxygen and carbon dioxide of the air occur in the native state. e.g. Au, Pt, and noble gases.

23. (a) (i) → (t), (ii) → (s), (iii) → (q), (iv) → (p)

24. (a) (i) → (r), (ii) → (s), (iii) → (p), (iv) → (q,t)

25. (5) The carbonate ores are siderite, azurite, calamine, calcite, magnesite.

26. (3) Fluorapatite is $3Ca_3(PO_4)_2 \cdot CaF_2$ or $Ca_9(PO_4)_6CaF_2$.

27. (5) The oxide ores are pyrolusite, rutile, cassiterite, haematite and cuprite.

28. (5) **Silicon is present in** Beryl = $3BeO \cdot Al_2O_3 \cdot 6SiO_2$;
Mica = $K_2O \cdot 3Al_2O_3 \cdot 6SiO_2 \cdot 2H_2O$
Zn_2SiO_4 = Willemite; Feldspar = $KAlSi_3O_8$.
China clay = $Al_2O_3 \cdot 2SiO_2 \cdot 2H_2O$ (or) $[Al_2(OH)_4 Si_2O_5]$
Silicon is not present in
Limonite = $2Fe_2O_3 \cdot 3H_2O$; Pyrargyrite (Ruby silver) = $3Ag_2S_3Sb_2S_3$.

29. (5) Chromite ($Fe_2Cr_2O_4$), Limonite ($Fe_2O_3 \cdot 3H_2O$), Cryolite (Na_3AlF_6), Caliche ($NaIO_3$), Diaspore ($Al_2O_3 \cdot H_2O$).

30. (c) Aluminium cannot be extracted from clay profitably.

DPP-2 Metallurgy Process of Concentration of Ores

1. The three steps involved in the extraction are :
(1) Concentration of the ore to remove impurities.
(2) Reduction of the ore to get the metal.
(3) Purification of the ore.

2. The ore is sulphide ore. Hence, it is concentrated by froth floatation process.

3. The term 'Gangue" is used for all the earthly impurities associated with the ore of the metal. These need to be removed before the extraction step. e.g. In iron ore, sand SiO_2 is the main gangue.
The term 'Flux' refers to the compound added during extraction, which reacts with a non-fusible (high melting point) gangue and forms a fusible product called slag.
This can then be removed by simple physical method. e.g. In the blast furnace for extraction of iron, the flux added is limestone $CaCO_3$. This decomposes to give CaO (a basic oxide) which reacts with sand (gangue - an acidic oxide) and forms a fusible slag $CaSiO_3$.
The term slag refers to the easily fusible product formed between gangue and flux.

$$\underset{\text{Gangue}}{SiO_2} + \underset{\text{Flux}}{CaO} \longrightarrow \underset{\text{Slag}}{CaSiO_3}$$

4. NaCN combines with ZnS to form a complex $Na_2[Zn(CN)_4]$ on the surface of ZnS and this prevents it from forming the froth.

$$ZnS + 4NaCN \longrightarrow Na_2[Zn(CN)_4] + Na_2S$$

5. Flux is a substance which combines with gangue (earthly impurities) still present in roasted ore to form a fusible product called slag.
(i) **Acid fluxes** For basic impurities like those of lime or oxides of iron present in the ore, acidic fluxes like silica (SiO_2), borax ($Na_2B_4O_7 \cdot 10H_2O$) are used.
e.g.

$$\underset{\text{Gangue}}{FeO} + \underset{\text{Flux}}{SiO_2} \longrightarrow \underset{\text{Slag}}{FeSiO_3}$$

(ii) **Basic fluxes** For acidic impurities like silica, phosphorus pentaoxide (P_4O_{10}) present in the ore, basic fluxes like lime stone ($CaCO_3$) magnesite ($MgCO_3$) haematite (Fe_3O_4) are used.

$$\underset{\text{Gangue}}{SiO_2} + \underset{\text{Flux}}{CaCO_3} \longrightarrow \underset{\text{Slag}}{CaSiO_3} + CO_2\uparrow$$

6. (b) $CaCO_3 \longrightarrow CaO + CO_2$
$CaO + SiO_2 \longrightarrow CaSiO_3$

7. (c) This method is based upon the fact that the surface of sulphide over is preferentially wetted by oils while that of gangue is preferentially wetted by water.

8. (d) In this process the finely divided powdered argentite or the native silver or gold is treated with a dilute solution (0.5%) of sodium or potassium cyanide while a current of air is continuously passed. As a result silver pass into solution forming their respective soluble complex cyanides while the impurities remain unaffected which are filtered off.

$$4Ag + 8NaCN + O_2 + 2H_2O \longrightarrow 4Na[Ag(CN)_2] + 4NaOH$$

or

$$Ag_2S + 4NaCN \longrightarrow \underset{\substack{\text{Sodium} \\ \text{dicyanoargentate(I)}}}{2Na[Ag(CN)_2]} + Na_2S$$

9. (a) The ore is crushed into a fine powder and mixed with water to form suspension.

10. (b) Electromagnetic separation method of concentration is employed when either the ore or the impurities associated with it are magnetic in nature. Tinstone or cassiterite (SnO_2) an ore of tin being non-magnetic can be separated from magnetic impurities ($FeWO_4$).

11. (a) $SiO_2 + CaO \longrightarrow CaSiO_3$

$$\underset{\text{Acidic flux}}{SiO_2} + \underset{\substack{\text{Basic} \\ \text{impurities}}}{FeO} \longrightarrow \underset{\text{Fusible slag}}{FeSiO_3}$$

12. (b) The mineral particle becomes wet by oils while the gangue particles by water. A rotating paddle agitates the mixtures and draws air in it. As a result froth is formed which carrier the mineral particles. The froth is light and is skimmed of. If is then dried for recovery of the ore particles.

13. (b) **Leaching** This process consists in treating the powdered ore with a suitable reagent (such as acids, bases or other chemicals) which can selectively dissolves the ore but not the impuritier. It is used to extract silver metals.

14. (a) $SiO_2 + CaO \longrightarrow CaSiO_3$

$$\underset{\substack{\text{Acidic} \\ \text{flux}}}{SiO_2} + \underset{\substack{\text{Basic} \\ \text{impurities}}}{FeO} \longrightarrow \underset{\text{Fusible slag}}{FeSiO_3}$$

15. (b) The ore particles preferentially wetted by the oil become lighter and thus rise to the surface along with the froth while the gangue particle wetted by wated become heavier and thus settle down at the bottom.

16. (a,b,c) Cinnabar, Galena, zinc blende are concetrated by froth floatation method.

17. (a,b) They are light and water repellent.

18. (a,b,c) Hoope's method is used to obtain aluminium of very high purity. The metal obtained in the Hall-Heroult process is about 99% pure and for most purposes it is taken as pure metal. However further purification of aluminium can be carried out by the Hoope's process. This is an electrolytic process.

19. (a,c) The common impurities present in bauxite are Fe_2O_3 and SiO_2.

20. (a,b,d) Flux is not involved in froth floatation process.

21. (a) The process by which lighter earthly particles are free from the heavier ore particles by washing with water is called levigation.

22. (b) In the Baeyer process Al_2O_3 is obtained from bauxite ore by treating the powered ore with a strong solution of NaOH when alumina (Al_2O_3) dissolves forming sodium *meta*-aluminate leaving behind impurities.

$$Al_2O_3 \cdot 2H_2O(s) + 2NaOH(aq) + H_2O(l) \longrightarrow \underset{\text{Sod. } meta \text{ aluminate}}{2Na[Al(OH)_4](aq) \text{ or } 2NAlO_2(aq) + 4H_2O(l)}$$

23. (a) (i) $\rightarrow$ (r), (ii) $\rightarrow$ (p), (iii) $\rightarrow$ (q), (iv) $\rightarrow$ (s)

24. (c) (i) $\rightarrow$ (q), (ii) $\rightarrow$ (r,t), (iii) $\rightarrow$ (p), (iv) $\rightarrow$ (s)

25. (3) Galena, zinc blende and copper iron pyrites are concentrated by froth floatation method.

26. (4) Bauxite, horn silver, native gold, tinstone.

27. (0) In this reaction oxidation state of silver is zero.

28. (3) Pitche blende, tinstone, chromite are purified by levigation.

29. (4) Wolframite, chromite, magnetite, rutile (contain iron impurities).

30. (b) Because cresol and aniline are froth stabilisers, but only ethyl xanthate, pine oil which enhance non-wettability of ore particles.

DPP-3 **Extraction of Crude Metals from Concentrated Ores**

1.

	Calcination		Roasting
1.	It is heating of the ore, in the absence of air to decompose it or drive volatile matter.	1.	It is heating of the ore in the presence of air to oxidise it.
2.	It occurs at temperature lower than the melting point of the ore.	2.	It occurs at higher temperatures, higher than the melting point of the ore.
	$ZnCO_3 \xrightarrow[\Delta]{\text{Calcination}} ZnO + CO_2\uparrow$		$2ZnS + 3O_2 \xrightarrow[\Delta]{\text{Roasting}} 2ZnO + 2SO_2\uparrow$ $4FeS + 7O_2 \xrightarrow{\text{Roasting}} 2Fe_2O_3 + 4SO_2\uparrow$

In both cases, the ore becomes porous for easy reduction.

2. (i) Roasting is a chemical process where a sulphide ore is strongly heated in a current of O_2 to oxidise the ore. Therefore, it is an oxidation process.

$$2ZnS + 3O_2 \xrightarrow{\text{Roasting}} 2ZnO + 2SO_2\uparrow$$

(ii) SO_2 gas is the byproduct formed and it is used for the manufacture of sulphuric acid by the contact process.

3. Al has greater affinity for oxygen. It acts as reducing agent when the metal having high melting point is to be extracted from its oxide. $Cr_2O_3 + 2Al \longrightarrow 2Cr + Al_2O_3$

4. Copper is comparatively less reactive metal as indicated by its SRP value i.e.

($E^\circ_{(Cu^{2+}/Cu)} = 0.3V$). It can be displaced from solutions of Cu^{2+} -ions by more active metals with E° value less than copper like $Zn(E^\circ_{(Zn/Zn^{2+})} = -0.76$ V).

In contrast to displace zinc from solutions of Zn^{2+} ions , a more reactive metal than zinc is required. But the more active metal readily react with water liberating hydrogen gas. Hence, copper can be extracted by hydrometallurgy but not zinc.

5. This process is used in the reduction of metal oxides in which the metals have very high melting point and those oxides for which carbon is not satisfactory as reducing agent.

A mixture of concentrated oxide ore and aluminium powder commonly called thermite , is taken in a steel crucible placed in a bed of sand.

The reaction is started by the use of an ignition mixture containing Mg powder and BaO_2.

$$Cr_2O_3 + 2Al \longrightarrow Al_2O_3 + 2Cr$$
$$3Mn_3O_4 + 8Al \longrightarrow 9Mn + 4Al_2O_3$$

Large amount of heat energy is released during reduction which fuses both the alumina and the metal.

6. (c) **Cyanide process** The extraction of gold from its ores by dissolving the gold in a solution of potassium cyanide, KCN. The resulting potassium aurocyanide, $KAu(CN)_2$, with zinc, filtering off, melting down and cupelling the metal.

$$4Au + 8KCN + O_2 + 2H_2O \longrightarrow \underset{\substack{\text{Pot. dicyanoaurate(I)} \\ \text{(soluble complex)}}}{4K[Au(CN)_2]} + 4KOH$$

7. (a) Cryolite (Na_3AlF_6) is added to increase the electrical conductivity of pure aluminium.

8. (d) $2Na[Ag(CN)_2] + Zn \longrightarrow Na_2[Zn(CN)_4] + 2Ag\downarrow$

9. (d) Sometimes the sulphide ore may be oxidised to sulphate which is then reduced to the free metal by the excess of the sulphide ore.

$$\underset{\text{Copper pyrites}}{2Cu_2S} + 3O_2 \longrightarrow 3Cu_2O + 2SO_2$$

$$2Cu_2O + Cu_2S \longrightarrow 6Cu + SO_2$$

10. (b) The nitrate of $AgNO_3$ leaves metallic globule on strongly heating.

11. (b) When the gold pass into the solution of sodium or potassium cyanide forming the respective soluble complex cyanides while the impurities remain unaffected which are filtered off.

$$4Au + 8KCN + O_2 + 2H_2O \longrightarrow \underset{\substack{\text{Pot. dicyanoaurate(I)} \\ \text{(Soluble complex)}}}{4K[Au(CN)_2]} + 4KOH$$

12. (b) The suphide ores of copper are heated in reverberatory furnace. If the ore contains iron. It is mined with silica before heating. Iron oxide slags of as iron silicate and copper produced in the form of copper matte, which contain Cu_2S and FeS.

$$FeO + SiO_2 \longrightarrow \underset{\text{(Slag)}}{FeSiO_3}$$

13. (a) $4Au(s) + 8CN^-(aq) + 2H_2O(aq) \xrightarrow{O_2} \underset{(X)}{4[Au(CN)_2]^-} + 4OH^-(aq)$

$2[Au(CN)_2]^-(aq) + Zn(s) \longrightarrow \underset{(Y)}{[Zn(CN)_4]^{2-}}(aq) + 2Au(s)$

14. (c) Iron is produced by the following reaction.

$$FeO + C \longrightarrow Fe + CO$$

15. (b) The reactions involve in roasting process.

$2ZnS + 3O_2 \longrightarrow 2ZnO + 2SO_2$

$2PbS + 3O_2 \longrightarrow 2PbO + 2SO_2$

$2Cu_2S + 3O_2 \longrightarrow 2Cu_2O + 2SO_2$

The SO_2 produced is utilised for manufacturing H_2SO_4.

16. (a,c,d) $2\,2ZnS + 3O_2 \longrightarrow 2ZnO + 2SO_2$,

$2Cu_2S + 3O_2 \longrightarrow 2Cu_2O + 2SO_2$

$Ag_2S + 2NaCl + O_2 \longrightarrow 2AgCl + Na_2SO_4$

17. (a,b,c) Gravity separation, smelting, pudding process.

18. (a,b,c) Pyrometallurgy = Fe, Zn and Cu
Electrometallurgy = Al, Mg and Na
Hydrometallurgy = Ag and Au

19. (a,b) Self reduction or auto-reduction or self electronation.

(i) $\underset{\text{Copper pyrites}}{2Cu_2S} + 3O_2 \longrightarrow 2Cu_2O + 2SO$

$2Cu_2O + Cu_2S \longrightarrow 6\,Cu + SO_2$

(ii) $\underset{\text{Galena}}{2PbS} + 3O_2 \longrightarrow 2PbO + 2SO_2$

$2PbO + PbS \longrightarrow 3Pb + SO_2$

(iii) $HgS(s) + O_2(g) \longrightarrow Hg(g) + SO_2(g)$ or

$2HgS(s) + 3O_2(g) \longrightarrow 2HgO(s) + 2SO_2(g)$

20. (b,c,d) The process of reduction of metal oxide to the metal with the help of aluminium powder as the reducing agent is called Goldschmidt thermite process.

$$2Al + Fe_2O_3 \longrightarrow Al_2O_3 + 2Fe$$

21. (a) Copper pyrites is $CuFeS_2$. Copper can be extracted from this by roasting of sulphide partially and reduction.

22. (d) Carbon (coal or coke) reduces the metal oxide to the free metal. e.g.

$ZnO + C \longrightarrow Zn + CO\uparrow$

$PbO + C \longrightarrow Pb + CO\uparrow$

$Fe_2O_3 + 3C \longrightarrow 2Fe + 3CO\uparrow$

The carbon monoxide obtained by the incomplete combustion of carbon can also bring about the reduction of metal oxides to free metal.

e.g. $FeO + CO \xrightarrow{1123K} Fe + CO_2\uparrow$

$Fe_2O_3 + CO \xrightarrow{823\,K} 2FeO + CO\uparrow$

23. (b) (i) → (q), (ii) → (s), (iii) → (r), (iv) → (p)

Low electropositive metal oxides like Fe_2O_3 and ZnO can be reduced by CO. Cr_2O_3 cannot be reduced by CO or C because of their high melting point, some metal sulphides self reduced during roasting like Cu_2S, PbS and HgS.

24. (c) (i) → (r), (ii) → (p,t), (iii) → (q), (iv) → (s)

25. (4) **Pyrometallurgy** It is a branch of extractive metallurgy. It involves the thermal treatment of minerals and metallurgical ores and concentrates to bring about physical and chemical transformation in the materials to enable recovery of valuable metals.

It include the extraction of oxide of less reactive elements like, Pb, Sn, Zn and Cu.

26. (6) Pb, Ag, Au, Zn, Cu, Al, Electrometallurgy is the process of extraction of metals by electrolysis of their fused salt.

27. (8) $4Au + 8CN^- + 2H_2O + O_2 \longrightarrow 4[Au(CN)_2]^- + 4OH^-$

28. (3) The molar ratio of Al and F_2 is 2 : 3.

29. (3) In thermite mixture contains 3 parts of Fe_2O_3 and one part of Al.

30. (c) Cast iron is obtained by remelting pig iron with scrap iron and coke using hot air blast.

In the extraction of silver. silver is extracted as anionic complex $[Ag(CN)_2]^-$.

DPP-4 Thermodynamic Principles of Metallurgy, Ellingham's Diagram and Its Applications

1. Ellingham diagrams are the particular graphical form of the principle that the thermodynamic feasibility of a reaction depends on the sign of ΔG, the Gibbs free energy change, which is equal to $\Delta H - T\Delta S$, where ΔH is the enthalpy change and ΔS is the entropy change. i.e. Ellingham diagrams are the plots between $\Delta_f G°$ of the formation of oxides of elements and temperature.

It provides a sound idea about selecting a reducing agent in the reduction of oxides. Such diagrams help in predicting the feasibility of a thermal reduction of an ore. ΔG must be negative at a given temperature for a reaction to be feasible.

Limitations

It does not take kinetics of reduction into consideration i.e. how fast reduction will occur cannot be determined.

2. (i) The reduction equations of C and CO are as follows :

$2C + O_2 \longrightarrow 2CO$

$C + O_2 \longrightarrow CO_2$

$2CO + O_2 \longrightarrow 2CO_2$

From the Ellingham diagram, it is clear that at the temperature 673K, $\Delta G_f°$ of CO_2 from CO is more negative than the formation of CO or CO_2 from carbon.

Hence, at 673 K CO is better reducing agent than C.

(ii) From the Ellingham diagram, it is clear that the free energy of formation of CO from C is lower at above 1120 K while that of CO_2 from carbon is lower above 1323 K than free energy of formation of ZnO. However the free energy of formation of CO_2 from CO is always higher than that of ZnO.Hence, C is a better reducing agent of ZnO.

3. (i) The temperature intersection of the Al $\longrightarrow Al_2O_3$ and Mg $\longrightarrow$ MgO curves in Ellingham diagram is 1665 K. At the point of intersection of the Al_2O_3 and MgO curves, the $\Delta G°$ becomes zero for the reaction : $\Delta G_f°$ of Al_2O_3 at temperature below 1665 K is less negative than $\Delta G_f°$ of MgO. Thus, below 1665 K Mg can reduce Al_2O_3 to Al.

(ii) Above 1665K this temperature $\Delta GG_f°$ of Al_2O_3 is more negative than $\Delta G_f°$ of MgO. Hence, above 1665K Al can reduce MgO into Mg. But the process will be uneconomical.

4. The entropy is higher if the metal is in liquid state than when it is in solid state. The value of entropy change (ΔS) of the reduction process is more on +ve side when the metal formed is in liquid state and the metal oxide being reduced is in solid state. Thus, the value of $\Delta G°$ becomes more on negative side and the reduction becomes easier.

5. The Ellingham curve for the reaction
$2C(s) + O_2(g) \longrightarrow 2CO(g)$ slopes down and falls below the curves for all the metals. Hence, carbon can normally act as a reducing agent for all metal oxides at very high temperatures. But chromium formed at these temperatures reacts with carbon to form its carbide, which gives undesirable properties to the chromium metal obtained. Hence, for high temperature reduction of chromic oxide, carbon cannot be used.

6. (b) Ellingham diagram represents a graph of $\Delta G°$ *vs* T.

7. (c) In the case of carbon reacting with dioxygen.

$$C + \frac{1}{2}O_2 \longrightarrow CO$$

two volumes of CO for every one volume of dioxygen used. Thus, ΔS is positive and hence ΔG becomes increasilngly negative as temperature increases. Consequently the line on Ellingham diagram slope downwards.

8. (a) ΔG_f° is less negative for (Cr_2O_3).

9. (d) Because the change in free energy i.e. ΔG is highly negative for this reaction.

10. (a) $3Fe_2O_3 + CO \longrightarrow 2Fe_3O_4 + CO_2$

$$Fe_3O_4 + CO \longrightarrow 3FeO + CO_2$$
$$FeO + CO \longrightarrow Fe + CO_2$$

11. (c) Because the $\Delta G°$ of the formation of CO is more negative than that of ZnO at 10000°C.

12. (d) Because $\Delta G°$ for reductions of rest three oxides [Al_2O_3, ZnO and MgO] are less negative than that of Cu_2O at lower temperature.

13. (b) In Hall -Heroult process the following reactions occur

$$3C + 2Al_2O_3 \longrightarrow 4Al + 3CO_2$$

$$\Delta G° = 3\Delta G_f^\circ(CO_2) - 2\Delta G_f^\circ(Al_2O_3) = 3(-394) - 2(-1520) = 1858 \text{ kJ}$$

$$\Delta G° = -nFE° \quad \text{or} \quad = -E° = \frac{\Delta G°}{nF} = \frac{1858 \times 1000}{12 \times 96500} = 1.60 \text{ V}$$

14. (b) $3Ca + Al_2O_3 \longrightarrow 2Al + 3CaO$

$$\Delta G° = 3 \times \Delta G_f^\circ(CaO) - \Delta G_f^\circ(Al_2O_3)$$
$$= 3 \times (-604.2) - (-1582)$$
$$= -1812 + 1582 = -230 \text{ kJ}$$

$\Delta G°$ is negative. Hence, this reaction is feasible.

15. (a,d) Calcined ore of haematite (Fe_2O_3) and zincite (ZnO) can be reduced by carbon.

16. (c,d) According to Ellingham diagram, ΔG_f° for Al_2O_3 and MgO is more negative than Cr_2O_3.

17. (a,b)

18. (b,c) $3CaO + 2Al \longrightarrow Al_2O_3 + 3Ca$

$$\Delta G° = \Delta G_f^\circ(Al_2O_3) - 3\Delta G_f^\circ(CaO)$$
$$= -1582 - 3(-604)$$
$$= -1582 + 1812 = +230 \text{ kJ}$$

$\Delta G°$ is positive. Hence this reaction is not spontaneous.

19. (c,d) In the extraction of iron, at low temperature CO is effective and at high temperature carbon is more effective.

20. (b) The temperature should be above the intersecting temperature .

21. (d) At 1100° C the $\Delta G°$ for
$ZnO + Mg \longrightarrow Zn + MgO$, is more negative then other three reactions.

22. (a) (i) → (q), (ii) → (r), (iii) → (s), (iv) → (p)

23. (b) (i) → (r,t), (ii) → (p), (iii) → (s), (iv) → (q)

24. (0) $\Delta G = \Delta H - T\Delta S$
At eqbm. ($\Delta G = 0$)

$$0 = \Delta H - T\Delta S \Rightarrow \Delta H = T\Delta S$$

25. (6)

26. (6) 6 electrons from 2 Al-atoms participate.

27. (3) PbO, ZnO and SnO_2 are reduced by carbon.

28. (2) $\Delta G° = -nFE°$, n = number of electrons involved

29. (d) Carbon is a good reducing agent for oxides. It is a poor reducing agent for sulphides. The reason why carbon reduces so many oxides at elevated temperature is that the $\Delta G°/T$ line for CO has a negative slope. There is no compound as analogous to CO with a steep negative $\Delta G°/T$ line. Thus sulphides are normally roasted in air to form oxides before reducing with carbon

$$2MS + 3O_2 \longrightarrow 2MO + 2SO_2$$

30. (d) Below 983 K iron having higher affinity towards oxygen than carbon, carbon cannot reduce Fe_2O_3 to Fe.

DPP-5 **Refining of Crude Metals**

1. Molten NaCl on electrolysis (Down's process) produce Na metal at cathode and Cl_2 gas at anode. NaCl aqueous solution on electrolysis (Nelson's process) produce hydrogen at cathode and chlorine at anode.

$$2NaCl(l) \xrightarrow{\text{Electrolysis}} 2Na(s) + Cl_2(g)$$

$$2NaCl(aq) + 2H_2O \xrightarrow{\text{Electrolysis}} H_2(g) + Cl_2(g) + 2NaOH(aq)$$

2. Zone refining is based on the difference in solubility of impurities in molten and solid state of the metal. A movable heater is fitted around the impure metal rod and slowly moved across the rod.
The metal melts at the point of heating and as the heater moves on, pure metal crystallizes while the impurities pass on the adjacent melted zone. e.g. Si, Ge, Ga etc., are purified.

3. (i) x = 330 – 350 K and y = 450 – 470 K
(ii) a = 523 K and b = 1700 K

4. (i) $AgNO_3$ electrolysis using Pt electrodes produce silver at cathode and oxygen at anode.
(ii) $AgNO_3$ electrolysis using Ag electrodes cause dissolution of silver metal at anode and deposition of silver at cathode.

5. The impure metal is heated in a furnace and a blast of compressed air is blown through the molten mass. The impurities get oxidised. e.g. the molten pig iron is taken in a Bessemer convertor and compressed air is passed which oxidises the impurities.

$$2Mn + O_2 \rightarrow 2MnO$$
$$Si + O_2 \rightarrow SiO_2$$
$$2C + O_2 \rightarrow 2CO$$
$$MnO + SiO_2 \rightarrow MnSiO_3(\text{slag})$$

6. (a) Poling (metallurgy) is a method employed in the purification of copper which contains cuprous oxide as an impurity.

7. (c) In this method, the impure metal is converted into a block which forms the anode while cathode is made up of a pure strip of the same metal. These electrodes are suspended in an electrolyte which is the solution of a soluble salt of the metal usually double salt of the metal.

8. (c) **Zone refining** This method is employed for preparing extremely pure metals. Germanium, silicon, boron, gallium and indium are purified by this method.

This method is based upon the principle that when a molten solution of the impure metal is allowed to cool the pure metal crystallises out while the impurities remain in the melt.

9. (d) The Hoope's process is a metallurgical process used to obtain the aluminium metal of very high purity. It gives about 99.99% pure aluminium.

10. (a) **Cupellation** This method is applied when the impure metal contains impurities of other metals which form volatile oxides.

Distillation process This method is employed for purification of volatile metals like zinc, mercury, cadmium etc. The impure metal is heated in an iron retort and the vapours are condensed in separate receivers.

11. (b) **van Arkel method** This method is used for preparation of ultra pure metals such as titanium and zirconium which are used in space technology. In this method, zirconium is heated in iodine vapours at about 870 K to form unstable volatile zirconium tetraiodide which when heated over a tungsten filament at 2075 K decompose to give pure zirconium.

$$Zr(s) + 2I_2(g) \xrightarrow{870K} ZrI_4(g) \xrightarrow{2075K} \underset{\text{Pure}}{Zr(s)} + 2I_2(g)$$

12. (d) The extraction of highly electropositive metals is carried out by electrolysis of their oxides, hydroxides etc., in fused state because their compounds are not reduced by ordinary reducing agents.

13. (c) **van Arkel method** This method is used in the production of ultra-pure titanium and zirconium. It primarily involves the formation of the metal iodides and their subsequent decompositon to yield pure metal.

14. (a) Since Mg is highly electropositive its extractive metallurgy is carried out by fused salt electrolysis.

15. (c) The process takes place are under.

Anode $M \longrightarrow M^{n+} + ne^-$

Cathode $M^{n+} + ne^- \longrightarrow M$

The net result is the transfer of pure metal from the anode to the cathode.

16. (a,b,c) **Mac Arthur Forrest process** It is a metallurgical technique for extracting gold from low grade ore by converting the gold to a water soluble coordination complex. It is the most commonly used process for gold extraction.

17. (a,c,d) Mond process is used for refining of nickel. When impure nickel is heated in a current of CO at 330-350 K. It forms volatile nickel carbonyl leaving behind the impurities.

$$Ni + 4CO \xrightarrow{330-350K} \underset{\text{Nickel carbonyl}}{Ni(CO)_4}$$

$$Ni(CO)_4 \xrightarrow{450-470K} \underset{\text{Pure}}{Ni} + 4CO$$

18. (a,b,c) The self reduction of the oxide and sulphide occurs, giving impure, blister copper which is 98-99% pure.

$$Cu_2S + 2Cu_2O \longrightarrow 6Cu + SO_2$$

The blister copper is cast into blocks and refined by electrolysis using Cu electrodes with an electrolyte of dilute H_2SO_4 and $CuSO_4$.

19. (a,b,d) Ag can be refined by electrolysis of $AgNO_3$ (containing HNO_3) using impure Ag as anode and pure Ag as cathode.

20. (a,c) Alumina (Al_2O_3) and silica gel used as adsorbet in column chromatography.

21. (b) The metal is obtained by thermal decomposition. Impure nickel is heated in a current of CO at 330-350 K. It forms volatile nickel carbonyl leaving behind the impurities.

$$\underset{\text{Impure}}{Ni} + 4CO \xrightarrow{330-350K} \underset{\text{Nickel carbonyl}}{Ni(CO)_4}$$

$$Ni(CO)_4 \xrightarrow{450-470K} \underset{\text{Pure}}{Ni} + 4CO$$

22. (c) Zone refining (Fractional crystallisation). This method is employed for preparing extremely pure metals. e.g. germanium, silicon, boron, gallium and indium.

23. (a) (i) → (q), (ii) → (r), (iii) → (s), (iv) → (p)

24. (b) (i) → (q), (ii) → (p), (iii) → (s), (iv) → (r)

25. (6) $\underset{\text{Copper pyrites}}{2Cu_2S} + 3O_2 \longrightarrow 2Cu_2O + 2SO_2$

$$2Cu_2O + Cu_2S \longrightarrow 6Cu + SO_2$$

26. (0) The complex formed in the Mond process is ($Ni(CO)_4$).

Let the oxidation state of N be x.

Therefore, $x + 4(0) = 0$

$x = 0$

27. (4) Nickel is purified by Mond's process.

$$\underset{\text{Impure}}{Ni} + 4CO \xrightarrow{330-350K} \underset{\text{Vapours}}{Ni(CO)_4} \xrightarrow{450-470K} \underset{\text{Pure}}{Ni} + 4CO$$

28. (4) $Mg^{2+} + Ca(OH)_2 \rightarrow Mg(OH)_2 + Ca^{2+}$

24g 74g

Slaked lime required for 1297.3g Mg^{2+} ion

$$= \frac{1297.3}{24} \times 74 = 4000 \text{ g} = 4 \text{ kg}$$

29. (4) 24 carat gold = 100% pure

$$23 \text{ carat gold} = \frac{23}{24} \times 100 = 95.83\ (96)$$

Hence, % impurity is nearly 4%.

30. (c) Blister copper contains 0 – 2% impurities.

DPP-6 Metallurgy of Some Important Metals-I (Cu,Fe, Ag)

1. Iron is more active metal with negative SRP value while Cu and Ag are noble metals with positive SRP values. Hence, Fe liberate hydrogen but Cu and Ag do not react with dilute acids.

2. Compound (*A*) that satisfy the given data is $AgNO_3$.

$$2AgNO_3 \longrightarrow Ag_2O + 2NO_2 + O_2$$
$$AgNO_3 + NaCl \longrightarrow AgCl\downarrow + NaNO_3$$
$$AgCl + 2NH_4OH \longrightarrow [Ag(NH_3)_2Cl] + 2H_2O$$

3. Photographic plates contain a light sensitive emulsion having AgBr. To dissolve the AgBr present on the plate exposed to light, it is treated with hypo solution. Hypo dissolves AgBr forming a soluble complex ,sodiumargentothiosulphate.

$$AgBr + 2Na_2S_2O_3 \longrightarrow Na_3[Ag(S_2O_3)_2] + NaBr$$

4. The inertness exhibited by metals under conditions when chemical activity is to be expected is called chemical passivity. The passivity of iron is believed to be due to the formation of an extremely thin film of oxide on the surface of iron. Cr, Ni, Co and Al also exhibit passivity.

5. At 300°C it becomes anhydrous. Anhydrous $FeSO_4$ is colourless. When strongly heated, it breaks into ferric oxide, SO_3 and SO_2.

$$\underset{\text{(green)}}{FeSO_4 \cdot 7H_2O} \xrightarrow[-7H_2O]{300°C} \underset{\text{(White)}}{FeSO_4}$$
$$2FeSO_4 \xrightarrow{\text{High temp}} Fe_2O_3 + SO_3 + SO_2$$

6. (a) It is an impure form of iron cast into blacks (Pigs) obtained from iron over by blast furnace process.

Graphite is the major impurity in pig iron. Pig-iron is later converted into cast iron or steel.

7. (a) Blast furnace is most suitable for smelting which is used for the manufacture of cast iron.

8. (d) **Cyanide process is used for gold** and silver as follow

$$4Ag + 8NaCN + O_2 + 2H_2O \longrightarrow 4Na[Ag(CN)_2] + 4NaOH$$
or $$Ag_2S + 4NaCN \longrightarrow \underset{\text{Sod. dicyanoargentate(1)}}{2Na[Ag(CN)_2]} + Na_2S$$

9. (b) Wrought iron is an iron with a very low carbon (less than 0.08%) content in contrast to cast iron (2.1 to 4%). Stainless steel contain maximum of 0.15 % carbon. Chrome steel contains high carbon content.

10. (a) The impurities of more electropositive metals remain as ions in the bath whereas impurities of less electropositive metals (Ag and Au) remain unionised and settle down under the anode as anode mud or anode sludge.

11. (a) It is natural basic copper (II) carbonate $2CuCO_3 \cdot Cu(OH)_2$. Having an intense blue colour it is used as gemstone.

12. (c) Slag zone is the zone of blast furnace where flux and gangue mix to form slag.

13. (b) In the conversion of cast iron into wrought iron, the carbon impurities present as Fe_3C are oxidised and escapes as carbon monoxide.

14. (d) Cuprous sulphide (Cu_2S) and cuprous oxide (Cu_2O) react with each other to form copper metal along with evolution of SO_2 gas.

15. (b) $Ag_2S + 4NaCN + 2O_2 \longrightarrow 2Na[Ag(CN)_2] + Na_2SO_4$

16. (b,c,d) $\underset{\text{Basic impurities}}{FeO} + \underset{\text{Acidic flux}}{SiO_2} \longrightarrow \underset{\text{Fusible slag}}{FeSiO_3}$

17. (a,c) Limonite ($Fe_2O_3 \cdot 3H_2O$), Cassiterite (SnO_2), Magnetite (Fe_3O_4), green vitriol ($FeSO_4 \cdot 7H_2O$)

18. (a,c,d) Collectors enhance the non-wettability of the ore particles and froth stabilises (e.g. cresols and aniline.)

19. (a,c) Poling is employed when the crude metal contains impurities of the oxide of metal itself.

20. (b,d) $CuSO_4 \cdot 5H_2O$ is blue vitriol.

21. (b) Alnico is Fe, Al, Ni, Co.

22. (a) Invar steel is Fe and Ni.

23. (a) (i) → (q), (ii) → (p), (iii) → (s), (iv) → (p)

24. (b) (i) → (r,s), (ii) → (p), (iii) → (r), (iv) → (t)

25. (4) Duralumin is an alloy having Al (95%), Cu(4%) and Mg (1%).

26. (0) German silver is the alloy of Cu + Zn + Ni.

27. (6) Mohr's salt formula $FeSO_4(NH_4)_2SO_4 \cdot 6H_2O$.

28. (0) $[Ag(CN)_2]^-$ complex having configuration of central atom (Ag) as $[Kr]4d^{10}$ have zero unpaired electron.

29. (3) Nessler's reagent ($K_2[HgI_4]$), Tollen's reagent $[Ag(NH_3)_2]^+$, Baeyer's reagent Alk. $KMnO_4$, Corrosive sublimate ($HgCl_2$), Horn silver (AgCl), Ruby silver ($3Ag_2SSb_2S_3$), Calomel (Hg_2Cl_2).

30. (b) Statement II does not explains the statement correctly.

DPP-7 Metallurgy of Some Important Metals-II (Al, Zn, Au)

1. When Zn is heated strongly in air, it burns and forming a white smoke which settles down to soft wooly flocks of zinc oxide. This is called philosophers wool or pompholyx.

2. The most important use of zinc is for the galvanisation of iron. Galvanising is a process of depositing a thin layer of zinc over the surface of iron. The zinc coating protects iron from rusting .The galvanisation is done either by spraying or the molten zinc or dipping in molten zinc or electrolytically.

3. When zinc oxide is heated with cobalt nitrate, a green mass is formed due to the formation of cobaltzincate .This is called Rinmann's green.

$$2Co(NO_3)_2 \longrightarrow 2CoO + 2NO_2 + O_2$$
$$ZnO + CoO \longrightarrow \underset{\text{(Rinmann's green)}}{CoZnO_2(CoO \cdot ZnO)}$$

4. The purity of gold is expressed in carats .Pure gold is 24 carats. Ornamental gold contain gold alloyed with silver ,copper etc., in certain percentage.

e.g. 18 carats golds has 75% gold and 25% other metals.

5. (i) When the surface of aluminium metal is exposed to air, a protective oxide coating forms almost instantaneously. This oxide layer is corrosion resistant and can be further enhanced with surface treatments such as anodising.

(ii) **Electrical and Thermal Conductivity of Aluminium**

Aluminium is an excellent conductor of both heat and electricity. The great advantage of aluminium is that by weight, the conductivity of aluminium is around twice that of copper. This means that aluminium is now the most commonly used material in large power transmission lines.

The best alternatives to copper are aluminium alloys. These can be used for all electrical conduction applications including domestic wiring.

6. (c) Electrodes are suspended in an electrolyte which is the solution of a soluble salt of the metal usually a double salt of the metal.

7. (a) ΔG° is higher for the formation of Cr_2O_3.

8. (a) $4Au + 8KCN + O_2 + 2H_2O \longrightarrow 4K[Au(CN)_2]$

$[4Au(CN)_2]^- + Zn \longrightarrow [Zn(CN)_4]^{2-} + Au$

9. (c) $[Au(CN)_2]^-$ is water soluble complex.

10. (c) In the Hoope's process the middle layer consists of cryolite and BaF_2.

11. (d) Thermite is a mixture of metallic oxide (Say. FeO, Cr_2O_3, Mn_3O_4) etc. and aluminium powder.

12. (d) Magnalium consists of aluminium with from 5% to 30% mangesium. Duralumin is a light hard aluminium alloy containing about 4% copper and small amount of magnesium.

13. (a) Aluminium is the metal which has the largest abundance in the earth's crust.

14. (a) The processes used in the refining of aluminium and zinc metals are Hoope's process and fractional distillation.

15. (b) Due to incomplete roasting $ZnSO_4$ is formed.

16. (a,b) Alumina (Al_2O_3) can be purified by Baeyer's process and Hall process.

17. (a,c,d) Argentite (silver glance). Ag_2S It is an important ore of silver. Argentite are sulphide and cuprite are oxides.

18. (a,c,d) Bell metal is an alloy of copper (60% – 85%) and tin, some time including zinc or lead.

It is the type of bronze used to cost bells. German silver (Zn + Cu + Ni), Brass (Cu + Zn).

19. (a,c) White vitriol = $ZnSO_4 \cdot 7H_2O$, Fool's gold = FeS_2;

Philosopher's wool = ZnO and calomel. mercury (I) chloride

20. (a,b) Spinel = MgO Al_2O_3, Feldspar = $KAlSi_3O_8$, Corundum Al_2O_3, Magnalium = Mg + Al alloy.

21. (a) Zinc oxide is converted into zinc by reduction with carbon.

22. (c) The third graph is the correct Ellingham diagram of Zn.

23. (a) (i) → (q), (ii)→ (r), (iii) → (p), (iv) → (s)

24. (c) (i) → (q), (ii) → (p), (iii) → (s), (iv) → (r)

25. (3) Gibbsite is $Al_2O_3 \cdot 3H_2O$.

26. (3) Gold dissolves in aqua regia i.e. HNO_3 and HCl (1:3) ratio.

27. (4) Duralumin is Al 95%, Cu 4%, Mg 0.5% and Mn 0.5%.

28. (3) 2 moles of Al = 6 equivalents Al. Therefore hydrogen produced is also 6 equivalents i.e. 6g = 3 moles.

29. (4) $6O^{2-} \longrightarrow 3O_2 + 12e^-$

and $4Al^{3+} + 12e^- \longrightarrow 4Al$.

This shows that for every 3 moles of oxygen 4 moles Al is formed.

30. (c) Many gemstones are impure forms of alumina (Al_2O_3) for e.g. The impurity present in ruby is Cr while sapphire contains Co as the impurity. In cyanide process dilute solution (0.5%) of sodium or potassium cyanide is used.

Revisal Problem (JEE Main)

1. (d) Metals like Na, K, Mg, Ca, Al extracted by the electrolysis of their salts in molten state. This is called electrometallurgy.

2. (d) **Leaching** This method consists in treating the powdered ore with a suitable reagent (acids, bases or other chemicals) which can selectively dissolve the ore but not the impurities.

3. (a) Chalcocite = Copper glance = Cu_2S.

4. (b) $\underset{\text{Slaked lime}}{Ca(OH)_2} + \underset{\text{Sea water}}{MgCl_2} \longrightarrow \underset{\text{Precipitate}}{Mg(OH)_2} + CaCl_2$

$Mg(OH)_2 + HCl \xrightarrow{\text{Heat}} MgCl_2$

5. (b) $CaCO_3 \longrightarrow CaO + CO_2$, $CaO + SiO_2$ (slag)

6. (c) Zone refining is based on the principle that the impurities are more soluble in the melt than in the solid state of the metal.

7. (b) **Manganese steel**, an alloy of exceptional durability that found uses in the construction of railroad rails and rock-crushing machinery. Manganese steel containing 12–14 per cent manganese, it provides a hard, wear-resistant etc.

8. (c) CuO, K_2O cannot be reduced to give respective metals.

9. (a) The reduction process is accompanied by an increase in volume and is therefore facilitated when implemented under reduced pressure condition. Hence, ΔG is positive.

10. (c) The major impurities in bauxite are iron oxides (haematite), silicon dioxide.

11. (c) Fluorspas (CaF_2) is used to lower the temperature of the melt. It is also used to increase the conductivity of the mixture.

12. (a) $2Al + Fe_2O_3 \longrightarrow Al_2O_3 + 2Fe$

13. (d) $M + 8CN^- + 2H_2O + O_2 \longrightarrow \underset{\text{Soluble complex}}{4M(CN)_2^-} + 4OH^-$ $\quad M$ = Au, Ag

14. (c) **Cyanide process** Extraction of Au, Ag

15. (c) Heavy spar is $BaSO_4$, Anglesite = $PbCO_3$

Cerussite = $PbSO_4$, Galena = PbS

16. (d) **Spiegeleisen** is a ferromanganese alloy containing approximately 15% manganese and small quantities of carbon and silicon. Historically, this was the standard form in which manganese was traded and used in steel making.

Being remarkably free from impurities, such as phosphorus, sulphur, and silica, it was much used in the Bessemer process for the purpose of reintroducing carbon. It reduces the iron oxide and is itself subject to oxidation.

17. (b) Argentite (Ag_2S) is sulphide and cuprite are oxides.

18. (c) The molten mixture of Al_2O_3 and Na_3AlF_6 used as an electrolyte.

19. (b) Aqueous solution of calcium chloride upon electrolysis give calcium hydroxide because calcium metal reacts with water to produce calcium hydroxide.

20. (a) Hydrometallurgy is the process of bringing the metal into solution by the action of suitable chemical reagents (NaCN solution or chlorine in water etc.) followed by recovery of the metal by the use of proper precipitating agent which is a more electropositive metal.

21. (b) Argentiferrous lead contain upto 2% silver in it. To remove such silver in lead in desilverisation methods are used like Parke's and Pattinson's process. In this zinc is added which combine with Ag forming zinc-silver alloy which is lighter and separated.

22. (c) Cinnabar (HgS) on roasting from HgO which is unstable (due to + ve ΔH°_f) at high temperature and decompose forming the metal.

23. (a) Except the first one all other reactions occur.

24. (b) Furnaces are lined with refactries which are resistant to different chemicals, withstand high temperature and loads etc.

25. (b) $4Na_2S + 5O_2 + 2H_2O \longrightarrow 2Na_2SO_4 + 4NaOH + S$

26. (d) The zinc ores are sphalerite (ZnS) and $ZnCO_3$.

27. (d) The sequence of steps should be roasting, reduction and electrolytic refining.

28. (a) The uppermost part of the furnace where the temperature is 250–700°C haematite is reduced by CO whereas carbon is effective as reducing agent at high temperature.

29. (b) To eliminate carbon impurities furnace is lined with heamatite.

30. (d) To extract aqueous solution should not be used, only molten chemical should be used.

Revisal Problem (JEE Advanced)

1. (d) Hall process
$$Al_2O_3 + Na_2CO_3 \longrightarrow 2NaAlO_2 + CO_2$$
Baeyer's process
$$Al_2O_3 + 2NaOH + 3H_2O \longrightarrow 2Na[Al(OH)_4]$$
Serpeck process
$$Al_2O_3 + 3C + N_2 \longrightarrow 2AlN + 3CO$$
and $\quad SiO_2 + 2C \longrightarrow Si + 2CO$

2. (b) Puddling process is the method of converting pig iron into wrought iron by subjecting it to heat and frequent stirring in a furnace in the presence of oxidizing substances.

3. (b) Ruby is a red form of corundum. Al_2O_3 and impurity present in it is chromium. Sapphire is a blue transparent corundum (alumina, Al_2O_3). Its colour is due to the tracer of cobalt.

4. (d) Cresols and aniline are the froth stabilisers which stabilises froth.

5. (b) From argentite,
$4Ag + 8NaCN + O_2 + 2H_2O \longrightarrow 4Na[Ag(CN)_2] + 4NaOH$
silver is extracted by leaching process.

6. (c) Poling method is employed when the crude metal contains impurities of the oxide of the metal itself.
$$3Cu_2O + CH_4 \longrightarrow 6Cu + 2H_2O + CO$$

7. (b) Ag_2O and HgO

8. (d) $2CuFeS_2 + 2SiO_2 + 4O_2 \longrightarrow Cu_2S + 2FeSiO_3 + 3SO_2$
$$Cu_2S + O_2 \longrightarrow 2Cu + SO_2$$

9. (a) Type metal is an alloy of lead, antimony and tin used to make type character for printing.

10. (b) $Ag_2S + 4NaCN \rightleftharpoons 2Na[Ag(CN)] + Na_2S$
$4Au + 8KCN + 2H_2O + O_2 \longrightarrow 4K[Au(CN)_2] + 4KOH$

11. (b,c) Depressants are used to prevent one type of sulphide ore particles from forming the froth with air bubbles. e.g. NaCN. It is used to separate lead sulphide (PbS) ore from zinc sulphide (ZnS) ore.
$$4NaCN + ZnS \longrightarrow Na_2[Zn(CN)_4] + Na_2S$$

12. (a,c,d) $Al_2O_3(s) + 2NaOH(aq) + 3H_2O(l) \xrightarrow{473\text{-}523K} \underset{\text{Sod. meta aluminate}}{2Na[Al(OH)_4]}$

13. (a,b) $\underset{\text{Dolomite}}{CaCO_3 \cdot MgCO_3} \xrightarrow{\Delta} CaO + MgO + 2CO_2 \uparrow$
$$\underset{\text{Calamine}}{ZnCO_3} \xrightarrow{\Delta} ZnO + CO_2 \uparrow$$

14. (b,d) **Mac Arthur Forest Process**
$4Au + 8NaCN + O_2 + 2H_2O \longrightarrow 4Na[Au(CN)_2] + 4NaOH$
Mond's Process $Ni + 4CO \xrightarrow[350K]{330-} Ni(CO)_4 \xrightarrow{\Delta} Ni + 4CO$

15. (c) Lapis lazuli is a deep blue semi-precious stone.
$$(Na,Ca)_8(AlSiO_4)_6(S,SO_4,Cl)_{1\text{-}2}$$

16. (a,b,d) The reaction $C + \frac{1}{2}O_2 \longrightarrow CO$ does not satisfy the upward slope for Ellingham diagram.

17. (b,c) Lithopone is $ZnS\text{-}BaSO_4$

18. (b,c,d) Corrosive sublimate $= HgCl_2$ and Hg_2Cl_2 is called calomel. Litharge = PbO, Lunar caustic $= AgNO_3$

19. (a,c) (a) Limonite $2Fe_2O_3 \cdot 3H_2O$, Cassiterite (SnO_2), Bauxite $= Al_2O_3$, Haematite $= Fe_2O_3$

20. (c,d) The points *A* and *B* show minimum and maximum value of ΔG° with melting and boiling points of metals.

21. (c) Magnesite ($MgCO_3$), Epsomite ($MgSO_4 \cdot 7H_2O$), Carnallite ($KCl \cdot MgCl_2 \cdot 6H_2O$), Spodumene $= LiAl(SiO_3)_3$

22. (b) The ore carnallite is $KCl \cdot MgCl_2 \cdot 6H_2O$. Due to potassium it gives lilac colour.

23. (b) (i) → (r), (ii) → (s), (iii) → (p), (iv) → (q)

24. (b) (i) → (q), (ii) → (r), (iii) → (p), (iv) → (s)

25. (3) Beryl is $Be_3Al_2Si_6O_{18}$. To balance the charge on all atoms $n = 3$
$(2n + 6 + 24 - 36 - 0)$

26. (4) In blue vitriol 4 water molecules are coordinated to Cu^{2+} and one water molecule is coordinated to SO_4^{2-} ion.

27. (6) Oxide ores are Rutile (TiO_2), Tinstone (SnO_2), Pyrolusite (MnO_2), Chromite ($FeCr_2O_4$), Corundum (Al_2O_3), Pitche blende (U_3O_8)
Beryl ($3BeO \cdot Al_2O_3 \cdot 6SiO_2$), Chalcosite ($Cu_2S$), Gypsum ($CaSO_4 \cdot 2H_2O$, Cryolite ($Na_3AlF_6$), Feldspar ($KAlSi_3O_8$)

28. (0) German silver (Cu, Zn and Ni)

29. (3) Among the given metals, Zr, Ti, Ni can be refined by vapour phase refining.

30. (c) F_2 produced, violently reacts with water.

JEE Main & AIEEE Archive

1. (d) This problem involves the concept of Ellingham diagram as this diagram predicts the basis of oxidation state and stability of iron compounds.

(a) $Fe \xrightarrow{\text{Dil. } H_2SO_4} FeSO_4 \xrightarrow{H_2SO_4,\, O_2} Fe_2(SO_4)_3 \xrightarrow{\text{Heat}} Fe$

The above given reaction series is incorrect. The correct reactions series is as follows :

$$Fe + H_2SO_4 \longrightarrow FeSO_4 + H_2$$

$$H_2SO_4 + 2FeSO_4 + \frac{1}{2}O_2 \longrightarrow Fe_2(SO_4)_3 + H_2O$$

$$Fe_2(SO_4)_3 \xrightarrow{\Delta} Fe_2O_3(s) + 3SO_3 \uparrow$$

(b) $Fe \xrightarrow{O_2,\ \text{heat}} FeO \xrightarrow{\text{dil. } H_2SO_4} FeSO_4 \xrightarrow{\text{Heat}} Fe$

This reaction series is also incorrect. The correct reaction series is

$F \xrightarrow[\Delta]{O_2} FeO$ [It could also be Fe_2O_3 or Fe_3O_4]

$$FeO + H_2SO_4 \longrightarrow FeSO_4 + H_2O$$

$$2FeSO_4 \xrightarrow{\Delta} Fe_2O_3 + SO_2 + SO_3$$

(c) $Fe \xrightarrow{Cl_2,\ \text{heat}} FeCl_3 \xrightarrow{\text{Heat, air}} FeCl_2 \xrightarrow{Zn} Fe$

The reaction of $FeCl_3$ is not possible. Thus, the above reaction series is incorrect.

(d) $Fe \xrightarrow[O_2]{\Delta} Fe_3O_4 \xrightarrow{CO,\ 600°C} FeO \xrightarrow{CO,\ 700°C} Fe$

This reaction series is correct.

2. (d) van-Arkel method is used for refining titanium and zirconium. This method is useful for removing volatile impurities like O_2, N_2, I_2 present in the metals of Zr and Ti.

In this method, metal iodide is formed by direct combination of metal and iodine.

The metal iodide undergo decomposition on a tungsten filament and thus, pure metal is deposited on the filament.

3. (c) $2MS + C \longrightarrow 2M + CS_2;\quad \Delta G_1 = +ve$

$$2MO + C \longrightarrow 2M + CO_2;\quad \Delta G_2 = -ve$$

The value of ΔG for the formation of CO_2 is negative, i.e. it is thermodynamically more stable than CS_2. Also, metal sulphides are thermodynamically more stable than CS_2.

Metal sulphides are more stable than the corresponding oxides, so they are roasted to convert into less stable oxides.

4. (b) During electrolysis, noble metals (inert metals) like Ag, Au and Pt are not affected and separate as anode mud from the impure anode.

5. (c) Froth floatation is used to concentrate sulphide ores [Galena (PbS)].

6. (d) Aluminium is extracted by electrolysis of a fused mixture of alumina (2–8%), synthetic cryolite Na_3AlF_6 (80–85%), AlF_3 and fluorspar. This makes alumina good conductor and lowers the fusion temperature also.

7. (b) Cyanide process or Mac Arthur-Forest cyanide process

$$2Ag_2S + 8NaCN + O_2 + 2H_2O \longrightarrow 4Na[Ag(CN)_2] + 4NaOH + 2S$$

Soluble silver complex is filtered and treated with zinc dust and silver gets precipitated.

$$2Na[Ag(CN)_2] + Zn \longrightarrow Na_2[Zn(CN)_4] + 2Ag \downarrow$$

8. (c) For purification, impurities like Se, Te, Ag, Au gets oxidised (deelectronation) falling into solution near anode and settle as anode mud.

JEE Advanced & IIT JEE Archive

1. (b,c,d) (b) $4CuO \xrightarrow{1100°C} 2Cu_2O + O_2$

$$2Cu_2O + Cu_2S \xrightarrow{\Delta} 6Cu + SO_2$$

(c) $Cu_2S + 2Cu_2O \xrightarrow{\Delta} 6Cu + SO_2$

(d) $CuSO_4 \xrightarrow{720°C} CuO + SO_2 + \frac{1}{2}O_2$

$$4CuO \xrightarrow{1100°C} 2Cu_2O + O_2$$

$$2Cu_2O + Cu_2S \xrightarrow{\Delta} 6Cu + SO_2$$

Reaction is believed to proceed as

$$Cu_2S \rightleftharpoons 2Cu^+ + S^{2-}$$

$$2Cu_2O \rightleftharpoons 4Cu^+ + 2O^{2-}$$

$$S^{2-} + 2O^{2-} \longrightarrow SO_2 + 6e^-$$

$$6Cu^+ + 6e^- \longrightarrow 6Cu;\ E^\circ_{cell} = 0.52$$

Here, copper sulphide is reduced to copper metal. Solidified copper has blistered appearance due to evolution of SO_2 and thus obtained copper is known as blister copper.

Other compounds which give Cu are

(i) CuO as $4CuO \xrightarrow{1100°C} 2Cu_2O + O_2$

$$2Cu_2O + Cu_2S \xrightarrow{\Delta} 6Cu + SO_2$$

(ii) $CuSO_4$ as $CuSO_4 \xrightarrow{720°C} CuO + SO_2 + \frac{1}{2}O_2$

$$4CuO \xrightarrow{\Delta} 2Cu_2O + O_2$$

$$2Cu_2O + Cu_2S \xrightarrow{\Delta} 6Cu + SO_2$$

While $CuFeS_2$ will not give Cu on heating. The heating in the presence of O_2 gives Cu_2S and FeS with the evolution of SO_2.

2. (c,d) Al has greater affinity for oxygen hence oxide is not reduced by carbon. MgO and CaO (formed in the calcination from carbonates) are stable species and not reduced by carbon.

During smelting $SnO_2 + C \xrightarrow{1300°C} Sn + CO$

$$2Fe_2O_3 + 3C \xrightarrow{\Delta} 4Fe + 3CO_2$$

3. (a)

Element	Ores	Name
Ag	Ag_2S	Argentite
Cu	$CuFeS_2$	Copper pyrites
Pb	PbS	Galena
Sn	SnO_2	Cassiterite
Mg	$MgCO_3 \cdot CaCO_3$	Dolomite
Al	$Al_2O_3 \cdot xH_2O$	Bauxite

4. (b) The reactions involved in the extraction of silver by cyanide process are

$Ag_2S + CN^- + O_2 \longrightarrow [Ag(CN)_2]^- + SO_2$ ...(i)

$[Ag(CN)_2]^- + Zn \longrightarrow [Zn(CN)_4]^{2-} + Ag$ ...(ii)

In reaction (i), sulphide is oxidised to SO_2 by oxygen. In the reaction (ii), silver ion (Ag^+) is reduced to Ag by Zn. Therefore, O_2 is oxidising agent and Zn is reducing agent.

5. (d) Haematite is Fe_2O_3, in which oxidation number of iron is III. Magnetite is Fe_3O_4 which is infact a mixed oxide ($FeO \cdot Fe_2O_3$), hence, iron is present in both II and III oxidation state.

6. (a,d) The important ore of tin is cassiterite (SnO_2). Tin is extracted from cassiterite ore by carbon reduction method in a blast furnace.

$$SnO_2 + 2C \longrightarrow Sn + 2CO$$

The product often contain traces of iron which is removed by blowing air through the melt to oxidise to FeO which then floats to the surface.

$$2Fe + O_2 \longrightarrow 2FeO$$

7. (b) $2CuFeS_2 + O_2 \longrightarrow Cu_2S + 2FeS + SO_2 \uparrow$

$2Cu_2S + 3O_2 \longrightarrow 2Cu_2O + 2SO_2 \uparrow$

$2FeS + 3O_2 \longrightarrow 2FeO + 2SO_2 \uparrow$

8. (d) $FeO + SiO_2 \longrightarrow FeSiO_3$ (slag)

9. (c) $Cu_2S + 2Cu_2O \longrightarrow 6Cu + SO_2$

In Cu_2S, sulphur is S^{2-} and in SO_2, sulphur is in +4 state.

Hence, S^{2-} is acting as reducing agent.

10. (d) (i) → (p,s), (ii) → (q,s), (iii) → (r,t), (iv) → (q,t)

(i) $3Cu + \underset{\text{dil.}}{8HNO_3} \longrightarrow 3Cu(NO_3)_2 + 2NO + 4H_2O$

(ii) $Cu + \underset{\text{conc.}}{4HNO_3} \longrightarrow Cu(NO_3)_2 + 2NO_2 + 2H_2O$

(iii) $4Zn + \underset{\text{dil}}{10HNO_3} \longrightarrow 4Zn(NO_3)_2 + N_2O + 5H_2O$

(iv) $Zn + \underset{\text{conc.}}{4HNO_3} \longrightarrow Zn(NO_3)_2 + 2NO_2 + 2H_2O$

11. (4) The complex has formula $[Cu(H_2O)_4]SO_4 \cdot H_2O$

12. (6) In crystalline state, $AlCl_3$ has rock-salt like structure with coordination number of Al = 6.

13. (c) (i) → (p), (ii) → (q), (iii) → (r,s), (iv) → (p,s)

(i) $PbS \xrightarrow{O_2} PbO + SO_2$ roasting

(ii) $CaCO_3 \longrightarrow CaO + CO_2 \uparrow$; calcination

(iii) $ZnS \longrightarrow Zn$, can be done by carbon reduction or self reduction

(iv) $Cu_2S \longrightarrow Cu$, roasting followed by self reduction

14. (b) A water soluble complex with silver and dilute aqueous solution of NaCN is $Na[Ag(CN)_2]$. In the cyanide process, the native silver is crushed and treated with aqueous NaCN solution and aerated.

$$4Ag + 8NaCN + 2H_2O + O_2 \longrightarrow 4Na[Ag(CN)_2] + 4NaOH$$

15. (b) Zinc blende contain ZnS which is first roasted partially and then subjected to reduction with carbon.

$ZnS + O_2 \longrightarrow ZnO + SO_2$ Roasting

$ZnO + C \xrightarrow{\Delta} Zn + CO \uparrow$ Carbon reduction

16. (b) (i) → (p,r), (ii) → (p), (iii) → (q), (iv) → (s)

	Extraction methods		Metals extracted
i.	Self reduction	p.	Copper, (r) Lead
ii.	Carbon reduction	q.	Lead
iii.	Complex formation and displacement by metal		Silver : $Ag_2S + NaCN \longrightarrow Na[Ag(CN)_2]$
iv.	Decomposition of iodide	r.	$Na[Ag(CN)_2] + Zn \longrightarrow Na_2[Zn(CN)_4] + Ag$
		s.	Boron : $BI_3 \xrightarrow{\Delta} B + \frac{3}{2}I_2$

17. (c) Chalcopyrite contain both iron and copper.

18. (a) $2AgBr + \underset{\text{Hydroquinone (developer)}}{C_6H_4(OH)_2} \longrightarrow 2Ag + 2HBr + \underset{\text{Quinone}}{C_6H_4O_2}$

$AgBr + 2Na_2S_2O_3 \longrightarrow Na_3[Ag(S_2O_3)_2] + NaBr$

(b) $Na_2S_2O_3 + 2H^+ \longrightarrow 2Na^+ + H_2SO_3 + \underset{\text{Colloidal sulphur}}{S \downarrow}$

5. Qualitative Analysis

DPP-1 Principles Involved in Qualitative Analysis of Inorganic Salts

1. (i) Barium (Ba^{2+}), beryllium (Be^{2+}), magnesium (Mg^{2+}) (as, Be and Mg have very high ionisation energy).

(ii) green, yellow

(iii) $NaBO_2$ and B_2O_3

(iv) $NaNH_4HPO_4 \cdot 4H_2O$ and $Na_2B_4O_7 \cdot 10H_2O$

(v) greater [$AB \rightleftharpoons A^+ + B^-$]

For precipitation, ionic product > K_{sp}

(vi) (> or greater than) $M(OH)_3 \rightleftharpoons M^{3+} + 3OH^-$

$$K_{sp} = [M^{3+}][OH^-]^3$$
$$[OH^-]^3 > 1 \times 10^{-9}$$
$$[OH^-] > 1 \times 10^{-3}$$

Maximum pOH = 3 and minimum pH = 9

i.e. pH > 9

2. (i) Precipitates are generally washed with a solution containing common ion to minimise the loss of precipitate. So,

(a) H_2O

(b) $CaCl_2$ (Ca^{2+} is in excess)

(c) KI

(d) KCl

(e) NaF

(ii) (a) pH increases (b) pH decreases

(c) pH increases (d) pH increases

3. (b) The principle involved for the classification of basic radicals is the solubility product.

Common ion effect is helpful only in decreasing or increasing the concentration of certain species in a reaction mixture but play no role in the classification of basic radicals.

4. (b) Al^{3+}, Fe^{3+} and Cr^{3+} all get precipitated as their hydroxides when treated with NH_4OH [or $NH_3(aq)$] in the presence of NH_4Cl.

5. (b) With NaOH, first hydroxide is obtained. Hydroxides of Al^{3+} and As^{3+} being amphoteric form soluble salt with excess of NaOH. $Cu(OH)_2$ is basic, so does not dissolve in NaOH.

6. (c) HCl and $BaCl_2$ both are strong electrolytes, so addition of HCl provides more [Cl^-]. It leads to $[Ba^{2+}][Cl^-]^2 > (K_{sp})_{BaCl_2}$, i.e. $BaCl_2$ precipitates out.

7. (c)
$$\underset{\text{(Colourless)}}{\underset{\text{Borax}}{Na_2B_4O_7}} \xrightarrow{\Delta} 2NaBO_2 + \underset{y}{B_2O_3} \xrightarrow{Cu^{2+}, \Delta} \underset{\text{(Blue)}}{Cu(BO_2)_2}$$

$$\underset{\text{(Colourless)}}{\underset{\text{Microcosmic salt}}{Na(NH_4)HPO_4}} \xrightarrow{\Delta} \underset{y}{NaPO_3} + H_2O\uparrow + NH_3\uparrow$$

$$NaPO_3 + Cu^{2+} \longrightarrow \underset{\text{(Blue bead in oxidising flame)}}{NaCuPO_4}$$

8. (a) Reduction of nitrate by Fe^{2+} ions.

$$NO_3^- + 3Fe^{2+} + 4H^+ \longrightarrow \underset{\text{Nitric oxide}}{NO} + 3Fe^{3+} + 2H_2O$$

$$Fe^{2+} + NO + 5H_2O \longrightarrow [Fe(H_2O)_5NO]^{2+}$$

9. (d) Silver chloride readily forms a soluble complex $[Ag(NH_3)_2]Cl$ with $aq.NH_3$ but insoluble in H_2O.

$AgNO_3$ or HCl suppresses its solubility by providing common ion.

10. (d) NH_4OH is a weak electrolyte, so its ionisation decreases in the presence of strong electrolyte having a common ion like NH_4Cl.

$$NH_4OH \rightleftharpoons NH_4^+ + OH^- \quad \ldots(i)$$
$$NH_4Cl \longrightarrow NH_4^+ + Cl^-$$
(NH_4^+: Common ion)

Due to excess of NH_4^+, the reaction (i) goes in backward direction.

11. (b) HCl being acidic neutralises ammonia, so more Zn^{2+} remains free, i.e. concentration of free Zn^{2+} increases in the solution.

12. (c) The colour of zinc oxide becomes yellow on heating due to metal excess defect. Being amphoteric, it forms salt with NaOH as well as HCl.

13. (a) Precipitation takes place only when ionic product > solubility product (K_{sp})

	Salt	Ionic product	K_{sp}
I	HgS	$10^{-3} \times 10^{-16} = 10^{-19}$	10^{-54}
II	ZnS	$10^{-3} \times 10^{-16} = 10^{-19}$	10^{-20}
III	MnS	$10^{-3} \times 10^{-16} = 10^{-19}$	10^{-15}
IV	FeS	$10^{-3} \times 10^{-16} = 10^{-19}$	10^{-23}

Since, K_{sp} of HgS is very low, so it will precipitate out first.

14. (d) In A : $ZnCl_2 + H_2S \longrightarrow ZnS + 2HCl$

HCl formed suppresses the degree of ionisation of H_2S (due to common ion effect) hence, only small concentration of S^{2-} ion is available, which is not sufficient for the precipitation of ZnS (as its K_{sp} is very high and for achieving ionic product $> K_{sp}$, S^{2-} should be high enough). So, ZnS remains soluble in HCl (i.e. no ppt is formed).

In B : $Zn(CH_3COO)_2 + H_2S \longrightarrow ZnS + 2CH_3COOH$

CH_3COOH formed in B, itself is a weak acid, so has no effect at the ionisation of H_2S. Thus, concentration of S^{2-} ions is high enough for the precipitation of ZnS.

15. (a,b,d) In common ion effect, the degree of ionisation of weak electrolyte decreases in the presence of a strong electrolyte having a common ion.

AgCl, NH_4OH and AgCN all are weak electrolytes, so their degree of dissociation is suppressed by adding NaCl, NH_4Cl or KCN respectively.

NaCl and HCl both are strong electrolytes, so do not exhibit common ion effect.

16. (a,b,d) For the analysis of radicals of group second, a very low concentration of S^{2-} is required, which is achieved by adding a strong electrolyte like HCl to provide common ion H^+.

$$H_2S \rightleftharpoons 2H^+ + S^{2-}$$
$$HCl \longrightarrow H^+ + Cl^-$$
(H^+: common ion)

Excess of H^+ shifts the reaction in backward direction, so concentration of S^{2-} decreases.

17. (c) $pK_a = -\log K_a \Rightarrow 4.74 = -\log K_a$

$$K_a = \text{antilog}(-4.74) = 1.8 \times 10^{-5}$$

When no HCl is present,

$$x = \sqrt{\frac{K_a}{C}} = \sqrt{\frac{1.8 \times 10^{-5}}{0.05}} = 0.019 = 1.9 \times 10^{-2}$$

$$AcOH \rightleftharpoons AcO^- + H^+$$
$$\begin{matrix} C & 0 & 0 \\ (C - Cx) & Cx & Cx \end{matrix}$$

In the presence of 0.01 M HCl,

$$\begin{matrix} HCl & \rightleftharpoons & H^+ & + & Cl^- \\ 0 & & 0.01 & & 0.01 \end{matrix}$$

So, total $[H^+] = Cx + 0.01$

But $[H^+]_{AcOH} << [H^+]_{HCl}$ so, $[H^+]_{AcOH}$, i.e. Cx is neglected.

$\therefore \quad [H^+]_{total} = 0.01\,M$

$$K_a = \frac{[AcO^-][H^+]}{[AcOH]}$$

$$1.8 \times 10^{-5} = \frac{(Cx')(0.01)}{C(1-x')} \quad \left[\begin{matrix}\because x <<< 1 \\ \text{hence } 1 - x = 1\end{matrix}\right]$$

$$\approx x'(0.01)$$

$\therefore$ $$x' = \frac{1.8 \times 10^{-5}}{0.01} = 1.8 \times 10^{-3}$$

Thus, in the presence of HCl, ionisation of AcOH (i.e. CH_3COOH) is decrease to one tenth (i.e. it is $\frac{x}{10}$).

18. (c) In the presence of 0.1 M AcONa

$$[AcO^-]_{total} = [AcO^-]_{from\ AcOH} + [AcO^-]_{from\ AcONa}$$
$$= Cx + 0.1\,M \approx 0.1\,M$$
$$K_a = \frac{[AcO^-][H^+]}{[AcOH]}$$
$$1.8 \times 10^{-5} = \frac{(0.1)(x)}{(1-x)} = \frac{0.1}{(1-x)} \approx 0.1\,x$$

or $$x = \frac{1.8 \times 10^{-5}}{0.1} = 1.8 \times 10^{-4}$$

19. (b) Basic radicals are classified on the basis of their solubility product as a basic radical forms ppt only when the ionic product of its salt > its K_{sp} value.
Salts of most of the alkali metals are soluble in water.

20. (d) Borax bead test is a dry test, not a wet test. Although, it is used for the identification of coloured cations.

$$Na_2B_4O_7 \xrightarrow{\Delta} 2NaBO_2 + B_2O_3$$
$$\underset{\text{(Coloured)}}{Cu^{2+}} + B_2O_3 \longrightarrow \underset{\text{(Blue bead)}}{Cu(BO_2)_2}$$

DPP-2 Qualitative Analysis of Acidic Radicals (Anions)

1. (i) (a) It is because ammonium acetate forms soluble lead acetate with white ppt of $PbSO_4$.

$$SO_4^{2-} + (CH_3COO)_2Pb \longrightarrow PbSO_4 \downarrow + 2CH_3COO^-$$
$$\underset{\text{Ammonium acetate}}{2CH_3COONH_4} + \underset{\text{(White ppt)}}{PbSO_4} \longrightarrow \underset{\text{(Soluble)}}{(CH_3COO)_2Pb} + (NH_4)_2SO_4$$

(b) In case of Ag_2SO_4 or $PbSO_4$, ppt of AgCl or $PbCl_2$ is formed with $BaCl_2$, so in place of $BaCl_2$, $Ba(NO_3)_2$ is used for testing SO_4^{2-} group here.

(ii) (a) First the mixture is treated with acidified $K_2Cr_2O_7$ solution then with lime water.
SO_2 evolved turned the acidified $K_2Cr_2O_7$ green.
CO_2 then turns the lime water milky.

(b) Add $BaCl_2$ and then conc. HCl. $BaSO_4$ and $BaSO_3$ are formed. The former one is insoluble in conc. HCl but latter remains soluble.

(c) Add chlorine water and $CHCl_3$ to the mixture of I^- and Br^-. The following reactions take place:

$$2I^- + Cl_2 \longrightarrow \underset{\text{(Violet)}}{I_2} + 2Cl^-$$
$$2Br^- + Cl_2 \longrightarrow \underset{\text{(Orange-yellow)}}{Br_2} + 2Cl^-$$

Since, the order of reducing power is $I^- > Br^- > Cl^-$, so Br_2 also oxidises I^- to I_2. Thus, first there will be violet colour in chloroform layer. Extract aqueous layer (containing Br^-) and again add chlorine water and chloroform. Orange-yellow colour in chloroform layer confirms the presence of Br^-.

2. (i) (a) When treated with dilute HCl, sulphite ion (SO_3^{2-}) gives SO_2 gas which turns acidified $K_2Cr_2O_7$ green but CO_2 evolved in case of CO_3^{2-} does not give this test.

(b) When treated with $AgNO_3$, Cl^- forms a white ppt of AgCl which is soluble in ammonia (*aq*) but I^- gives yellow ppt, which is insoluble in $NH_3(aq)$.

(c) Only NO_2^- gives NO_2 gas (brown fumes) with dilute acid. NO_3^- does not give this reaction with dilute acid.

(ii) Arsenate (AsO_4^{3-})

$$Mg^{2+} + NH_4^+ + AsO_4^{3-} \longrightarrow \underset{\text{(White ppt)}}{MgNH_4AsO_4}$$

(iii) CO and CO_2 gases are produced. CO (molar mass 28) diffuses at a faster rate than CO_2, thus, it burns. After further heating, CO_2 gas upwards which puts off the burning CO.

3. (i) (a) CO_3^{2-} (b) Basic (c) Alkaline

(d) When CO_2 (obtained by the reaction of CO_3^{2-} with H_2SO_4) is passed into above pink coloured solution, HCO_3^- (colourless) is formed.

$$CO_3^{2-} + CO_2 + H_2O \longrightarrow HCO_3^-$$

(ii) The reactions are related with the chromyl chloride test of Cl^- ion.
Thus, $A = CrCl_3$, $B = CrO_2Cl_2$, $C = Na_2CrO_4$, $D = PbCrO_4$

$$CrCl_3 + 2NaOH \xrightarrow[\text{water}]{Br_2} \underset{C}{Na_2CrO_4}$$

4. (c) H_2SO_4 being a stronger oxidising agent than HBr or HI, oxidises them into Br_2 or I_2 respectively.

$$I^- + H_2SO_4 \longrightarrow HSO_4^- + \underset{\text{Colourless}}{HI}$$
$$2HI + H_2SO_4 \longrightarrow \underset{\text{(Violet colour)}}{I_2 \uparrow} + SO_2 + 2H_2O$$

But it cannot oxidised HCl (colourless solution) formed in case of Cl^- ion.

$$Cl^- + H_2SO_4 \longrightarrow HSO_4^- + \underset{\text{(Colourless)}}{HCl}$$

5. (b) $$SO_3^{2-} + H_2SO_4 \longrightarrow SO_4^{2-} + H_2O + \underset{\text{(Colourless pungent gas)}}{SO_2}$$

$$Cr_2O_7^{2-} + 2H^+ + 3SO_2 \longrightarrow 3SO_4^{2-} + \underset{\text{(Green)}}{2Cr^{3+}} + H_2O$$

6. (b) The given reactions show that X contains a nitrite ion.

$$2NO_2^- + H_2SO_4 \longrightarrow SO_4^{2-} + 2HNO_2$$
$$3HNO_2 \longrightarrow HNO_3 + 2NO \uparrow + H_2O$$
$$6FeSO_4 + 2HNO_3 + 3H_2SO_4 \longrightarrow 3Fe_2(SO_4)_3 + 4H_2O + 2NO \uparrow$$
$$\underset{\text{(aqueous)}}{FeSO_4} + NO + H_2O \longrightarrow [Fe(H_2O)_5NO]SO_4$$

or

$$[Fe(H_2O)_5NO]^{2+}$$

7. (d) Formation of deep brown colouration on passing Cl_2 indicates that the salt is a iodide. Moreover, treatment with $BaCl_2$ or NaOH results in the formation of a white ppt, which shows the presence of sulphate group.

$$\underset{X}{MgI_2} + Cl_2(g) \longrightarrow MgCl_2 + I_2$$

$$\underset{Y}{MgSO_4} + BaCl_2 \longrightarrow MgCl_2 + \underset{\text{(White ppt)}}{BaSO_4}$$

$$\underset{Y}{MgSO_4} + 2NaOH \longrightarrow Na_2SO_4 + \underset{\text{(White ppt)}}{Mg(OH)_2}$$

8. (a) Since, the order of reducing power is $I^- > Br^- > Cl^-$, so Br_2 (formed) and Cl_2 (added) both oxidise iodide to iodine, hence the colour of organic layer (i.e. $CHCl_3$) will become violet.

$$2I^- + Cl_2 \longrightarrow \underset{\text{(Violet)}}{I_2} + 2Cl^-$$

$$2Br^- + Cl_2 \longrightarrow \underset{\text{(Orange-yellow)}}{Br_2} + 2Cl^-$$

$$2I^- + Br_2 \longrightarrow \underset{\text{(Violet)}}{I_2} + 2Br^-$$

9. (b) $$BO_3^{3-} + \underset{\text{(conc.)}}{3H_2SO_4} \longrightarrow 3HSO_4^- + \underset{\text{Boric acid}}{H_3BO_3}$$

$$H_3BO_3 + 3C_2H_5OH \longrightarrow \underset{\substack{\text{Ethyl borate} \\ \text{(Volatile)} \\ \text{burning with green} \\ \text{edged flame } (Y)}}{(C_2H_5)_3BO_3} + 3H_2O$$

10. (b) All the given statements are true, as NO_2^- (nitrite) and NO_3^- (nitrate) both give brown ring test and evolves NO_2 (brown gas) with conc. H_2SO_4.

However, only NO_3^- gives ammonia when treated with NaOH/Cu.

11. (d) S^{2-} ions react with sodium nitroprusside to give purple coloured compound, sodium thionitroprusside.

Thus, X = sodium nitroprusside, $Na_2[Fe(CN)_5NO]$

$$2 \times 1 + x + (-1)5 + 1 = 0$$

$$2 + x - 4 = 0 \Rightarrow x = +2$$

Y = sodium thionitroprusside, $Na_2[Fe(CN)_5NOS]$

$$2 \times 1 + x + (-1)5 + 1 = 0$$

$$2 + x - 4 = 0 \Rightarrow x = 2$$

12. (a) Since, X occupies 0.35 L per g at NTP.

So, its molecular mass = 64 g mol^{-1}

Further, it reduces MnO_4^-/H^+

$$MnO_4^- + 8H^+ + 5e^- \longrightarrow Mn^{2+} + 4H_2O$$

which indicates it may be SO_2 (sulphur dioxide).

Production of white turbidity on passing H_2S to the aqueous solution of X confirms that X is SO_2.

$$SO_2 + 2H_2S \longrightarrow 2H_2O + \underset{\text{(Colloidal)}}{3S}$$

13. (c) (i) → (q), (ii) → (p), (iii) → (r), (iv) → (p)

In chromyl chloride test, complex involved is CrO_2Cl_2.

$$\therefore \quad x + (-2)2 + (-1)2 = 0$$

$$x - 4 - 2 = 0 \Rightarrow x = +6$$

In ring test of NO_3^-, the complex involved is $[Fe(H_2O)_5NO]SO_4$.

$$x + (0)5 + 1 = +2$$

$$x = +2 - 1 = +1$$

Complex involved in sodium nitroprusside test are

$Na_2[Fe(CN)_5NO]$ and $Na_2[Fe(CN)_5NOS]$

In both the cases oxidation state is + 2.

In $AgNO_3$ test of Br^-, the complex involved is $[Ag(NH_3)_2]Cl$.

$$x + (0)2 = +1$$

$$\Rightarrow \quad x = +1$$

14. (a) (i) → (p,q,r,t), (ii) → (s), (iii) → (t), (iv) → (r,t), (v) → (q,r)

$AgNO_3$ (silver nitrate) gives white ppt of AgCl (with Cl^-), pale yellow ppt of AgBr, soluble in $NH_3(aq)$ (with Br^-), yellow ppt of AgI (with I^-) and red ppt of Ag_2CrO_4 (with CrO_4^{2-}).

$FeCl_3$ gives blood red colour with CH_3COO^-.

$$BaCl_2 + CrO_4^{2-} \longrightarrow \underset{\text{ppt}}{BaCrO_4\downarrow} + 2Cl^-$$

$$Pb(CH_3COO)_2 \xrightarrow{I^-} PbI_2\downarrow$$

$$Pb(CH_3COO)_2 \xrightarrow{CrO_4^{2-}} PbCrO_4\downarrow$$

Chlorine being stronger oxidising agent oxidises Br^- and I^- into Br_2 and I_2 respectively.

15. (a) (i) → (q,r,s), (ii) → (r,s), (iii) → (t), (iv) → (p)

$$\text{(i) } \underset{\text{(Green)}}{2FeSO_4} \xrightarrow{\Delta} Fe_2O_3 + SO_2 + SO_3$$

$SO_2 \xrightarrow{Cr_2O_7^{2-}/H^+}$ Green; $SO_3 \xrightarrow{H_2SO_4} \underset{\text{Oleum}}{H_2S_2O_7}$

$$\text{(ii) } Na_2SO_3 \xrightarrow{\text{dil. HCl}} 2NaCl + SO_2 + H_2O$$

$SO_2 \xrightarrow{Cr_2O_7^{2-}/H^+}$ Green

$$\text{(iii) } 2NaCl + \underset{\text{(Conc.)}}{H_2SO_4} \longrightarrow 2HCl + Na_2SO_4$$

$$AlCl_3 + \underset{\substack{\text{(Moist} \\ \text{air)}}}{3H_2O} \longrightarrow Al(OH)_3 + 3HCl$$

$$\text{(iv) } NaNO_2 \xrightarrow{\text{dil. HCl}} NaCl + HNO_2$$

$$3HNO_2 \longrightarrow HNO_3 + 2NO\uparrow + H_2O$$

$$2NO \xrightarrow{\text{Air } (O_2)} 2NO_2 \text{ or } N_2O_4$$

$$HNO_2 + HNO_3 \xrightarrow[-H_2O]{} \underset{\text{anhydride of } HNO_2 \text{ and } HNO_3}{N_2O_4}$$

16. (a)

$$\underset{\text{Mixture}}{CO_3^{2-} + SO_3^{2-}} \xrightarrow{\text{dil.} H_2SO_4} \underset{Y}{SO_2} + \underset{Z}{CO_2} \xrightarrow{\text{Lime water}} \text{Turns milky}$$

$SO_2 \xrightarrow{Cr_2O_7^{2-}/H^+}$ Green (Cr^{3+}); $CO_2 \xrightarrow{Cr_2O_7^{2-}/H^+}$ No reaction

$$\underset{A}{O_2} \xrightarrow{\text{Pyrogallol}} \underset{\text{(Black solution)}}{\text{Soluble}}$$

17. (c) The order of analysis is SO_2, CO_2, O_2.

18. (c) $$NO_2^- + H_2SO_4 \longrightarrow HSO_4^- + HNO_2$$

$$\underset{\text{Ulrea}}{NH_2CONH_2} + 2HNO_2 \longrightarrow 2N_2\uparrow + 3H_2O + CO_2$$

Thus, nitrite is decomposed by urea. Its decomposition is necessary as it interferes in the test of nitrate.

19. (c) $$\underset{\text{Carbonate ion}}{CO_3^{2-}} + 2HCl \longrightarrow 2Cl^- + H_2O + \underset{\substack{\text{Colourless} \\ \text{gas}}}{CO_2\uparrow}$$

$$\underset{\text{Bicarbonate ion}}{HCO_3^-} + HCl \longrightarrow Cl^- + H_2O + \underset{\substack{\text{Colourless} \\ \text{gas}}}{CO_2}$$

In group VI, Mg^{2+} is present and it forms a white ppt only with carbonate but not with bicarbonate.

20. (3) The chlorides of mercury, owing to their slight ionisation do not respond to chromyl chloride test.

Only partial conversion to CrO_2Cl_2 takes place with the chlorides of Pb, Ag, Sb and Sn, so these also not give this test.

Thus, among the given only HCl, NaCl and $BaCl_2$ give this test.

DPP-3 **Qualitative Analysis of Basic Radicals** (Cations)

1. (i) (a) Dil. HCl; as only Ag^+ reacts with it to give white ppt of AgCl, Cu^{2+} gives no ppt with it.

(b) H_2S in the presence of HCl. (Here, only Cd^{2+} gets precipitated as CdS but ZnS does not because only low concentration of S^{2-} ions is available,which is not enough for the precipitation of ZnS).

(c) HCl followed by hot water; treatment with HCl result in precipitation of both Ag^+ and Pb^{2+} as AgCl and $PbCl_2$. But only $PbCl_2$ is soluble in hot water.

(d) NH_4OH in NH_4Cl; here, only Al^{3+} gets precipitated as $Al(OH)_3$, but Mg^{2+} (because of low concentration of $O\bar{H}$) does not.

(e) Same as (iv), i.e. NH_4OH in NH_4Cl; here, Fe^{3+} gets precipitated as $Fe(OH)_3$ but Ni^{2+} does not [because of high K_{sp} of $Ni(OH)_2$].

(ii) (a) Potassium chromate (K_2CrO_4)

$$2Ag^+ + CrO_4^{2-} \longrightarrow \underset{\text{(Red)}}{Ag_2CrO_4}\downarrow$$

$$Ba^{2+} + CrO_4^{2-} \longrightarrow \underset{\text{(Yellow)}}{BaCrO_4}\downarrow$$

(b) Ammonium thiocyanide (NH_4SCN)

$$Co^{2+} + 4SCN^- \longrightarrow \underset{\text{(Blue)}}{[Co(SCN)_4]^{2-}}$$

$$Cu^{2+} + 2SCN^- \longrightarrow \underset{\text{(Black)}}{Cu(SCN)_2}\downarrow$$

(c) K_2CrO_4

$$Cu^{2+} + CrO_4^{2-} \longrightarrow \underset{\text{(Chocolate)}}{CuCrO_4}\downarrow$$

$$Pb^{2+} + CrO_4^{2-} \longrightarrow \underset{\text{(Yellow)}}{PbCrO_4}\downarrow$$

(d) NH_4SCN

$$Fe^{3+} + SCN^- \longrightarrow \underset{\text{(Red)}}{[Fe(SCN)]^{2+}}$$

$$Cu^{2+} + 2SCN^- \longrightarrow \underset{\text{(Black)}}{Cu(SCN)_2}\downarrow$$

2. (i) Concentrated HCl results in high concentration of Cl^-, which increases the solubility of precipitate of chlorides of group I by the formation of soluble complex ions.

$$AgCl + 2Cl^- \longrightarrow \underset{\text{(Soluble)}}{AgCl_3^{2-}}$$

$$PbCl_2 + 2Cl^- \longrightarrow \underset{\text{(Soluble)}}{PbCl_4^{2-}}$$

(ii) The group reagent of first group is dilute HCl. It should be in excess as its excess not only causes complete precipitation of chlorides of this group but also prevents the formation of white ppt of BiOCl and SbOCl along with chlorides of group I.

$$BiOCl + 2HCl \rightleftharpoons \underset{\text{Soluble}}{BiCl_3} + H_2O$$

$$SbOCl + 2HCl \rightleftharpoons \underset{\text{(Soluble)}}{SbCl_3} + H_2O$$

(iii) Because if H_2S is present in alkaline medium, its dissociation increases and more S^{2-} are generated, which are enough to cause the precipitation of basic radicals of group IV.

(iv) Although, $Al(OH)_3$ and $Fe(OH)_3$ both are insoluble in water but only the former is soluble in NaOH, forming $NaAlO_2$. Hence, these two can be separated by using NaOH.

$$Al(OH)_3 + NaOH \longrightarrow NaAlO_2 + 2H_2O$$

$$Fe(OH)_3 + NaOH \longrightarrow \text{No reaction}$$

(v) $NH_4Cl \rightleftharpoons NH_4^+ + Cl^-$

$$2NH_3 \longrightarrow NH_2^- + NH_4^+$$

In ammonia, NH_4^+ (ammonium ion) is formed, which have a tendency to give H^+ (or accept e^-) and hence behaves like an acid.

3. (c) The complex is soluble in $HNO_3(aq)$ to give AgCl back.

$$[Ag(NH_3)_2]Cl + 2HNO_3 \longrightarrow 2NH_4NO_3 + AgCl\downarrow$$

4. (c) Lead salt is the only salt, which gives white turbidity with HCl/H_2O and black ppt with H_2S/HCl.

Thus, the black ppt *Y* is PbS and it is soluble in nitric acid.

5. (d) Since, HCl is not present, so common ion effect does not operate and H_2S ionises to give enough S^{2-} ions. Hence, all Zn^{2+}, Cu^{2+} and Cd^{2+} get precipitated as sulphides.

6. (b) $Cr_2O_7^{2-}$ is orange in colour, while Fe^{2+} is colourless. When acidified, $Cr_2O_7^{2-}$ gets reduced into green coloured Cr^{3+} and Fe^{2+} is oxidised to Fe^{3+}.

$$\underset{\text{(Orange)}}{K_2Cr_2O_7} + 7H_2SO_4 + 6FeSO_4 \longrightarrow K_2SO_4 + \underset{\text{(Green)}}{Cr_2(SO_4)_3} + 3Fe_2(SO_4)_3 + 7H_2O.$$

7. (d) The given reactions suggest that *A* is Hg_2Cl_2 (also called calomel).

$$\underset{(A)}{Hg_2Cl_2} + 2NH_3 \longrightarrow \underset{\text{(Black)}}{HgNH_2Cl + Hg} + NH_4Cl$$

$$\underset{(A)}{Hg_2Cl_2} + SnCl_2 \longrightarrow \underset{\text{(Grey)}}{2Hg}\downarrow + SnCl_4$$

8. (c) Solubility of *X* in hot water and of black ppt in hot nitric acid suggests that *X* is $PbCl_2$ (white substance).

$$\underset{\substack{(X)\\\text{(Soluble in hot water)}}}{PbCl_2} + H_2S \longrightarrow \underset{\substack{\text{Black ppt}\\(\text{Soluble in } HNO_3)}}{PbS\downarrow} + 2HCl$$

$$PbS \xrightarrow{\text{conc. } H_2SO_4} \underset{\text{(White ppt)}}{PbSO_4}\downarrow$$

9. (c) Precipitate of $Fe(OH)_3$ is not formed by just exposing the iron salt to air. All other statements are true.

10. (b) Formation of red ppt with KI which dissolves in excess of KI suggests that the metal ion is Hg^{2+}.

$$Hg^{2+} + 2KI \longrightarrow \underset{\text{Reddish orange}}{HgI_2}\downarrow + 2K^+$$

$$HgI_2 + 2KI \longrightarrow \underset{\substack{\text{II}\\\text{Soluble}\\\text{(colourless)}}}{K_2HgI_4}$$

When Hg^{2+} is treated with Co(III) thiocyanate, a blue ppt of cobalt tetrathiocyanatomercurate (II) is formed.

$$Hg^{2+} + Co^{2+} + 4SCN^- \longrightarrow \underset{\text{(Blue)}}{Co[Hg(SCN)_4]}\downarrow$$

11. (a,b,c) When Cu reacts with conc. H_2SO_4, there is formation of $CuSO_4$ (by oxidation of Cu to Cu^{2+}) and SO_2 (by reduction of SO_4^{2-} to SO_2).

$$Cu \longrightarrow Cu^{2+} + 2e^-$$

$$SO_4^{2-} + 2e^- + 4H^+ \longrightarrow SO_2 + 2H_2O$$

$$SO_4^{2-} + 8H^+ + 8e^- \longrightarrow S^{2-} + 4H_2O$$

This reaction is favourable due to low solubility of CuS and High $[H^+]$.

12. (a,b) Yellow ammonium sulphide is added to dissolve the sulphide of radicals of group II, B (i.e. As^{3+}, Sb^{3+}, Sn^{4+}) by converting them into thio compound, e.g.

$$As_2S_3 + 3(NH_4)_2S \longrightarrow 2(NH_4)_3AsS_3$$

13. (a,b,c,d) Fe^{2+} gives no colouration with $K_4[Fe(CN)_6]$ but Fe^{3+} can give.

All the given reagents oxidise Fe^{2+} into Fe^{3+}, which gives blue colouration with $[K_4Fe(CN)_6]$.

$$Fe^{2+} \xrightarrow[\text{(iii) } H_2O_2/H^+ \text{ or (iv) } HNO_3/H_2SO_4]{\text{(i) } MnO_4^-/H^+ \text{ or (ii) } Cr_2O_7^{2-}/H^+ \text{ or}} Fe^{3+}$$

$$4Fe^{3+} + 3K_4[Fe(CN)_6] \longrightarrow \underset{\text{Prussian blue}}{Fe_4[Fe(CN)_6]_3} + 12K^+$$

14. (d) Ammonium nitrate cannot be used instead of NH_4Cl as NO_3^- ion will oxidise Mn^{2+} ion to Mn^{3+} ions and hence, $Mn(OH)_3$ will be precipitated in group III. NH_4OH is a weak base and HNO_3 or HCl are strong acids.

15. (a) ZnO is an amphoteric oxide, so reacts with aqueous alkali as:

$$ZnO + 2NaOH \longrightarrow Na_2ZnO_2 + H_2O$$

CuO, being basic, does not react with the NaOH so ZnO and CuO can be separated from the mixture by boiling with aqueous alkali (like NaOH, KOH etc.)

16. (d) Cu^{2+} forms complex with KCNS but Cd^{2+} does not.

$$Cu^{2+} + 2KCNS \longrightarrow \underset{\text{Black}}{Cu(SCN)_2} + 2K^+$$

$[Cu(CN)_4]^{3-}$ being stable, remains unaffected from H_2S gas.

17. (c) I. $Fe^{2+} + K_3[Fe(CN)_6] \longrightarrow Fe^{3+} + K_4[Fe(CN)_6]$

$$4Fe^{3+} + 3[Fe(CN)_6]^{4-} \longrightarrow \underset{\text{Turnbull's blue}}{Fe_4[Fe(CN)_6]_3\downarrow}$$

II. $4Fe^{3+} + 3K_4[Fe(CN)_6]^{4-} \longrightarrow \underset{\text{Prussian blue}}{Fe_4[Fe(CN)_6]_3} + 12K^+$

III. $Ni^{2+} + DMG \longrightarrow \underset{\text{(Ni-DMG complex)}}{\text{Cherry red ppt}}$

IV. $Co^{2+} + 4SCN^- \longrightarrow \underset{\text{Blue}}{[Co(SCN)_4]^{2-}}$

(I) → (iv,B), (II) → (iii,C), (III) → (ii,A), (IV) → (i,D)

18. (a) (i) → (3,s), (ii) → (4,p), (iii) → (2,t), (iv) → (1,q), (v) → (5,r)

Zn^{2+}, Ni^{2+} belong to group IV, and their group reagents are H_2S/NH_4OH, NH_4Cl.

Ag^{2+}, Hg_2^{2+} – group I (so dil. HCl)

Ca^{2+}, Ba^{2+} – group V [so, $(NH_4)_2CO_3/NH_4OH$]

Cu^{2+}, Cd^{2+} – group II (so H_2S/H^+)

Al^{3+}, Cr^{3+} – group III (so NH_4Cl/NH_4OH)

19. (a) (i) → (r), (ii) → (r), (iii) → (p,q), (iv) → (p,q,s)

MnO_4^-/H^+ or $Cr_2O_7^{2-}/H^+$ both show a colour change (i.e. get reduced to colourless Mn^{2+} or green Cr^{3+} respectively) with Fe^{2+} and oxidises it to Fe^{3+}.

$$2I^- + Pb^{2+} \longrightarrow \underset{\text{Yellow}}{PbI_2} \xrightarrow{I^- \text{ excess}} \underset{\text{Soluble}}{[PbI_4]^{2-}}$$

$$Ag^+ + I^- \longrightarrow \underset{\text{Yellow}}{AgI\downarrow}$$

$$CrO_4^{2-} + Pb^{2+} \longrightarrow \underset{\text{Yellow}}{PbCrO_4}$$

$$2Ag^+ + CrO_4^{2-} \longrightarrow \underset{\text{Red}}{Ag_2CrO_4\downarrow}$$

$$Ba^{2+} + CrO_4^{2-} \longrightarrow \underset{\text{Yellow}}{BaCrO_4\downarrow}$$

20. (5) (i) $\underset{\text{Insoluble}}{AgCl} + KCN \longrightarrow \underset{\text{Soluble}}{K[Ag(CN)_2]}$

(ii) $AgCl + NH_3 \longrightarrow \underset{\text{Soluble}}{[Ag(NH_3)_2]Cl}$

(iii) $\underset{\text{(Insoluble)}}{HgI_2} + KI \longrightarrow \underset{\text{Soluble}}{K_2HgI_4}$

(iv) $\underset{\text{Soluble}}{BiCl_3} + H_2O \xrightarrow{HCl} \underset{\text{Insoluble}}{BiOCl}$

(v) $\underset{\text{Soluble}}{SbCl_3} + H_2O \xrightarrow{HCl} \underset{\text{Insoluble}}{SbOCl}$

(vi) $NH_3 + \underset{\text{Insoluble}}{AgI} \longrightarrow \underset{\text{Insoluble}}{[Ag(NH_3)]I}$

DPP-4 Qualitative Analysis of Inorganic Salts (Acidic and Basic Radicals)

1. (i) Hydroxide of Na, i.e. NaOH is soluble in water but hydroxide of Fe^{3+} does not. So, water can be used to separate them.

(ii) AgCl is soluble in aqueous NH_3 but AgI does not.

$$AgCl + 2NH_3(aq) \longrightarrow \underset{\text{Soluble}}{[Ag(NH_3)_2]Cl}$$

(iii) MnS is soluble in dil. HCl but CoS does not.

$$MnS + 2HCl \longrightarrow MnCl_2 + H_2S\uparrow$$

(iv) $Ni(OH)_2$ is soluble in NH_4OH but $Fe(OH)_3$ does not dissolve in it.

$$Ni(OH)_2 + 4NH_4OH \longrightarrow \underset{\text{Soluble}}{[Ni(NH_3)_4]^{2+}}$$

(v) Cr_2O_3 gets dissolved in excess of NaOH but $Fe(OH)_3$ does not.

$$Cr_2O_3 + 3H_2O \longrightarrow 2Cr(OH)_3$$

$$Cr(OH)_3 + NaOH \longrightarrow \underset{\text{Soluble}}{Na[Cr(OH)_4]}$$

2. *A* turns red litmus blue so *A* is basic in nature. *A* imparts golden yellow colour in flame so *A* has Na^+. *A* gives H_2 gas with Zn or Al. *A* is NaOH.

$$2NaOH + Zn \longrightarrow Na_2ZnO_2 + H_2\uparrow$$

$$2NaOH + 2H_2O + 2Al \longrightarrow 2NaAlO_2 + 3H_2\uparrow$$

$$ZnCl_2 + 2NaOH \longrightarrow \underset{\text{White ppt}}{Zn(OH)_2\downarrow} + 2NaCl$$

$$Zn(OH)_2 + 2NaOH \longrightarrow Na_2[Zn(OH)_4]$$

or $\underset{\text{Sodium zincate}}{Na_2ZnO_2}$

$$AlCl_3 + 3NaOH \longrightarrow \underset{\text{White ppt}}{Al(OH)_3\downarrow} + 3NaCl$$

$$Al(OH)_3 + NaOH \longrightarrow Na[Al(OH)_4]$$

or $\underset{\text{Sodium metaaluminate}}{NaAlO_2}$

3. (i) $A \xrightarrow{NaOH,\ \Delta} B \xleftarrow{H_2O} Mg_3N_2$

Since, $Mg_3N_2 + 6H_2O \longrightarrow 3Mg(OH)_2 + \underset{'B'}{2NH_3}$

Thus, gas *B* is ammonia and hence, *A* should be an ammonium salt (i.e. having NH_4^+ as basic radical). When the reaction is completed, the mixture becomes ammonia free.

$$NH_4^+ + NaOH \longrightarrow Na^+ + NH_3 + H_2O$$

Now, all the NH_4^+ is replaced by Na^+ and anion of (*A*) gives ring test, i.e. it is a nitrate.

Thus, *A* is ammonium nitrate (NH_4NO_3).

The reaction involved are as follows :

$$\underset{(A)}{NH_4NO_3} + NaOH \xrightarrow{\Delta} \underset{(B)}{NH_3\uparrow} + NaNO_3 + H_2O$$

$$NaNO_3 + H_2SO_4 \longrightarrow NaHSO_4 + HNO_3$$

$$3H_2SO_4 + 6FeSO_4 + 2HNO_3 \longrightarrow 3Fe_2(SO_4)_3 + 2NO + 4H_2O$$

$$[Fe(H_2O)_6]SO_4 + NO \longrightarrow \underset{\text{Brown coloured ring }(C)}{[Fe(H_2O)_5NO]SO_4} + H_2O$$

$$\underset{(A)}{NH_4NO_3} \xrightarrow{\Delta} \underset{(E)}{N_2O} + \underset{(D)}{2H_2O}$$

Thus, F = Hyponitrous acid ($H_2N_2O_2$)

(ii)

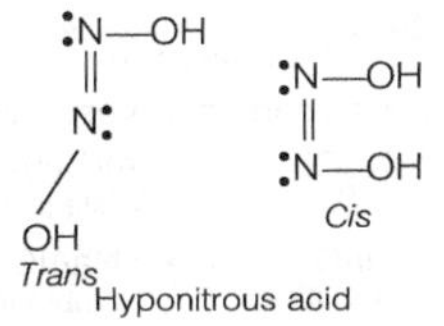

Hyponitrous acid

4. Gas 'C' turns lead acetate black, so 'C' must be H_2S and A must have S^{2-} as acidic radical.

B gives blue colouration with $[Fe(CN)_6]^{3-}$, so it must contain Fe^{2+} ion and hence, A must be FeS.

The reactions are as follows :

$$\underset{'A'}{FeS} + H_2SO_4 \longrightarrow \underset{'B'}{FeSO_4} + \underset{'C'}{H_2S} \xrightarrow{(CH_3COO)_2Pb} \underset{\text{Black ppt}}{PbS\downarrow}$$

$$\underset{'B'}{FeSO_4} \xrightarrow{2NaOH} \underset{\substack{\text{green ppt}\\ 'D'}}{Fe(OH)_2\downarrow} \xrightarrow[\text{Atm.}]{O_2}$$

$$\underset{\substack{\text{Brown}\\ 'E'}}{2Fe_2O_3\cdot H_2O} \xrightarrow{HCl} \underset{\substack{\text{Deep yellow}\\ 'G'}}{2FeCl_3}$$

$$\underset{'G'}{FeCl_3} \xrightarrow{CNS^-} \underset{\text{Red}}{[Fe(CNS)]^{2+}}$$

$$\underset{'B'}{FeSO_4} \xrightarrow{[Fe(CN)_6]^{3-}} \underset{\substack{\text{Turnbull's blue}\\ 'F'}}{Fe[Fe(CN)_6]^-}$$

5. (d) Mix I and II in aqueous solution. If there is formation of white ppt, then these are $Al_2(SO_4)_3\cdot 18H_2O$ and $BaCl_2\cdot 2H_2O$.

$$Ba^{2+} + SO_4^{2-} \longrightarrow \underset{\text{White ppt}}{BaSO_4\downarrow}$$

III is certainly KOH.

Add III into I, if there is white ppt and ppt dissolves in excess of III then I is $Al_2(SO_4)_3\cdot 18H_2O$.

$$Al_2(SO_4)_3 + 6\,KOH \longrightarrow \underset{\text{White ppt}}{2\,Al(OH)_3\downarrow} + 2K_2SO_4$$

$$Al(OH)_3 + KOH \longrightarrow \underset{\text{Soluble}}{KAl(OH)_4}$$

Add III into II and there is no ppt formation then II is certainly $BaCl_2\cdot 2H_2O$.

Thus, I is $Al_2(SO_4)_3\cdot 18\,H_2O$, II is $BaCl_2\cdot 2H_2O$ and III is KOH.

6. (c) $PbCl_2 + 2KI \longrightarrow \underset{A}{PbI_2} + 2KCl$

$$\downarrow KI$$

$$\underset{\substack{(B)\\ \text{Soluble}}}{K_2[PbI_4]}$$

7. (c) (i) $2CuSO_4 + 4KI \longrightarrow \underset{\text{White ppt}}{Cu_2I_2\downarrow} + I_2 + 2K_2SO_4$

$$I_2 + \underset{\text{Hypo}}{2Na_2S_2O_3} \longrightarrow 2NaI + Na_2S_4O_6$$

(ii) $I^- + Ag^+ \longrightarrow \underset{\substack{\text{Yellow}\\ \text{(insoluble in}\\ NH_4OH)}}{AgI\downarrow}$

8. (d) Sulphides of Ni and Co do not precipitate in the presence of acid ($\because$ In aqueous solution Cl^- combines with H^+ to give HCl). Between Cu and Cd, later one give this reaction in excess of KCN as:

$$CdSO_4 + 4\,KCN \longrightarrow \underset{\text{Unstable (I)}}{K_2[Cd(CN)_4]} + K_2SO_4$$

$$CuSO_4 + 8\,KCN \longrightarrow \underset{\substack{\text{Stable (II)}\\ \text{(very less dissociation)}}}{2K_3[Cu(CN)_4]} + K_2SO_4$$

Since, I being unstable dissociates to give Cd^{2+}, hence, it will give the ppt of CdS when treated with H_2S.

9. (b) $X \xrightarrow{HNO_3/AgNO_3}$ White ppt [soluble in NH_3 (aq)]

The reaction shows that white ppt is of AgCl and X contains Cl^- ion as acidic radical.

$$X \xrightarrow{NH_4OH/H_2S} \text{White ppt} \xleftarrow{H_2S}$$

$$X \xrightarrow{NaOH} \text{White ppt} \xrightarrow{NaOH} \text{dissolve}$$

In basic medium, i.e. in IV group, Zn^{2+} forms white ppt of ZnS with H_2S.

$$ZnCl_2 + H_2S \longrightarrow \underset{\text{White ppt}}{ZnS\downarrow} + 2HCl$$

$$\therefore\; ZnCl_2 + 2NaOH \xrightarrow[-\,2NaCl]{} Zn(OH)_2\downarrow \xrightarrow[2H_2O]{2NaOH} Na_2ZnO_2$$

$$\xrightarrow[-\,2NaOH]{H_2S} \underset{\substack{\text{White ppt}\\ (Y)}}{ZnS}$$

10. (b) $PbCrO_4 + 2HCl \longrightarrow PbCl_2 + H_2CrO_4$

$$\downarrow \text{Cooling}$$

$$\text{White ppt} \xrightarrow{\Delta} \text{Dissolves}$$

Thus, compound is lead chromate.

11. (a) Since, C turns lime water milky, so C must be CO_2 or SO_2.

B burns with blue flame, so it must be CO.

CO_2 and CO are produced when oxalate salt is heated, so A is an oxalate.

The metal present in lime water, i.e. E is Ca, so A is calcium oxalate.

$$CaC_2O_4 \xrightarrow{\Delta} \underset{B}{CO} + \underset{'C'}{CO_2} + \underset{'D'}{CaO} \xrightarrow{H_2O} \underset{\substack{'E'\\ \text{Lime water}}}{Ca(OH)_2}$$

$$CO_2 + \underset{[Ca(OH)_2]}{E} \longrightarrow \text{milky } (CaCO_3)$$

12. (a) Formation of white ppt with $AgNO_3$ suggests the presence of Cl^- ion, i.e. A is a chloride.

Formation of deep blue colour in the non-luminous flame of borax bead test indicates that the cation present in Co^{2+}. Thus, the salt A is $CoCl_2$.

Since, $Co(OH)_2$ is a weak base and HCl is a strong acid, so aqueous solution of A is acidic.

$$Co^{2+} + 2Cl^- + H_2O \longrightarrow Co(OH)_2 + H^+ + Cl^-$$

$$Co^{2+} + H_2O \longrightarrow Co(OH)_2 + H^+$$

i.e. cationic hydrolysis.

13. (b) X gives yellow dirty ppt with H_2S soluble in YAS (yellow ammonium sulphide) so

X has Sn^{2+}.

X gives white ppt with $HNO_3/AgNO_3$, so has Cl^-.

X has Cl^- ion and thus X is $SnCl_2$.

The reaction involved are as follows :

(i) $SnCl_2 + 2HgCl_2 \longrightarrow \underset{\text{White}}{Hg_2Cl_2\downarrow} + SnCl_4$

$$SnCl_2 + Hg_2Cl_2 \longrightarrow \underset{\text{Grey}}{2Hg\downarrow} + SnCl_4$$

(ii) $\underset{\text{Yellow}}{2FeCl_3} + SnCl_2 \longrightarrow \underset{\text{Green}}{2FeCl_2} + SnCl_4$

(iii) $SnCl_2 + 2NaOH \longrightarrow \underset{\text{White}}{Sn(OH)_2\downarrow} + 2NaCl$

$$Sn(OH)_2 + 2NaOH \longrightarrow \underset{\substack{\text{Soluble}\\ \text{(sodium stannite)}}}{Na_2SnO_2} + 2H_2O$$

(iv) $SnCl_2 + H_2S \longrightarrow 2HCl + \underset{\text{Yellow}}{SnS\downarrow}$

$$\underset{\text{YAS}}{SnS + (NH_4)_2S_2} \longrightarrow \underset{\text{Soluble}}{(NH_4)_2SnS_3}$$

14. (b) Molar mass of $7H_2O = 7[1 \times 2 + 16] = 7 \times 18 = 126$

$\therefore$ Molar mass of $M_2X = 126$

[$\because$ Water and M_2X are in 1 : 1 by weight]

Since, salt gives a gas with dil. H_2SO_4 which turns $Cr_2O_7^{2-}/H^+$ green, so gas is SO_2 and salt is a SO_3^{2-} (sulphite salt).

Thus, $X = SO_3^{2-} \Rightarrow M_2SO_3 = 126$

$$2x + 32 + 16 \times 3 = 126$$
$$2x + 32 + 48 = 126$$
$$2x = 126 - 80$$
$$2x = 46$$
$$x = 23 \quad \text{(atomic mass of Na)}$$

So, M = Na

Thus, formula of the salt in $Na_2SO_3 \cdot 7H_2O$.

15. (a,b,c,d) $Cu^{2+} \xrightarrow{4CN^-} [Cu(CN)_4]^{3-}$

$$HgCl_2 + 2KI \xrightarrow[-2KCl]{} HgI_2\downarrow \xrightarrow{KI} \underset{\text{Soluble}}{K_2HgI_4}$$

$$NiCl_2 + 4KCN \longrightarrow K_2[Ni(CN)_4] + KCl$$

$$As_2S_3 + 3(NH_4)_2S \longrightarrow 2(NH_4)_3AsS_3$$

16. (a,d) Only HgS is the black sulphide which is insoluble in YAS (yellow ammonium sulphide) as well as sulphuric acid but dissolves in aqua-regia.

The nitrate ion of aqua-regia in the presence of H^+ ion oxidise the S^{2-} ions to free S and the Cl^- ion reacts with Hg^{2+} to form soluble complex ion $HgCl_4^{2-}$.

$$3HgS + 2NO_3^- + 8H_3O^+ + 12Cl^- \longrightarrow 3HgCl_4^{2-} + 2NO + 3S + 12H_2O$$

17. (a,b,d) A gives white ppt with NH_4Cl and NH_4OH, so A contains Al^{3+} ion and the ppt is of $Al(OH)_3$, which is soluble in NaOH. So, B is NaOH, and C is sodium metaaluminate ($NaAlO_2$).

Thus, (a) B (NaOH) turns red litmus blue.

(b) $NaOH + ZnO \longrightarrow \underset{\text{Sodium zincate}}{Na_2ZnO_2} + H_2O$

(c) The white fumes are of HCl, not of $AlCl_3$.

$$AlCl_3 + 3H_2O \longrightarrow Al(OH)_3 + 3HCl\uparrow \text{ (fumes)}$$

(d) Being electron deficient, $AlCl_3$ acts as a Lewis acid and used as a catalyst in Friedel-Crafts reaction.

$$2AlCl_3 \xrightarrow[\text{Dimerisation}]{} Al_2Cl_6$$

The reactions involved can be represented as

$$\underset{(A)}{AlCl_3} + \underset{(B)}{3NaOH} \longrightarrow \underset{\text{White ppt}}{Al(OH)_3} + 3NaCl$$

$$Al(OH)_3 + \underset{\text{(excess)}}{NaOH} \longrightarrow \underset{\text{Soluble }(C)}{NaAlO_2} + 2H_2O$$

$$AlCl_3 + 3H_2O \longrightarrow Al(OH)_3 + 3HCl\uparrow \text{ (fumes)}$$

$$AlCl_3 + 3NH_4OH \xrightarrow{NH_4Cl} Al(OH)_3\downarrow + 3NH_4Cl$$

Explanation for Q. Nos. (18-20)

Conversion of blue litmus red suggests that A is an acidic salt. Treatment with NH_4SCN gives red colouration which suggest that A has Fe^{3+}. Chromyl chloride test suggests the presence of Cl^-. Thus, A is $FeCl_3$.

The reactions are as follows :

$$FeCl_3 \xrightarrow{\text{blue litmus}} \text{Red}$$

$$FeCl_3 \xrightarrow{NH_4SCN} \underset{\substack{\text{Red}\\ E}}{Fe(SCN)_3}$$

$$FeCl_3 \xrightarrow{Cr_2O_7^{2-}/H^+} \underset{\substack{\text{Red vapours}\\ B}}{CrO_2Cl_2} \xrightarrow{NaOH} \underset{C}{Na_2CrO_4} \xrightarrow[(AcO)_2/Pb]{AcONa} PbCrO_4$$

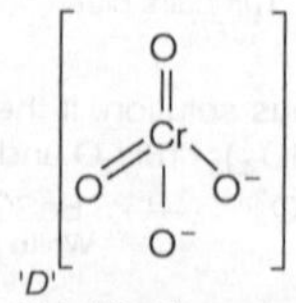

'D'

Hybridisation = sp^3

structure = tetrahedral

Acidic radical of A is Cl^-

$$MnO_2 + 4Cl^- \longrightarrow MnCl_2 + Cl_2 + 2O^{2-}$$

$$Cl_2 + 2Fe^{2+} \longrightarrow 2Fe^{3+} + 2Cl^-$$

$$\underset{\text{Excess}}{Cl_2} + NH_3 \longrightarrow \underset{\text{Explosive}}{NCl_3} + 3HCl$$

18. (b) **19.** (d) **20.** (b)

DPP-5 Organic Qualitative Analysis

1.

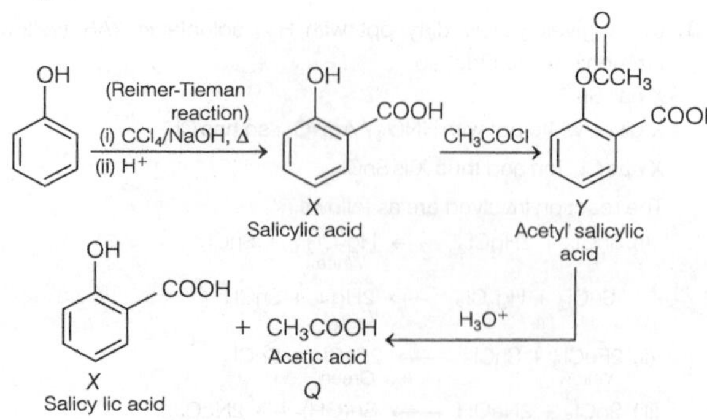

(i) $NaHCO_3$, (both give effervescence of CO_2).

(ii) $FeCl_3$ test in which X gives violet colour as

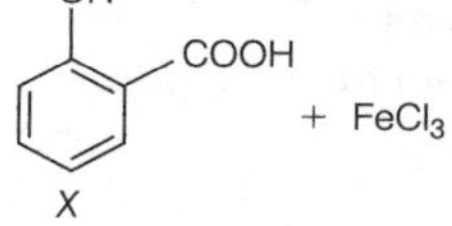

Esterification (i.e. treatment with acid) etc.

(iii) Y is also called aspirin (acetyl salicylic acid).

2. $C_3H_6O_2 \xrightarrow[H^+]{H_2O} A + B$

The molecular formula and the above reaction suggests that $C_3H_6O_2$ is an ester. So, *A* and *B* are acid and alcohol.

Since, *A* gives positive test with Fehling solution as well as with Tollen's reagent, so *A* should be formic acid (HCOOH, the only acid with reducing nature).

Since, *B* gives iodoform (*C*) with I_2 and *aq*. NaOH, so *B* should be ethanol (C_2H_5OH).

Thus, ester (organic compound) is $HCOOC_2H_5$.

The reactions are as follows :

$$\underset{\text{Ethyl formate}}{HCOOC_2H_5} \xrightarrow[H^+]{HOH} \underset{'A'}{HCOOH} + \underset{'B'}{C_2H_5OH}$$

$$\underset{(A)}{HCOOH} + \underset{\text{Tollen's reagent}}{Ag_2O} \longrightarrow H_2O + CO_2 + 2Ag\downarrow$$

$$\underset{(A)}{HCOOH} + \underset{\text{Fehling solution}}{2CuO} \longrightarrow H_2O + CO_2 + Cu_2O\downarrow$$

$$\underset{(B)}{C_2H_5OH} + 4I_2 + 6NaOH \longrightarrow \underset{C}{CHI_3} + \underset{\text{Sodium formate}}{HCOONa} + 5NaI + 5H_2O$$

$$\underset{(C)}{2CHI_3} + 6Ag \longrightarrow \underset{\text{Acetylene}}{CH\equiv CH} + 6AgI$$

3. (a) The possible isomers of the formula C_7H_8O are as follows :

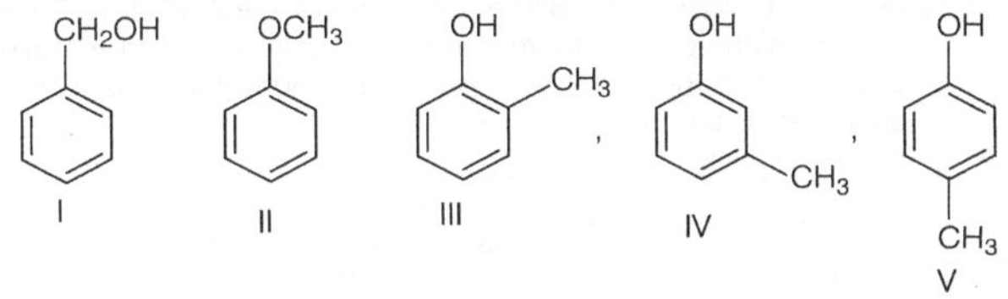

Benzyl alcohol (I) can be distinguished from III, IV and V by following tests.

(i) I decolourises alkaline $KMnO_4$ and give a brownish black ppt of MnO_2.

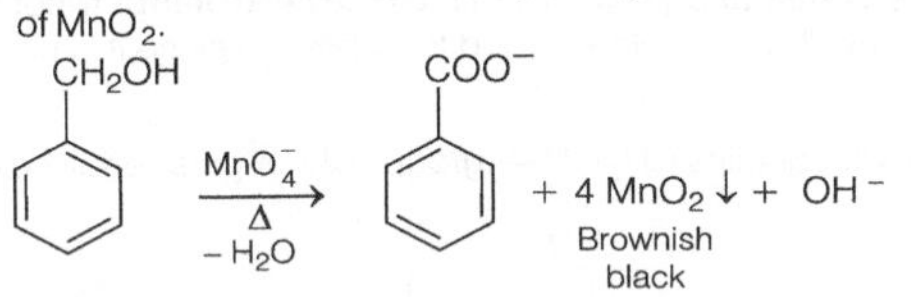

This reaction is not given by phenols.

(ii) Phenols give violet colouration with $FeCl_3$ but this test is not given by benzyl alcohol.

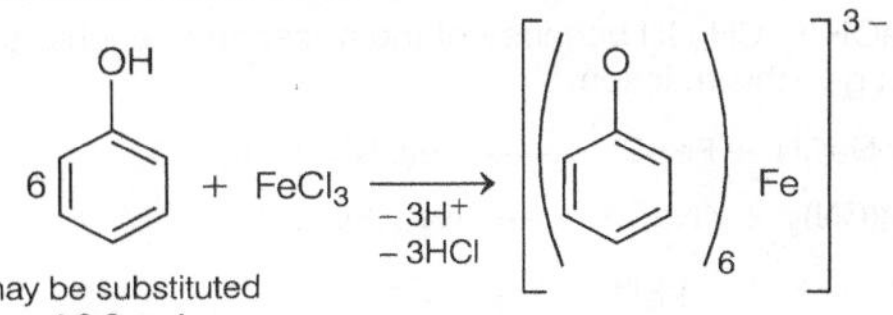

(may be substituted at 2,3 or 4 position)

Further II is neutral towards litmus and does not liberate H_2 with metal (Na) while other can liberate.

(b) Benedict's solution (Cu^{2+} + citric acid + base) and Fehling solution (Cu^{2+} + tartaric acid + base) are weaker oxidising agents than Tollen's reagent, so they can oxidise the aliphatic aldehydes but not capable of oxidising aromatic aldehydes.

4. (a) Benzyl amine is an aliphatic amine, so does not form stable diazonium salt and hence, does not give dye test.

Phthalic acid does not give fluorescence test and acetamide is not a primary amine, so does not give carbylamine test. Phenol however gives phthalein test.

5. (d) Although fructose contains $\gt C{=}O$ group but in alkaline medium, it gets converted into its enomers like mannose, glucose etc, so gives positive Fehling's solution test, Tollen's test and Benedict's test. Thus, these tests cannot be used to distinguish between the glucose and fructose. However, optical rotation of glucose is opposite of fructose, so these two can be distinguished on this basis.

6. (c) Organic compound '*X*' C_5H_8O does not react appreciably with Lucas reagent at room temperature so, it contains a primary alcohol ($-CH_2OH$) group. Further, it gives white ppt with ammoniacal $AgNO_3$, which indicates the presence of a terminal alkyne linkage. On hydrogenation followed by reaction with HI, pentane is formed, which indicates that '*X*' is a straight chain compound.

Thus, the possible structure of *X* is

$$\underset{OH}{CH_2}CH_2CH_2C\equiv CH$$

X contains two acidic H-atoms thus, 1 mole of *X* will give 2 moles of CH_4.

$$\underset{'X'}{\underset{OH}{CH_2}CH_2CH_2C\equiv CH} + 2CH_3MgBr \longrightarrow 2CH_4 + BrMgOCH_2CH_2CH_2C\equiv \underset{OMgBr}{C}$$

$\because$ 84g *X* gives 2×22.4 L CH_4

$\therefore$ 0.42g *X* will give $\dfrac{2 \times 22.4 \times 0.42}{84} = 224$ mL CH_4

7. (b) Liberation of CO_2 with $NaHCO_3$ indicates the presence of a $-COOH$ group while colour with neutral $FeCl_3$ shows the presence of $-OH$ group in the form of phenol (i.e. attached directly with the benzene nuclei).

8. (c) (i) $\rightarrow$ (q), (ii) $\rightarrow$ (r), (iii) $\rightarrow$ (p), (iv) $\rightarrow$ (s)

(i) Ethyne because of the presence of acidic H^+, gives ppt with ammoniacal Cu_2Cl_2 but ethene due to the absence of acidic hydrogen does not give this reaction.

$$CH\equiv CH \xrightarrow[NH_4OH]{Cu_2Cl_2} CCu\equiv CCu$$

(ii) $FeCl_3$ (neutral) can distinguish between HCOOH and CH_3COOH.

(iii) Biuret gives blue colour with $CuSO_4$ but glycine does not give this test.

(iv) *o*-toluidine being an aromatic amine forms dye with $NaNO_2$/HCl, phenol but benzylamine being aliphatic amine does not give this test.

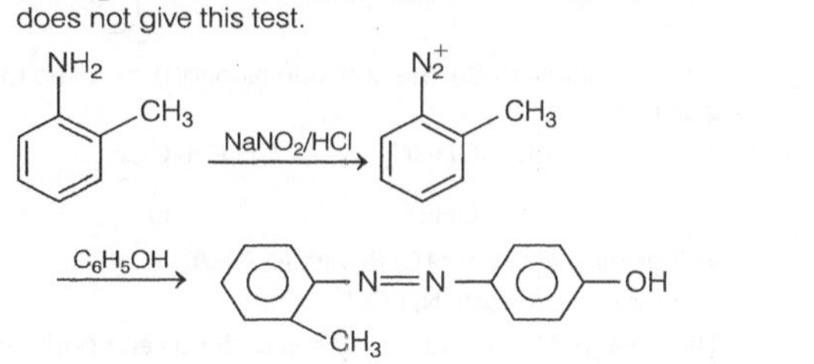

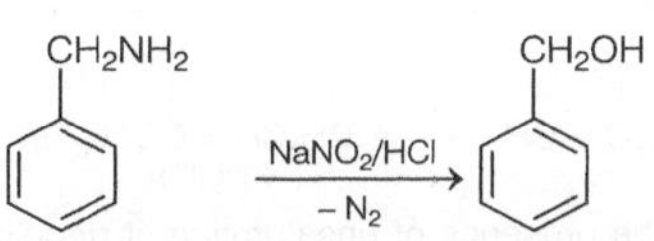

9. (a) (i) $\rightarrow$ (s), (ii) $\rightarrow$ (q,r), (iii) $\rightarrow$ (r), (iv) $\rightarrow$ (p)

Carboxylic acid gives effervescence of CO_2 with $NaHCO_3$.

Amides are hydrolysed by NaOH and diazotised with HNO_2 (However, the salt form is unstable).

Amines also get diazotised with HNO_2. However the aliphatic diazo compounds are highly unstable.

—OH(enolic) group gives characteristic colour with $FeCl_3$.

10. (b) (i) → (q), (ii) → (q,s), (iii) → (p,r), (iv) → (p)

Fructose in alkaline medium converts into its epimers and hence, reduces Tollen's reagent to silver mirror.

Benzaldehyde because of the presence of —CHO group gives silver mirror test and also give malachite green test.

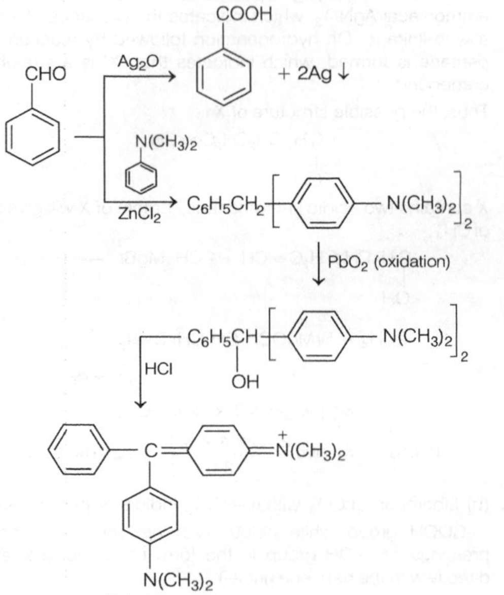

Malachite green

Phenols give dye test as well as Libermann's nitroso test while aniline being a 1° amine gives only dye test.

11. (a, c, d) The carbocation formed in case of (a), (c) and (d) are stabilised by resonance, so they readily give Cl^- in solution. Hence, there is no need in case of them to convert this chloride into sodium extract.

12. (a, b, c, d) To prepare butanol-2 by using CH_3MgBr, the aldehyde required is propanal. The reaction takes place in the following manner:

$$CH_3CH_2CHO \xrightarrow[\text{(ii)}NH_4Cl/H_2O]{\text{(i)}CH_3MgBr} CH_3CH_2\underset{\displaystyle OH}{\underset{|}{C}}HCH_3$$

In the presence of Cu, the obtained alcohol is oxidised to carbonyl group.

$$CH_3CH_2\underset{\displaystyle OH}{\underset{|}{C}}HCH_3 \longrightarrow CH_3CH_2\underset{\displaystyle O}{\underset{\|}{C}}CH_3$$

⇒ It gives yellow ppt of CHI_3 with $NaOH/I_2$.

⇒ It forms oxime with NH_2OH.

The obtained ketone tautomerises under given conditions to give following alcohol.

$$CH_3CH_2\overset{\displaystyle O}{\overset{\|}{C}}—CH_3 \longrightarrow \underset{\text{enol (C)}}{CH_3CH=\overset{\displaystyle OH}{\overset{|}{C}}CH_3}$$

⇒ Because of the presence of unsaturation, it decolourises Br_2 water.

⇒ Because of enol group, it gives colour with neutral $FeCl_3$.

13. (d) CCl_4 because of the presence of covalent bond, does not give free Cl^- ions and hence, does not give ppt with $AgNO_3$.

14. (a) Benedict's solution is a mixture of copper sulphate, sodium carbonate and sodium citrate. Here, Cu^{2+} is complexed with citrate ion which prevent the precipitation of Cu^{2+} as $Cu(OH)_2$.

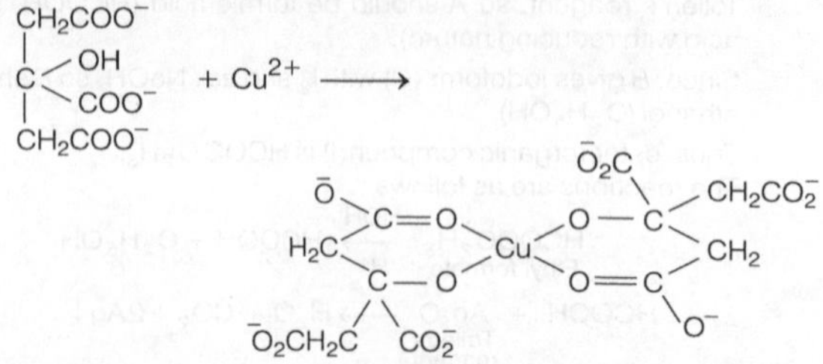

Copper-citrate complex (Benedict's solution)

15. (c) CH_3COCH_2OH (α-hydroxy ketone) can reduce Tollen's reagent but all the carbonyl compounds do not give this reaction. It is a characteristic of only aldehydes, not of ketones.

16. (b) By HCl a basic compound like 4-methyl aniline can be extracted but an acidic compound like phenol does not. So, it can be used to separate them.

17. (b) Z is an isocyanide. Since, oxidation of Z followed by heating with sodalime gives a hydrocarbon, so only one carbon chain is attached with the benzene ring and NH_2 group is present in side chain. i.e. it is benzylamine (II)

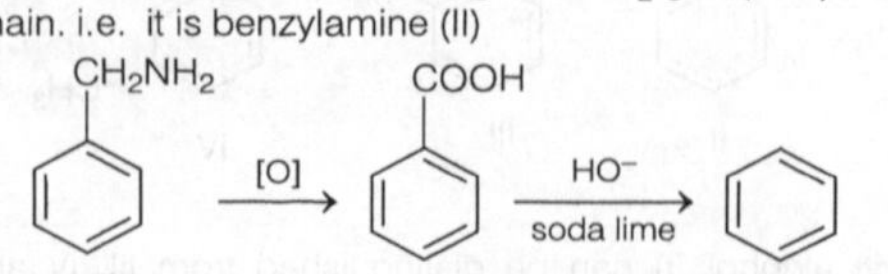

(II) Hydrocarbon B

The compound present in NaOH layer gives buff coloured ppt with $FeCl_3$ and is also a good food preserver, so it is benzoic acid (V).

18. (a) VIII contains $CH_3\underset{\displaystyle OH}{\underset{|}{C}}H—$ group, which gets readily oxidised to methyl ketone group $\left(CH_3\underset{\displaystyle O}{\underset{\|}{C}}—\right)$ and hence, forms iodoform with $NaOH/I_2$. CH_3OH because of the absence of such a group does not give this reaction.

19. (5) $NaCN + FeSO_4 \longrightarrow Fe(CN)_2 + Na_2SO_4$

$Fe(CN)_2 + 4NaCN \longrightarrow Na_4[Fe(CN)_6]$

$$Fe^{2+} \xrightarrow{OH^-} Fe^{3+}$$

$$4Fe^{3+} + 3Na_4Fe(CN)_6 \longrightarrow 12Na^+ + \underset{\text{Prussian blue}}{Fe_4[Fe(CN)_6]_3}$$

O.S of Fe in $Fe_4[Fe(CN)_6]_3$ = +2 and +3

So, sum of OS = + 5

20. (3) During acetylation, one —OH group is replaced by a —$COCH_3$ group, i.e. acetylation of one —OH shows an increase of C_2H_2O unit in alcohol molecule.

Since, the reaction shows an increase of 3 units of C_2H_2O (i.e. $C_6H_6O_3$) during acetylation of the given organic compound, so it has three —OH groups. Since, two —OH group cannot exist at the same C atom. Thus, one —OH is present at each carbon. Thus, organic compound is glycerol.

DPP-6 Quantitative Analysis

1. Normality of caustic soda $= \frac{20}{40} N = \frac{N}{2}$

Normality of $Na_2CO_3 = \frac{20}{53} = \frac{N}{2.65}$

Normality of $NaHCO_3 = \frac{20}{84} = \frac{N}{4.2}$

The volume of N—HCl used for the complete neutralisation of various substances can be calculated as

For NaOH, $N_1V_1 = N_2V_2$

$$\frac{N}{2} \times 50 = N \times V_2$$

$$V_2 = 25 \text{ mL}$$

For Na_2CO_3, $N_1V_1 = N_2V_2$

$$\frac{N}{2.65} \times 50 = N \times V_2$$

$$V_2 = 18.8 \text{ mL}$$

For $NaHCO_3$, $N_1V_1 = N_2V_2$

$$\frac{N}{4.2} \times 50 = NV_2$$

$$V_2 = 11.9 \text{ mL}$$

Thus,

(i) **When first titrated with phenolphthalein** (HPh)

HPh gives full titre value of NaOH and only half titre value of Na_2CO_3

So, titre value $= 25 + \frac{18.8}{2} = 34.4$ mL

(ii) **When MeOH is added after the first end point**

MeOH will give half titre value of Na_2CO_3 and full titre value of $NaHCO_3$

i.e. $\frac{18.8}{2} + 11.9 = 21.3$ mL

(iii) **When MeOH is added from the very beginning**

Here, it gives full titre value of all

i.e. $= 25 + 18.8 + 11.9 = 55.7$ mL

2. $\underset{KMnO_4}{N_1V_1} = \underset{\text{oxalic acid}}{N_2V_2}$

$$N_1 \times 37.5 = \frac{N}{10} \times 25$$

$$N_1 = \frac{25}{10 \times 37.5} = \frac{N}{15}$$

Before reduction,

$$24 \text{ mL of } \frac{N}{15} KMnO_4 = 24 \text{ mL of } \frac{N}{15} Fe^{2+}$$

$$= \frac{24 \times 56}{15 \times 1000} \text{ g/25 mL}$$

$$= \frac{24 \times 56}{15 \times 1000} \times \frac{1000}{25} \text{ g/L}$$

$$= 3.58 \text{ g/L}$$

After reduction,

$$36.75 \text{ mL of } \frac{N}{15} KMnO_4 = 36.75 \text{ mL of } \frac{N}{15} Fe^{2+} \text{ion}$$

$$= \frac{36.75 \times 56}{15 \times 1000} \text{ g/50 mL}$$

$$= \frac{36.75 \times 56}{15 \times 1000} \times \frac{250}{50} \text{ g/250 mL}$$

$$= 0.6861 \text{ g/250 mL}$$

But 0.6861 g of total Fe^{2+} are originally present in 100 mL of solution.

$\therefore$ Amount of Fe^{2+}/L = 6.861 g

$\therefore$ Fe^{3+} ion = 6.861 − 3.58 = 3.281 g

Amount of Fe^{2+} ion = 3.580 g and of Fe^{3+} ion = 3.281 g

3. (a) The solution is heated to increase the rate of reaction. Because with increase in temperature number of collisions (hence, chances of collisions) increases thus, rate increases.

4. (b) In the titration (reaction) of $KMnO_4$ with oxalic acid, in the presence of acid (mineral acid), HNO_3 cannot be used because it itself is an oxidising agent, so interfere with the oxidising action of $KMnO_4$.

5. (a) For H_2CO_3 $pK_{a_1} = -\log (K_{a_1})$

$$= -\log (2.5 \times 10^{-4}) = 3.6$$

$$pK_{a_2} = -\log (K_{a_2})$$

$$= -\log (2.4 \times 10^{-8}) = 7.619$$

The pH range of HPh (phenolphthalein) lies in between 8.4 to 10. So, it cannot be used here.

pH range of MeOH varies from 3.4 to 4.4, so it can be used for this titration.

6. (b) Amount of Mohr's salt, $w = \frac{ENV}{1000}$

$$= \frac{392 \times N \times 250}{20 \times 1000} = 4.9 \text{ g}$$

7. (a) Equivalent of $KMnO_4$ used $= \frac{50}{1000} \times \frac{1}{10} = 0.005$

$\therefore$ Equivalent of Mohr's salt reacted = 0.005

$\therefore$ Molecular mass of Mohr's salt required = 0.005 × 392 = 1.96 g

Thus, percentage purity of Mohr's salt

$$= \frac{1.96}{3.92} \times 100 = 50\%$$

8. (d) $$5C_2O_4^{2-} + \underset{\text{Purple}}{2MnO_4^-} + 16H^+ \longrightarrow \underset{\substack{\text{Colourless} \\ \text{(autocatalyst)}}}{2Mn^{2+}} + 8H_2O + 10CO_2$$

9. (c) pH of I > II and pH $\propto \frac{1}{\text{acidity}}$ thus I is a weak acid than II.

Although graph indicates that both the acids have same concentration.

10. (c) Equivalent mass of $MnO_4^- = \frac{\text{molar mass}}{7-2} = \frac{\text{molar mass}}{5}$

Similarly equivalent mass of $C_2O_4^{2-}$

$$= \frac{\text{Molar mass}}{2\,(4-3)} = \frac{\text{Molar mass}}{2}$$

Milliequivalent of $KMnO_4 = 50 \times 5 \times 0.04 = 10$

$\therefore$ Milliequivalent of $H_2C_2O_4 = 50 \times 2 \times 0.1 = 10$

11. (b) Suppose x g of potassium dichromate is present in the mixture.

So, $KMnO_4$ present = $(0.5 - x)$ g

$$\therefore \quad \left(\frac{x}{49} + \frac{0.5 - x}{31.6}\right) = \frac{100 \times 0.15}{1000}$$

(Here, 49 = equivalent weight of potassium dichromate and 31.6 = Equivalent weight of $KMnO_4$)

$\therefore$ $x = 0.0732$ g

$$\% \text{ of } K_2Cr_2O_7 = \frac{0.0732 \times 100}{0.5} = 14.64\%$$

12. (c) pH at the intermediate range $= \frac{3 + 6.3}{2} = 4.65$

$$NH_4^+ + H_2O \rightleftharpoons NH_4OH + H^+$$

[Aqueous solution of NH_4Br is acidic due to cationic hydrolysis.]

Hence, $pH = 7 - \frac{pK_b}{2} - \frac{\log C}{2} = 5.17$

Since, the order of pH is

$$3.1 < 5.17 < 6.3$$

So, the colour of indicator is orange.

13. (b,c,d) In redox titrations, loss and gain of electrons, i.e. oxidation and reduction take place simultaneously. Thus, titrations of oxalic acid *vs* $KMnO_4$, Mohr's salt *vs* $KMnO_4$ and As_2O_3 *vs* I_2 are the examples of redox titrations.

As_2O_3 *vs* I_2 however is also a iodimetric titration, a special type of redox titration.

14. (a,b) Molecular weight of oxalic acid = 126 g mol^{-1}

$\therefore$ Moles of oxalic acid

$$= \frac{126\ \text{g}}{126\ \text{g mol}^{-1}} = 1\ \text{mol}$$

$$2MnO_4^- + 5C_2O_4^{2-} + 16H^+ \longrightarrow 2Mn^{2+} + 10CO_2 + 8H_2O$$

5 moles of $C_2O_4^{2-}$ require $MnO_4^- = 2$ moles

$\therefore$ 1 mole of $C_2O_4^{2-}$ will require $MnO_4^- = \frac{2}{5}$ mol

Similarly,

$$Cr_2O_7^{2-} + 3C_2O_4^{2-} + 14H^+ \longrightarrow 2Cr^{3+} + 6CO_2 + 7H_2O$$

3 moles $C_2O_4^{2-}$ require $Cr_2O_7^{2-} = 1$ mol

$\therefore$ 1 mole $C_2O_4^{2-}$ will require $Cr_2O_7^{2-} = \frac{1}{3}$ mol

15. (d) $6Fe^{2+} \longrightarrow 6Fe^{3+} + 6e^-$
(From Mohr's salt)

$$Cr_2O_7^{2-} + 14H^+ + 6e^- \longrightarrow 2Cr^{3+} + 7H_2O$$

$N(Cr_2O_7^{2-}) = 6 \times M(Cr_2O_7^{2-}) = 0.6$ N

$$5Fe^{2+} \longrightarrow 5Fe^{3+} + 5e^-$$

$N(MnO_4^-) = 5 \times M(MnO_4^-) = 0.5$ N

Volume of $Fe^{3+} = V$ mL

and molarity of $Fe^{2+} \equiv 1$ M = 1 N

$$6Fe^{2+} \equiv Cr_2O_7^{2-}$$

or $$1Fe^{2+} \equiv \frac{Cr_2O_7^{2-}}{6}$$

$$5Fe^{2+} \equiv MnO_4^-$$

or $$1Fe^{2+} \equiv \frac{MnO_4^-}{5}$$

In all cases 1 equivalent of $X \equiv$ 1 equivalent of Y.

16. (a) (For MnO_4^-) $N_1V_1 = N_2V_2$

$$1 \times V = 0.5 \times X$$
$$X = 2V$$

Further, for $Cr_2O_7^{2-}$, $N_1V_1 = N_2V_2$

$$1 \times V = 0.6 \times Y$$
$$Y = 1.66\ V$$

Thus, $X > Y$

17. (d) At pH 3-4, phenolphthalein exists in its benzenoid form but methyl orange exists in its quinonoid form. The colour of quinonoid form is usually deeper than the benzenoid form.

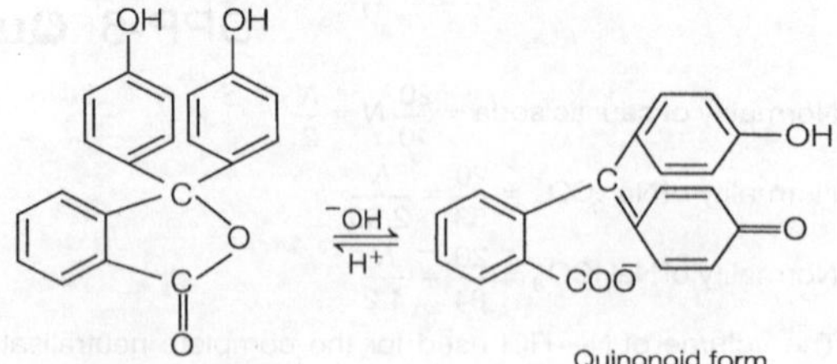

Benzenoid form Phenolphthalein

Quinonoid form at pH 9.7

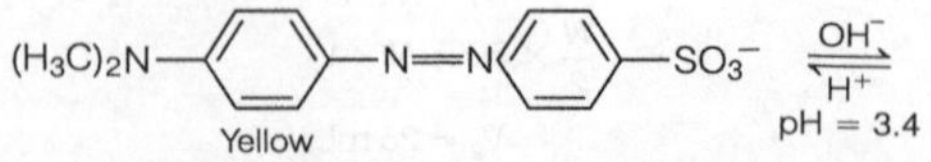

Yellow (Benzenoid form)

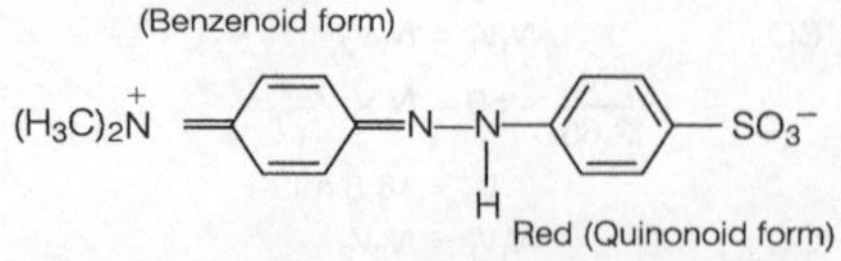

Red (Quinonoid form)

18. (a) (i)→(p,r), (ii)→(p, r), (iii)→(q,r,s), (iv)→(p,r)

All the given reactions involve oxidation of one species and reduction of other, so these all are the examples of redox titrations.

(i) $10FeSO_4 + 2KMnO_4 + 8H_2SO_4 \longrightarrow K_2SO_4 + 2MnSO_4 + 5Fe_2(SO_4)_3 + 8H_2O$

$$2KMnO_4 \equiv 5O \equiv 10H$$

or $8H^+ + MnO_4^- + 5e^- \longrightarrow Mn^{2+} + 4H_2O$

So equivalent weight $= \frac{M}{5}$

Equivalent weight of $Fe^{2+} = M$

(ii) $2CuSO_4 + 4KI \longrightarrow Cu_2I_2 + I_2 + 2K_2SO_4$

$$2CuSO_4 \equiv 2 \times I_2$$

i.e. 2×249.71 g $CuSO_4 = 2 \times 127$ g I_2

$$249.71\ \text{g}\ CuSO_4 = 127\ \text{g}\ I_2$$

Thus, equivalent weight of $CuSO_4 = M$

(iii) $As_2O_3 + 2I_2 + 2H_2O \longrightarrow As_2O_5 + 4HI$

$\therefore As_2O_3 \equiv 4I = 4H$ equivalent weight of $As_2O_3 = \frac{M}{4}$

(iv) $K_2Cr_2O_7 + 7H_2SO_4 + 6FeSO_4 \longrightarrow K_2SO_4 + Cr_2(SO_4)_3 + 3Fe_2(SO_4)_3 + 7H_2O$

$\therefore$ Equivalent weight of $K_2Cr_2O_7 = \frac{M}{6}$

Equivalent weight of $Fe^{2+} = M$

19. (3) External indicators are those which are not added to the volumetric flask in which the titration is carried out.

Among the given, only potassium ferricyanide, uranyl acetate and uranyl nitrate are the examples of external indicators.

20. (5) $MnO_4^- + 8H^+ + 5e^- \longrightarrow Mn^{2+} + 4H_2O$

$[Fe^{2+} \longrightarrow Fe^{3+} + e^-] \times 5$
(From Mohr's salt)

$$MnO_4^- + 8H^+ + 5Fe^{2+} \longrightarrow Mn^{2+} + 5Fe^{3+} + 4H_2O$$

DPP-7 Principles Involved in Synthesis, Neutralisation and Kinetic Studies

1. (a) Because potassium reacts with water.

2. (d) Number of water of crystallisation in alum = 24

Number of water of crystallisation in Mohr's salt = 6

$$\text{Ratio} = \frac{24}{6}$$

$$= 4:1$$

3. (c) This is because, this mixture also oxidises amino group of the aniline.

4. (c) Enthalpy of neutralisation of weak acid (acetic acid) = heat released during neutralisation + heat absorbed during ionisation

$\therefore$ Enthalpy of dissociation $= -50.6 - (-55.9)$

$$= +5.3 \text{ kJ mol}^{-1}$$

5. (b) When H_2S is passed through HNO_3, colloidal sol of S is obtained (as yellow turbidity).

6. (b) $AgNO_3 + KI \longrightarrow AgI + KNO_3$

Since, $AgNO_3$ is in excess, so AgI absorbs Ag^+ to form AgI : Ag^+

7. (a) When *p*-nitroacetanilide is subjected to hydrolysis in the presence of alkali, then there is high possibility of activated NS of —$NHCOCH_3$ group by —OH. However, when hydrolysis is carried out by using dilute acid, such side reaction is not possible, so dilute acid is used to hydrolysed it.

8. (c)

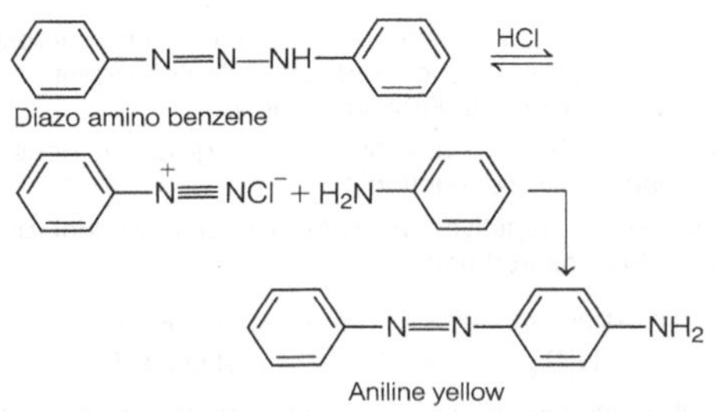

9. (b) In the preparation of Mohr's salt, small amount of dil. H_2SO_4 is added to prevent the hydrolysis of ferrous sulphate, i.e. to prevent the conversion of $FeSO_4$ into $Fe(OH)_2$.

10. (b) The main purpose of using leveling bulb is to assure that pressure within the reaction vessel is same as that of the room.

11. (b) Lower (negative value) is the heat of neutralisation, more is the dissociation energy and weaker is the acid.

12. (a) Heat produced by 200 Milliequivalent of

$$KOH + HCl = \frac{-13.7 \times 200}{1000} \text{ kcal}$$

This value raises the temperature of 400 mL solution by T_1

Thus, $$\frac{13.7 \times 200}{1000} = 400 \times T_1$$

Similarly, heat produced by 100 Milliequivalent of

$$KOH + HCl = \frac{-13.7 \times 100}{1000}$$

Thus, $$\frac{13.7 \times 100}{1000} = 200 \times T_2$$

Thus, $$T_1 = T_2$$

13. (b) The sodium salt of sulphanilic acid cannot be acetylated by acetic acid (AcOH) because acetic acid mostly converts it into a dipolar Zwitter ionic form.

Since, N atom of Zwitter ion has no free electron pair, so acetylation is not possible.

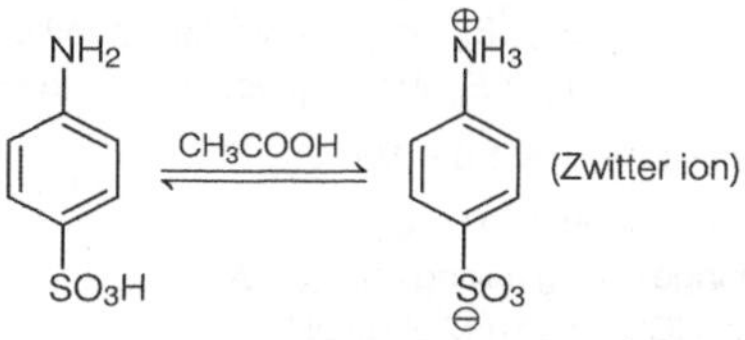

However, in the presence of $(CH_3CO)_2O$ lone pair of electrons are free and Zwitter ionic character is lost, so acetylation takes place resulting in the formation of acetanilide derivative.

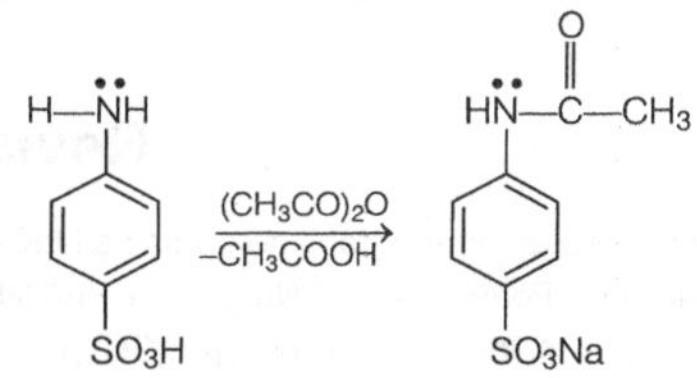

14. (a,b,d) It is a carcinogenic compound and basic in nature, (i.e. basic dye).

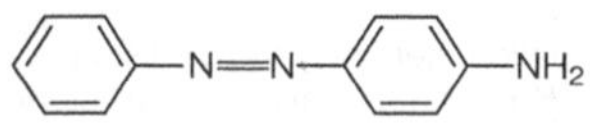

Aniline yellow or *p*-aminoazobenzene or 4-phenylazoaniline

15. (b,c,d)

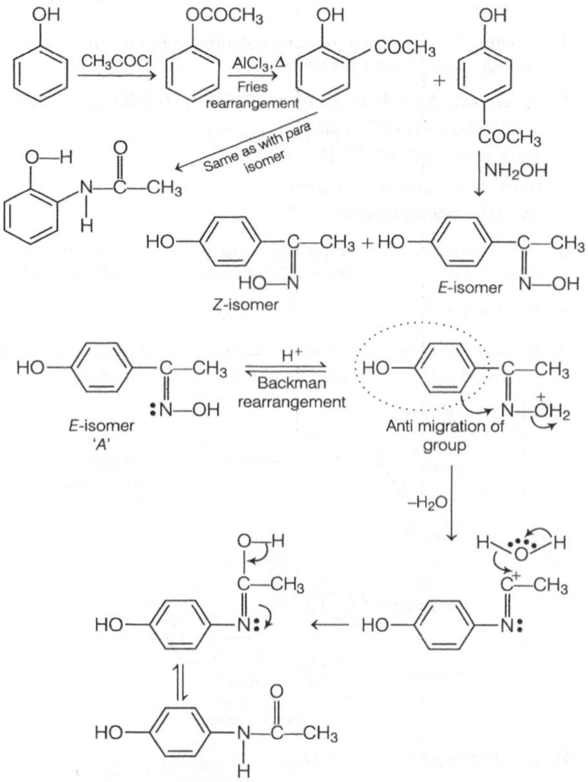

16. (c) For strong acid and strong base, enthalpy of neutralisation = – 57.3. In case of HF, it is greater than this value, due to extensive solvation of F^- ion on account of its smaller size.

In non-polar protic solvent, $\Delta H_{neu} < 57.3$.

17. (a) Dissolution of a substance may be exothermic or endothermic depending upon the value of E_h and E_i. Here, E_h = enthalpy of hydration and E_i = enthalpy of ionisation.

$$\Delta H_{solution} = E_h + E_i$$

If $E_i > E_h$, $\Delta H_s = +$ ve i.e. endothermic

If $E_i < E_h$, $\Delta H_s = -$ ve, i.e. exothermic

18. (c) mM of HCl = $50 \times 1.0 = 50$

mM of NaOH = $50 \times 1.0 = 50$

Heat change during mixing = $m\,C_s \cdot \Delta T$

50 mM of HCl and 50 mM of NaOH

$= -(100 \times 1) \times 4.18 \times 6.5$

$= -2717$ J $= -2.72$ kJ

19. (c) $\because$ On mixing 50 moles of HCl and 50 moles of NaOH, heat produced = 2.72 kJ

$\therefore$ 1000 moles of HCl (or 1000 milliequivalent) and 1000 moles of NaOH when mixed, the heat produced

$$= -\frac{2.72 \times 1000}{50} = -54.4 \text{ kJ}$$

20. (6) With NaOI, triiodomethane, (i.e. iodoform) is given by only those compounds which contains either $CH_3-\underset{\underset{O}{\|}}{C}-$ (methyl ketone) or $CH_3-\underset{\underset{OH}{|}}{C}H-$ (2-ol) group.

Thus, butanol ($CH_3CH_2CH_2CH_2OH$), $C_6H_5COC_6H_5$ (benzophenone) and formic acid (HCOOH) do not give this reaction.

Revisal Problems (JEE Main)

1. (a) The extract obtained from aniline contains NaCN.

$$2NaCN + FeSO_4 \longrightarrow Na_2SO_4 + Fe(CN)_2$$

$$Fe(CN)_2 + 4NaCN \longrightarrow Na_4[Fe(CN)_6]$$

$$3Na_4[Fe(CN)_6] + 4FeCl_3 \longrightarrow \underset{\substack{\text{Prussian blue}\\ \text{(Ferricferrocyanide)}}}{Fe_4[Fe(CN)_6]_3} + 12NaCl$$

2. (a) In redox titration, generally the oxidising agent (like $KMnO_4$, $K_2Cr_2O_7$) is taken as a standard because, its standard solution, is stable and free of impurities.

3. (b) Because lead sulphate is insoluble but lead chloride this soluble in hot water.

4. (c) Here, Zn is added to remove coloured impurities as well as to protect it from being oxidised.

5. (d) General formula of an alum is $M_2SO_4 \cdot M_2'(SO_4)_3 \cdot 12H_2O$ or $M_2M_2'(SO_4)_4 \cdot 12H_2O$ or $MM'(SO_4)_2 \cdot 12H_2O$

Thus, ratio of $M : M' : SO_4^{2-} = 1:1:2$

Here, M = monovalent cation

and M' = trivalent cation

6. (a) The colour of permanganate is violet and in acidic medium it oxidises Fe^{2+} to Fe^{3+} and itself gets reduced to Mn^{2+} (yellow).

7. (c) $As_2O_3 + 3H_2S \longrightarrow As_2S_3 + 3H_2O$

8. (a) Schiff's reagent is a solution of *p*-rosaniline hydrochloride in saturated solution of SO_2 in water (i.e. sulphurous acid.).

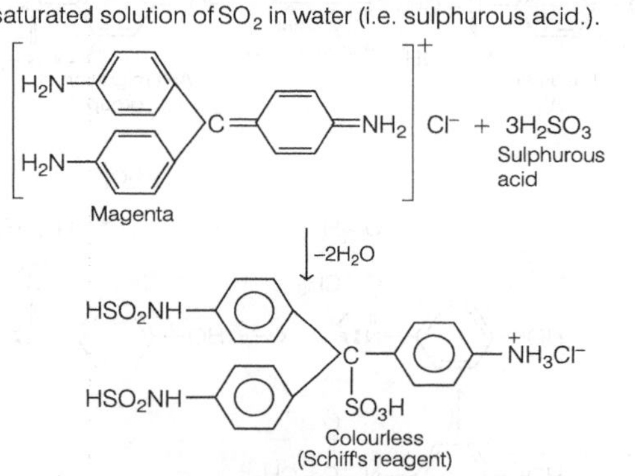

9. (c) Chromyl chloride is CrO_2Cl_2.

Its cation part is CrO_2^{2+} and anionic part is Cl^-.

10. (c) $CoS \xrightarrow{\text{Aqua regia}} CoCl_2$

$$CoCl_2 + NaHCO_3 + NaOH \xrightarrow{\Delta} 2NaCl + CoCO_3 + H_2O$$

$$2NaHCO_3 \xrightarrow{\Delta} Na_2CO_3 + H_2O + CO_2$$

$$CoCO_3 + 2Na_2CO_3 \xrightarrow{\Delta} Na_4[Co^{II}(CO_3)_3]$$

$$2Na_4[Co^{II}(CO_3)_3] + 2NaHCO_3 + [O] \xrightarrow{\Delta}$$

$$\underset{\substack{\text{Green}\\ \text{(sodium cobalticarbonate)}}}{2Na_3[Co^{III}(CO_3)_3]} + 2Na_2CO_3 + H_2O$$

11. (a) Fe^{3+} gives blood red colour with SCN^- in acidic media.

$$Fe^{3+} + 3SCN^- \longrightarrow \underset{\text{Blood red}}{[Fe(SCN)_3]}$$

12. (a) Because of epimerisation in alkaline medium, sucrose acts as reducing sugar and hence, reduces Tollen's reagent (give silver mirror) and Fehling's reaction.

Because of the absence of phenolic group, it does not give Libermann's nitroso reaction.

It also gives negative test with Baeyer's reagent due to the absence of unsaturation.

13. (b) This experiment involves following reactions:

$$H_2O_2 + 2I^- + 2H^+ \longrightarrow 2H_2O + I_2$$

The liberated iodine reacts with hypo and gets reduced to I^- ion.

$$I_2 + \underset{\text{Thiosulphate ion}}{2S_2O_3^{2-}} \xrightarrow{\text{Fast}} \underset{\text{Tetrathionate ion}}{S_4O_6^{2-}} + 2I^-$$

14. (c) As it acts as interfering radical here.

15. (b) The ppt of $Al(OH)_3$ is adsorbed by litmus, which gave lake like appearance.

16. (b) A + tap water $\longrightarrow$ white turbidity soluble in aqueous NH_3. Tap water has Cl^- and turbidity is soluble in aqueous NH_3.

$\Rightarrow$ turbidity is of AgCl and thus A has Ag^+.

A also gives **ring test** of NO_3^-.

$\Rightarrow$ A has NO_3^- and thus A is $AgNO_3$.

$$2AgNO_3 \xrightarrow{\Delta} \underset{B}{2Ag} + \underset{C}{2NO_2} + O_2$$

NO_2 is paramagnetic due to one unpaired electron as shown and thus forms dimer by using unpaired electron.

$$2NO_2 \rightleftharpoons \underset{\text{dimer (VD = 46)}}{N_2O_4}$$

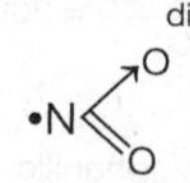

17. (b) Sulphuric acid can form insoluble sulphates with certain basic radicals like Pb^{2+}, Ba^{2+}, etc., so it is not used to prepare original solution for the analysis of basic radicals.

18. (d) Mohr's salt is highly stable and does not lose water of crystallisation, so it is generally used in place of Fe^{2+} salt for its titration.

19. (c) More the volume of acid as well as base, more energy is released and higher is the temperature.

20. (b) Lyophilic sols being stable can be prepared by just dissolving the substance in a suitable solvent but peptisation (or other specific methods) are used for the preparation of lyophobic sols.

21. (c) $HCl(g) + (aq.) \longrightarrow H^+(aq) + Cl^-(aq); \quad \Delta H^\circ = ?$

$$\Delta H^\circ_{HCl(g)} = [\Delta H^\circ_{f\,Cl^-(aq)} + \Delta H^\circ_{f\,H^+(aq)}] - [\Delta H^\circ_{f\,HCl} + \Delta H^\circ_{f(aq)}]$$
$$= [-167.44 + 0.0] - [-92.30 + 0.0] = -75.14\ \text{kJmol}^{-1}$$

22. (b) This is because $KMnO_4$ oxidises HCl to Cl_2 which itself is a good oxidant.

$$[2Cl^- \longrightarrow Cl_2 + 2e^-] \times 5$$
$$MnO_4^- + 8H^+ + 5e^- \longrightarrow Mn^{2+} + 4H_2O] \times 2$$
$$2MnO_4^- + 16H^+ + 10Cl^- \longrightarrow 2Mn^{2+} + 5Cl_2 + 8H_2O$$

23. (c) As the pH increases, dissociation of H_2S increases leading to the complete precipitation of metal ion as *MS*.

24. (b) The given characteristics matches with it.

$$Cl^- + AgNO_3 \longrightarrow \underset{\text{White ppt}}{AgCl} + NO_3^-$$

25. (a) $pK_a = -\log K_a = 4$

Thus, at pH 7 (when $pH > pK_a$) its colour is yellow.

26. (a) $CuSO_4 + 5H_2O \longrightarrow CuSO_4 \cdot 5H_2O; \ \Delta H = ?$

Given that

$$CuSO_4(\text{anhy}) + 5H_2O\ (aq) \longrightarrow CuSO_4 \cdot 5H_2O(aq);\ \Delta H_1 = -66.4\text{kJ} \quad \ldots\text{(i)}$$

and $$CuSO_4 \cdot 5H_2O + (aq) \longrightarrow CuSO_4 \cdot 5H_2O(aq),\ \Delta H_2 = +11.7\ \text{kJ} \quad \ldots\text{(ii)}$$

From the above Eq.

$$\Delta H = \Delta H_1 - \Delta H_2 = -66.4 - 11.7 = -78.1\ \text{kJ}$$

27. (b) First of all H_2S is passed in acidified solution, so that Cu^{2+} gets precipitated and Al^{3+} and Ni^{2+} remains in the filtrate.

The filtrate is then boil to remove H_2S and then NH_4Cl/NH_4OH is added. This will precipitate Al^{3+} as $Al(OH)_3$.

The filtrate contains only Ni^{2+}. It is then treated with H_2S/OH^-, which results in the precipitation of NiS.

28. (a) *A* burns with sooty flame, so it must be an aromatic compound. Formation of yellow orange ppt with Brady's reagent indicates the presence of carbonyl group (aldehyde or ketone). Negative Tollen's test suggests that it is not an aldehyde. So, it is acetophenone (an aromatic ketone).

29. (c) $$Fe^{2+} \longrightarrow Fe^{3+} + e^-] \times 6$$

From Mohr's salt

$$Cr_2O_7^{2-} + 14H^+ + 6e^- \longrightarrow 2Cr^{3+} + 7H_2O$$
$$6Fe^{2+} + Cr_2O_7^{2-} + 14H^+ \longrightarrow 6Fe^{3+} + 2Cr^{3+} + 7H_2O$$
$$\text{m moles of } Fe^{2+} = 750 \times 0.6 = 450$$
$$\text{moles of } Fe^{2+} = \frac{450}{1000} = 0.45\ \text{mol}$$
$$6 \text{ moles of } Fe^{2+} \equiv 1 \text{ mole } Cr_2O_7^{2-}$$
$$0.45 \text{ mole } Fe^{2+} = \frac{0.450}{6} = 0.075 \text{ mole } Cr_2O_7^{2-}$$
$$= 0.075 \times 294\ \text{g}$$
$$= 22.05\ \text{g}$$

30. (a) $Zn(CH_3COO)_2 + H_2S \longrightarrow ZnS + 2CH_3COOH$

(it does not dissolve ZnS)

31. (b) $PbCl_2$ is soluble in hot water but Hg_2Cl_2 does not. So, these can be separated by hot water, Hg_2Cl_2 is blackened by *aq* NH_3 as

$$Hg_2Cl_2 + 2NH_3 \longrightarrow \underset{\text{black}}{HgNH_2\,Cl;} + Hg + NH_4Cl$$

32. (a) $$8H^+ + MnO_4^- + Br_2 \longrightarrow \underset{\text{Colourless}}{2Br^-} + Mn^{2+} + 4H_2O$$

So, it does not interfere in the chromyl chloride test.

33. (b) Benzaldehyde does not reduce Fehling's solution but acetaldehyde can. So, these two can be separated by Fehling's solution.

All the carbonyl compounds form bisulphite adduct with $NaHSO_3$.

Revisal Problems (JEE Advanced)

1. (a) The order of oxidising power is

$$Cl^- > Br^- > I^-$$

Thus, Cl^- oxidises Br^- as well as I^- but Br^- further oxidise I^- to I_2, colour of which is violet.

2. (b) $Cr_2O_7^{2-}$ is orange in colour while I^- is colourless. When treated with acid, $Cr_2O_7^{2-}$ oxidise I^- to brown solution.

$$Cr_2O_7^{2-} + 14H^+ + 6e^- \longrightarrow 2Cr^{3+} + 7H_2O$$
$$[2I^- \longrightarrow I_2 + e^-]\,6$$

3. (a) When equilibrium is achieved, the Mg^{2+} concentration will become the same.

4. (a) $$\underset{A}{AgBr} + Na_2S_2O_3 \longrightarrow \underset{\substack{\text{Soluble}\\ B}}{Na_3[Ag(S_2O_3)_2]}$$

$$\underset{\substack{\text{Black}\\ D}}{Ag_2S\downarrow} \xleftarrow[-SO_2]{\text{Decompose}} \underset{\substack{\text{White ppt}\\ C}}{Ag_2S_2O_3} \xleftarrow[-NaNO_3]{AgNO_3} \underset{A}{Na_2S_2O_3}$$

5. (b) The colour of $PbCrO_4$ and $BaCrO_4$ is yellow but of Ag_2CrO_4 is red.

6. (a) Among the given only $Zn(OH)_2$ dissolves in NH_4OH/NH_4Cl and hence, its volume decreases.

7. (b) $1\,M\ H_2SO_4 \equiv 2$ equivalent H_2SO_4

$1\,M\ HCl \equiv 1$ equivalent HCl

Thus, for equal volume of two acids to be neutralised separately with NaOH, heat evolved will be twice in case of H_2SO_4 to that of HCl.

8. (c) Turnbull's blue – $Fe^{II}[Fe^{III}(CN)_6]^-$

Prussian's blue – $Fe^{III}[Fe^{II}(CN)_6]^-$

9. (c) NH_4Cl, $(NH_4)_2CS$ (thiourea) and NH_2SO_3H (sulphamic acid) all decompose the NO_2^- into N

e.g.

$$(NH_2)_2C{=}S + 2HNO_2 \xrightarrow[-CO_2]{-SO_2} N_2 + 3H_2O$$

$$NaNO_2 + NH_4Cl \longrightarrow NaCl + NH_4NO_2 \xrightarrow{\Delta} N_2 + H_2O$$

10. (b) Ppt of Hg_2^{2+} is black, of Ag^+ is brown and of Pb^{2+} is white, which dissolves in the excess of reagent.

11. (d) Brown fumes means the gas is NO_2 and O_2, relights the glowing splinter.

12. (b) $[Ba(OH)_2]_{mix} = \frac{80}{100} \times 1.0 \times 10^{-5} M = 0.80 \times 10^{-5} M$

Further $[Fe_2(SO_4)_3]_{mix} = \frac{20}{100} \times 1.0 \times 10^{-5} M = 0.20 \times 10^{-5} M$

$$\underset{}{Ba(OH)_2} \rightleftharpoons \underset{0.80 \times 10^{-5} M}{Ba^{2+}} + \underset{2 \times 0.80 \times 10^{-5} M}{2OH^-}$$

$$Fe_2(SO_4)_3 \rightleftharpoons \underset{2 \times 0.20 \times 10^{-5} M}{2Fe^{3+}} + \underset{3 \times 0.20 \times 10^{-5} M}{3SO_4^{2-}}$$

$BaSO_4$ is precipitated if $[Ba^{2+}][SO_4^{2-}] > K_{sp}$

$$[Ba^{2+}][SO_4^{2-}] = (0.80 \times 10^{-5})(0.60 \times 10^{-5}) = 0.48 \times 10^{-10} < K_{sp}$$

So, $BaSO_4$ is not precipitated.
$Fe(OH)_3$ is precipitated if $[Fe^{3+}][OH^-]^3 > K_{sp}$

$$[Fe^{3+}][OH^-]^3 = (0.40 \times 10^{-5})(1.60 \times 10^{-5})^3 = 1.638 \times 10^{-20} > K_{sp}$$

Hence, $Fe(OH)_3$ gets precipitated.

13. (b) $pH = -\log[H_3O^+]$, or $\log[H_3O^+] = -pH$

$\therefore$ $[H_3O^+] = \text{antilog of } (-pH)$

For pH = 3.1 $[H_3O^+]_1 = \text{antilog of } (-3.1) = \text{antilog of } (\bar{4}.9) = 7.94 \times 10^{-4}$

For pH = 4.5 $[H_3O^+]_2 = \text{antilog of } (-4.5) = \text{antilog of } (\bar{5}.5) = 3.16 \times 10^{-5}$

Since, neutral point lies at the centre of the hydrogen ion concentration in the given pH range, hence $[H_3O^+]$ at the neutral point,

$$[H_3O^+] = \frac{[H_3O^+]_1 + [H_3O^+]_2}{2} = \frac{7.94 \times 10^{-4} + 3.16 \times 10^{-5}}{2} = 4.13 \times 10^{-4} M$$

Let indicator be HIn behaving as weak acid, then

$$HIn + H_2O \rightleftharpoons H_3O^+ + In^-$$

$$K_{In} = \frac{[H_3O^+][In^-]}{[HIn]}$$

[Ionisation constant of indicator is K_{In}]

$= [H_3O^+]$ {since at neutral point $[In^-] = [HIn]$}

$= 4.13 \times 10^{-4}$

14. (c) In case I, II and IV, the acids and base involved are strong and enthalpy of neutralisation of a strong acid by a strong base or *vice-versa* is always the same i.e. – 57.3 kJ.

15. (b) $HNO_3 + H_2SO_4 \longrightarrow \underset{\text{Nitronium ion}}{NO_2^+} + H_2O + HSO_4^-$

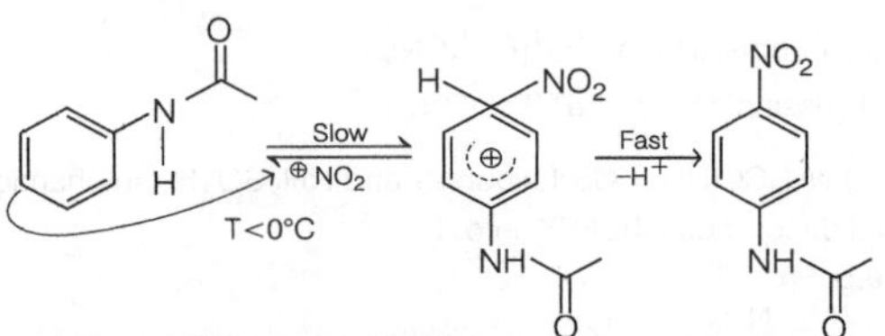

Major

$H^+ + HSO_4^- \longrightarrow H_2SO_4$

16. (c, d) PbS is black and CdS is yellow. No reaction with $CaCO_3$.

17. (a, b) In the removal of PO_4^{3-}, CH_3COOH / CH_3COO^- behaves as a buffer solution and also exert common ion effect.

18. (a, b) Y is SO_2.

$$SO_2 + O_2 \longrightarrow \underset{Z}{SO_3} + H_2O$$

$$Fe_2(SO_4)_3 \xrightarrow{\Delta} Fe_2O_3 + SO_2 + SO_3$$

It is also used as a drying agent. It is an anhydride of H_2SO_4. Marshall's acid is $H_2S_2O_8$.

19. (a,b,c)

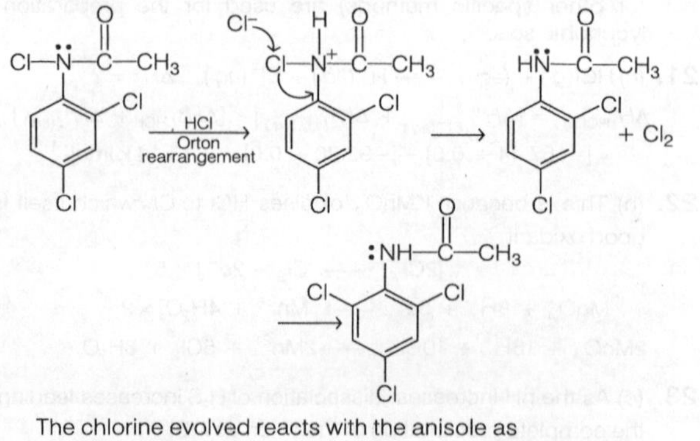

The chlorine evolved reacts with the anisole as

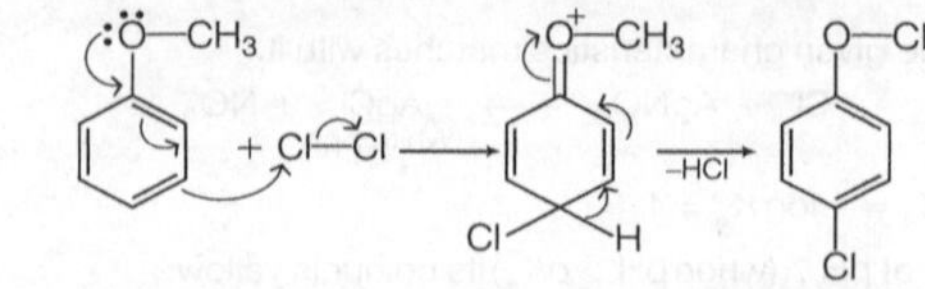

Similar attack at *ortho* position gives *o*-chloro anisole.

20. (a,c,d) $K_3\overset{+3}{[Fe(CN)_6]} + \overset{-1}{KI} \text{ (excess)} \longrightarrow K_4\overset{+2}{[Fe(CN)_6]} + \overset{1,0}{KI_3}$
Brownish yellow solution

It is a redox reaction.

$$K_4[Fe(CN)_6] + ZnSO_4 \longrightarrow \underset{\text{White ppt}}{K_2Zn[Fe(CN)_6]}$$

$$\underset{\text{Brownish yellow filtrate}}{I_3^-} + 2Na_2S_2O_3 \longrightarrow \underset{\text{Clear solution}}{Na_2S_4O_6} + 2NaI + \underset{\text{Turns starch solution blue}}{I_2}$$

$K_2Zn[Fe(CN)_6]$ reacts with NaOH as

$$K_2Zn[Fe(CN)_6] + 4NaOH \longrightarrow [Zn(OH)_4]^{2-} + [Fe(CN)_6]^{4-}$$

21. (a, b, c) Only primary aromatic amines give azo dye test. Thus, such amines with molecular formula C_7H_9N are

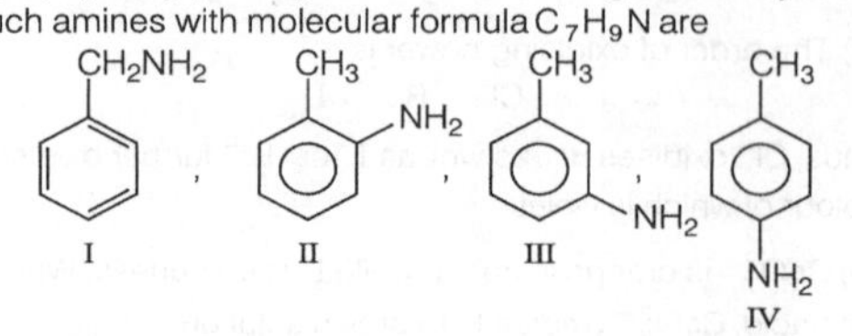

I is oxidised to benzoic acid while in case of II, III and IV aniline is the product.

22. (d)

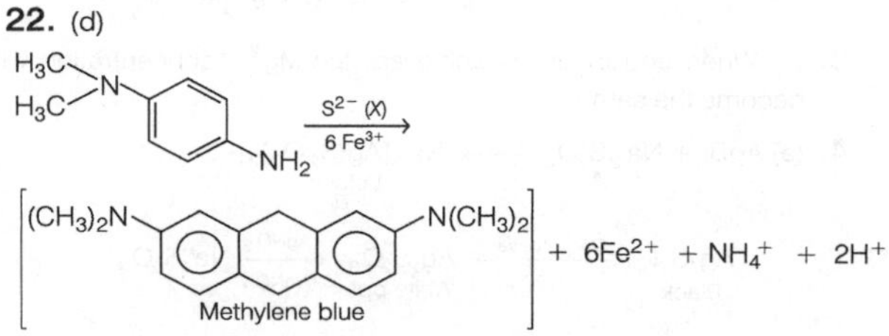

Thus X is Na_2S.

23. (c) Y is $FeCl_3$.

24. (b) $Fe^{3+} + [Fe(CN)_6]^{3-} \longrightarrow \underset{\text{Brown}}{Fe[Fe(CN)_6]}$

Explanations for Q. Nos. (25 and 26)

A must be an aromatic methyl ketone.

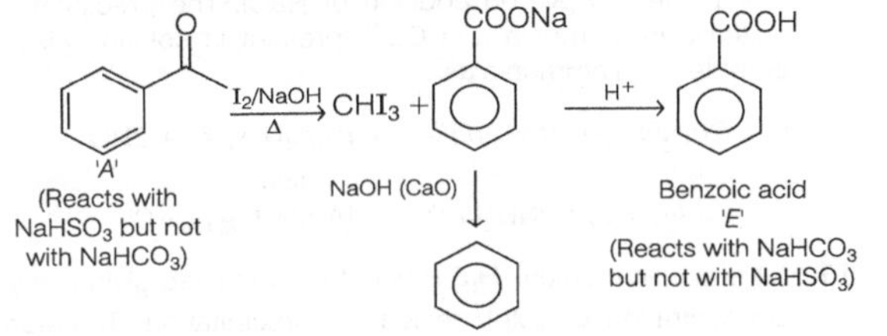

Acid group is identified by $NaHCO_3$ test.

25. (c) **26.** (b)

27. (c) NH_3 can be used to separate $PbCl_2$ from AgCl as it forms soluble complex $[Ag(NH_3)_2]Cl$ with AgCl.

28. (c) A chocolate colour ppt is obtained during this reaction but oxidation state of iron remains the same.

$$2CuSO_4 + K_4\overset{+2}{[Fe}(CN)_6] \longrightarrow \underset{\text{Chocolate coloured ppt}}{Cu_2\overset{+2}{[Fe}(CN)_6]} + 2K_2SO_4$$

29. (c) Ascorbic acid contains no —COOH group, but gives CO_2 with $NaHCO_3$ due to the presence of highly acidic H.

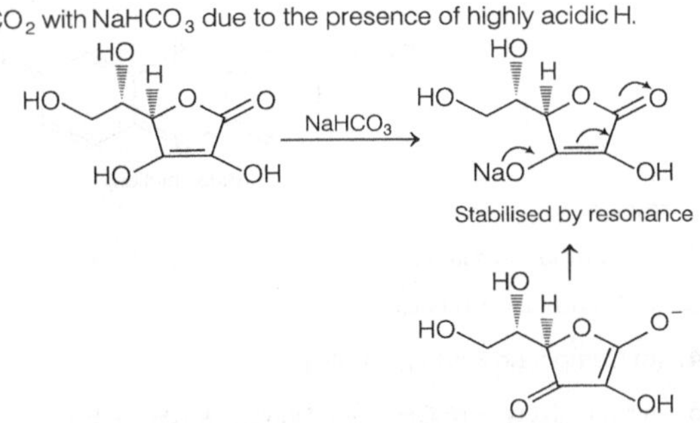

30. (b) (i)→ (p), (ii)→ (p,q), (iii)→ (p,r), (iv)→ (p,s)

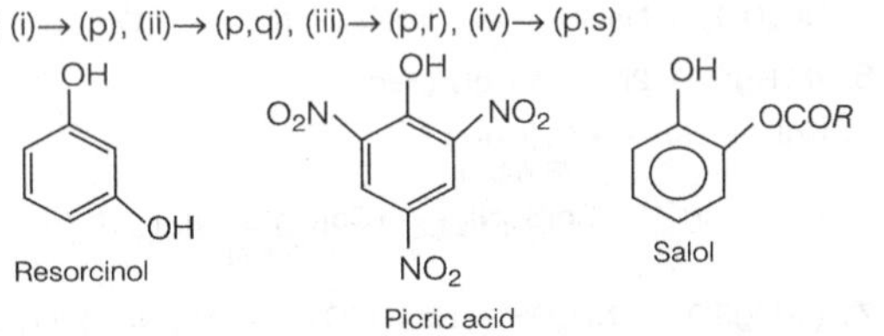

Ascorbic acid contains —OH enolic as well as —OH alcoholic.

31. (a) (i)→ (q), (ii)→ (q), (iii)→ (p,r), (iv)→ (p,r,s)

32. (d) (i)→ (q,r,s), (ii)→ (q,s), (iii)→ (r), (iv)→ (p)

(i) HCl and NaOH are the pair of strong acid and strong base, so methyl orange, phenolphthalein and thymolphthalein can be used as the indicator.

(ii) NaOH is a strong base but CH_3COOH is a weak acid so in this case, indicator used are phenolphthalein and thymolphthalein.

(iii) Since acid is strong, so only methyl orange can be used.

(iv) Both acid and base are weak, so phenol red is used.

33. (c) $[Fe(CNS)_3]$ is red, Prussian blue is blue in colour, Ni-DMG complex is cherry red, $PbCrO_4$ formed by reaction of $CrO_2Cl_2|NaOH|Pb^{2+}$ is yellow and Mn^{2+} reacts with PbO_2 in H^+ to give pink coloured compound.

34. (4) Only phosphate, fluoride, oxalate and borate act as interfering radicals.

35. (9) Enthalpy of solution, $\Delta H_{\text{solution}}$ = Enthalpy change per mole on dissolving 1 mole solute in excess of solvent.

∴ For metal chloride

$$\Delta H_{\text{solution}} = \frac{74.5}{8.27} \times X = 9X \text{ kcal}$$

36. (9) Total heat produced in acid-base reaction

= Heat taken by solution ($\Delta S, \Delta T$) + heat taken by calorimeter ($m_2S_2\Delta T$)

$$m_2S_2 = 12 \text{ g}$$
$$V = 400 + 400 = 800 \text{ mL}$$

$$\therefore \quad \Delta H_{\text{reaction}} = -(800 \times 1 \times 4.4) + (12 \times 4.4) = -3572.8 \text{ cal}$$

Further,

$$\text{meq } HX = 1 \times 1 \times 400 = 400 \text{ meq}$$
$$\text{meq of } BOH = 1 \times 2 \times 400 = 800 \text{ meq}$$

Thus, − 3572.8 cal heat is released by 400 meq of acid and base while 400 meq of base is in excess.

Thus, enthalpy of neutralisation = Heat released by 1000 meq of acid-base in a reaction

$$= \frac{1000}{400} \times 3572.8 = -8.9 \text{ kcal} \approx -9 \text{ kcal}$$

JEE Advanced & IIT JEE Archive

1. (6 or 7) From qualitative analysis of the different metal ions it is found that

PbS, CuS, HgS, Ag_2S, NiS, CoS are black coloured

MnS–dirty pink/buff coloured, SnS_2–yellow coloured

Bi_2S_3–brown/black (brownish black) coloured

Hence, correct integer value is 6 or 7.

2. (b) **Plan** This problem can be solved by using concept of chemical reactions of transition metal ions (colour and structure of transition metal compounds).

Here, among given four options Ni^{2+} and Zn^{2+} has ability to form tetrahedral as well as square planar complex depending upon types of reagent used.

Ni^{2+} on reaction with KCN forms square planar complex $[Ni(CN)_4]^{2-}$ due to strong field strength of CN.

$$Ni^{2+} + KCN \longrightarrow \underset{\text{Square planar}}{[Ni(CN)_4]^{2-}}$$

While on reaction with HCl, Ni^{2+} forms stable tetrahedral complex $[Ni(Cl)_4]^{2-}$.

Zn^{2+}, on the other hand, on reaction with KCN as well as HCl produces tetrahedral complex because of its d^{10} electronic configuration.

$$Zn^{2+} \xrightarrow{HCl} [ZnCl_4]^{2-}$$
$$Zn^{2+} \xrightarrow{KCN} [Zn(CN)_4]^{2-}$$
$$Zn^{2+} \xrightarrow[excess]{KOH} \underset{\text{White ppt}}{[Zn(OH)_4]^{2-}}$$

Complete reaction sequence can be shown as

$$\underset{\text{Tetrahedral}}{[NiCl_4]^{2-}} \xleftarrow[excess]{HCl(Q)} \underset{(M_1)}{Ni^{2+}} \xrightarrow[excess]{KCN(R)} \underset{\text{Square planar}}{[Ni(CN)_4]^{2-}}$$

$$\underset{\text{Tetrahedral}}{[ZnCl_4]^{2-}} \xleftarrow[excess]{HCl(Q)} \underset{(M_2)}{Zn^{2+}} \xrightarrow[excess]{KCN(R)} \underset{\text{Tetrahedral}}{[Zn(CN)_4]^{2-}}$$

$$Zn^{2+} \xrightarrow{KOH\,(s)} \underset{\text{White ppt}}{Zn(OH)_2} \xrightarrow[excess]{KOH} \underset{\text{Soluble}}{[Zn(OH)_4]^{2-}}$$

3. (d) Zn^{2+} on treatment with excess of KOH produces $[Zn(OH)_4]^{2-}$

Explanations for Q. Nos. (4-5)

Plan $PbCl_2$ is soluble in hot water.

In ammoniacal medium, cations of group III and IV may be precipitated as hydroxide or sulphide and dissolved in H_2O_2 due to oxidation.

$$\text{Mixture of two inorganic salts} \xrightarrow[\text{(ii) Filter}]{\text{(i) HCl}} \text{ppt.}$$

$$\underset{\substack{\text{Yellow}\\(S)}}{CrO_4^{2-}} \xleftarrow[NaOH]{H_2O_2} \underset{\text{(ppt)}}{Cr(OH)_3} \xleftarrow{NH_4OH/H_2S} \text{Filtrate}$$

(as CH (NaCl) soluble in hot water)

4. (a) **5.** (d)

6. (d) **Plan** $K_{sp}(ZnS)$ is very high and Zn^{2+} is precipitated as ZnS by high concentration of S^{2-} formed when H_2S is passed in ammoniacal solution.

$$H_2S \rightleftharpoons 2H^+ + S^{2-} \quad (I)$$
$$H^+ + OH^- \rightleftharpoons H_2O \quad (II)$$

Reaction (I) is favoured in forward side if H^+ is removed immediately by OH^- (NH_4OH).

$$Zn^{2+} + S^{2-} \longrightarrow \underset{\text{(White ppt)}}{ZnS\downarrow}$$

Fe^{3+} and Al^{3+} are precipitated as hydroxide.

7. (a,c,d) $K_3\overset{+3}{[Fe}(CN)_6] + KI \text{ (excess)} \longrightarrow$

$$K_4\overset{+2}{[Fe}(CN)_6] + \underset{\text{(Brownish yellow solution)}}{KI_3\text{(Redox)}}$$

$$K_4[Fe(CN)_6] + ZnSO_4 \longrightarrow K_2Zn_3[Fe(CN)_6]_2$$

or $\underset{\text{(White ppt)}}{K_2Zn[Fe(CN)_6]}$

$$\underset{\text{(Brownish yellow filtrate)}}{I_3^-} + 2Na_2S_2O_3 \longrightarrow \underset{\text{(Clear solution)}}{Na_2S_4O_6} + 2NaI + \underset{\text{(Turns starch solution blue)}}{I_2}$$

$K_2Zn[Fe(CN)_6]$ reacts with NaOH as

$$K_2Zn[Fe(CN)_6] + NaOH \longrightarrow [Zn(OH)_4]^{2-} + [Fe(CN)_6]^{4-}$$

8. (b) Appearance of blue colour on addition of the metal rod M to the solution of N is an indication that metal may be Cu. Later formation of white precipitate on addition of NaCl and dissolution of this precipitate in aqueous ammonia confirm that white precipitate is of AgCl. This implies that the solution N is of $AgNO_3$. This confirms that the metal M is only Cu.

9. (a) The compound N is $AgNO_3$, explained above.

10. (c) Since in the beginning, concentrated $AgNO_3$ solution was taken, some $AgNO_3$ remain in solution which gives white precipitate of AgCl on addition of NaCl. The precipitate finally dissolve in ammonia and Cu^{2+} present in solution also forms complex with ammonia as

$$Cu(NO_3)_2 + 4NH_3(aq) \longrightarrow \underset{\text{(Blue)}}{[Cu(NH_3)_4]^{2+}} + 2NO_3^-$$

$$AgCl(s) + 2NH_3(aq) \longrightarrow [Ag(NH_3)_2]^+ + Cl^-$$

11. (a) In acidic medium, H_2S is very feebly ionised giving very small concentration of sulphide ion for precipitation. Therefore, the most insoluble salts CuS and HgS are precipitated only.

Explanations for Q. Nos. (12-14)

The comprehension describing methylene blue test.

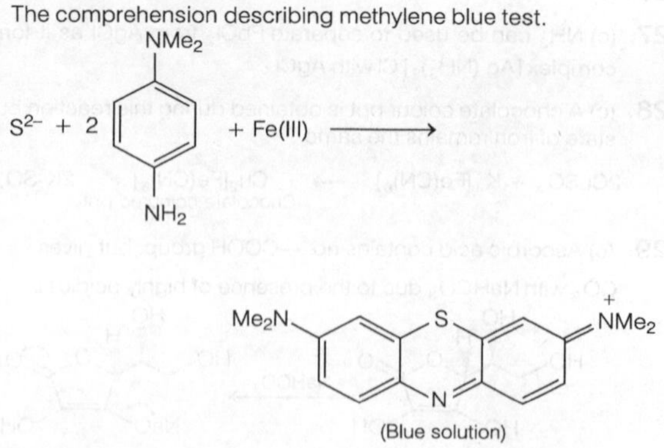

Therefore,

12. (d) Compound X is Na_2S.

13. (c) Compound Y is $FeCl_3$.

14. (b) Compound Z is $Fe[Fe(CN)_6]$.

15. (a,b) $NH_4NO_3 + NaOH \longrightarrow NaNO_3 + NH_3 + H_2O$

$NH_4NO_2 + NaOH \longrightarrow NaNO_2 + NH_3 + H_2O$

16. (b) $Hg^{2+} + 2I^- \longrightarrow \underset{\text{(Red)}}{HgI_2}$

$$HgI_2 + 2KI \longrightarrow \underset{\text{(Soluble)}}{K_2[HgI_4]}$$

$$Hg^{2+} + Co(SCN)_2 \longrightarrow \underset{\text{(Blue)}}{Co[Hg(SCN)_4]}$$

17. (a) $MgSO_4 + NH_4OH + Na_2HPO_4 \longrightarrow \underset{\text{(White)}}{Mg(NH_4)PO_4\downarrow} + Na_2SO_4 + H_2O$

18. (b) $CuSO_4 + 2KCN \longrightarrow Cu(CN)_2 + K_2SO_4$

$$\underset{\text{(Unstable)}}{2Cu(CN)_2} \longrightarrow 2Cu(CN) + \begin{matrix}CN\\|\\CN\end{matrix} \text{ (Cyanogen)}$$

$$CuCN + 3KCN \longrightarrow 3K^+ + [Cu(CN)_4]^{3-}$$

19. (a) $Zn^{2+} + 2H_2O \longrightarrow \underset{\text{(White)}}{Zn(OH)_2\downarrow} + 2H^+$

20. (b) $Bi^{3+} + 3I^- \longrightarrow \underset{\text{(Black)}}{BiI_3\downarrow} \xrightarrow[excess]{I^-} \underset{\text{(Orange solution)}}{[BiI_4]^-}$

Last 3 Years' Questions
JEE Main & Advanced
(2016-2018)

JEE Main

2016

1. Which of the following compounds is metallic and ferromagnetic? **[*d* and *f*-block Elements]**

(a) CrO_2 (b) VO_2
(c) MnO_2 (d) TiO_2

2. The species in which the N-atom is in a state of *sp* hybridisation is **[*p*-block Elements]**

(a) NO_2^- (b) NO_3^-
(c) NO_2 (d) NO_2^+

3. The pair having the same magnetic moment is [at. no. Cr = 24, Mn = 25, Fe = 26 and Co = 27] **[Coordination Compounds]**

(a) $[Cr(H_2O)_6]^{2+}$ and $[Fe(H_2O)_6]^{2+}$ (b) $[Mn(H_2O)_6]^{2+}$ and $[Cr(H_2O)_6]^{2+}$
(c) $[CoCl_4]^{2-}$ and $[Fe(H_2O)_6]^{2+}$ (d) $[Cr(H_2O)_6]^{2+}$ and $[CoCl_4]^{2-}$

4. Which one of the following ores is best concentrated by froth floatation method? **[Metallurgy]**

(a) Siderite (b) Galena
(c) Malachite (d) Magnetite

5. The pair in which phosphorus atoms have a formal oxidation state of +3 is **[*p*-block Elements]**

(a) pyrophosphorous and hypophosphoric acids
(b) orthophosphorous and hypophosphoric acids
(c) pyrophosphorous and pyrophosphoric acids
(d) orthophosphorous and pyrophosphorous acids

6. Which one of the following complexes shows optical isomerism? **[Coordination Chemistry]**

(a) *cis* $[Co(en)_2Cl_2]Cl$ (b) *trans* $[Co(en)_2Cl_2]Cl$
(c) $[Co(NH_3)_4Cl_2]Cl$ (d) $[Co(NH_3)_3Cl_3]$

7. The reaction of zinc with dilute and concentrated nitric acid, respectively, produce **[*p*-block Elements]**

(a) NO_2 and NO (b) NO and N_2O
(c) NO_2 and N_2O (d) N_2O and NO_2

8. Galvanisation is applying a coating of **[*d* and *f*-block Elements]**

(a) Cr (b) Cu
(c) Zn (d) Pb

2017

9. In the following reactions, ZnO is respectively acting as a/an **[*d* and *f*-block Elements]**

(i) $ZnO + Na_2O \longrightarrow Na_2ZnO_2$
(ii) $ZnO + CO_2 \longrightarrow ZnCO_3$

(a) base and acid (b) base and base
(c) acid and acid (d) acid and base

10. The products obtained when chlorine gas reacts with cold and dilute aqueous NaOH are **[p-block Elements]**

(a) ClO^- and ClO_3^- (b) ClO_2^- and ClO_3^-
(c) Cl^- and ClO^- (d) Cl^- and ClO_2^-

11. On treatment of 100 mL of 0.1 M solution of $CoCl_3 \cdot 6H_2O$ with excess of $AgNO_3$; 1.2×10^{22} ions are precipitated. The complex is **[Coordination Compounds]**

(a) $[Co(H_2O)_4Cl_2]Cl \cdot 2H_2O$ (b) $[Co(H_2O)_3Cl_3] \cdot 3H_2O$
(c) $[Co(H_2O)_6]Cl_3$ (d) $[Co(H_2O)_5Cl]Cl_2 \cdot H_2O$

2018

12. The oxidation states of Cr, in $[Cr(H_2O)_6]Cl_3$, $[Cr(C_6H_6)_2]$, and $K_2[Cr(CN)_2(O)_2(O_2)(NH_3)]$ respectively are **[Coordination Compounds]**

(a) +3, +4 and +6 (b) +3, +2 and +4
(c) +3, 0 and +6 (d) +3, 0 and +4

13. The compound that does not produce nitrogen gas by the thermal decomposition is **[p-block Elements]**

(a) $Ba(N_3)_2$ (b) $(NH_4)_2Cr_2O_7$
(c) NH_4NO_2 (d) $(NH_4)_2SO_4$

14. Consider the following reaction and statements :

$$[Co(NH_3)_4Br_2]^+ + Br^- \longrightarrow [Co(NH_3)_3Br_3] + NH_3$$

I. Two isomers are produces if the reactant complex ion is a *cis*-isomer.
II. Two isomers are produced if the reactant complex ion is a *trans*-isomer.
III. Only one isomer is produced if the reactant complex ion is a *trans*-isomer.
IV. Only one isomer is produced if the reactant complex ion is a *cis*-isomer.

The correct statements are **[Coordination Compounds]**

(a) (I) and (II) (b) (I) and (III)
(c) (III) and (IV) (d) (II) and (IV)

JEE Advanced

2016

1. Among $[Ni(CO)_4]$, $[NiCl_4]^{2-}$, $[Co(NH_3)_4Cl_2]Cl$, $Na_3[CoF_6]$, Na_2O_2 and CsO_2, the total number of paramagnetic compounds is **[Coordination Compounds, Type 1]**

(a) 2 (b) 3
(c) 4 (d) 5

2. The possible number of geometrical isomers for the complex $[CoL_2Cl_2]^-$ ($L = H_2NCH_2CH_2O^-$) is (are) ... **[Coordination Compounds, Type 3]**

3. In neutral or faintly alkaline solution, 8 moles of permanganate anion quantitative oxidise thiosulphate anions to produce X moles of a sulphur containing product. The magnitude of X is ... . **[d and f-block Elements, Type 3]**

4. In the following reaction sequence in aqueous solution, the species X, Y and Z, respectively, are

$$S_2O_3^{2-} \xrightarrow{Ag^+} \underset{\text{(Clear solution)}}{X} \xrightarrow{Ag^+} \underset{\text{(White ppt.)}}{Y} \xrightarrow{\text{With time}} \underset{\text{(Black ppt.)}}{Z}$$

[Qualitative Analysis, Type 3]

(a) $[Ag(S_2O_3)_2]^{3-}$, $Ag_2S_2O_3$, Ag_2S (b) $[Ag(S_2O_3)_3]^{5-}$, Ag_2SO_3, Ag_2S
(c) $[Ag(SO_3)_2]^{3-}$, $Ag_2S_2O_3$, Ag (d) $[Ag(SO_3)_3]^{3-}$, Ag_2SO_4, Ag

Note ***Type-1*** *Only one correct option,* ***Type-2*** *More then one correct option,* ***Type-3*** *Single digit Answer,* ***Type-4*** *Numerical value based*

5. The geometries of the ammonia complexes of Ni^{2+}, Pt^{2+} and Zn^{2+}, respectively, are **[Coordination Compounds, Type 1]**

(a) octahedral, square planar and tetrahedral
(b) square planar, octahedral and tetrahedral
(c) tetrahedral, square planar and octahedral
(d) octahedral, tetrahedral and square planar

6. Extraction of copper from copper pyrite ($CuFeS_2$) involves **[Metallurgy, Type 2]**

(a) crushing followed by concentration of the ore by froth-floatation
(b) removal of iron as slag
(c) self reduction step to produce 'blister copper' following evolution of SO_2
(d) refining of 'blister copper' by carbon reduction

7. The nitrogen containing compound produced in the reaction of HNO_3 with P_4O_{10} **[p-block Element, Type 2]**

(a) can also be prepared by reaction of P_4 and HNO_3
(b) is diamagnetic
(c) contains one N—N bond
(d) reacts with Na metal producing a brown gas

2017

8. Addition of excess aqueous ammonia to a pink coloured aqueous solution of $MCl_2 \cdot 6H_2O(X)$ and NH_4Cl gives an octahedral complex Y in the presence of air. In aqueous solution, complex Y behaves as 1 : 3 electrolyte. The reaction of X with excess HCl at room temperature results in the formation of a blue colured complex Z. The calculated spin only magnetic moment of X and Z is 3.87 B.M., whersas it is zero for complex Y.

Among the following options, which statement(s) is (are) correct? **[Coordination Compounds, Type 2]**

(a) The hybridisation of the central metal ion in Y is d^2sp^3
(b) Addition of silver nitrate to Y given only two equivalents of silver chloride
(c) When X and Y are in equilibrium at 0°C, the colour of the solution is pink
(d) Z is a tetrahedral complex

9. The correct statement(s) about the oxoacids, $HClO_4$ and HClO, is (are) **[p-block Elements, Type 1]**

(a) The central atom in both $HClO_4$ and HClO is sp^3-hybridised
(b) $HClO_4$ is formed in the reaction between Cl_2 and H_2O
(c) The conjugate base of $HClO_4$ is weaker base than H_2O
(d) $HClO_4$ is more acidic than HClO because of the resonance stabilisation of its anion

10. The colour of the X_2 molecules of group 17 elements changes gradually from yellow to violet down the group. This is due to **[p-block Elements, Type 2]**

(a) decrease in $\pi^* - \sigma^*$ gap down the group
(b) decrease in ionisation energy down the group
(c) the physical state of X_2 at room temperature changes from gas to solid down the group
(d) decreases in HOMO-LUMO gap down the group

11. The order of the oxidation state of the phosphorus atom in H_3PO_2, H_3PO_4, H_3PO_3 and $H_4P_2O_6$ is **[p-block Elements, Type 1]**

(a) $H_3PO_4 > H_3PO_2 > H_3PO_3 > H_4P_2O_6$
(b) $H_3PO_4 > H_4P_2O_6 > H_3PO_3 > H_3PO_2$
(c) $H_3PO_2 > H_3PO_3 > H_4P_2O_6 > H_3PO_4$
(d) $H_3PO_3 > H_3PO_2 > H_3PO_4 > H_4P_2O_6$

Paragraph (Q. Nos. 12-13)

Upon heating $KClO_3$ in presence of catalytic amount of MnO_2, a gas W is formed. Excess amount of W reacts with white phosphorus to give X. The reaction of X with pure HNO_3 gives Y and Z.

12. Y and Z are, respectively **[p-block Elements]**

(a) N_2O_4 and HPO_3
(b) N_2O_4 and H_3PO_3
(c) N_2O_3 and H_3PO_4
(d) N_2O_5 and HPO_3

13. *W* and *X* are, respectively **[p-block Elements, Type 2]**

(a) O_2 and P_4O_{10} (b) O_2 and P_4O_6
(c) O_3 and P_4O_6 (d) O_3 and P_4O_{10}

2018

14. **The compound(s) which generate (s) N_2 gas upon thermal decomposition below 300°C is (are)** **[Qualitative Analysis, Type 2]**

(a) NH_4NO_3 (b) $(NH_4)_2Cr_2O_7$
(c) $Ba(N_3)_2$ (d) Mg_3N_2

15. **The correct statement (s) regarding the binary transition metal carbonyl compounds is (are) (Atomic numbers : Fe = 26, Ni = 28)** **[Coordination Compounds, Type 2]**

(a) Total number of valence shell electrons at metal centre in $Fe(CO)_5$ or $Ni(CO)_4$ is 16
(b) These are predominantly low spin in nature
(c) Metal-carbon bond strengthens when the oxidation state of the metal is lowered
(d) The carbonyl C—O bond weakens when the oxidation state of the metal is increased

16. **The correct option(s) regarding the complex $[Co(en)(NH_3)_3(H_2O)]^{3+}$ ($en = H_2NCH_2CH_2NH_2$) is (are)** **[Coordination Compounds, Type 2]**

(a) It has two geometrical isomers
(b) It will have three geometrical isomers, if bidentate 'en' is replaced by two cyanide ligands
(c) It is paramagnetic
(d) It absorbs light at longer wavelength as compared to $[Co(en)(NH_3)_4]^{3+}$

17. **The correct option(s) to distinguish nitrate salts to Mn^{2+} and Cu^{2+} taken separately is (are)** **[Qualitative Analysis, Type 2]**

(a) Mn^{2+} shows the characteristic green colour in the flame test
(b) Only Cu^{2+} shows the formation of precipitate by passing H_2S in acidic medium
(c) Only Mn^{2+} shows the formation of precipitate by passing H_2S in faintly basic medium
(d) Cu^{2+}/Cu has higher reduction potential than Mn^{2+}/Mn (measured under similar conditions)

18. **The total number of compounds having at least one bridging oxo group among the molecules given below is ....** **[p-block Elements, Type 4]**

$N_2O_3, N_2O_5, P_4O_6, P_4O_7, H_4P_2O_5, H_5P_3O_{10}, H_2S_2O_3, H_2S_2O_5$

19. **Match each set of hybrid orbitals from List–I with complexes given in List-II.**

	List–I		List–II
P.	dsp^2	1.	$[FeF_6]^{4-}$
Q.	sp^3	2.	$[Ti(H_2O)_3Cl_3]$
R.	sp^3d^2	3.	$[Cr(NH_3)_6]^{3+}$
S.	d^2sp^3	4.	$[FeCl_4]^{2-}$
		5.	$[Ni(CO)_4]$
		6.	$[Ni(CN)_4]^{2-}$

The correct option is **[Coordination Compounds]**

(a) $P \to 5$; $Q \to 4, 6$; $R \to 2, 3$; $S \to 1$ (b) $P \to 5,6$; $Q \to 4$; $R \to 3$; $S \to 1,2$
(c) $P \to 6$; $Q \to 4, 5$; $R \to 1$; $S \to 2, 3$ (d) $P \to 4,6$; $Q \to 5, 6$; $R \to 1,2$; $S \to 3$

20. **Based on the compounds of group 15 elements, the correct statement(s) is (are)** **[p-Block Elements, Type 2]**

(a) Bi_2O_5 is more basic than N_2O_5
(b) NF_3 is more covalent than BiF_3
(c) PH_3 boils at lower temperature than NH_3
(d) The N — N single bond is stronger than the P — P single bond

Answer *with* Explanations

JEE Main

1. *(a)* Only three elements iron (Fe), cobalt (Co) and nickel (Ni) show ferromagnetism at room temperature. CrO_2 is also a metallic and ferromagnetic compound which is used to make magnetic tapes for cassette recorders.

2. *(d)*

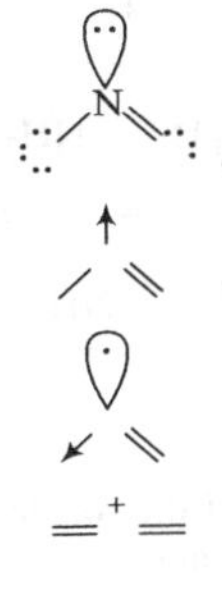

sp^2

3. *(a)*

Complex ion	Electronic configuration of metal ion	Number of unpaired electrons (n)
$[Cr(H_2O)_6]^{2+}$	Cr^{2+}; [Ar] $3d^4$	↑ ↑ ↑ ↑ □ ; 4
$[Fe(H_2O)_6]^{2+}$	Fe^{2+}; [Ar] $3d^6$	↑↓ ↑ ↑ ↑ ↑ ; 4
$[Mn(H_2O)_6]^{2+}$	Mn^{2+}; [Ar] $3d^5$	↑ ↑ ↑ ↑ ↑ ; 5
$[CoCl_4]^{2-}$	Co^{2+}; [Ar] $3d^7$	↑↓ ↑↓ ↑ ↑ ↑ ; 3

4. *(b)* Sulphide ores are concentrated by froth floatation method e.g. Galena (PbS).

5. *(d)* Orthophosphorous acid, H_3PO_3 :

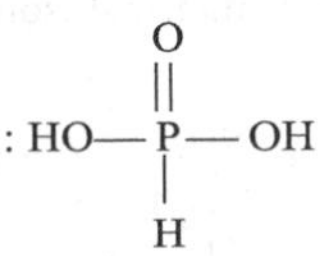

$$H_3\overset{x}{P}O_3 = 3 + x + 3(-2) = 0$$

or $$x = +3$$

Pyrophosphorous acid, $H_4P_2O_5$:

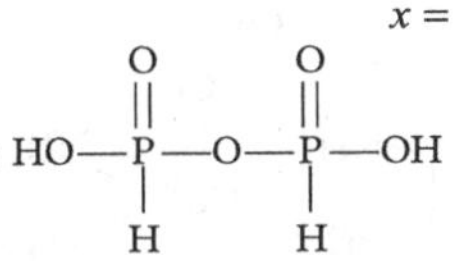

$$H_4\overset{x}{P_2}O_5 = 4 + 2x + 5(-2) = 0$$

$$4 + 2x - 10 = 0, \; x = +3$$

6. *(a)*

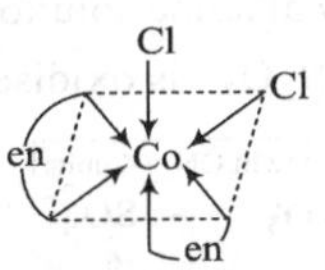

cis-$[Co(en)_2Cl_2]Cl$ (optically active)

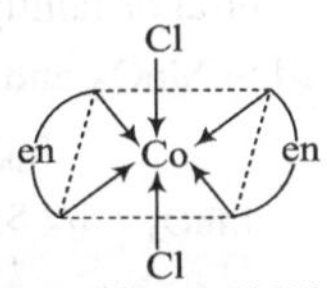

trans-$[Co(en)_2Cl_2]Cl$ (optically inactive due to plane of symmetry)

$[Co(NH_3)_4Cl_2]Cl$ can exist in both *cis* and *trans* forms that are given below:

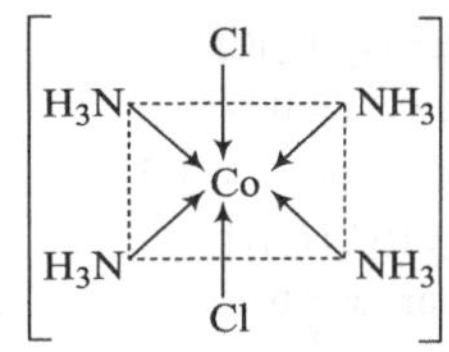

trans-$[Co(NH_3)_4Cl_2]Cl$ (optically inactive)

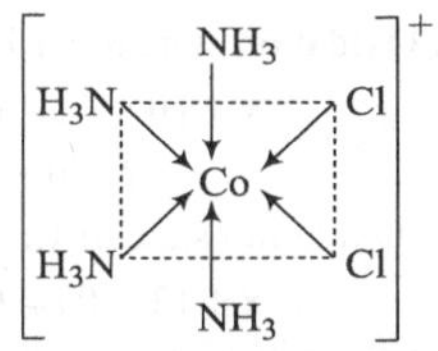

cis-$[Co(NH_3)_4Cl_2]Cl$ (optically inactive)

$[Co(NH_3)_3Cl_3]$ exists in *fac* and *mer*-isomeric forms and both are optically inactive.

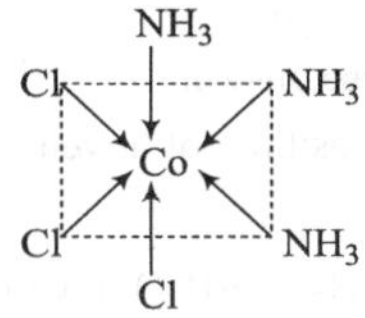

fac-isomer (optically inactive)

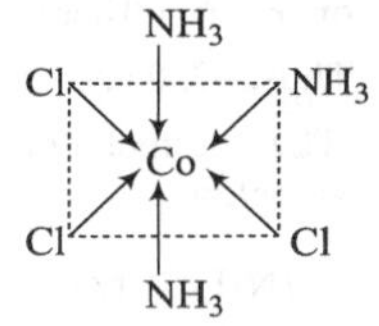

mer-isomer (optically inactive)

7. *(d)* $$\underset{\text{(Conc.)}}{Zn + 4HNO_3} \longrightarrow Zn(NO_3)_2 + 2H_2O + 2NO_2$$

$$\underset{\text{(Dil.)}}{4Zn + 10HNO_3} \longrightarrow 4Zn(NO_3)_2 + N_2O + 5H_2O$$

8. *(c)* Zinc metal is the most stable metal to cover iron surfaces. The process of coating the iron surface by zinc is called galvanisation.

9. *(d)* Zinc oxide (ZnO) when react with Na_2O it act as acid while with CO_2 it act as base. Therefore, it is an amphoteric oxide.

$$\underset{\text{Acid}}{ZnO} + \underset{\text{Base}}{Na_2O} \longrightarrow \underset{\text{Salt}}{Na_2ZnO_2}$$

$$\underset{\text{Base}}{ZnO} + \underset{\text{Acid}}{CO_2} \longrightarrow \underset{\text{Salt}}{ZnCO_3}$$

10. *(c)* Cl_2, Br_2 and I_2 form a mixture of halide and hypohalites when react with cold dilute alkalies while a mixture of halides and haloate when react with concentrated cold alkalies.

$$\underset{\text{Cold and dilute}}{Cl_2 + 2NaOH} \longrightarrow NaCl + NaClO + H_2O$$

$\therefore$ Cl^- and ClO^- are obtained as products when chlorine gas reacts with cold and dilute aqueous NaOH.

11. *(d)* $$\text{Molarity }(M) = \frac{\text{Number of moles of solute}}{\text{Volume of solution (in L)}}$$

$\therefore$ Number of moles of complex

$$= \frac{\text{Molarity} \times \text{volume (in mL)}}{1000} = \frac{0.1 \times 100}{1000} = 0.01 \text{ mole}$$

Number of moles of ions precipitate

$$= \frac{1.2 \times 10^{22}}{6.02 \times 10^{23}} = 0.02 \text{ moles}$$

$\therefore$ Number of Cl^- present in ionisation sphere

$$= \frac{\text{Number of moles of ions precipitated}}{\text{Number of moles of complex}} = \frac{0.02}{0.01} = 2$$

$\therefore$ $2Cl^-$ are present outside the square brackets, i.e. in ionisation sphere. Thus, the formula of complex is $[Co(H_2O)_5Cl]Cl_2 \cdot H_2O$.

12. *(c)* Let the oxidation state of Cr in all cases is 'x'

(i) Oxidation state of Cr in $[Cr(H_2O)_6]Cl_3$

$$x + (0 \times 6) + (-1 \times 3) = 0$$

or $\quad x + 0 - 3 = 0 \quad$ or $\quad x = +3$

(ii) Oxidation state of Cr in $[Cr(C_6H_6)_2]$

$$x + (2 \times 0) = 0 \quad \text{or} \quad x = 0$$

(iii) Oxidation state of Cr in

$$K_2[Cr(CN)_2(O)_2(O_2)(NH_3)]$$

$$1 \times 2 + x + (-1 \times 2) + (-2 \times 2) + (-2) + 0 = 0$$

or $\quad 2 + x - 2 - 4 - 2 = 0$

or $x - 6 = 0$ hence $x = +6$

Thus, +3, 0 and +6 is the answer.

13. *(d)* The thermal decomposition of given compounds is shown below

$$(NH_4)_2Cr_2O_7 \xrightarrow{\Delta} N_2 + 4H_2O + Cr_2O_3$$

$$NH_4NO_2 \xrightarrow{\Delta} N_2 + 2H_2O$$

$$(NH_4)_2SO_4 \xrightarrow{\Delta} 2NH_3 + H_2SO_4$$

$$Ba(N_3)_2 \longrightarrow Ba + 3N_2$$

Thus, only $(NH_4)_2SO_4$ does not gives N_2 on heating (It give NH_3). While rest of the given compounds gives N_2 on their thermal decomposition.

14. *(b)* If the reactant is *cis* isomer than following reaction takes place.

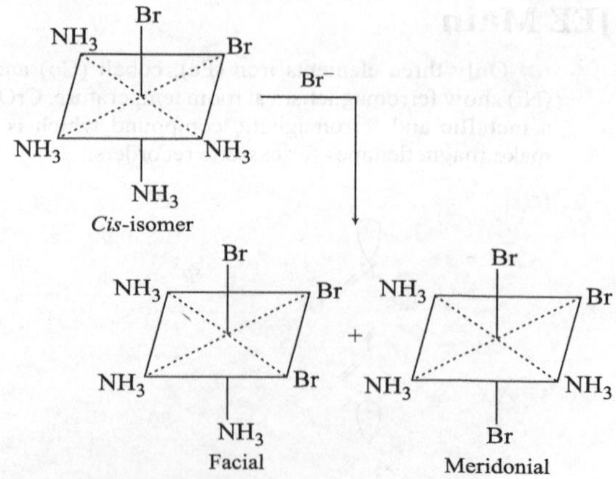

i.e. two isomers are produced. If the reactant is *trans* isomer than following reaction takes place.

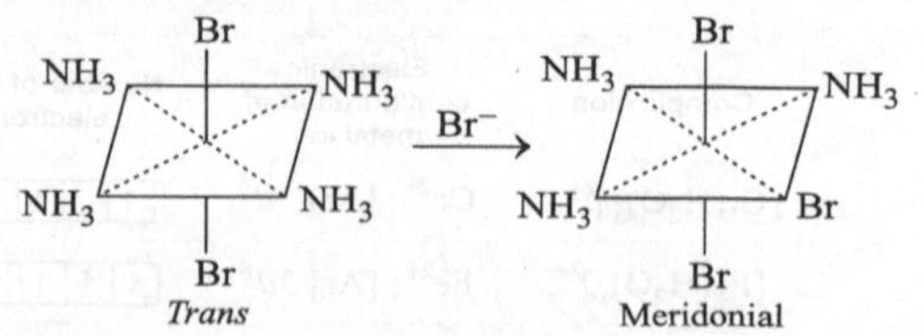

i.e. only 1 isomer is produced. Thus, statement (I) and (III) are correct resulting to option (b) as the correct answer.

JEE Advanced

1. *(b)*

Compounds	Hybridisation	Unpaired electron (s)	Magnetic character
$Ni(CO)_4$	sp^3	No	Diamagnetic
$[NiCl_4]^{2-}$	sp^3	two	Paramagnetic
$[Co(NH_3)_4Cl_2]Cl$	sp^3d^2	No	Diamagnetic
$Na_3[CoF_6]$	sp^3d^2	three	Paramagnetic
Na_2O_2	—	No	Diamagnetic (O_2^{2-})
CsO_2	—	One	Paramagnetic
			O_2^- (superoxide ion is paramagnetic)

2. **(5)** Ligand is $\begin{matrix} CH_2 - NH_2 \\ | \\ CH_2\bar{O} \end{matrix}$

Geometrical isomers are

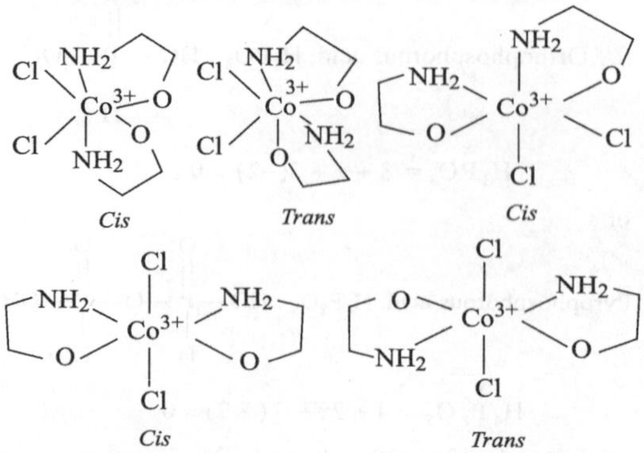

3. **(6)** In neutral or faintly alkaline solution, MnO_4^- is reduced to MnO_2 and $S_2O_3^{2-}$ is oxidised to SO_4^{2-}.

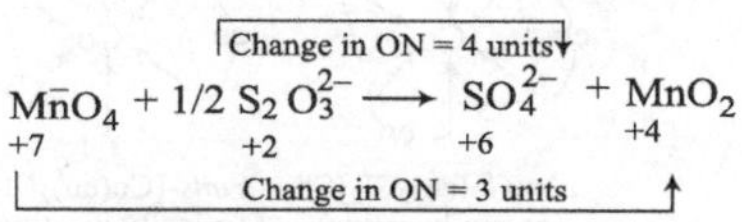

Thus, $4MnO_4^- + \frac{3}{2}S_2O_3^{2-} \longrightarrow 3SO_4^{2-} + 4MnO_2$

or $\quad 8MnO_4^- + 3S_2O_3^{2-} \longrightarrow 6SO_4^{2-} + 8MnO_2$

Thus, moles of SO_4^{2-} formed by 8 moles of $MnO_4^- = 6$

4. *(a)* $2S_2O_3^{2-} + Ag^+ \longrightarrow \underset{\text{(Clear solution)}}{[Ag(S_2O_3)_2]^{3-}}$

$[Ag(S_2O_3)_2]^{3-} + 3Ag^+ \longrightarrow \underset{\text{(White ppt.)}}{2Ag_2S_2O_3}$

$Ag_2S_2O_3 + H_2O \longrightarrow \underset{\text{(Black ppt.)}}{Ag_2S} + H_2SO_4$

5. *(a)* $[Ni(NH_3)_6]^{2+}$ sp^3d^2 octahedral

$[Pt(NH_3)_4]^{2+}$ dsp^2 square planar

$[Zn(NH_3)_4]^{2+}$ sp^3 tetrahedral

6. *(a, b, c)* $CuFeS_2$ (copper pyrite) is converted into copper into following steps:

Step I Crushing (grinding) followed by concentration by froth-floatation process.

Step II Roasting of ore in the presence of SiO_2 which removes iron as slag ($FeSiO_3$).

$$2CuFeS_2 + O_2 \longrightarrow Cu_2S + 2FeS + SO_2$$
$$2FeS + 3O_2 \longrightarrow 2SO_2 + 2FeO$$
$$FeO + SiO_2 \longrightarrow FeSiO_3\,(\text{Slag})$$

Step III Self-reduction in Bessemer converter

$$2Cu_2S + 3O_2 \longrightarrow 2Cu_2O + 2SO_2$$
$$2Cu_2O + Cu_2S \longrightarrow 6Cu + SO_2$$

Copper obtained is blister copper (98% pure).

Step IV Refining of blister copper is done by electrolysis

Impure copper—Anode

Pure copper— Cathode

At anode : $Cu \longrightarrow Cu^{2+} + 2e^-$

At cathode : $Cu^{2+} + 2e^- \longrightarrow Cu$

Carbon-reduction method is not used. Thus, (d) is incorrect.

7. *(b, d)* P_4O_{10} is a dehydrating agent and converts HNO_3 into N_2O_5

$$2HNO_3 \longrightarrow N_2O_5 + H_2O$$
$$P_4O_{10} + 6H_2O \longrightarrow 4H_3PO_4$$

(a) $P_4 + 20HNO_3 \longrightarrow 4H_3PO_4 + 20NO_2 + 4H_2O$
Thus, (a) is incorrect.

(b) N_2O_5 has no unpaired electron and is thus, diamagnetic thus, (b) is correct.

(c)

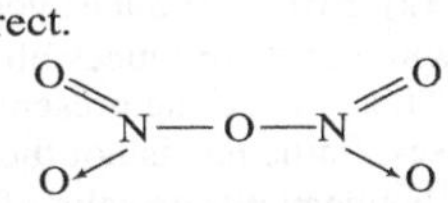

There is no N—N bond, thus, (c) is incorrect.

(d) $N_2O_5 + Na \longrightarrow NaNO_3 + NO_2$
N_2O_5 vapours are of brownish colour. Thus, (d) is correct.

8. *(a, b, d)* $\underset{\text{Pink }(X)}{[Co(H_2O)_6]Cl_2} \xrightarrow[O_2\text{(Air)}]{\text{Excess }NH_4OH/NH_4Cl} \underset{Y}{Co(NH_3)_6]Cl_3}$

$\underset{X}{[Co(H_2O)_6]^{2+}} + \underset{\text{(Excess)}}{4Cl^-} \longrightarrow \underset{\text{blue }Z}{[CoCl_4]^{2-}}$

(a) Since NH_3 is moderately strong ligand, hybridisation of cobalt in Y is d^2sp^3.

(b) Cobalt is sp^3-hybridised in $[CoCl_4]^{2-}$.

(c) $\underset{Y}{[Co(NH_3)_6]Cl_3} + 3AgNO_3(aq) \longrightarrow 3AgCl\downarrow$

(d) $\underset{\text{Blue}}{[CoCl_4]^{2-}} + 6H_2O \rightleftharpoons \underset{\text{Pink}}{[Co(H_2O)_6]^{2+}} + 4Cl^-; \Delta H < 0$

9. *(a, c, d)*

(a) ClO_4^- is more stable than ClO^-.

(b) Incorrect : $Cl_2 + H_2O \longrightarrow HCl + HOCl$

(c)

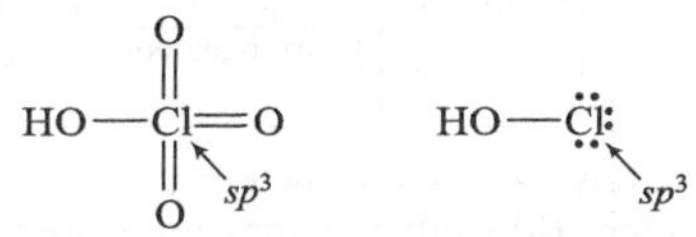

(d) $HClO_4$ is stronger acid than H_2O.

10. *(b,c)* Colour of halogen arises due to transition from HOMO to LUMO in the visible region. On moving down a group, the difference in energy between HOMO and LUMO decreases electronic transition occur more easily and colour intensity increases.

11. *(b)* $H_3\overset{+5}{P}O_4 > H_4\overset{+4}{P_2}O_6 > H_3\overset{+3}{P}O_3 > H_3\overset{+1}{P}O_2$

Paragraph-1

$$KClO_3 \xrightarrow[\Delta]{MnO_2} KCl + \underset{W}{O_2}$$

$$O_2 + P_4 \xrightarrow{\Delta} \underset{X}{P_4O_{10}} \xrightarrow{HNO_3} \underset{Y}{N_2O_5} + \underset{Z}{HPO_3}$$

12. *(a)*

13. *(b)*

14. *(b, c)* Among the given compounds, those which generate N_2 on thermal decomposition below 300°C are **ammonium dichromate**, i.e. $(NH_4)_2Cr_2O_7$ and **barium azide** or nitride i.e., $Ba(N_3)_2$. Reactions of their thermal decomposition are given below

(i) $(NH_4)_2Cr_2O_7 \xrightarrow[\text{Below 300°C}]{\Delta} N_2\uparrow + Cr_2O_3 + 4H_2O$

(ii) $Ba(N_3)_2 \xrightarrow[\text{Around 160° and above}]{\Delta} Ba + 3N_2\uparrow$

Ammonium nitrate (NH_4NO_3) on heating below 300°C gives N_2O as

$$NH_4NO_3 \xrightarrow[\text{below 300°C}]{\Delta} N_2O + 2H_2O$$

However, on rapid heating or explosion (i.e. above 300°C) it gives off nitrogen as

$$2NH_4NO_3 \xrightarrow[\text{or explosion}]{\text{Rapid heating}} 2N_2 + O_2 + 4H_2O$$

Magnesium nitride (Mg_3N_2) does not decompose at lower temperatures being comparatively more stable. Its thermal decomposition requires a minimum temperature of 700°C and proceeds as

$$Mg_3N_2 \xrightarrow{700\text{–}1500°C} 3Mg + N_2\uparrow$$

15. *(b, c)* Statement wise explanation is

(i) **Statement (a)** The total number of valence shell electrons at metal centre in $Fe(CO)_5$ or $Ni(CO)_4$ is 8 instead of 16 as shown below

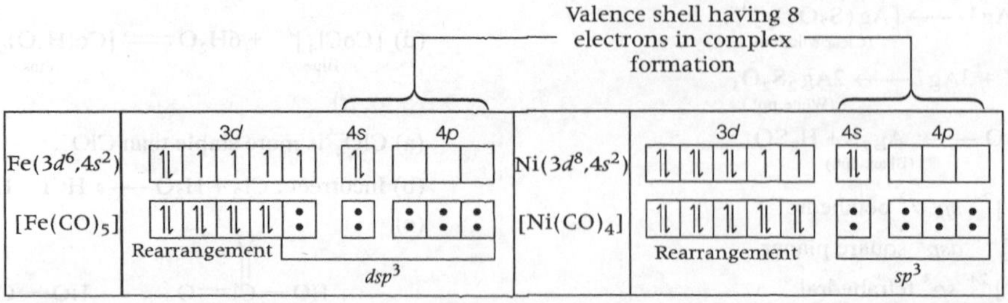

Hence, this statement is incorrect.

(ii) **Statement (b)** Carbonyl complexes are predominantly low spin complexes due to strong ligand fields. Hence, this statement is correct.

(iii) **Statement (c)** For central metal lowering of oxidation state results to increase in electron density on it. This in turn results to increase in extent of synergic bonding. Thus, we can say "metal carbonyl bond strengthens, when oxidation state of metal is lowered".
Hence, it is a correct statement.

(iv) **Statement (d)** Increase in positive charge on metal (i.e., increase in oxidation state) results to decrease in synergic bonding strength.
This in turn makes C—O bond stronger instead of weaker. Hence this statement is also incorrect.

16. *(a, b, d)* Statement wise explanation is

Statement (a) $[Co(en)(NH_3)_3H_2O]^{3+}$ have following 2 geometrical isomers.

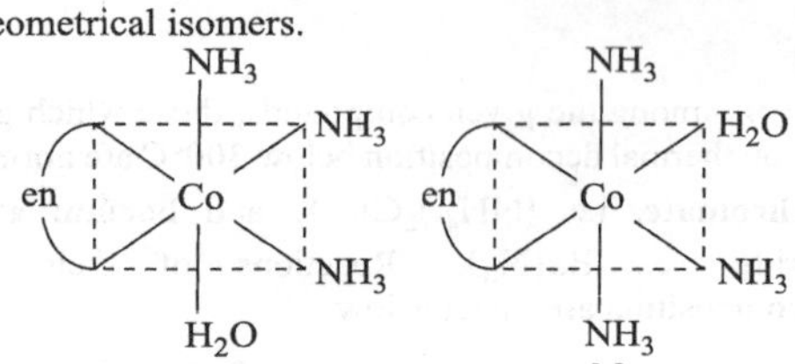

Fac *Mer*

$[M(AA)b_3c]$ type complex

Hence, this is correct statement.

Statement (b) If bidentate ligand 'en' is replaced by two cyanide ligands then $[Co(NH_3)_3(H_2O)(CN)_2]^+$ is formed.

It is $[Ma_3b_2c]$ type complex which has following 3 geometrical isomers.

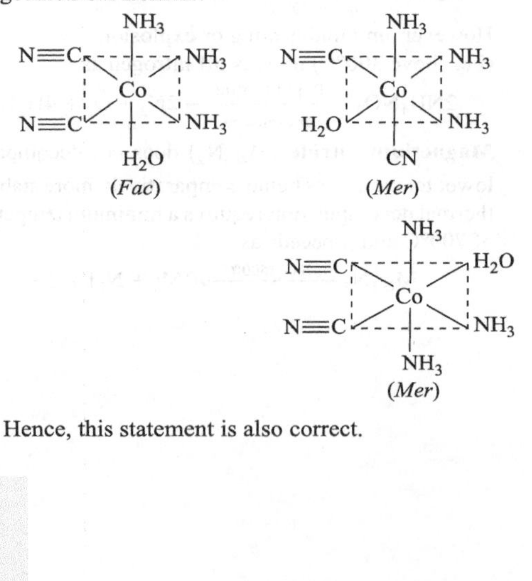

Hence, this statement is also correct.

Statement (c)

Co metal has $[Ar]3d^7 4s^2$ configuration while in $[Co(en)(NH_3)_3(H_2O)]^{3+}$ it is in +3 oxidation state. Thus, Co^{3+} has $[Ar]3d^6$ configuration, and d^2sp^3 in complex.

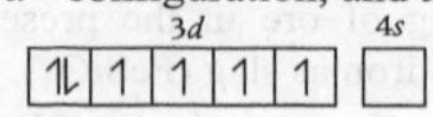

As *en* is a strong ligand, so pairing will occur

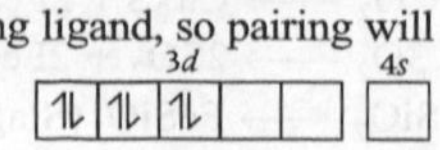

Due to the presence of all paired electrons it show diamagnetic behaviour rather than paramagnetic.

Hence, this statement is incorrect.

Statement (d)

According to CFT, absorption of light by coordination complexes depends upon CFSE i.e., crystal field splitting energy (Δ_0)as

$$\Delta_0 \propto \frac{1}{\lambda}$$

Among the complexes given $[Co(en)(NH_3)_4]^{3+}$ has more Δ_0 value as compared to complex $[Co(en)(NH_3)_3(H_2O)]^{3+}$. Thus, $[Co(en)(NH_3)_3(H_2O)]^{3+}$ absorbs the light at longer wavelength for *d-d* transition.

Hence, this statement is also correct.

Note : *For any complex, the value of Δ_0 can be calculated via the difference or gap between e_g and t_{2g} values.*

17. *(b,d)* Statement wise explanation is

Statement (a) Mn^{2+} produces yellow-green colour in flame test while Cu^{2+} produces bluish-green colour in flame test. Thus, due to the presence of green colour in both the cases, flame test is not the suitable method to distinguish between nitrate salts of Cu^{2+} and Mn^{2+}.

Hence this statement is wrong.

Statement (b) Cu^{2+} belong to group II of cationic or basic radicals. It gives black ppt. of CuS if H_2S is passed through it in the presence of acid (e.g HCl). Mn^{2+} does not show this property hence this can be considered as a suitable method to distinguish between Mn^{2+} and Cu^{2+}.

Hence, this statement is correct.

Statement (c) In faintly basic medium when H_2S is passed both Cu^{2+} and Mn^{2+} forms precipitates. Thus, it is not suitable method to distinguish between them.

Hence, this statement is incorrect

Statement (d) The standard reduction potential of Cu^{2+}/Cu is +0.34 V while that of Mn^{2+}/Mn is –1.18V.

This can be used to distinguish between Cu^{2+} and Mn^{2+}.
In general less electropositive metals have higher SRP.

Hence, this statement is correct.

18. *(6)* The structures of various molecules given in problem are discussed below—

1. **N_2O_3** It is the tautomeric mixture of following two structures—

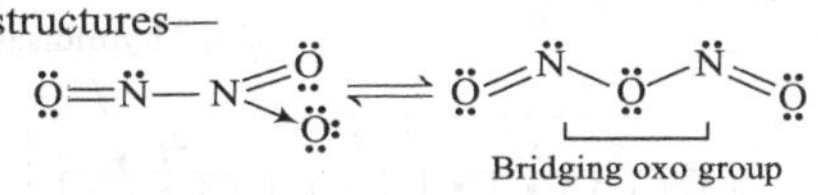

Conclusion 1 bridging oxo group is present in the compound.

2. **N_2O_5** It has following structure.

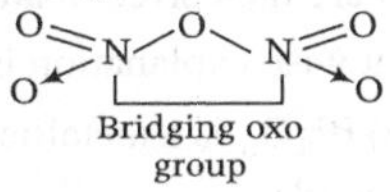

Conclusion 1 bridging oxo group is present in the compound.
where, p_i = initial pressure, p_t = total pressure
y = number of gaseous products per mole of reactant

3. **P_4O_6**

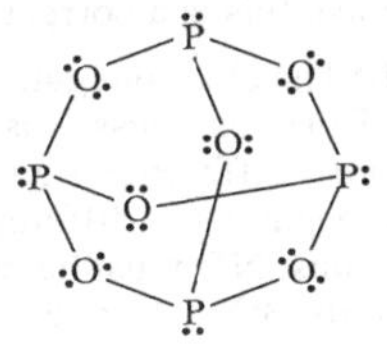

Conclusion 6 bridging oxo groups are present in the compound.

4. **P_4O_7**

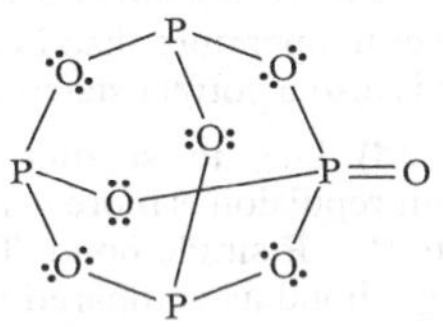

Conclusion 6 bridging oxo groups are present in the compound.

5. **$H_4P_2O_5$**

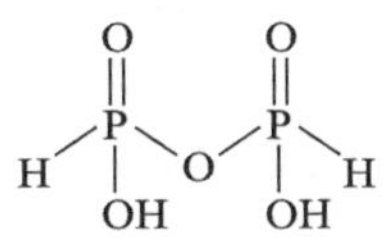

Conclusion 1 bridging oxo group is present in the compound.

6. $H_5P_3O_{10}$

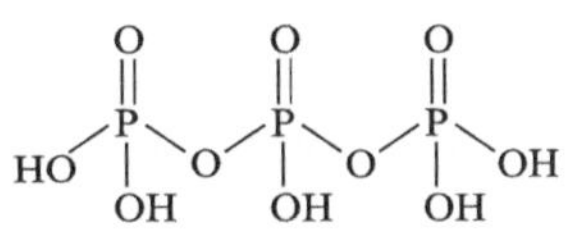

Conclusion 2 bridging oxo groups are present in the compound.

7. **$H_2S_2O_3$**

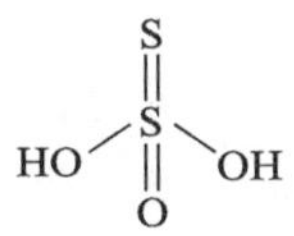

Conclusion This compound does not contain any bridging oxo group.

8. **$H_2S_2O_5$**

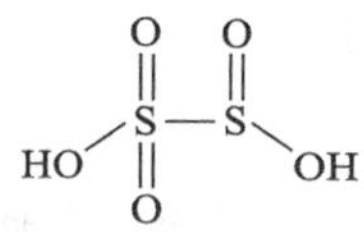

Conclusion This compound also does not contain any bridging oxo group.

19. *(c)* **For, *P* i.e. dsp^2,** It is seen in $[Ni(CN)_4]^{2-}$

$$Ni—[Ar]3d^8 4s^2$$
$$Ni^{2+}—[Ar]3d^8$$

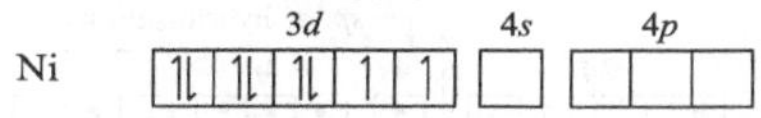

as CN^- is a strong ligand so when it approaches towards central metal pairing of unpaired electrons takes place.

Thus, in $[Ni(CN)_4]^{2-}$

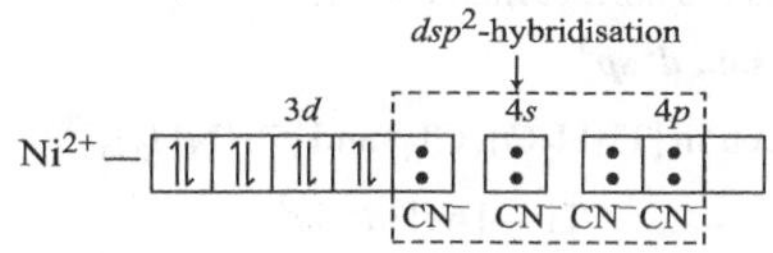

Structure : Square planar
So correct match for P is 6.

For *Q* i.e., sp^3

It is seen in $[FeCl_4]^{2-}$ and $Ni(CO)_4$

$$Fe - [Ar]3d^6 4s^2$$
$$Fe^{2+} - [Ar]3d^6$$

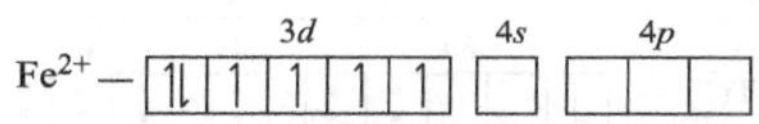

As Cl^- is a weak ligand so when it approaches towards central metal pairing of unpaired electrons does not take place.

Thus, in $[FeCl_4]^{2-}$

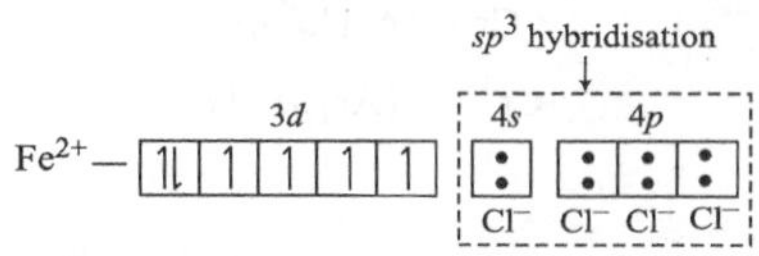

Structure-Tetrahedral
Likewise in $Ni(CO)_4$
$Ni—[Ar]3d^8 4s^2$

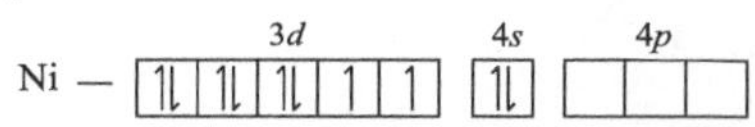

As CO is a strong ligand, hence when it approaches towards central metal atom pairing of unpaired electron of central atom takes place.

Thus, in $Ni(CO)_4$

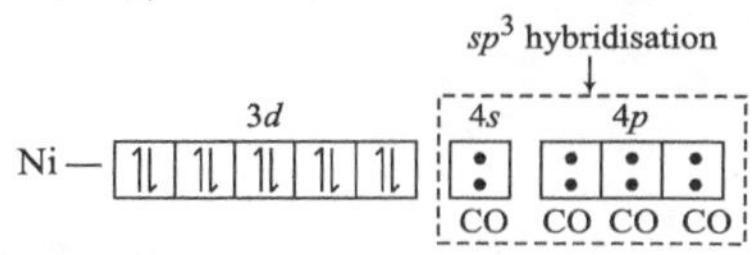

Structure Tetrahedral
So, for *Q*-4 and 5 are correct match.

For *R* i.e., sp^3d^2

It is seen in $[FeF_6]^{4-}$

Fe—$[Ar]3d^6 4s^2$

Fe^{2+}—$[Ar]3d^6$

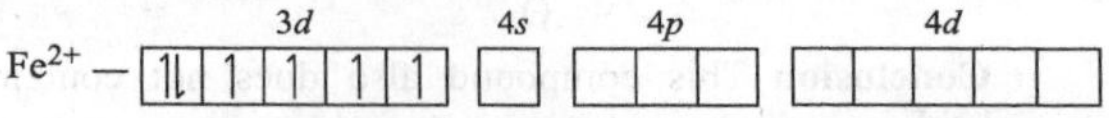

As F^- is a weak field ligand hence, when it approaches towards central metal atom, pairing of its electrons does not take place.

Thus, in $[FeF_6]^{4-}$

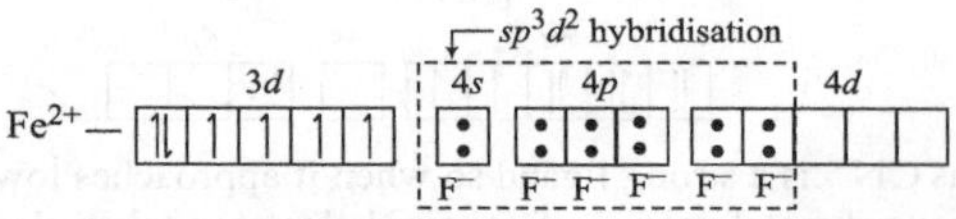

Structure : Octahedral

So, 1 is the correct match for *R*.

For *S* i.e., d^2sp^3

It is seen in $[Ti(H_2O)_3Cl_3]$ and $[Cr(NH_3)_6]^{3+}$

Ti—$[Ar]3d^2 4s^2$

Ti^{3+}—$[Ar]3d^1$

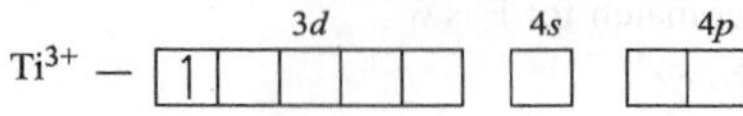

Here, both H_2O and Cl are weak ligands

So, in $[Ti(H_2O)_3Cl_3]$

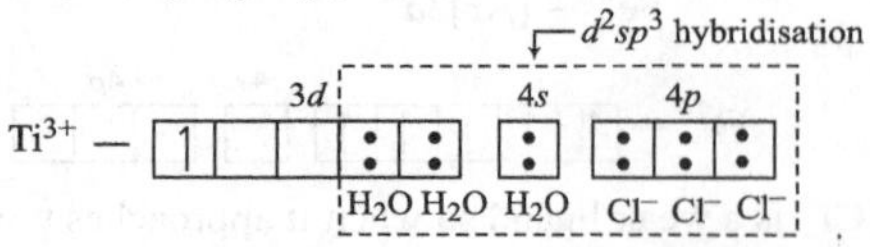

Structure Octahedral

Likewise in $[Cr(NH_3)_6]^{3+}$

Cr—$[Ar]3d^5 4s^1$

Cr^{3+}—$[Ar]3d^3 4s^0$

Here, NH_3 is also a weak field ligand so due to its approach no pairing takes place in Cr.

Thus, In $[Cr(NH_3)_6]^{3+}$

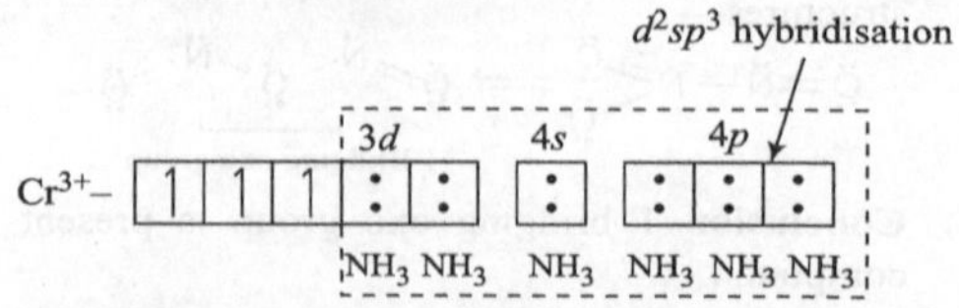

So for, S-2 and 3 are the correct match.

20. *(a,b,c)* Statement wise explanation is

(i) **Statement(a)** Bi_2O_5 is a metallic oxide while N_2O_5 is a non-metallic oxide.
Metallic oxides being **ionic** are **basic in nature** while non metallic oxides being **covalent** are **acidic in nature.** This confirms more basic nature of Bi_2O_5 in comparison to N_2O_5. Hence, this is a correct statement.

(ii) **Statement (b)** The electronegativity difference between N(3) and F(4) is less as compared to the electronegativity difference between Bi(1.7) and F(4). More electronegativity difference leads to ionic compounds. Thus, NF_3 must be more covalent in nature as compared to BiF_3. Hence, this statement is also correct.

(iii) **Statement (c)** In NH_3 intermolecular hydrogen bonding is present, which is altogether absent in PH_3. Thus, PH_3 boils at lower temperature than NH_3.
Hence, this is also a correct statement.

(iv) **Statement (d)** Due to smaller size of N the lone pair-lone pair repulsion is more in N—N single bond as compared to P—P single bond. This results to weaker N—N single bond as compared to P—P single bond. Hence, this statement is incorrect.